I0049692

Molecular Advances in Wheat and Barley

Molecular Advances in Wheat and Barley

Special Issue Editor

Manuel Martinez

MDPI • Basel • Beijing • Wuhan • Barcelona • Belgrade

MDPI

Special Issue Editor
Manuel Martinez
Universidad Politecnica de Madrid
Spain

Editorial Office
MDPI
St. Alban-Anlage 66
4052 Basel, Switzerland

This is a reprint of articles from the Special Issue published online in the open access journal *International Journal of Molecular Sciences* (ISSN 1422-0067) from 2018 to 2019 (available at: https: //www.mdpi.com/journal/ijms/special_issues/Wheat_Barley)

For citation purposes, cite each article independently as indicated on the article page online and as indicated below:

LastName, A.A.; LastName, B.B.; LastName, C.C. Article Title. *Journal Name* **Year**, *Article Number*, Page Range.

ISBN 978-3-03921-371-9 (Pbk)
ISBN 978-3-03921-372-6 (PDF)

© 2019 by the authors. Articles in this book are Open Access and distributed under the Creative Commons Attribution (CC BY) license, which allows users to download, copy and build upon published articles, as long as the author and publisher are properly credited, which ensures maximum dissemination and a wider impact of our publications.

The book as a whole is distributed by MDPI under the terms and conditions of the Creative Commons license CC BY-NC-ND.

Contents

About the Special Issue Editor

Manuel Martinez is a senior researcher in the Plant Genomics and Biotechnology Center (CBGP). He received a PhD from the Universidad Complutense de Madrid working in cereal cytogenetics (1999). Then, he moved to the Biotechnology Department at the Universidad Politécnica de Madrid, with a Post-Doc fellowship from 1999 to 2002 to work in plant molecular biology. At the same Institution, he received a Ramón y Cajal research contract in 2003, a contract as Assistant Professor from December 2005, and a permanent contract as Associate Professor in 2012. In September 2008, he moved to the Plant Genomics and Biotechnology Center to carry out his research activity. At present, he belongs to the UPM research group "plant-pest molecular interactions". His research activity is focused on the molecular characterization of defense genes against pathogens and pests, the relationships between proteases and their inhibitors with senescence and germination mechanisms in cereals, and the bioinformatic analysis of the evolution and composition of gene families. He has coauthorized more than 70 research articles in international journals and participates as Guarantor Researcher in the Severo Ochoa grant awarded by the CBGP. Currently, he carries out his teaching activity in different courses related to plant biotechnology and bioinformatics.

International Journal of
Molecular Sciences

MDPI

Editorial

Editorial for Special Issue "Molecular Advances in Wheat and Barley"

Manuel Martinez [1,2]

[1] Centro de Biotecnologia y Genomica de Plantas (CBGP, UPM-INIA), Universidad Politecnica de, Madrid (UPM)- Instituto Nacional de Investigacion y Tecnología Agraria y Alimentaria (INIA), Campus, Montegancedo, Pozuelo de Alarcon, 28223 Madrid, Spain; m.martinez@upm.es; Tel.: +34-910679149

[2] Departamento de Biotecnologia-Biologia Vegetal, Escuela Tecnica Superior de Ingenieria Agronomica, Alimentaria y de Biosistemas, UPM, 28040 Madrid, Spain

Received: 10 July 2019; Accepted: 15 July 2019; Published: 16 July 2019

Along with maize and rice, allohexaploid bread wheat and diploid barley are the most cultivated crops in the world (FAOSTAT database, http://www.fao.org/faostat, accessed on 22 June 2019). Their economic importance and close relationship supports a parallel study of both cereals. Nowadays, analyses based on high-throughput sequencing have become a key approach in genome-wide biology for crop improvement. Advances in genomics have resulted in the development of new technologies and strategies that give support to experimental research. In this context, the release of the genomic sequences of wheat and barley has permitted the application of genome-scale approaches, such as those related to metabolomics, proteomics, transcriptomics, and phenomics analyses. Additionally, new tools for gene identification, such are Genome-Editing and Genome-Wide Association Studies, are being developed. Modern research based in this new technological scenario is focused on understanding regulatory systems in order to improve crop productivity. The final goal should be the functional genomic analysis of genes and regulatory networks that control important agronomic traits and biological processes, such as yield, grain quality, disease and pest resistance, nutrient-use efficiency, and abiotic stress resistance. This Special Issue aimed to report novel molecular research and reviews related to wheat and barley biology using these new technologies. The Special Issue presents a total of 18 articles (Table 1).

Five articles are reviews covering different aspects of both wheat and barley crops. Two of these reviews are focused on understanding the role of phytases and proteases in the grain using novel technologies with an agronomical goal. In the case of phytases, single-stomached animals and humans depend on phytase supplied through the diet to hydrolyze phytate and make associated nutrients, such as phosphorous, iron, and zinc, bioavailable. This review highlights advances in the understanding of the molecular basis of the phytase activity and how understanding the function and regulation of the *PAPhy_a* gene may support the development of improved wheat and barley with even higher phytase activity [1]. Proteases are crucial for the continuous release of nutrients from the endosperm to the embryo to achieve the correct development of the new plant and to avoid agronomical losses due to the absence of seed germination. Many advances have been made in understanding the role of proteases in the grain due to their potential value for the brewing industry and their relationship with celiac disease. Novel technologies have permitted the application of genome-scale approaches, such as those used in functional genomics and proteomics, to increase the repertoire and knowledge on the barley and wheat proteases involved in germination [2].

Table 1. Contributors to the Special Issue "Molecular Advances in Wheat and Barley".

Publication	Species	Topic/Molecular Advance	Reference
Review	Wheat/barley	Phytases in grains	Madsen et al. [1]
		Proteases in grain germination	Martinez et al. [2]
		Cas endonuclease technology	Koeppel et al. [3]
		Resistance to nematodes	Ali et al. [4]
		Host-induced gene silencing	Qi et al. [5]
Article	Wheat	Over-expression of genes for abiotic resistance	Pigolev et al. [6]
		Over-expression of genes for abiotic resistance	Ayadi et al. [7]
		GWAS for minerals in grains	Bhatta et al. [8]
		Bulked segregant analyses for markers	Nishijima et al. [9]
		BSR-seq for biotic resistance markers	Hu et al. [10]
		Fine-mapping of biotic resistance genes	Gill et al. [11]
		FISH for instability of alien introgressions	Perničková et al. [12]
		Wheat–rye small translocations for biotic resistance	Du et al. [13]
		Proteomic analysis in biotic resistant lines	Liang et al. [14]
		DArTseq technology for population structure	Robbana et al. [15]
	Barley	Transcriptomics in grain development	Bian et al. [16]
	Wheat/barley	Sequencing of Long tandem repeats	Kapustová et al. [17]
Brief report	Wheat	ND-FISH probes for chromosome identification	Xi et al. [18]

GWAS: genome-wide association study; BSR-seq: bulked segregant analysis-RNA-Seq; FISH: fluorescence in situ hybridization; DArTseq: diversity array technology sequencing; ND-FISH: non-denaturing fluorescence in situ hybridization.

Three reviews are related to the development of novel techniques with a strong potential to be used as biotechnological tools, and their specific use against cereal cyst nematodes. One of these reviews explores the possibilities of the newly emerging Cas endonuclease technology, which allows for the induction of mutations at user-defined positions in the plant genome. Current trends in the development of this technology and its biotechnological application in wheat and barley are reviewed [3]. Likewise, recent studies on host-induced gene silencing (HIGS) technology employing RNA silencing mechanisms in wheat and barley are reviewed [5]. RNA silencing mechanisms provide a transgenic approach for disease management. This approach has been successfully applied in crop disease prevention by silencing the targets of invading pathogens, being a valuable tool to protect wheat and barley from diseases in an environmentally friendly way. Finally, the use of modern tools for the enhancement of cereal cyst nematode resistance in wheat and barley is examined [4]. Besides genome-wide association studies, the application of various transgenic strategies has been exploited, including host-induced gene silencing, nematode effector genes, proteinase inhibitors, or chemodisruptive peptides, with an emphasis on the future applicability of Cas endonuclease technology.

Research articles cover most of the modern molecular approaches used to further advance wheat and barley knowledge. Two articles are focused on transgenic engineering. The overexpression in wheat of the Arabidopsis *AtOPR3* gene, one of the key genes in the jasmonic acid (JA) biosynthesis pathway, affected wheat development and altered tolerance to environmental stresses [6]. Transgenic durum wheat overexpressing the wheat plasma membrane aquaporin TdPIP2;1 gene exhibited improved germination rates and biomass production and retained low Na+ and high K+ concentrations in their shoots under high salt and osmotic stress conditions [7].

Four articles tried to identify single nucleotide polymorphism (SNP) markers in order to perform molecular marker-assisted selection in wheat breeding. Synthetic hexaploid wheat was used to quantify 10 grain minerals by an inductively coupled mass spectrometer for a genome-wide association study (GWAS). For this analysis, 92 marker-trait associations (MTAs) were identified, of which 60 were novel and 40 were within genes, and the genes underlying 20 MTAs had annotations suggesting a potential role in grain mineral concentration [8]. Likewise, synthetic hexaploid wheat lines were used to perform RNA sequencing (RNA-seq)-based bulked segregant analysis (BSA). This analysis permitted the identification of several SNP markers around the *Net2* gene, a causative locus to hybrid necrosis [9]. The bulked segregant analysis-RNA-Seq technique was also used to find new single

SNPs, kompetitive allele specific polymorphisms (KASPs), and simple sequence repeat (SSR) markers to saturate the genetic linkage map for *Pm61*, a gene that confers powdery mildew resistance. The newly saturated genetic linkage map will be useful in molecular marker-assisted selection of *Pm61* in breeding for disease-resistant cultivars [10]. With a similar goal, the gene *Lr42*, which confers effective resistance against leaf rust, was fine-mapped by using recombinant inbred lines (RILs). The identified region included nine nucleotide-binding domain leucine-rich repeat genes, and two KASP markers flanking *Lr42* were developed to facilitate marker-assisted selection for rust resistance in wheat breeding programs [11].

In three papers, chromosomal imaging techniques were used. Somatic nuclei of wheat with rye introgressions were analyzed by tridimensional fluorescence in situ hybridization (3D-FISH). While introgressed rye chromosomes or chromosome arms occupied discrete positions similar to chromosomes of the wheat host, their telomeres frequently occupied improper positions. This feature probably impacts the ability of introgressed chromosomes to migrate into the telomere bouquet at the onset of meiosis, leading to their gradual elimination over generations [12]. Non-denaturing fluorescence in situ hybridization (ND-FISH) was used to identify small segment translocations after irradiation in a wheat-rye 6RL^{Ku} minichromosome addition line. A translocated chromosome 6DL/6RL^{Ku} included the powdery mildew resistance from rye, supporting the practical utilization of the resistance gene on 6RL^{Ku} [13]. Additionally, ND-FISH technology provided suitable positive oligo probes for distinguishing alien *Thinopyrum* chromosomes in wheat backgrounds [18]. These oligo probes could be a convenient tool for the utilization of *Thinopyrum* germplasms in wheat breeding programs.

Finally, four articles are good examples of the suitability of advanced molecular techniques to delve into different wheat and barley issues. Proteomics techniques led to the identification of proteins that were up- and downregulated after powdery mildew inoculation of the wheat line L699, which includes the *Pm40* resistance gene. The identified proteins were predicted to be associated with the defense response as well as with other physiological processes [14]. The transcriptional dynamics of barley grain development was investigated through RNA sequencing at four developmental time points. Transcriptome profiling found notable shifts in the abundance of transcripts involved in both primary and secondary metabolism during grain development and highlighted the existence of numerous RNA editing events [16]. Population genetics on durum wheat lines were assessed using diversity array technology sequencing (DArTseq). Cluster analysis and discriminant analysis of principal components allowed five distinct groups to be distinguished, thus supporting the importance of genomic characterization for enhancing knowledge on population structure [15]. Genomic sequencing was also used to identify missing tandemly organized repetitive sequences in wheat and barley genomes, which are underrepresented in genome assemblies generated from short-read sequence data. The authors demonstrated that this missing information may be added to the pseudomolecules with the aid of nanopore sequencing of individual bacterial artificial chromosome (BAC) clones and optical mapping [17].

Overall, the 18 contributions published in this Special Issue (Table 1) illustrate research advances in wheat and barley knowledge using modern molecular techniques. These molecular approaches at genomic, transcriptomic, proteomic, and phenomic levels, together with new tools for gene identification and the development of new molecular markers, have contributed to developing a further understanding of regulatory systems in order to improve wheat and barley performance. In the near future, the development of novel techniques will permit us to increase our knowledge on the regulation of important agronomic traits, which will facilitate the breeding of improved wheat and barley varieties.

Acknowledgments: The financial support from the Ministerio de Ciencia y Universidades of Spain (project BIO2017-83472-R) is gratefully acknowledged.

Conflicts of Interest: The author declares no conflict of interest.

References

1. Madsen, C.K.; Brinch-Pedersen, H. Molecular Advances on Phytases in Barley and Wheat. *Int. J. Mol. Sci.* **2019**, *20*, 2459. [CrossRef] [PubMed]
2. Diaz-Mendoza, M.; Diaz, I.; Martinez, M. Insights on the Proteases Involved in Barley and Wheat Grain Germination. *Int. J. Mol. Sci.* **2019**, *20*, 2087. [CrossRef] [PubMed]
3. Koeppel, I.; Hertig, C.; Hoffie, R.; Kumlehn, J. Cas Endonuclease Technology-A Quantum Leap in the Advancement of Barley and Wheat Genetic Engineering. *Int. J. Mol. Sci.* **2019**, *20*, 2647. [CrossRef] [PubMed]
4. Ali, M.A.; Shahzadi, M.; Zahoor, A.; Dababat, A.A.; Toktay, H.; Bakhsh, A.; Nawaz, M.A.; Li, H. Resistance to Cereal Cyst Nematodes in Wheat and Barley: An Emphasis on Classical and Modern Approaches. *Int. J. Mol. Sci.* **2019**, *20*, 432. [CrossRef] [PubMed]
5. Qi, T.; Guo, J.; Peng, H.; Liu, P.; Kang, Z. Host-Induced Gene Silencing: A Powerful Strategy to Control Diseases of Wheat and Barley. *Int. J. Mol. Sci.* **2019**, *20*, 206. [CrossRef] [PubMed]
6. Pigolev, A.V.; Miroshnichenko, D.N.; Pushin, A.S.; Terentyev, V.V.; Boutanayev, A.M.; Dolgov, S.V.; Savchenko, T.V. Overexpression of Arabidopsis *OPR3* in Hexaploid Wheat (*Triticum aestivum* L.) Alters Plant Development and Freezing Tolerance. *Int. J. Mol. Sci.* **2018**, *19*, 3989. [CrossRef] [PubMed]
7. Ayadi, M.; Brini, F.; Masmoudi, K. Overexpression of a Wheat Aquaporin Gene, TdPIP2;1, Enhances Salt and Drought Tolerance in Transgenic Durum Wheat cv. Maali. *Int. J. Mol. Sci.* **2019**, *20*, 2389. [CrossRef] [PubMed]
8. Bhatta, M.; Baenziger, P.S.; Waters, B.M.; Poudel, R.; Belamkar, V.; Poland, J.; Morgounov, A. Genome-Wide Association Study Reveals Novel Genomic Regions Associated with 10 Grain Minerals in Synthetic Hexaploid Wheat. *Int. J. Mol. Sci.* **2018**, *19*, 3237. [CrossRef] [PubMed]
9. Nishijima, R.; Yoshida, K.; Sakaguchi, K.; Yoshimura, S.I.; Sato, K.; Takumi, S. RNA Sequencing-Based Bulked Segregant Analysis Facilitates Efficient D-genome Marker Development for a Specific Chromosomal Region of Synthetic Hexaploid Wheat. *Int. J. Mol. Sci.* **2018**, *19*, 3749. [CrossRef] [PubMed]
10. Hu, J.; Li, J.; Wu, P.; Li, Y.; Qiu, D.; Qu, Y.; Xie, J.; Zhang, H.; Yang, L.; Fu, T.; et al. Development of SNP, KASP, and SSR Markers by BSR-Seq Technology for Saturation of Genetic Linkage Map and Efficient Detection of Wheat Powdery Mildew Resistance Gene. *Int. J. Mol. Sci.* **2019**, *20*, 750. [CrossRef] [PubMed]
11. Gill, H.S.; Li, C.; Sidhu, J.S.; Liu, W.; Wilson, D.; Bai, G.; Gill, B.S.; Sehgal, S.K. Fine Mapping of the Wheat Leaf Rust Resistance Gene *Lr42*. *Int. J. Mol. Sci.* **2019**, *20*, 2445. [CrossRef]
12. Perničková, K.; Koláčková, V.; Lukaszewski, A.J.; Fan, C.; Vrána, J.; Duchoslav, M.; Jenkins, G.; Phillips, D.; Šamajová, O.; Sedlářová, M.; et al. Instability of Alien Chromosome Introgressions in Wheat Associated with Improper Positioning in the Nucleus. *Int. J. Mol. Sci.* **2019**, *20*, 1448. [CrossRef]
13. Du, H.; Tang, Z.; Duan, Q.; Tang, S.; Fu, S. Using the 6RL^{Ku} Minichromosome of Rye (*Secale cereale* L.) to Create Wheat-Rye 6D/6RL^{Ku} Small Segment Translocation Lines with Powdery Mildew Resistance. *Int. J. Mol. Sci.* **2018**, *19*, 3933. [CrossRef] [PubMed]
14. Liang, Y.; Xia, Y.; Chang, X.; Gong, G.; Yang, J.; Hu, Y.; Cahill, M.; Luo, L.; Li, T.; He, L.; et al. Comparative Proteomic Analysis of Wheat Carrying *Pm40* Response to *Blumeria graminis* f. sp. tritici Using Two-Dimensional Electrophoresis. *Int. J. Mol. Sci.* **2019**, *20*, 933. [CrossRef] [PubMed]
15. Robbana, C.; Kehel, Z.; Ben Naceur, M.; Sansaloni, C.; Bassi, F.; Amri, A. Genome-Wide Genetic Diversity and Population Structure of Tunisian Durum Wheat Landraces Based on DArTseq Technology. *Int. J. Mol. Sci.* **2019**, *20*, 1352. [CrossRef] [PubMed]
16. Bian, J.; Deng, P.; Zhan, H.; Wu, X.; Nishantha, M.D.L.C.; Yan, Z.; Du, X.; Nie, X.; Song, W. Transcriptional Dynamics of Grain Development in Barley (*Hordeum vulgare* L.). *Int. J. Mol. Sci.* **2019**, *20*, 962. [CrossRef] [PubMed]

17. Kapustová, V.; Tulpová, Z.; Toegelová, H.; Novák, P.; Macas, J.; Karafiátová, M.; Hřibová, E.; Doležel, J.; Šimková, H. The Dark Matter of Large Cereal Genomes: Long Tandem Repeats. *Int. J. Mol. Sci.* **2019**, *20*, 2483. [CrossRef] [PubMed]

18. Xi, W.; Tang, Z.; Tang, S.; Yang, Z.; Luo, J.; Fu, S. New ND-FISH-Positive Oligo Probes for Identifying Thinopyrum Chromosomes in Wheat Backgrounds. *Int. J. Mol. Sci.* **2019**, *20*, 2031. [CrossRef] [PubMed]

© 2019 by the author. Licensee MDPI, Basel, Switzerland. This article is an open access article distributed under the terms and conditions of the Creative Commons Attribution (CC BY) license (http://creativecommons.org/licenses/by/4.0/).

International Journal of
Molecular Sciences

MDPI

Review

Molecular Advances on Phytases in Barley and Wheat

Claus Krogh Madsen and Henrik Brinch-Pedersen *

Department of Molecular Biology and Genetics, Research Center Flakkebjerg, Aarhus University, 4200-Slagelse, Denmark; ClausKrogh.Madsen@mbg.au.dk
* Correspondence: hbp@mbg.au.dk

Received: 20 March 2019; Accepted: 15 May 2019; Published: 18 May 2019

Abstract: Phytases are pro-nutritional enzymes that hydrolyze phytate and make associated nutrients, such as phosphorous, iron, and zinc, bioavailable. Single-stomached animals and humans depend on phytase supplied through the diet or the action of phytase on the food before ingestion. As a result, phytases—or lack thereof—have a profound impact on agricultural ecosystems, resource management, animal health, and public health. Wheat, barley and their Triticeae relatives make exceptionally good natural sources of phytase. This review highlights advances in the understanding of the molecular basis of the phytase activity in wheat and barley, which has taken place over the past decade. It is shown how the phytase activity in the mature grains of wheat and barley can be ascribed to the *PAPhy_a* gene, which exists as a single gene in barley and in two or three homeologous copies in tetra- and hexaploid wheat, respectively. It is discussed how understanding the function and regulation of *PAPhy_a* may support the development of improved wheat and barley with even higher phytase activity.

Keywords: phytase; wheat; barley; purple acid phosphatase phytase; PAPhy; mature grain phytase activity (MGPA)

1. Introduction

Phytases (myo-inositol hexakisphosphate 3-,6- and 5-phosphohydrolase, EC 3.1.3.8, EC 3.1.3.26 and EC 3.1.3.72) are phosphatases that can initiate the stepwise hydrolysis of phytate (IP6, myoinositol-(1,2,3,4,5,6)-hexakisphosphate) and thereby provide phosphate (P), inositol phosphates, and inositol for a range of cellular activities [1]. In addition to purely scientific inquiries, phytase research has for many years been driven by the urgent need for improving utilization of phytate-phosphorus in diets for single-stomached animals, such as pigs and poultry, and to reduce the anti-nutritional effect of non-digested IP6 chelating micronutrients in the digestive tracts of humans and animals. As such, phytases can be regarded as tools for managing global phosphate resources and for alleviating human micro-nutrient deficiencies mainly in the developing world.

IP6 is the main storage form of phosphate in plants, typically amounting 2/3 of the total P content in the seed (Table 1). IP6 is a strong chelator and exists in the plant seeds as an insoluble mixed salt with cations called phytin. In cereals and many other plant seeds, phytin forms spherical crystalloid inclusions called globoids inside protein storage vacuoles. The globoids are the principal site of phosphorous (P), potassium (K) and magnesium (Mg) in the mature cereal grain but they also contain calcium (Ca), iron (Fe), zinc (Zn), copper (Cu), manganese (Mn), sodium (Na), sulfur (S), and protein [2,3].

Table 1. Total P, IP6 bound P, proportion of IP6 bound P, and phytase activity in wheat, barley and other cereals seeds. The number of samples is n and ± denotes the standard deviation. The range is given if $n \geq 2$. Data from [4] [1], [5] [2], [6] [3].

Cereal	*n*	Total P (% of Dry Matter)	Total IP6 P (% of Dry Matter)	Percent IP6 out of Total P	Phytase Activity (FTU/kg) *
Wheat [1]	13	0.33 ± 0.02	0.22 ± 0.02	67 ± 4.8	1193 ± 223
Wheat [2]	18	0.40 ± 0.04	0.29 ± 0.04	73 ± 8.1	2886 ± 645
Wheat [3]	30	0.29 ± 0.03	0.23 ± 0.03	79 ± 0.07	1637 ± 275
Barley [1]	9	0.37 ± 0.02	0.22 ± 0.01	60 ± 2.4	582 ± 178
Barley [2]	15	0.42 ± 0.4	0.26 ± 0.03	63 ± 3.5	2323 ± 648
Barley [3]	21	0.31 ± 0.03	0.19 ± 0.02	61 ± 0.04	1016 ± 330
Rye [1]	2	0.36 (0.35-0.36)	0.22 (0.20-0.23)	61 ± (56 -66)	5130 (4132-6127)
Rye [2]	13	0.36 ± 0.02	0.24 ± 0.2	67 ± 5.0	6016 ± 1578
Rye [3]	6	0.34 ± 0.03	0.20 ± 0.01	59 ± 0.02	5147 ± 649
Triticale [1]	6	0.37 ± 0.02	0.25 ± 0.02	67 ± 3.7	1688 ± 227
Triticale [2]	12	0.40 ± 0.03	0.28 ± 0.03	70 ± 5.4	2799 ± 501
Oats [1]	6	0.36 ± 0.03	0.21 ± 0.04	59 ± 11	42 ± 50
Oats [2]	6	0.37 ± 0.01	0.25 ± 0.02	67 ± 5.4	496 ± 35
Oats [3]	9	0.29 ± 0.02	0.17 ± 0.03	59 ± 0.07	84 ± 39
Maize [1]	11	0.28 ± 0.03	0.19 ± 0.03	68 ± 5.9	15 ± 18
Maize [3]	7	0.32 ± 0.01	0.18 ± 0.01	78 ± 0.01	70 ± 7
Rice [4]	1				72

* One FTU is the amount of enzyme that liberates 1 µmol of inorganic phosphorus per minute from sodium phytate at pH 5.5 and 37 °C.

Micronutrients are also chelated by phytate in food and feed, and hydrolysis is most wanted for improving micronutrient bioavailability. The anti-nutritional effect of phytate is in particular regarded as critical for Fe and Zn, where phytate is considered the single most important anti-nutritional compound for the bioavailability of these two micronutrients [7]. Iron deficiency is the primary cause of anemia and ranks among the most widespread nutrient deficiencies, estimated to affect 1.6 billion people worldwide [8]. Iron deficiency anemia has been linked to maternal and prenatal mortality, and to impairment of cognitive skills and physical activity [9]. For zinc, around 800,000 child deaths worldwide per year are attributable to Zn deficiency [10] because it significantly increases the risk of diarrhea, pneumonia, and malaria. Moreover, Zn deficiency has been linked to the morbidity and mortality of children younger than five [11].

Humans and single-stomached animals have insufficient phytase activity in their digestive tract unless it is provided by the diet. Unfortunately, major food and feed components like rice, maize and soybeans contribute with phytate but negligible phytase [4]. Because of the missing phytase activity, the phytate passes largely un-digested through the single-stomached animals' digestive tract and enters the environment when their manure is spread on agricultural fields. Moreover, to ensure that farm animals get the phosphate needed, bio-available mined P is added to the feed. However, this strategy has become critical in many regions of the world where intense livestock production and spreading of manure with high levels of undigested phytate P on oversupplied agricultural soil leads to run-off of phosphorus to aquatic ecosystems. The resulting eutrophication is a severe environmental risk [12]. However, also from a resource perspective, inefficient utilization of plant phytate P is inappropriate. P is a non-renewable resource, essential for efficient agricultural production, and complete depletion of mined P will have unmanageable consequences for global food production [13].

2. Plant and Microbial Phytases

IP6 is resistant to most phosphatases whereas the lower inositol phosphates can be degraded by a wider range of phosphatases. Phosphatases that can initiate the dephosphorylation of IP6 are classified as phytases. So far, four classes of phytases have been identified: (1) Histidine acid phosphatase (HAP), (2) purple acid phosphatase (PAP), (3) cysteine phosphatase (CP) and (4) β-propeller phytase (BPP). Each phytase type has unique structural features due to their distinct catalytic apparatus that allows them to utilize phytate as a substrate in various environments [14].

For decades, applied phytase research was focusing mainly on microbial phytases and to our knowledge, all commercial phytases currently used for feed supplementation are microbial enzymes belonging to the HAP class [14]. Similarly, until recently, microbial HAP phytases were used exclusively for increasing plant seeds phytase activity through transgenesis. However, scientific achievements in recent years have led to a substantially increased knowledge based on the complements of phytases, in particular barley and wheat, and have demonstrated significant potentials of their phytases as highly stable and potent enzymes with potentials both in feed and food (see later).

3. Mature Grain Phytase Activity

When hydrated, the mature seed tissues activate a battery of preformed hydrolytic enzymes that degrades the large internal pool of IP6 but also storage compounds like lipids, carbohydrates, and proteins. When ungerminated seeds are used as feedstuffs, this battery of enzymes constitutes all plant-derived hydrolytic activities. We refer to this as the first wave of activity and for phytase, it constitutes what is called the mature grain phytase activity (MGPA). In parallel with imbibition, the embryo synthesizes and secretes the plant hormone gibberellic acid. The aleurone and the scutellum layer of the embryo are thereby turned into secretory tissues where a wide range of hydrolytic enzymes are synthesized and secreted into the endosperm for degradation of cell walls, starch grains and storage proteins—the second wave of hydrolysis [2].

Cereals generally express phytases to assist in their IP6 metabolism but the MGPA varies several orders of magnitude. This is in strong contrast to the modest variation in total and proportional content of seed IP6 (Table1).

4. Classes of Phytases in Barley and Wheat

In wheat and barley, two types of phytases have been described, phytases belonging to the HAP class and phytases belonging to the PAP class of phosphatases [1]. The HAP phytases belong to the multiple inositol polyphosphate phosphatase (MINPP) group [15]. MINPP phytases have been reported to be expressed both during grain development, and thereby, potentially contribute to the MGPA, and during germination contributing to the second wave of phytate hydrolysis. The PAP phytases are represented by the TaPAPhy_a/bs from wheat and the HvPAPhy_a/bs from barley, respectively. Expression analysis showed that *PAPhy_a* genes are preferentially expressed during grain filling whereas the *PAPhy_b* genes are preferentially expressed during germination [16]. The Km value with phytate as a substrate for recombinant wheat MINPP rTaPhyIIa2 phytase and barley rHvPhyIIb phytase is around ten-fold higher than for the rTaPAPhy_a/b and rHvPAPhy_a/b PAP phytases, indicating PAPhys to be more potent phytases than the HAPhys (Table 2).

Table 2. Kinetic parameters from HAP and PAPhy phytases from wheat, barley, and *Aspergillus ficuum*. Data from [1] [15], [2] [16], [3] [17].

Class	Enzyme	Km (μM)	$Vmax$ (μmol/(min × mg))	$Kcat$ (s^{-1})	$Kcat/Km$ ($s^{-1}M^{-1}$)	pH Optimum
PAP Phytases	rTaPAPhy_a1 [2]	35	223	279	796×10^4	5.5
	rTaPAPhy_b1 [2]	45	216	270	600×10^4	5
	rHvPAPhy_a [2]	36	208	260	722×10^4	
	rHvPAPhy_b [2]	46	202	253	550×10^4	
HAP Phytases	rTaPhyIIa2 [1]	246				4.5
	rHvPhyIIb [1]	334				4.5
	A. ficuum phytase [3]	27	-	348	129×10^5	5.5

The contribution of HvPAPhy_a and HvPAPhy_b to the total MGPA in barley was recently evaluated using CRISPR/Cas and TALEN [18]. TALEN- and CRISPR/Cas9 were used for introducing targeted mutations in the promoter of the barley phytase gene *HvPAPhy_a*. Barley lines with substantial deletions in the *HvPAPhy_a* promoter and 5′CDS retained <5% normal MGPA. This confirms that the barley PAPhy_a enzyme is the main contributor to the MGPA and can be regarded as the main target for modulating MGPA.

5. Biochemical Properties and Storage of the PAPhys

Wheat and barley mainly store phytate in the protein storage vacuoles (PSVs) of the aleurone layer. PAPhy accumulated during grain filling is localized in the same organelles [16]. This suggests that some mechanism protects the phytate from hydrolysis during grain filling. The PSV's are rapidly acidified in response to gibberellic acid as germination commences. A decrease from pH 6.6 to 5.9 was reported in the PSV's of barley protoplasts incubated with 5 μM GA_3 and the authors speculated that pH might play a crucial role in regulating vacuolar hydrolases [19]. Recombinant wheat phytases rTaPAPhy_a1 and rTaPAPhy_b1 showed pH optima of 5.5 and 5, respectively [16]. Optima of pH 5 and 6 were measured for seed purified barley phytases P1 (=HvPAPhy_b) and P2 (=HvPAPhy_a) respectively [20]. This shows that the PAPhy's are most active when the PSV is in the acidified lytic, state and the higher pH optimum of preformed PAPhy_a may even be an adaptation, which enhances activity in the earliest stages of germination. Nevertheless, rTaPAPhy_a retained some activity up to pH 7.5 so pH regulation of the enzyme alone does not offer a satisfying explanation for the protection of phytate against premature hydrolysis. The second layer of pH-dependent protection is provided by the substrate's organization into globoids, which provides some degree of water exclusion and steric hindrance [3]. A membrane surrounds the globoids and immuno-gold localization suggests that PAPhy is located outside this membrane [16,21]. The membrane, therefore, seems to provide an additional layer of protection, by physical separation. The temperature optimum of the recombinant wheat enzymes was 50 and 55 °C, respectively. The temperature curves showed a broad peak with 50% activity already at 30–35 °C but decreasing sharply around 60 °C. Optima of 55 and 45 °C were reported for the seed purified HvPAPhy_a and HvPAPhy_b respectively [20]. The kinetic parameters of recombinant wheat and barley phytases at 36 °C and pH 5 are summarized in Table 2. The corresponding values for *Aspergillus ficuum* phytase are given for comparison [17]. The values for the PAPhys are very similar.

6. PAPhy Genetics

The *PAPhy_a* and *PAPhy_b* genes are paralogs, which originate from gene duplication in a common ancestor of wheat, barley, and rye (i.e., the Triticeae tribe) [22]. Rice, maize, and sorghum diverged from the Triticeae earlier and carry only one PAPhy gene whereas *Brachypodum distachyon* has the duplication but lack the conserved *PAPhy_a* promoter of the Triticeae (Figure 1) [22]. Allopolyploidzation has united the A and B genomes to form tetraploid wheat (e.g., durum wheat). *Triticum urartu* and

Aegilops speltoides are the closest living relatives of the A and B genomes, respectively. Additional hybridizations have added the *Ae. tauschii* derived D genome to produce hexaploid wheat (bread wheat, spelt) and the rye derived R genome to produce triticale (Figure 1) [23]. Allopolyploidization results in large-scale gene duplication because most genes in one parent will have a homolog in the other parent species. In allopolyploids, such sets of genes are termed homeologs. Some homeologs may be lost or translocated as the polyploid species continues to evolve but the Triticeae *PAPhy* gene copy number and chromosomal localization are highly conserved. A single locus of *PAPhy_a* and *PAPhy_b* resides on chromosome 5 and 3, respectively in barley and on the homologous chromosomes on the three subgenomes of wheat [22]. The sequenced diploid members of the *Triticum* and *Aegilops*, i.e., *Ae. tauschii, Ae. speltoides, Ae. sharonensis, T. Urartu,* and *T. monococcum* also have one *PAPhy_a* and one *PAPhy_b* [24]. The chromosomal localization in these relatives has not been determined but it is reasonable to expect a conserved synteny between wheat and its ancestors *T. urartu, Ae. Speltoides,* and *Ae. tauschii. Secale* provides an exception since some members of this tribe have two *PAPhy_a* loci. In the case of domesticated rye, one and two *PAPhy_a* variants were isolated from the cultivars Imperial and Picasso, respectively [25]. Rye is an outbreeding species, unlike wheat and barley, and tends to have higher allele heterogeneity. Therefore, it cannot be excluded that the two variants from Picasso are alleles of the same gene even though phylogeny suggests that one allele may have been introgressed from *S. strictum* [25]. Thus, it is not known with certainty if domesticated rye has one or two *PAPhy_a* loci or whether it is cultivar-dependent.

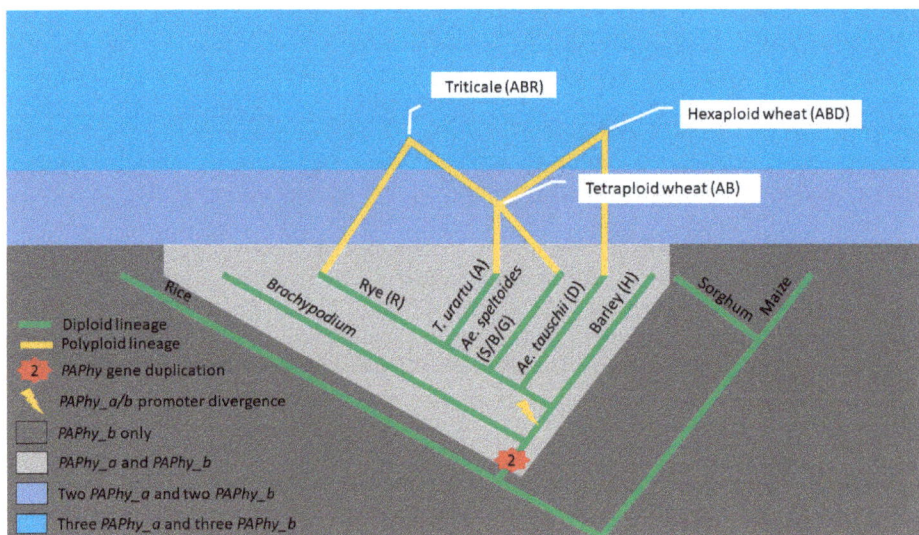

Figure 1. Key events in *PAPhy* evolution. Gene duplication and divergence of the PAPhy_a/b promoters and further duplications through polyploidization. For simplicity, rye is assumed to have just one *PAPhy_a* locus, and phylogenetic distance is not drawn to scale.

The intron/exon structure and the respective promoters of *PAPhy_a* and *PAPhy_b* are highly conserved. *PAPhy_a* has four introns and *PAPhy_b* has five. The position and, to a large extent, the size of the introns is conserved between Triticeae species and between the two paralogs [22]. Both genes have a core promoter of 3–400 base pairs which is conserved in all studied Triticeae. Both promoters have two TATA-boxes and upstream of those reside cis-acting regulatory elements consistent with the differential expression pattern of the two genes (Figure 2). For *PAPhy_b*, they are ABRE (abscisic-acid-responsive), TGACG (methyl-jasmonate-responsive) and—conserved in all examined *PAPhy_b* promoters—GARE (gibberellic acid responsive) [22]. The *PAPhy_a* promoter lacks these hormone responsive elements, except TGACG which is found at the very beginning of the conserved sequence. Instead, the most notable feature is a composite element with the consensus 5' **GAACATG**AGTC*ATGCATG* 3' which is made of the GAMYB binding motif (bold) [26], the odd base palindrome/GCN4 (underlined) [27,28] and the RY element (italic) [29]. These are all elements associated with seed development and storage proteins. Between the composite element and the first TATA box is a G-box motif [22]. Deletion of the odd base palindrome and the RY-element reduce barley MGPA by approximately 40% whereas deletion of the whole element and an additional 10 base pairs 3' reduce MGPA by 75%. Deletions immediately 3' of the composite element have even more severe effects [14]. It is not clear if this is caused by an unknown cis-acting regulatory element at that position or by the change in distance between the composite element and elements further downstream.

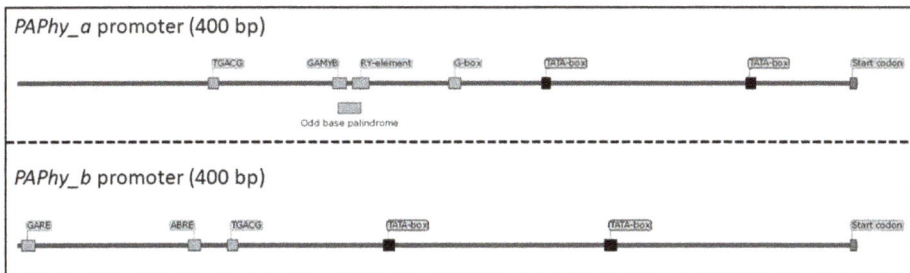

Figure 2. Consensus promoters of *PAPhy_a* and *PAPhy_b*, respectively, showing the most conserved cis-acting regulatory elements.

7. Applied Potentials of the PAPhys

Although the PAPhy phytases appear to be somewhat less active than fungal HAP phytases, they certainly have technological potential. Transformation of barley with a genomic clone of *HvPAPhy_a*, including its native promoter (cisgenesis), more than doubled the MGPA [30]. Moreover, overexpressing *HvPAPhy_a* using the 35S promoter resulted in a line with impressive 40.000 FTU [31]. This level of MGPA represents more than a 20 fold increase and provides so much activity that <2% of the recombinant grains in a feed mixture could theoretically replace conventional phytase supplementation. Moreover, the enzyme activity was highly stable, mature leaves and straw accumulated PAPhy_a and high phytase activity remained after three years of storage. The PAPhy_a was easily extracted in water and could be added to feed or other processes.

The efficacy of a phytase in the digestive tract depends in part on how well intrinsic biochemical parameters, such as pH optimum, match the environment but also on resistance to proteolysis and inhibition. Feeding experiments are, therefore, necessary to prove efficacy. Early evidence for the efficacy of PAPhy_a was provided by comparing phosphorous and calcium utilization in pigs fed maize and triticale, respectively [32]. In this case, triticale-based feed performed better on both parameters, indicating an effect of the higher MGPA associated with triticale. In humans, an intervention study compared wheat bran with or without phytase (native vs. heat treated) in ileostomy patients. This resulted in recoveries of 40% vs. 95% phytate in the ileostomy content [33]. It is also possible to utilize the MGPA to achieve dephytination during food or feed preparation. Up to 96% of the phytate in whole grain wheat bread could be removed by simply adjusting the pH during proofing to five [34]. Similarly, fonio (*Digitaria exilis*) porridge could be dephytinized in one hour at 50 °C and pH 5 when 25% of wheat was included. This was demonstrated to significantly increase iron absorption in West African women [35]. Taken together, these studies demonstrate a lot of potential for the utilization of endogenous Triticeae PAPhy_a in food and feed.

8. Achieving Higher MGPA with the PAPhys

Given the high conservation of the *PAPhy_a* coding sequence and highly similar enzymatic properties of all examined PAPhys, including PAPhy_b as well as rice and maize PAPhy, it seems unlikely that alternative *PAPhy_a* alleles, encoding decisively better enzymes, exist. On the other hand, increasing *PAPhy_a* expression has proven a viable strategy, as discussed. Efforts to increase MGPA by conventional breeding should, therefore, focus on the discovery of more highly expressed *PAPhy_a* alleles and the elimination of defective alleles.

Polyploids (e.g., wheat) are prone to acquire defective alleles because homeologs on the other subgenomes provide functional redundancy. Nevertheless, defective *PAPhy_a* alleles may reduce MGPA because the gene copies appear to function in an additive manner as demonstrated by cisgenic gene duplication [30]. Further support for this hypothesis is provided by the discovery of a defective *TaPAPhy_a2* gene in the wheat cultivars Skagen and Bob White that correlated with a lower MGPA compared to Chinese Spring, which has three normal loci [22,25]. A similar reduction of phytase activity has been observed in wheat lines with induced deleterious mutations in the *PAPhy_a* genes (author's unpublished results). It is not known how commonly and how many different defective alleles exist in the elite wheat gene pool. A systematic investigation of this combined with the development of markers for the defective alleles would be helpful in ensuring a minimum MGPA in new cultivars.

More active alleles might be found in crop relatives. Wild emmer (*T. dicocoides*) can be crossed directly with other tetraploid wheat (e.g., durum wheat) and with hexaploid wheat trough synthetic hexaploids. Alleles from *Ae. tauscii* can also be introduced by this route or through the "octo-amphiploid bridge" [36]. A higher allele diversity is expected for these species because they have not gone through the bottlenecks of domestication and hexaploidization [37]. A small sampling of wheat and the closest relatives support this assumption [25]. Furthermore, it was found that at least one exotic allele had already been introduced in the elite wheat gene pool in the pursuit of other breeding goals—from *T. timopheevii*. In addition to the abovementioned, einkorn (wild and domesticated), *T. urartu*, rye, and *Ae. speltoides* could be used but here the introgression is more complicated.

Another strategy is mutagenesis. A wheat mutant with two to three times increased MGPA has been patented by the authors and co-workers [38]. This "HighPhy" wheat has an SNP in the composite element discussed above. Incremental replacement (1/3, 2/3, and 3/3) of conventional wheat with HighPhy all improved calcium and phosphorous digestibility compared to the control diet in a comprehensive feeding experiment with broiler chicken. Complete replacement with HighPhy wheat even significantly outperformed a control to which the standard dose of microbial phytase was added [39].

With the exception of the Triticeae cereals, none of the major food and feed grains accumulate nutritionally sufficient amounts of phytase during grain development [4,5]. Rather, phytase is synthesized de novo during germination in response to gibberellic acid. Introducing a transgene driven by a seed development specific promoter has, therefore, been considered the only realistic approach to introduce preformed phytase in major grains, such as maize, rice, and soybean. However, the recent advances in genome editing may change this because they enable targeted modification of endogenous genes [40]. In theory, this can introduce preformed phytase activity in one of two ways (a) by modifying the promoter of endogenous phytase genes to become active during grain development or (b) by modifying the active site of phosphatases already expressed during grain development so they become phytases. It seems possible to pursue option "a" by combining the lessons of the natural evolution of the *PAPhy_a* promoter from the *PAPhy_b* counterpart with target species-specific information on cis-acting regulatory elements involved in seed development-specific expression. The pursuit of option "b" would have to be guided by structural biology but has the advantage that mutations could be tested in a heterologous system before the more laborious plant gene editing. It is also likely that this approach would require smaller changes to the target gene sequence because a single base mutation is enough to change one critical amino acid.

Int. J. Mol. Sci. **2019**, *20*, 2459

Proteinaceous inhibitors are known for many hydrolases including amylases, proteases, and xylanases. To our knowledge, such an inhibitor has never been reported for a phytase of microbial or plant origin. However, it was recently discovered that barley grain extracts do inhibit *Aspergillus* phytase. The inhibition could be reduced by the addition of pepstatin A. This suggests that the apparent inhibition is caused by an aspartic acid proteolytic activity [41]. It would represent a new and important variable, if plants use specialized proteases rather than conventional inhibitors to counteract phytases, e.g., from pathogens.

In conclusion, preformed wheat and barley phytase has the potential to counter the negative effects of phytate in the nutrition of single-stomached farm animals and humans alike. Realizing this potential depends on awareness because minor adjustments to processing may be needed. Generally, these adjustments should ensure that the phytase is not inactivated before ingestion or, alternatively, that the phytase has optimal conditions to perform the dephytination before inactivation. The higher the MGPA, the better the chances of a successful dephytination. Characterization of the *PAPhy_a* gene over the past decade should be most helpful for plant breeders who make MGPA a priority.

Funding: This research received no external funding.

Conflicts of Interest: The authors are co-inventors of patent WO 2012/146597Al.

References

1. Brinch-Pedersen, H.; Madsen, C.K.; Holme, I.B.; Dionisio, G. Increased understanding of the cereal phytase complement for better mineral bio-availability and resource management. *J. Cereal Sci.* **2014**, *59*, 373–381. [CrossRef]
2. Bethke, P. From Storage Compartment to Lytic Organelle: The Metamorphosis of the Aleurone Protein Storage Vacuole. *Ann. Bot.* **1998**, *82*, 399–412. [CrossRef]
3. Bohn, L.; Josefsen, L.; Meyer, A.S.; Rasmussen, S.K. Quantitative Analysis of Phytate Globoids Isolated from Wheat Bran and Characterization of Their Sequential Dephosphorylation by Wheat Phytase. *J. Agric. Food Chem.* **2007**, *55*, 7547–7552. [CrossRef] [PubMed]
4. Eeckhout, W.; De Paepe, M. Total phosphorus, phytate-phosphorus and phytase activity in plant feedstuffs. *Anim. Feed Sci. Technol.* **1994**, *47*, 19–29. [CrossRef]
5. Steiner, T.; Mosenthin, R.; Zimmermann, B.; Greiner, R.; Roth, S. Distribution of phytase activity, total phosphorus and phytate phosphorus in legume seeds, cereals and cereal by-products as influenced by harvest year and cultivar. *Anim. Feed Sci. Technol.* **2007**, *133*, 320–334. [CrossRef]
6. Viveros, A.; Centeno, C.; Brenes, A.; Canales, R.; Lozano, A. Phytase and Acid Phosphatase Activities in Plant Feedstuffs. *J. Agric. Food Chem.* **2000**, *48*, 4009–4013. [CrossRef]
7. Bouis, H.E. Special Issue on Improving Human Nutrition Through Agriculture. *Food Nutr. Bull.* **2000**, *21*, 351–578.
8. De Benoist, B.; Cogswell, M.; Egli, I.; McLean, E. Worldwide prevalence of anaemia 1993–2005. In *WHO Global Database of Anaemia*; WHO: Geneva, Switzerland, 2008.
9. Stoltzfus, R.J.; Mullany, L.; Black, R.E. Iron deficiency anemia. In *Comparative Quantification of Health Risks: Global and Regional Burden of Disease Attribution to Selected Major Risk Factors*; Ezzati, M., Lopez, A.D., Rodgers, A., Murray, C.J., Eds.; World Health Organization: Geneva, Switzerland, 2004; ISBN 9241580313.
10. Brown, K.H.; Rivera, J.A.; Bhutta, Z.A.; Gibson, R.S.; King, J.C.; Lonnerdal, B. Chapter 2: Assessment of the Risk of Zinc Deficiency in Populations. *Food Nutr. Bull.* **2004**, *25*, S130.
11. Yakoob, M.Y.; Theodoratou, E.; Jabeen, A.; Imdad, A.; Eisele, T.P.; Ferguson, J.; Jhass, A.; Rudan, I.; Campbell, H.; Black, R.E.; et al. Preventive zinc supplementation in developing countries: impact on mortality and morbidity due to diarrhea, pneumonia and malaria. *BMC Public Health* **2011**, *11*, S23. [CrossRef] [PubMed]
12. Schindler, D.W.; Carpenter, S.R.; Chapra, S.C.; Hecky, R.E.; Orihel, D.M. Reducing Phosphorus to Curb Lake Eutrophication is a Success. *Environ. Sci. Technol.* **2016**, *50*, 8923–8929. [CrossRef] [PubMed]
13. Cordell, D.; Drangert, J.-O.; White, S. The story of phosphorus: Global food security and food for thought. *Glob. Environ. Chang.* **2009**, *19*, 292–305. [CrossRef]

14. Lei, X.G.; Porres, J.M.; Mullaney, E.J.; Brinch-Pedersen, H. Phytase: Source, Structure and Application. In *Industrial Enzymes: Structure, Function and Applications*; Polaina, J., MacCabe, A., Eds.; Springer: Dordrecht, The Netherlands, 2007; pp. 505–530. ISBN 978-1-4020-5376-4.

15. Dionisio, G.; Holm, P.B.; Brinch-Pedersen, H. Wheat (*Triticum aestivum* L.) and barley (*Hordeum vulgare* L.) multiple inositol polyphosphate phosphatases (MINPPs) are phytases expressed during grain filling and germination. *Plant Biotechnol. J.* **2007**, *5*, 325–338. [CrossRef]

16. Dionisio, G.; Madsen, C.K.; Holm, P.B.; Welinder, K.G.; Jørgensen, M.; Stoger, E.; Arcalis, E.; Brinch-Pedersen, H. Cloning and characterization of purple acid phosphatase phytases from wheat, barley, maize, and rice. *Plant Physiol.* **2011**, *156*. [CrossRef] [PubMed]

17. Ullah, A.H.J.; Phillippy, B.Q. Substrate selectivity in Aspergillus ficuum phytase and acid phosphatases using myo-inositol phosphates. *J. Agric. Food Chem.* **1994**, *42*, 423–425. [CrossRef]

18. Holme, I.B.; Wendt, T.; Gil-Humanes, J.; Deleuran, L.C.; Starker, C.G.; Voytas, D.F.; Brinch-Pedersen, H. Evaluation of the mature grain phytase candidate HvPAPhy_a gene in barley (Hordeum vulgare L.) using CRISPR/Cas9 and TALENs. *Plant Mol. Biol.* **2017**, *95*, 111–121. [CrossRef]

19. Swanson, S.J.; Jones, R.L. Gibberellic Acid Induces Vacuolar Acidification in Barley Aleurone. *Plant Cell* **1996**, *8*, 2211, LP-2221. [CrossRef] [PubMed]

20. Greiner, R.; Jany, K.-D.; Larsson Alminger, M. Identification and Properties of myo-Inositol Hexakisphosphate Phosphohydrolases (Phytases) from Barley (Hordeum vulgare). *J. Cereal Sci.* **2000**, *31*, 127–139. [CrossRef]

21. Jiang, L.; Phillips, T.E.; Hamm, C.A.; Drozdowicz, Y.M.; Rea, P.A.; Maeshima, M.; Rogers, S.W.; Rogers, J.C. The protein storage vacuole. *J. Cell Biol.* **2001**, *155*, 991, LP-1002. [CrossRef]

22. Madsen, C.K.; Dionisio, G.; Holme, I.B.; Holm, P.B.; Brinch-Pedersen, H. High mature grain phytase activity in the Triticeae has evolved by duplication followed by neofunctionalization of the purple acid phosphatase phytase (PAPhy) gene. *J. Exp. Bot.* **2013**, *64*. [CrossRef]

23. IWGSC A chromosome-based draft sequence of the hexaploid bread wheat (Triticum aestivum) genome. *Science* **2014**, *345*, 1251788. [CrossRef]

24. IWGSC IWGSC-Blast. Available online: https://urgi.versailles.inra.fr/blast_iwgsc/blast.php (accessed on 17 May 2019).

25. Madsen, C.K.; Petersen, G.; Seberg, O.; Brinch-Pedersen, H. Evolution and diversity of PAPhy_a phytase in the genepool of wheat (Triticum aestivum L., Poaceae). *Genet. Resour. Crop Evol.* **2017**, *64*. [CrossRef]

26. Guo, W.; Yang, H.; Liu, Y.; Gao, Y.; Ni, Z.; Peng, H.; Xin, M.; Hu, Z.; Sun, Q.; Yao, Y. The wheat transcription factor TaGAMyb recruits histone acetyltransferase and activates the expression of a high-molecular-weight glutenin subunit gene. *Plant J.* **2015**, *84*, 347–359. [CrossRef]

27. De Pater, S.; Katagiri, F.; Kijne, J.; Chua, N.-H. bZIP proteins bind to a palindromic sequence without an ACGT core located in a seed-specific element of the pea lectin promoter. *Plant J.* **1994**, *6*, 133–140. [CrossRef] [PubMed]

28. Müller, M.; Knudsen, S. The nitrogen response of a barley C-hordein promoter is controlled by positive and negative regulation of the GCN4 and endosperm box. *Plant J.* **1993**, *4*, 343–355. [CrossRef]

29. Bäumlein, H.; Nagy†, I.; Villarroel, R.; Inzé, D.; Wobus, U. Cis-analysis of a seed protein gene promoter: the conservative RY repeat CATGCATG within the legumin box is essential for tissue-specific expression of a legumin gene. *Plant J.* **1992**, *2*, 233–239.

30. Holme, I.B.; Dionisio, G.; Brinch-Pedersen, H.; Wendt, T.; Madsen, C.K.; Vincze, E.; Holm, P.B. Cisgenic barley with improved phytase activity. *Plant Biotechnol. J.* **2012**, *10*. [CrossRef]

31. Holme, I.B.; Dionisio, G.; Madsen, C.K.; Brinch-Pedersen, H. Barley HvPAPhy_a as transgene provides high and stable phytase activities in mature barley straw and in grains. *Plant Biotechnol. J.* **2017**, *15*. [CrossRef]

32. Pointillart, A.; Fourdin, A.; Fontaine, N. Importance of Cereal Phytase Activity for Phytate Phosphorus Utilization by Growing Pigs Fed Diets Containing Triticale or Corn. *J. Nutr.* **1987**, *117*, 907–913. [CrossRef]

33. Sandberg, A.-S.; Andersson, H. Effect of Dietary Phytase on the Digestion of Phytate in the Stomach and Small Intestine of Humans. *J. Nutr.* **1988**, *118*, 469–473. [CrossRef] [PubMed]

34. Türk, M.; Carlsson, N.-G.; Sandberg, A.-S. Reduction in the Levels of Phytate During Wholemeal Bread Making; Effect of Yeast and Wheat Phytases. *J. Cereal Sci.* **1996**, *23*, 257–264. [CrossRef]

35. Koréissi-Dembélé, Y.; Fanou-Fogny, N.; Moretti, D.; Schuth, S.; Dossa, R.A.M.; Egli, I.; Zimmermann, M.B.; Brouwer, I.D. Dephytinisation with intrinsic wheat phytase and iron fortification significantly increase iron absorption from fonio (Digitaria exilis) meals in West African women. *PLoS ONE* **2013**, *8*, e70613. [CrossRef] [PubMed]

36. Singh, N.; Wu, S.; Tiwari, V.; Sehgal, S.; Raupp, J.; Wilson, D.; Abbasov, M.; Gill, B.; Poland, J. Genomic Analysis Confirms Population Structure and Identifies Inter-Lineage Hybrids in Aegilops tauschii. *Front. Plant Sci.* **2019**, *10*, 9. [CrossRef]

37. Feuillet, C.; Langridge, P.; Waugh, R. Cereal breeding takes a walk on the wild side. *Trends Genet.* **2008**, *24*, 24–32. [CrossRef] [PubMed]

38. Brinch-Pedersen, H.; Madsen, C.K.; Dionisio, G.; Holm, P.B. High expression cereal phytase gene. WO 2012/146597 Al. 2012. Available online: https://patents.google.com/patent/WO2012146597A1/en (accessed on 17 May 2019).

39. Scholey, D.; Burton, E.; Morgan, N.; Sanni, C.; Madsen, C.K.; Dionisio, G.; Brinch-Pedersen, H. P and Ca digestibility is increased in broiler diets supplemented with the high-phytase HIGHPHY wheat. *Animal* **2017**, *11*. [CrossRef]

40. Yin, K.; Gao, C.; Qiu, J.-L. Progress and prospects in plant genome editing. *Nat. Plants* **2017**, *3*, 17107. [CrossRef] [PubMed]

41. Bekalu, Z.E.; Madsen, C.K.; Dionisio, G.; Brinch-Pedersen, H. Aspergillus ficuum phytase activity is inhibited by cereal grain components. *PLoS ONE* **2017**, *12*. [CrossRef]

© 2019 by the authors. Licensee MDPI, Basel, Switzerland. This article is an open access article distributed under the terms and conditions of the Creative Commons Attribution (CC BY) license (http://creativecommons.org/licenses/by/4.0/).

International Journal of
Molecular Sciences

MDPI

Review

Insights on the Proteases Involved in Barley and Wheat Grain Germination

Mercedes Diaz-Mendoza [1], Isabel Diaz [1,2] and Manuel Martinez [1,2,*

[1]   Centro de Biotecnologia y Genomica de Plantas (CBGP, UPM-INIA), Universidad Politecnica de Madrid (UPM)- Instituto Nacional de Investigacion y Tecnología Agraria y Alimentaria (INIA), Campus Montegancedo, 28223 Pozuelo de Alarcon, Madrid, Spain; mercedes.diaz.mendoza@upm.es (M.D.-M.); i.diaz@upm.es (I.D.)

[2]   Departamento de Biotecnologia-Biologia Vegetal, Escuela Tecnica Superior de Ingenieria Agronomica, Alimentaria y de Biosistemas, UPM, 28040-Madrid, Spain

*   Correspondence: m.martinez@upm.es; Tel.: +34-910679149

Received: 26 March 2019; Accepted: 24 April 2019; Published: 28 April 2019

Abstract: Seed storage proteins must be hydrolyzed by proteases to deliver the amino acids essential for embryo growth and development. Several groups of proteases involved in this process have been identified in both the monocot and the dicot species. This review focuses on the implication of proteases during germination in two cereal species, barley and wheat, where proteolytic control during the germination process has considerable economic importance. Formerly, the participation of proteases during grain germination was inferred from reports of proteolytic activities, the expression of individual genes, or the presence of individual proteins and showed a prominent role for papain-like and legumain-like cysteine proteases and for serine carboxypeptidases. Nowadays, the development of new technologies and the release of the genomic sequences of wheat and barley have permitted the application of genome-scale approaches, such as those used in functional genomics and proteomics. Using these approaches, the repertoire of proteases known to be involved in germination has increased and includes members of distinct protease families. The development of novel techniques based on shotgun proteomics, activity-based protein profiling, and comparative and structural genomics will help to achieve a general view of the proteolytic process during germination.

Keywords: barley; wheat; protease; germination; grain

---

1. Introduction

Barley is considered a model organism for the investigation of the cereal germination process. Along with maize and rice, allohexaploid bread wheat and diploid barley are the most cultivated crops in the world (FAOSTAT database, http://www.fao.org/faostat, access on 22 April 2019). Their economic importance and close relationship support a parallel study of both cereals. The role of plant proteases in the mobilization of storage proteins that have accumulated in seeds has been largely established in both the dicotyledonous and the monocotyledonous species [1–3]. Storage proteins must be degraded to sustain embryo growth and development until an autotrophic growth is reached. Thus, a controlled proteolysis is crucial for the accurate delivery of amino acids in the initial stages of seed germination. Several protease families are involved in the germination process. Cysteine proteases (CysProt) of the C1A family, which are known as papain-like, and the C13 family, alternatively called legumains or vacuolar processing enzymes (VPEs), are the main proteases involved in the germination of both dicot and monocot species [1,2,4]. In dicot species, storage proteins are placed in the mesophyll of the cotyledons and in the embryonic axis. Members of the papain-like, legumain-like, and subtilisin-like (S8) families have been demonstrated to participate in the breakdown and mobilization of reserve proteins from seeds to cotyledons during germination [5–7].

Monocot seeds include proteins with many different functions. Around 80% of these proteins are storage proteins, packed in the endosperm together with starch and lipids. These proteins are synthesized during grain development and maturation and consequently are involved in germination. Among the proteases involved in the germination process, CysProt are responsible for around 90% of the proteolytic activity [8]. Other than CysProt from the papain family (C1A) and the legumain family (C13), members of the S10 serine carboxypeptidases (SCP) have also been implicated in the germination process in cereal grains. Papain-like CysProt participating in different stages of the germination process include the cathepsin L-like proteases identified in rice (oryzains α and β) and triticale (EP8), the cathepsin H-like proteases (oryzain y) from rice, and the cathepsin B-like proteases (BdCathB) from *Brachypodium distachyon* [9–12]. Among legumains, the OsVPE-1 protease was described in the degradation of stored proteins in the rice grain [13], and the REP-2 rice legumain was suggested as an activator of other CysProt during rice germination [14]. In this process, the SCP46 serine carboxypeptidase from rice regulates grain filling and seed germination upon hormonal induction [15,16]. Besides, serine carboxypeptidases I and III from triticale grains effectively degraded storage proteins that were proteolytically modified by the cathepsin L-like protease EP8 [17,18].

The participation of proteases in the germination processes of barley and wheat will be widely described in following sections.

2. Mobilization of Stored Proteins During the Germination of Barley and Wheat

Monocot species like barley and wheat have caryopses or cereal grains as propagation units. Cereal grains are endospermic seeds, meaning that the storage proteins are accumulated in the endosperm tissue (Figure 1A). This tissue consists of the starchy endosperm, which is a dead storage tissue, and the aleurone layer, which is formed by living cells. The main tissues found in the barley embryo are the coleoptile, the scutellum, and the radicle. During seed development and maturation, deposition of reserves within the storage tissue takes place. Starch, proteins, and lipids are mainly accumulated in the endosperm tissue but are also found in axis organs like the radicle and the embryonic shoot, or in the outer aleurone layer [19]. Cereal grains contain relatively little protein, with an average stored amount of 10–12% of the dry weight [19]. This storage fraction represents 80–85% of the total protein content. The main seed storage proteins are classified as albumins, globulins, and prolamins, on the basis of their solubility [20]. Most storage proteins in barley and wheat are prolamins [19], which are named hordeins in barley and gliadins in wheat. The antagonism between gibberellins (GA) and abscisic acid (ABA) is an important factor regulating the developmental transition from seed maturation to seed germination [21]. In terms of physiological and morphological changes, seed germination typically begins with dry mature seed imbibition and ends with radicle protrusion (Figure 1B). The embryo is responsible for the synthesis of GA after imbibition in water. This hormone reaches the aleurone layer via the scutellum, where it induces the expression of genes encoding α-amylases and proteases. The combination of the CysProt stored in the protein bodies and de novo-formed proteolytic enzymes, which are spread into the endosperm, triggers the hydrolysis of most proteins in the storage tissue. The resulting amino acids and small peptides are absorbed by the scutellum, which delivers them to the growing seedling (Figure 1A). Once the radicle breaks and protrudes from the seed coat, the germination process is accomplished [22].

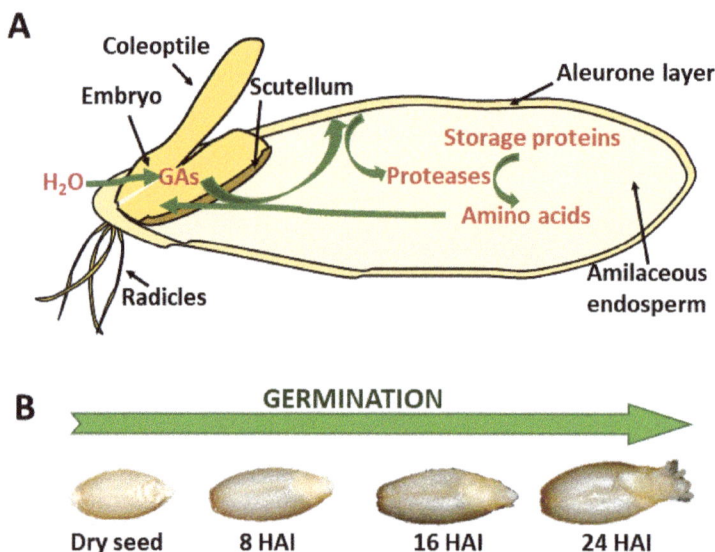

Figure 1. Cereal seed germination. (**A**) Schematic representation of the main events during storage protein remobilization in cereal seeds. (**B**) Photographs of the germination process of barley grains. HAI, hours after imbibition.

3. Investigation of Proteases in the Germination of Barley and Wheat before High-Throughput Technologies

Historically, the participation of proteases during grain germination was inferred from reports on proteolytic activities, the expression of individual genes during the process, or the presence of individual proteins in the germinating grain (Figure 2). Zhang and Jones [8], using bidimensional gel analyses and class-specific protease inhibitors, found 42 different spots with protease activity in the barley germinating grain, putatively corresponding to 27 cysteine proteases, 8 serine proteases, 4 aspartic proteases, and 3 metalloproteases. Similarly, Dominguez and Cejudo [23] used polyacrylamide gels copolymerized with gelatin to detect proteolytic activities in germinating wheat grains. Again, putative CysProt activity was preferentially observed, although serine, aspartic, and metalloprotease activities were also present. In parallel, individual proteases were isolated from barley and wheat germinating grains and identified by sequencing the corresponding cDNAs. The first barley CysProt described participating in the proteolytic degradation of the storage proteins in the grains [24–27] were the cathepsin L-like proteases EP-A and EP-B from the C1A family. Besides, a cathepsin H-like protease, aleurain, and a cathepsin B-like protein, HvCathB, were detected in the aleurone of the barley grain [28,29]. In wheat, several cathepsin L-like CysProt, WEB-1, WEP-2, EP-A, WCP-2, and gliadain, and the cathepsin B-like proteases Al16, Al20, and Al21 were described as participants in the proteolysis processes of the germinating grain [30–34]. In addition, members of the S10 serine carboxypeptidase family were implicated in the germination of barley and wheat [35–37], and some aspartic proteases were identified in barley and wheat seeds [38,39].

Technique	Subject	Barley/wheat references
Former techniques		
Zymograms	Proteolytic activities	[8], [23]
cDNA isolation	Individual protease identification	[24-29] [35] [38], [30-34] [36-37] [39]
Functional genomics		
Genomic searches	Members of each protease family	[44-45] [48-49]
Microarray assays	Proteases differentially expressed	[41-42], [53-54]
RNA-seq assays	Proteases differentially expressed	[51], [57]
RT-qPCR assays	Individual protease expression	[46-47] [49] [52]
Reverse genetics	Phenotypic/proteolytic features	[65]
Proteomics		
2D-gel electrophoresis/ Mass spectrometry	Protease identification	[58-59], [3] [60-61]
Shotgun/ Mass spectrometry	Protease identification	[63]
Activity based protein profiling	Active proteases	[73]
Purified recombinant protease	Storage protein hydrolysis	[46-47] [64-65], [34] [68-70]

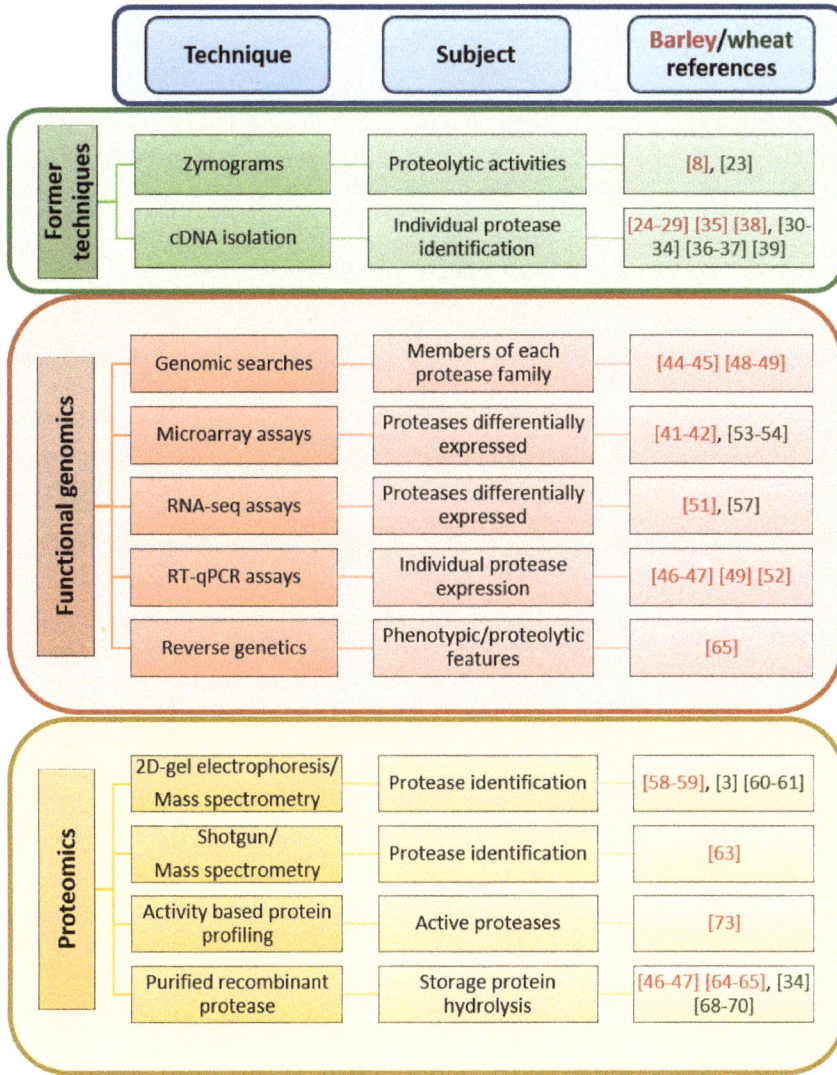

Figure 2. Global scheme of the main techniques used to analyze the implication of proteases during the germination of barley and wheat grains.

Nowadays, advances in high-throughput sequencing have resulted in the development of new technologies and strategies that give support to experimental research. The release of the genomic sequences of wheat and barley has permitted the application of genome-scale approaches, such as those used in functional genomics and proteomics. These technologies have been employed, mainly in barley, in the study of proteases during the germination process (Figure 2).

4. Functional Genomic-Based Advances in the Identification of Proteases in the Germination of Barley and Wheat Grains

The extremely large number of expressed sequence tags (EST) and cDNA sequences in barley permitted the first approximation of the protease repertoire. Using the Affymetrix Barley1 GeneChip

22K [40], the expression of about 20 C1A proteases could be checked. This GeneChip has been used in several studies on gene expression during germination and seedling elongation (reviewed in [41]). Among these studies, a detailed transcriptome analysis of barley grain germination can be highlighted [42]. In this analysis, several C1A papain-like CysProt, such as *EP-A*, *EP-B1*, and *EP-B2*, were expressed in both the aleurone and the embryo during grain germination. In addition, transcripts of cathepsin B-like (*HvPap-19* and *-20*), cathepsin H-like (*aleurain*), cathepsin F-like (*HvPap-1*), and some cathepsin L-like proteases (*HvPap-4* and *-6*) were abundant during seed germination in both the aleurone and the embryo, but were also expressed during seed maturation. Besides, transcripts for several serine carboxypeptidases were highly detected during germination in both the embryo and the aleurone [42].

A second step in the identification process came from the public release of the barley genome sequence [43]. Previous analysis from EST collections showed the presence of 32 C1A papain-like members [44], which increased to 41 members upon genomic mining [45]. A former analysis of the expression of cathepsin-L like proteases from different phylogenetic groups showed that four of the five selected cathepsin L-like genes (*HvPap-4*, *HvPap-6*, *HvPap-10*, and *HvPap-17*) were primarily transcribed in germinating embryos [46]. The most abundant transcripts were those of *HvPap-10*, which had a seed-specific pattern of expression, followed by *HvPap-6* and *HvPap-4*. Whereas a lower expression of the *HvPap-17* gene was detected in germinating embryos, *HvPap-16* was exclusively transcribed in barley leaves. No expression of any of the cathepsin L-like genes studied was detected in the developing barley endosperms. During germination, GA treatment of the aleurone layer increased the quantity of transcripts from *HvPap-6* and *HvPap-10* but had no effect on the expression level of the *HvPap-4* gene. Likewise, de-embryonated barley grains showed that the *HvPap-1* gene was expressed during grain germination and that GA treatment induced a remarkable increase in its expression [47].

The genomic content of C13 legumain-like genes has also been addressed. Formerly, five legumains were reported from EST collections [44], two additional legumains were later detected [48], and finally, a novel protein was added after an in-depth search of the barley genome [49]. Regarding the barley legumains group, it has been suggested that legumains are able to process other CysProt in order to activate them to take part in the proteolytic degradation of the storage proteins [50]. For example, the HvLeg-2 legumain of barley, which is highly expressed during germination, could be involved in the mobilization of storage proteins, either by direct proteolytic degradation or by the processing and activation of other CysProt [49,50]. In fact, the capacity of HvLeg-2 to degrade storage seed globulins was demonstrated, confirming its role as a hydrolytic enzyme against storage proteins. Likewise, HvLeg-2 could participate in the processing of other peptidases, such as the papain-like CysProt induced by GA, which degrade hordeins [46,47]. This is similar to the action of the legumain REP-2 on the papain-like peptidase REP-1 in germinating rice seeds [14].

Recently, a new methodology has been developed for isolating fragments of aleurone, starchy endosperm, embryo, scutellum, pericarp–testa, husk, and crushed cell layers from barley germinated grain [51]. This method is based on rapid fixation of the intact grain, followed by dissection for subsequent transcriptomic analyses. Using this technology, the expression profiles of many genes were precisely defined during the first 24 h of germination [51]. Interestingly, an analysis of the differential expressed genes (DEG) in the aleurone fragment nearest the embryo after 24 h of germination showed the induction of many different proteases, including papain-like CysProt, legumain-like CysProt, and serine carboxypeptidases, and also several aspartic proteases, metalloproteases, and subtilisin-like serine proteases. Many of these proteases were exclusively upregulated in this proximal part of the aleurone layer, and none of them were overexpressed in the distal fragment of the aleurone. Following 24 h of germination, the set of upregulated proteases in the embryo and scutellum comprised several proteases exclusively upregulated in these tissues and some of the proteases overexpressed in the aleurone. Recently, the upregulation of several subtilisin-like serine proteases during grain germination has been confirmed by RT-qPCR assays [52].

In contrast to wide knowledge on the gene expression in the barley germinating grain, studies on wheat are scarce. To date, transcriptome expression profiles during seed germination have been performed using the GeneChip® Wheat Genome Array [53,54]. In whole germinating grains, the expression of several cysteine, aspartic, and serine proteases increased significantly between 24 h and 48 h after imbibition, with a peak at 36 h for the most expressed class, the CysProt [54]. Efforts on optimizing RNA-seq analyses clashed with the poor quality of the first draft of the wheat genome released by the International Wheat Genome Sequencing Consortium [55] (IWGSC, 2014). However, a tremendous effort using novel approaches generated the recently published high-quality linear reference assembly, IWGSC RefSeq v1.0 [56]. Likewise, many RNA-seq analyses have been performed and implemented in different portals, such as the Wheat Expression Browser (www.wheat-expression.com, [57]) which includes 850 wheat RNA-sequencing samples derived from 32 tissues at different growth stages and/or challenged by different stress treatments. Unfortunately, these analyses did not cover the germinating grain.

5. Proteomic-Based Advances on Proteases in the Germination of Barley and Wheat Grains

Wide proteomic analyses of barley and wheat grains were formerly based on two-dimensional gel electrophoresis, which was subsequently coupled with mass spectrometry. In the absence of a genome sequence, barley gene and EST sequences, combined with information from other cereals, facilitated the identification of barley proteins. From this wide range of proteomic analyses, many barley seed proteins were identified (reviewed by [58]). Besides, to avoid the masking of low-abundant proteins by large amounts of starchy endosperm storage proteins, isolated aleurone layers were used in proteomic analyses. When treated with GA, several C1A cysteine proteases, S10 serine proteases, and A1 aspartic proteases were identified in this tissue [59].

In contrast, relatively few proteomic studies have been performed during wheat seed germination. Several analyses combined two-dimensional electrophoresis (2-DE) with MALDI –TOF/TOF MS to explore the proteomic changes in the embryo and endosperm that occur throughout the germination period. From these analyses, distinct differentially expressed proteins were found in the seed embryo and endosperm, which presumably cooperate in seed germination [3,60,61]. Although the authors claimed that some of these proteins were related to storage protein metabolism, no individual proteases were identified. Besides, the proteome of isolated aleurone layers has only been addressed during seed development [62]. Recently, a gel-free proteomics approach was performed to obtain a dynamic proteome survey during barley malting. This shotgun proteomic technique entails the in-solution tryptic digestion of precipitated proteins and an analysis of peptides by nanoLC–MS/MS. A high number of proteins were identified [63], including several aspartic, cysteine, metallo, and serine peptidases, which demonstrates the strength of this technique for identifying low-expressed proteins. Although similar proteomics approaches have been performed in wheat, no one has analyzed grain germination.

On the other hand, protease activity in the barley grain has been analyzed using two different approaches. A classical approach is based on the capacity of recombinant proteases to degrade storage compounds. In a first analysis, whereas recombinant HvPap-10 protease was able to completely degrade all electrophoretic bands corresponding to B, C, and D hordeins from grain extracts, HvPap-6 only partially reduced the presence of those bands [46]. In addition, the capacity of HvPap-10 to hydrolyze the recombinant hordeins (B1, B3, and γ1) expressed in *Escherichia coli* was tested. After 2 h of incubation with HvPap-10, an almost complete degradation of γ1 and a partial digestion of hordein B1 and B3 were observed [64]. Besides, recombinant HvPap-1 was able to degrade different barley proteins (hordeins, albumins, and globulins) stored in the barley endosperm [47]. Further insight on the role of proteases in grain germination in transgenic plants was provided by silencing or overexpressing a specific protease. Only the silencing or overexpressing of the cathepsin F-like *HvPap-1* protease by barley plants has been described. These plants showed differential accumulation of storage molecules such as starch, proteins, and free amino acids in the grain, as well as disturbed electrophoretic patterns of hordeins, globulins, and albumins during the germinating process. Silencing lines showed a drastic delay in the grain germination process. Remarkably, this phenotypic feature could not be directly related to cathepsin F-like deficiencies, as alterations in the cathepsin L/F-like proteolytic activities were also accompanied by changes in cathepsin B-like and trypsin-like proteolytic activities [65].

In wheat, the C1A protease gliadain, purified from *E. coli*, was able to hydrolyze the storage α, β, and γ gliadins, but not the glutenins from grain extracts [34]. Gluten is a heterogeneous mixture of insoluble storage proteins, called gliadins, which contain proline-rich and glutamine-rich repetitive sequences. The fact that several of the peptides derived from gluten are toxic for humans with celiac disease led to the identification of proteases with the ability to degrade it [66,67]. Proteases from protein extracts of germinated barley and wheat grains showed the ability to degrade gliadin-derived toxic peptides [68]. In particular, the C1A protease Triticain-α, formerly thought to participate in seed germination by digesting storage proteins [69], was shown to possess glutenase activity in vitro [70]. Triticain-α cleavage sites were found in the majority of the previously identified gluten-derived toxic peptides, including the major 33-mer α-gliadin-derived peptide. These findings support the potential of Triticain-α as a basic compound for the development of drugs against celiac disease [70].

The second approach is based on the development of activity-based proteomics, also known as activity-based protein profiling (ABPP). This method uses molecular probes which bind irreversibly to the reactive site of members of specific groups of enzymes. The results provided information on enzyme activity, not just protein abundance [71], allowing differentiation between the inactive plant proteases synthesized as zymogens and the active proteases, after proteolytic processing. As the specificity of many commercial protease inhibitors is inaccurate, specific fluorescent probes were developed for ABPP. When applied on *Arabidopsis* germinating seeds, the fluorescent activity-based probes specifically targeted three distinct cysteine protease subfamilies, revealing the dynamic activities of aleurain-like proteases, cathepsin B-like proteases, and vacuolar processing enzymes during the remobilization of stored proteins [72]. This technology has recently been applied to monitor the activity of different enzymes in the germination process of the barley grain. Using specific probes and ABPP to detect the active enzymes extracted from the aleurone layers of a commercial malting barley variety, several active proteases were found to be induced by GA, such as putative aleurains, cathepsin- B-like proteases, and serine hydrolases [73].

6. Conclusions and Future Perspectives

Germination is a key process in the life cycle of plants. During this period, a new plant starts its development from the embryo with the help of the compounds stored in the seed endosperm. The continuous release of nutrients from the endosperm to the embryo is crucial to achieve the correct development of the new plant and to avoid agronomical losses due to the absence of seed germination in the field. Therefore, knowledge on the regulatory mechanisms that take part during this process must be improved to establish the best conditions for a correct germination. In the case of barley and wheat, many advances have been made in understanding the role of proteases in the grain due to their potential value for the brewing industry and their relationship with the celiac disease. Using different technologies, many proteases have been identified and closely associated to the grain germination. The development of novel techniques based on shotgun proteomics, ABPP, and comparative and structural genomics will lead to a more comprehensive understanding of this process and the specific roles of the proteases involved. For example, the identification by ABPP of active proteases during the barley grain germination process will increase the selection of efficient prolamin-degrading proteases. To reduce costs, the brewing industry is replacing barley malt with unmalted grains [74]. Thus, these proteases may be included in the commercial brewing enzyme cocktails used to improve the wort obtained from unmalted barley grains. Likewise, technical advances will allow the selection of high gliadin-degrading efficient proteases, which could be combined with the low-gliadin wheat plants obtained by genome edition [75], as therapeutic alternatives in the treatment of celiac disease. Furthermore, silencing or overexpressing a specific protease will contribute to the knowledge on the role of the protease and also in understanding the intricate network of proteolytic reactions combining

Int. J. Mol. Sci. **2019**, *20*, 2087

different classes of proteases and protease inhibitors involved in the remobilization of storage proteins. The required technology has recently been developed, and the challenge is to combine research efforts to address key questions concerning the control of proteolysis during seed germination.

Author Contributions: M.M. conceived the review. M.D-M., I.D., and M.M. wrote the manuscript.

Acknowledgments: The financial support from the Ministerio de Economía, Industria y Competitividad (project BIO2014-53508-R) is highly acknowledged.

Conflicts of Interest: The authors declare no conflict of interest.

References

1. Grudkowska, M.; Zagdańska, B. Multifunctional role of plant cysteine proteinases. *Acta Biochim. Pol.* **2004**, *51*, 609–624. [PubMed]
2. Szewińska, J.; Simińska, J.; Bielawski, W. The roles of cysteine proteases and phytocystatins in development and germination of cereal seeds. *J. Plant Physiol.* **2016**, *207*, 10–21. [CrossRef] [PubMed]
3. Liu, Y.; Han, C.; Deng, X.; Liu, D.; Liu, N.; Yan, Y. Integrated physiology and proteome analysis of embryo and endosperm highlights complex metabolic networks involved in seed germination in wheat (*Triticum aestivum* L.). *J. Plant Physiol.* **2018**, *229*, 63–76. [CrossRef]
4. Tan-Wilson, A.L.; Wilson, K.A. Mobilization of seed protein reserves. *Physiol. Plant* **2012**, *145*, 140–153. [CrossRef]
5. Fischer, J.; Becker, C.; Hillmer, S.; Horstmann, C.; Neubohn, B.; Schlereth, A.; Senyuk, V.; Shutov, A.; Müntz, K. The families of papain- and legumain-like cysteine proteinases from embryonic axes and cotyledons of Vicia seeds: Developmental patterns, intracellular localization and functions in globulin proteolysis. *Plant Mol. Biol.* **2000**, *43*, 83–101. [CrossRef]
6. Schlereth, A.; Becker, C.; Horstmann, C.; Tiedemann, J.; Müntz, K. Comparison of globulin mobilization and cysteine proteinases in embryonic axes and cotyledons during germination and seedling growth of vetch (*Vicia sativa* L.). *J. Exp. Bot.* **2000**, *51*, 1423–1433. [PubMed]
7. Toyooka, K.; Okamoto, T.; Minamikawa, T. Mass transport of proform of a KDEL-tailed cysteine proteinase (SH-EP) to protein storage vacuoles by endoplasmic reticulum-derived vesicle is involved in protein mobilization in germinating seeds. *J. Cell Biol.* **2000**, *148*, 453–464. [CrossRef]
8. Zhang, N.; Jones, B.L. Characterization of germinated barley endoproteolytic enzymes by two dimensional gel electrophoresis. *J. Cereal Sci.* **1995**, *21*, 145–153. [CrossRef]
9. Watanabe, H.; Abe, K.; Emori, Y.; Hosoyama, H.; Arai, S. Molecular cloning and gibberellin-induced expression of multiple cysteine proteinases of rice seeds (oryzains). *J. Biol. Chem.* **1991**, *266*, 16897–16902. [PubMed]
10. Prabucka, B.; Bielawski, W. Purification and partial characteristic of a major gliadin-degrading cysteine endopeptidase from germinating triticale seeds. *Acta Physiol. Plant.* **2004**, *26*, 383. [CrossRef]
11. Prabucka, B.; Drzymała, A.; Grabowska, A. Molecular cloning and expression analysis of the main gliadin-degrading cysteine endopeptidase EP8 from triticale. *J. Cereal Sci.* **2013**, *58*, 284–289. [CrossRef]
12. González-Calle, V.; Iglesias-Fernández, R.; Carbonero, P.; Barrero-Sicilia, C. The BdGAMYB protein from Brachypodium distachyon interacts with BdDOF24 and regulates transcription of the BdCathB gene upon seed germination. *Planta* **2014**, *240*, 539–552. [CrossRef]
13. Wang, Y.; Zhu, S.; Liu, S.; Jiang, L.; Chen, L.; Ren, Y.; Han, X.; Liu, F.; Ji, S.; Liu, X.; et al. The vacuolar processing enzyme OsVPE1 is required for efficient glutelin processing in rice. *Plant J.* **2009**, *58*, 606–617. [CrossRef]
14. Kato, H.; Sutoh, K.; Minamikawa, T. Identification, cDNA cloning and possible roles of seed-specific rice asparaginyl endopeptidase, REP-2. *Planta* **2003**, *217*, 676–685. [CrossRef] [PubMed]
15. Washio, K.; Ishikawa, K. Organ-specific and hormone-dependent expression of genes for serine carboxypeptidases during development and following germination of rice grains. *Plant Physiol.* **1994**, *105*, 1275–1280. [CrossRef]
16. Li, Z.; Tang, L.; Qiu, J.; Zhang, W.; Wang, Y.; Tong, X.; Wei, X.; Hou, Y.; Zhang, J. Serine carboxypeptidase 46 Regulates Grain Filling and Seed Germination in Rice (*Oryza sativa* L.). *PLoS ONE* **2016**, *11*, e0159737. [CrossRef]

17. Drzymała, A.; Bielawski, W. Isolation and characterization of carboxypeptidase III from germinating triticale grains. *Acta Biochim. Biophys. Sin.* **2009**, *41*, 69–78. [CrossRef]
18. Drzymała, A.; Prabucka, B.; Bielawski, W. Carboxypeptidase I from triticale grains and the hydrolysis of salt-soluble fractions of storage proteins. *Plant Physiol. Biochem.* **2012**, *58*, 195–204. [CrossRef]
19. Shewry, P.R.; Halford, N.G. Cereal seed storage proteins: Structures, properties and role in grain utilization. *J. Exp. Bot.* **2002**, *53*, 947–958. [CrossRef]
20. Shewry, P.R.; Napier, J.A.; Tatham, A.S. Seed storage proteins: Structures and biosynthesis. *Plant Cell* **1995**, *7*, 945–956. [CrossRef] [PubMed]
21. Finch-Savage, W.E.; Leubner-Metzger, G. Seed dormancy and the control of germination. *New Phytol.* **2006**, *171*, 501–523. [CrossRef] [PubMed]
22. Bewley, J.D. Seed Germination and Dormancy. *Plant Cell* **1997**, *9*, 1055–1066. [CrossRef]
23. Dominguez, F.; Cejudo, F.J. Characterization of the Endoproteases Appearing during Wheat Grain Development. *Plant Physiol.* **1996**, *112*, 1211–1217. [CrossRef] [PubMed]
24. Poulle, M.; Jones, B.L. A Proteinase from Germinating Barley: I. Purification and Some Physical Properties of a 30 kD Cysteine Endoproteinase from Green Malt. *Plant Physiol.* **1988**, *88*, 1454–1460. [CrossRef] [PubMed]
25. Koehler, S.M.; Ho, T.H. A major gibberellic Acid-induced barley aleurone cysteine proteinase which digests hordein: Purification and characterization. *Plant Physiol.* **1990**, *94*, 251–258. [CrossRef] [PubMed]
26. Koehler, S.; Ho, T.H. Purification and characterization of gibberellic Acid-induced cysteine endoproteases in barley aleurone layers. *Plant Physiol* **1988**, *87*, 95–103. [CrossRef] [PubMed]
27. Mikkonen, A.; Porali, I.; Cercos, M.; Ho, T.H. A major cysteine proteinase, EPB, in germinating barley seeds: Structure of two intronless genes and regulation of expression. *Plant Mol. Biol.* **1996**, *31*, 239–254. [CrossRef]
28. Rogers, J.C.; Dean, D.; Heck, G.R. Aleurain: A barley thiol protease closely related to mammalian cathepsin H. *Proc. Natl. Acad. Sci. USA* **1985**, *82*, 6512–6516. [CrossRef]
29. Martinez, M.; Rubio-Somoza, I.; Carbonero, P.; Diaz, I. A cathepsin B-like cysteine protease gene from Hordeum vulgare (gene CatB) induced by GA in aleurone cells is under circadian control in leaves. *J. Exp. Bot.* **2003**, *54*, 951–959. [CrossRef]
30. Cejudo, F.J.; Murphy, G.; Chinoy, C.; Baulcombe, D.C. A gibberellin-regulated gene from wheat with sequence homology to cathepsin B of mammalian cells. *Plant J.* **1992**, *2*, 937–948.
31. Jivotovskaya, A.V.; Horstmann, C.; Vaintraub, I.A. Detection of the isoenzymes of wheat grain proteinase A. *Phytochemistry* **1997**, *45*, 1549–1553. [CrossRef]
32. Sutoh, K.; Kato, H.; Minamikawa, T. Identification and possible roles of three types of endopeptidase from germinated wheat seeds. *J. Biochem.* **1999**, *126*, 700–707.
33. Tsuji, A.; Tsuji, M.; Takami, H.; Nakamura, S.; Matsuda, Y. Molecular cloning and expression analysis of novel wheat cysteine protease. *Biochim. Biophys. Acta* **2004**, *1670*, 84–89. [CrossRef]
34. Kiyosaki, T.; Matsumoto, I.; Asakura, T.; Funaki, J.; Kuroda, M.; Misaka, T.; Arai, S.; Abe, K. Gliadain, a gibberellin-inducible cysteine proteinase occurring in germinating seeds of wheat, *Triticum aestivum* L., specifically digests gliadin and is regulated by intrinsic cystatins. *FEBS J.* **2007**, *274*, 1908–1917. [CrossRef]
35. Dal Degan, F.; Rocher, A.; Cameron-Mills, V.; von Wettstein, D. The expression of serine carboxypeptidases during maturation and germination of the barley grain. *Proc. Natl. Acad. Sci. USA* **1994**, *91*, 8209–8213. [CrossRef]
36. Dominguez, F.; Cejudo, F.J. Patterns of starchy endosperm acidification and protease gene expression in wheat grains following germination. *Plant Physiol.* **1999**, *119*, 81–88. [CrossRef]
37. Domínguez, F.; González, M.C.; Cejudo, F.J. A germination-related gene encoding a serine carboxypeptidase is expressed during the differentiation of the vascular tissue in wheat grains and seedlings. *Planta* **2002**, *215*, 727–734. [CrossRef]
38. Törmäkangas, K.; Runeberg-Roos, P.; Ostman, A.; Tilgmann, C.; Sarkkinen, P.; Kervinen, J.; Mikola, L.; Kalkkinen, N. Aspartic proteinase from barley seeds is related to animal cathepsin D. *Adv. Exp. Med. Biol.* **1991**, *306*, 355–359.
39. Tamura, T.; Terauchi, K.; Kiyosaki, T.; Asakura, T.; Funaki, J.; Matsumoto, I.; Misaka, T.; Abe, K. Differential expression of wheat aspartic proteinases, WAP1 and WAP2, in germinating and maturing seeds. *J. Plant Physiol.* **2007**, *164*, 470–477. [CrossRef]

40. Close, T.J.; Wanamaker, S.I.; Caldo, R.A.; Turner, S.M.; Ashlock, D.A.; Dickerson, J.A.; Wing, R.A.; Muehlbauer, G.J.; Kleinhofs, A.; Wise, R.P. A new resource for cereal genomics: 22K barley GeneChip comes of age. *Plant Physiol.* **2004**, *134*, 960–968. [CrossRef]

41. Daneri-Castro, S.N.; Svensson, B.; Roberts, T.H. Barley germination: Spatio-temporal considerations for designing and interpreting 'omics' experiments. *J. Cereal Sci.* **2016**, *70*, 29–37. [CrossRef]

42. Sreenivasulu, N.; Usadel, B.; Winter, A.; Radchuk, V.; Scholz, U.; Stein, N.; Weschke, W.; Strickert, M.; Close, T.J.; Stitt, M.; et al. Barley grain maturation and germination: Metabolic pathway and regulatory network commonalities and differences highlighted by new MapMan/PageMan profiling tools. *Plant Physiol.* **2008**, *146*, 1738–1758. [CrossRef]

43. Mayer, K.F.; Waugh, R.; Brown, J.W.; Schulman, A.; Langridge, P.; Platzer, M.; Fincher, G.B.; Muehlbauer, G.J.; Sato, K.; Close, T.J.; et al. A physical, genetic and functional sequence assembly of the barley genome. *Nature* **2012**, *491*, 711–716. [CrossRef]

44. Martinez, M.; Diaz, I. The origin and evolution of plant cystatins and their target cysteine proteinases indicate a complex functional relationship. *BMC Evol. Biol.* **2008**, *8*, 198. [CrossRef]

45. Diaz-Mendoza, M.; Velasco-Arroyo, B.; Gonzalez-Melendi, P.; Martinez, M.; Diaz, I. C1A cysteine protease-cystatin interactions in leaf senescence. *J. Exp. Bot.* **2014**, *65*, 3825–3833. [CrossRef]

46. Martinez, M.; Cambra, I.; Carrillo, L.; Diaz-Mendoza, M.; Diaz, I. Characterization of the Entire Cystatin Gene Family in Barley and Their Target Cathepsin L-Like Cysteine-Proteases, Partners in the Hordein Mobilization during Seed Germination. *Plant Physiol.* **2009**, *151*, 1531–1545. [CrossRef]

47. Cambra, I.; Martinez, M.; Dader, B.; Gonzalez-Melendi, P.; Gandullo, J.; Santamaria, M.E.; Diaz, I. A cathepsin F-like peptidase involved in barley grain protein mobilization, HvPap-1, is modulated by its own propeptide and by cystatins. *J. Exp. Bot.* **2012**, *63*, 4615–4629. [CrossRef]

48. Radchuk, V.; Weier, D.; Radchuk, R.; Weschke, W.; Weber, H. Development of maternal seed tissue in barley is mediated by regulated cell expansion and cell disintegration and coordinated with endosperm growth. *J. Exp. Bot.* **2011**, *62*, 1217–1227. [CrossRef]

49. Julian, I.; Gandullo, J.; Santos-Silva, L.K.; Diaz, I.; Martinez, M. Phylogenetically distant barley legumains have a role in both seed and vegetative tissues. *J. Exp. Bot.* **2013**, *64*, 2929–2941. [CrossRef]

50. Cambra, I.; Garcia, F.J.; Martinez, M. Clan CD of cysteine peptidases as an example of evolutionary divergences in related protein families across plant clades. *Gene* **2010**, *449*, 59–69. [CrossRef]

51. Betts, N.S.; Berkowitz, O.; Liu, R.; Collins, H.M.; Skadhauge, B.; Dockter, C.; Burton, R.A.; Whelan, J.; Fincher, G.B. Isolation of tissues and preservation of RNA from intact, germinated barley grain. *Plant J.* **2017**, *91*, 754–765. [CrossRef] [PubMed]

52. Galotta, M.F.; Pugliese, P.; Gutiérrez-Boem, F.H.; Veliz, C.G.; Criado, M.V.; Caputo, C.; Echeverria, M.; Roberts, I.N. Subtilase activity and gene expression during germination and seedling growth in barley. *Plant Physiol. Biochem.* **2019**, *139*, 197–206. [CrossRef] [PubMed]

53. Yu, Y.; Zhen, S.; Wang, S.; Wang, Y.; Cao, H.; Zhang, Y.; Li, J.; Yan, Y. Comparative transcriptome analysis of wheat embryo and endosperm responses to ABA and H2O2 stresses during seed germination. *BMC Genom.* **2016**, *17*, 97. [CrossRef] [PubMed]

54. Yu, Y.; Guo, G.; Lv, D.; Hu, Y.; Li, J.; Li, X.; Yan, Y. Transcriptome analysis during seed germination of elite Chinese bread wheat cultivar Jimai 20. *BMC Plant Biol.* **2014**, *14*, 20. [CrossRef] [PubMed]

55. The International Wheat Genome Sequencing Consortium (IWGSC). A chromosome-based draft sequence of the hexaploid bread wheat (*Triticum aestivum*) genome. *Science* **2014**, *345*, 1251788. [CrossRef] [PubMed]

56. Appels, R.; Eversole, K.; Feuillet, C.; Keller, B.; Rogers, J.; Stein, N.; Pozniak, C.J.; Choulet, F.; Distelfeld, A.; Poland, J.; et al. Shifting the limits in wheat research and breeding using a fully annotated reference genome. *Science* **2018**, *361*, aar7191. [CrossRef]

57. Ramírez-González, R.H.; Borrill, P.; Lang, D.; Harrington, S.A.; Brinton, J.; Venturini, L.; Davey, M.; Jacobs, J.; van Ex, F.; Pasha, A.; et al. The transcriptional landscape of polyploid wheat. *Science* **2018**, *361*, aar6089. [CrossRef]

58. Finnie, C.; Svensson, B. Barley seed proteomics from spots to structures. *J. Proteom.* **2009**, *72*, 315–324. [CrossRef]

59. Finnie, C.; Andersen, B.; Shahpiri, A.; Svensson, B. Proteomes of the barley aleurone layer: A model system for plant signalling and protein secretion. *Proteomics* **2011**, *11*, 1595–1605. [CrossRef]

60. Dong, K.; Zhen, S.; Cheng, Z.; Cao, H.; Ge, P.; Yan, Y. Proteomic Analysis Reveals Key Proteins and Phosphoproteins upon Seed Germination of Wheat (*Triticum aestivum* L.). *Front. Plant Sci.* **2015**, *6*, 1017. [CrossRef]

61. He, M.; Zhu, C.; Dong, K.; Zhang, T.; Cheng, Z.; Li, J.; Yan, Y. Comparative proteome analysis of embryo and endosperm reveals central differential expression proteins involved in wheat seed germination. *BMC Plant Biol.* **2015**, *15*, 97. [CrossRef]

62. Nadaud, I.; Tasleem-Tahir, A.; Chateigner-Boutin, A.L.; Chambon, C.; Viala, D.; Branlard, G. Proteome evolution of wheat (*Triticum aestivum* L.) aleurone layer at fifteen stages of grain development. *J. Proteom.* **2015**, *123*, 29–41. [CrossRef] [PubMed]

63. Mahalingam, R. Temporal Analyses of Barley Malting Stages Using Shotgun Proteomics. *Proteomics* **2018**, *18*, e1800025. [CrossRef]

64. Rosenkilde, A.L.; Dionisio, G.; Holm, P.B.; Brinch-Pedersen, H. Production of barley endoprotease B2 in Pichia pastoris and its proteolytic activity against native and recombinant hordeins. *Phytochemistry* **2014**, *97*, 11–19. [CrossRef] [PubMed]

65. Diaz-Mendoza, M.; Dominguez-Figueroa, J.D.; Velasco-Arroyo, B.; Cambra, I.; Gonzalez-Melendi, P.; Lopez-Gonzalvez, A.; Garcia, A.; Hensel, G.; Kumlehn, J.; Diaz, I.; et al. HvPap-1 C1A Protease and HvCPI-2 Cystatin Contribute to Barley Grain Filling and Germination. *Plant Physiol.* **2016**, *170*, 2511–2524. [CrossRef] [PubMed]

66. Bethune, M.T.; Khosla, C. Oral enzyme therapy for celiac sprue. *Methods Enzymol.* **2012**, *502*, 241–271. [CrossRef] [PubMed]

67. Scherf, K.A.; Wieser, H.; Koehler, P. Novel approaches for enzymatic gluten degradation to create high-quality gluten-free products. *Food Res. Int.* **2018**, *110*, 62–72. [CrossRef] [PubMed]

68. Hartmann, G.; Koehler, P.; Wieser, H. Rapid degradation of gliadin peptides toxic for coeliac disease patients by proteases from germinating cereals. *J. Cereal Sci.* **2006**, *44*, 368–371. [CrossRef]

69. Kiyosaki, T.; Asakura, T.; Matsumoto, I.; Tamura, T.; Terauchi, K.; Funaki, J.; Kuroda, M.; Misaka, T.; Abe, K. Wheat cysteine proteases triticain alpha, beta and gamma exhibit mutually distinct responses to gibberellin in germinating seeds. *J. Plant Physiol.* **2009**, *166*, 101–106. [CrossRef] [PubMed]

70. Savvateeva, L.V.; Gorokhovets, N.V.; Makarov, V.A.; Serebryakova, M.V.; Solovyev, A.G.; Morozov, S.Y.; Reddy, V.P.; Zernii, E.Y.; Zamyatnin, A.A.; Aliev, G. Glutenase and collagenase activities of wheat cysteine protease Triticain-α: Feasibility for enzymatic therapy assays. *Int. J. Biochem. Cell Biol.* **2015**, *62*, 115–124. [CrossRef]

71. Cravatt, B.F.; Wright, A.T.; Kozarich, J.W. Activity-based protein profiling: From enzyme chemistry to proteomic chemistry. *Annu. Rev. Biochem.* **2008**, *77*, 383–414. [CrossRef]

72. Lu, H.; Chandrasekar, B.; Oeljeklaus, J.; Misas-Villamil, J.C.; Wang, Z.; Shindo, T.; Bogyo, M.; Kaiser, M.; van der Hoorn, R.A. Subfamily-Specific Fluorescent Probes for Cysteine Proteases Display Dynamic Protease Activities during Seed Germination. *Plant Physiol.* **2015**, *168*, 1462–1475. [CrossRef] [PubMed]

73. Daneri-Castro, S.N.; Chandrasekar, B.; Grosse-Holz, F.M.; van der Hoorn, R.A.; Roberts, T.H. Activity-based protein profiling of hydrolytic enzymes induced by gibberellic acid in isolated aleurone layers of malting barley. *FEBS Lett.* **2016**, *590*, 2956–2962. [CrossRef] [PubMed]

74. Kok, Y.J.; Ye, L.; Muller, J.; Ow, D.S.; Bi, X. Brewing with malted barley or raw barley: What makes the difference in the processes? *Appl. Microbiol. Biotechnol.* **2019**, *103*, 1059–1067. [CrossRef] [PubMed]

75. Sánchez-León, S.; Gil-Humanes, J.; Ozuna, C.V.; Giménez, M.J.; Sousa, C.; Voytas, D.F.; Barro, F. Low-gluten, nontransgenic wheat engineered with CRISPR/Cas9. *Plant Biotechnol. J.* **2018**, *16*, 902–910. [CrossRef]

© 2019 by the authors. Licensee MDPI, Basel, Switzerland. This article is an open access article distributed under the terms and conditions of the Creative Commons Attribution (CC BY) license (http://creativecommons.org/licenses/by/4.0/).

International Journal of
Molecular Sciences

MDPI

Review

Cas Endonuclease Technology—A Quantum Leap in the Advancement of Barley and Wheat Genetic Engineering

Iris Koeppel [†], Christian Hertig [†], Robert Hoffie [†] and Jochen Kumlehn *

Plant Reproductive Biology, Leibniz Institute of Plant Genetics and Crop Plant Research (IPK) Gatersleben, 06466 Seeland, Germany; koeppel@ipk-gatersleben.de (I.K.); hertig@ipk-gatersleben.de (C.H.); hoffier@ipk-gatersleben.de (R.H.)
*   Correspondence: kumlehn@ipk-gatersleben.de
†   These authors contributed equally to this work.

Received: 8 May 2019; Accepted: 24 May 2019; Published: 29 May 2019

Abstract: Domestication and breeding have created productive crops that are adapted to the climatic conditions of their growing regions. Initially, this process solely relied on the frequent occurrence of spontaneous mutations and the recombination of resultant gene variants. Later, treatments with ionizing radiation or mutagenic chemicals facilitated dramatically increased mutation rates, which remarkably extended the genetic diversity of crop plants. However, a major drawback of conventionally induced mutagenesis is that genetic alterations occur simultaneously across the whole genome and at very high numbers per individual plant. By contrast, the newly emerging Cas endonuclease technology allows for the induction of mutations at user-defined positions in the plant genome. In fundamental and breeding-oriented research, this opens up unprecedented opportunities for the elucidation of gene functions and the targeted improvement of plant performance. This review covers historical aspects of the development of customizable endonucleases, information on the mechanisms of targeted genome modification, as well as hitherto reported applications of Cas endonuclease technology in barley and wheat that are the agronomically most important members of the temperate cereals. Finally, current trends in the further development of this technology and some ensuing future opportunities for research and biotechnological application are presented.

Keywords: cereals; CRISPR; crops; genetic engineering; genome editing; plant; *Triticeae*

1. Introduction

1.1. Historical View of Genetic Modification in Crop Plants

For more than 10,000 years, plants have been cultivated to feed humans and their livestock, as a source of raw materials and to generate energy. Over time, the driving forces of evolution, domestication, and breeding of cultivated plants have changed according to environmental conditions and societal demands. To cope with the growth of the world's population and the consequent increase in demand for food, the total production needs to be substantially increased during the next decades [1,2]. Not only is the growing world population a challenge, but also the expected climatic changes [3,4]. Thus, the plants of the future also have to be better adapted to harsh weather conditions [5,6]. In addition, the demands of modern society are increasing; food should be of better and consistent quality, healthier, and more diverse. At the same time, society calls for a substantially reduced application of fertilizers and pesticides to render agriculture more sustainable [2].

In nature, mutations occur spontaneously and alterations that provide an advantage under the given conditions are likely to prevail over time. For modern plant breeding, however, the rate of spontaneous mutations is too low to keep up with the demands for crop improvements. Moreover, it is in the nature of spontaneous mutations that they are unpredictable in terms of both position and resultant nucleobase sequence. In the 1930s, mutation breeding emerged, in which mutations are deliberately induced with chemicals such as ethylmethanesulfonate (EMS) or with ionizing radiation. Induced mutagenesis breeding has so far produced over 3000 approved varieties [7] and for many further cultivars it has not been documented which induced mutations they have inherited from germplasms they derive from. Using this technology, however, thousands of mutations occur at different sites in a single plant's genome at a time. Therefore, a vast number of undesirable mutations have to be eliminated via cumbersome and time-consuming back-crossing procedures [5]. The method of Targeted Induced Local Lesions in Genomes (TILLING) represented a significant progress in the detection of mutations. With TILLING or Eco-TILLING, respectively, it is possible to readily identify induced and spontaneous mutations in known genes of interest [8,9].

The method of gene transfer using *Agrobacterium tumefaciens* was introduced in the 1980s [10]. Ever since, genes derived from unrelated organisms or genetically modified gene variants can be introduced into plant genomes. This technique makes it possible to modify or introduce performance-determining genes into cultivated plants more quickly and in a more targeted manner. However, there are some limitations, such as the fact that the transferred gene is integrated at a random position in the genome and that changes to the plant's own genes are only possible to a very limited extent. The insertion site in the genome then also determines whether and how effectively the introduced gene is expressed, and thus to what extent the respective trait will be modified. In addition, not all plants can be transformed equally well by the use of *Agrobacterium*, with cereal species being a particular challenge in this respect [11].

With the new methods for targeted genome modification that are based upon customizable endonucleases, it is possible to modify DNA sequences at a previously defined site of choice in the host genome. The standard application of this technology is site-directed mutagenesis, in which the location of the mutation can be precisely determined, whereas the resulting nucleobase sequence is random [12]. However, the precision can be increased by a variety of approaches. The most sophisticated one involves the use of an artificial DNA repair template implemented via homology-directed DNA repair. By this still very challenging principle, any sequence of choice can be created at any predefined genomic locus.

Wheat (*Triticum aestivum*) and barley (*Hordeum vulgare*) are among the most important cereals in the world. Among all crops, wheat occupies the largest cultivated area in the world, and barley is one of the oldest domesticated crops, currently being the fourth most frequently cultivated cereal (http://faostat/fao.org). With a size of about 17 Gbp, wheat has a very large and complex genome. Only recently has the wheat genome been almost completely read out and this data has been made publicly available [13]. Hexaploid bread wheat evolved via hybridizations from several ancestors. This process involved allopolyploidization, resulting in a total of three diploid subgenomes [14]. In contrast to wheat, barley has a genome with a size of 5.1 Gbp. Already in 2012, a detailed draft of the barley genome was published [15], which was recently complemented by much improved data, including genomic sequences of a large number of representative gene bank accessions [16,17].

1.2. Platforms of Customizable Endonucleases

In plant research and biotechnology, four platforms of customizable endonucleases have been used so far; meganucleases, zinc-finger nucleases (ZFNs), transcription activator-like effector nucleases (TALENs), and the RNA-guided, clustered, regularly interspaced, short palindromic repeats (CRISPR)-associated (Cas) endonucleases.

Meganucleases are naturally occurring endonucleases that recognize and cut comparatively long (>12 bp) DNA sequences. The most commonly used meganuclease is the I-*Sce*I from *Saccharomyces cerevisiae* [18,19]. The ability of this endonuclease to induce double-strand breaks in plants was first demonstrated by Puchta et al. [20] using *Nicotiana* protoplasts. Meganucleases are very specific and efficient. It is, however, very difficult to reengineer these enzymes for target sequences other than their native ones [21]. Hence, their use is fairly restricted as compared to the alternative endonuclease platforms [22,23]. As a consequence, meganucleases have been used in plants almost exclusively for basic research, in particular for DNA repair mechanisms [24].

Zinc-finger nucleases are hybrid proteins with a DNA recognition domain consisting of at least three zinc-fingers combined with a *Fok*I restriction endonuclease domain [25]. Each zinc-finger specifically interacts with three base pairs (bp) of the genomic target sequence and several zinc-fingers can be consecutively assembled to recognize and bind a total of 9 to 12 bp of DNA [26]. Zinc-finger nucleases must always be used in pairs, since their *Fok*I endonuclease domain is only catalytically active if it is present as a dimer [25]. The target motifs on the DNA are selected in such a way that the two zinc-finger nuclease monomers bind to the target DNA in antiparallel orientation and at a suitable distance to each other. A DNA double-strand break (DSB) is then catalyzed in the interspace between the two binding sites [27,28]. In addition to the particularly complex production process of zinc-finger nucleases, there are limitations in the selection of possible binding sites, as well as unpredictable neighbor effects between adjacent zinc-fingers on the DNA binding specificity [22].

In 1989, it was shown for the first time how transcription activator-like effector (TALE) proteins are transferred from a pathogenic bacterium of the genus *Xanthomonas* into a host plant [29]. The TALE-based nucleases (TALENs) act, similarly to zinc-finger nucleases, as dimers [30]. The DNA-binding domains consist of up to 30 copies of highly conserved repeats of typically 34 amino acids. Only the amino acid positions 12 and 13 (called repeat-variable di-residues) of each repeat are not uniform, because they specify the different DNA base pairs to bind [31]. A *Fok*I endonuclease bound to such a TALE binding domain induces the double-strand break in the target motif [30]. Since each repeat recognizes just one nucleobase of the DNA target, the design and assembly of the binding domain is more straightforward and versatile than with zinc-finger nucleases [31].

The platform of RNA-guided Cas endonucleases is derived from the CRISPR/Cas adaptive immune system of microbes. The most commonly used Cas endonuclease is from *Streptococcus pyogenes* (Sp). The two-component system consists of the Cas restriction enzyme and an artificial guide RNA (gRNA) that navigates the Cas protein to a cognate DNA sequence motif [12]. The gRNA consists of a structurally functional and a variable part. The 3′-tail forms a spatial structure required for binding with Cas to form a ribonucleoprotein complex. The variable part of the gRNA located at the 5′-end usually comprises 20 nucleotides and defines the DNA-binding specificity of the gRNA according to the principle of complementary base pairing. The genomic target motif of such ribonucleoprotein complex includes the same nucleotide sequence as the variable part of the gRNA, which enables the gRNA to specifically bind the opposite DNA strand. This part of the target DNA is often referred to as the protospacer, a term adopted from the microbial immune system these molecules originate from. In addition, this major part of the target motif is complemented by the so-called protospacer-adjacent motif (PAM), which is recognized and bound by the Cas protein. In the case of SpCas9, the nucleobase triplet NGG, in which N stands for any of the four nucleobases, represents the PAM site [12]. The high versatility of the gRNA 5′-end allows a wide variety of target sequences of choice to be addressed. The comparatively simple application and the efficiency and reliability achievable in higher organisms have rendered this platform the most popular and frequently used tool for site-directed genome modification today [32,33].

2. Methodological Aspects of Cas Endonuclease Technology

2.1. System Components

When employing Cas endonuclease technology in plants, there are a number of opportunities and some specific requirements in terms of the construction of transformation vectors. Cas9 endonucleases have been modified in various ways for use in plants. Importantly, the coding sequences were complemented by one or two nuclear localization signals (NLSs) and the codon usage of the protein biosynthesis was optimized for use in plants in general, for monocots and dicots, or even more specifically for individual species, such as wheat and barley [34–36].

Depending on the host organism, also various promoters have been used to drive endonuclease expression. In crop plants, the doubled enhanced cauliflower mosaic virus (2x35S) promoter has been mainly used for test systems [34,37,38] and UBIQUITIN promoters for the generation of heritable modifications. Accordingly, the maize POLYUBIQUITIN1 promoter (ZmUBI1) has been commonly used for cas9 expression to generate heritable changes in barley and wheat [37,39,40]. Alternatively, Cas endonuclease-encoding genes can also be expressed by self-replicating virus particles in the plant cell [41]. This expression system ensures a comparatively high gene dosage, which may be particularly useful for homology-based approaches that are described in more detail below. Gil-Humanes et al. [42] have shown that the Wheat Dwarf Virus, a replicon-based geminivirus, can be modified to express gRNA and cas9 in wheat, albeit this approach has not yet been shown to be applicable for the generation of plants with heritable modifications.

The expression cassette for a gRNA typically consists of plant-derived RNA polymerase III (Pol III)-processed promoters and terminators derived from small nuclear RNA (snRNA)-encoding genes. Somehow surprisingly, it also proved to be sufficient in plants to use the transcriptional termination signal of bacterial CRISPR RNAs, which solely consists of a stretch of five to seven thymidines [34,43]. A comparative test in maize protoplasts had shown that wheat and rice U3 promoters were more efficient than the *Arabidopsis* U6 promoter that has been preferentially used in dicots [44]. The wheat U6 promoter has hitherto been mostly used for barley and wheat [37,39,40]. More recently, a study of Kumar et al. [45] indicated that the barley U3 promoter might be more efficient in site-directed mutagenesis of barley than the frequently used rice U3 promoter.

The preference of the U3 and U6 promoters for A and G as the first transcribed nucleotides limits the selection of target motifs to GN20GG and AN20GG, respectively [43]. However, there are thus far no examples where gRNAs were directly driven by Polymerase II-compatible promoters. The mRNAs otherwise expressed by Pol II promoters are processed at both ends and these modifications significantly alter the structure of the gRNA, and thus compromise the ability to bind both Cas9 and the target motif that is to be processed with sufficient efficiency [43].

Several systems have been developed for the expression of multiple gRNAs. Mostly, each gRNA is expressed by a separate Pol III promoter [44,46,47]. Alternatively, multiple gRNAs can be located on only one transcript and post-transcriptionally separated, as has already been demonstrated with wheat [48]. However, such expression systems have not yet led to improved efficiency of the technology with regard to individual target motifs in plant genomes [43,49,50].

2.2. Criteria for Target Motif Selection and in Silico gRNA Design

Thanks to the opportunity to equip the gRNA with any target-specific 5′-sequence, the plant genomic target site of the gRNA/Cas complex can, in principle, be chosen at will. However, the Cas9 endonuclease requires the aforementioned two guanines that are part of the protospacer-adjacent motif (PAM) at the 3′-end of the target sequence. In addition, there are further preferences of the target-specific part of the gRNA, some of which can have a considerable impact on the functionality of the gRNA/Cas9 complex. For instance, it was determined for the so-called seed region, that is, the six nucleotides residing immediately upstream of the PAM site, that a GC content above 50 percent increases the probability of sufficient gRNA functionality [51]. With regards to the entire target motif, an enrichment of guanines and low adenine content was shown to cause increased binding stability and activity of the gRNA/Cas9 complex [52].

The target sequence-specific part of the gRNA usually has a length of 20 nucleotides [12,53]. As previously stated, the transcriptional start defined by the U3 and U6 promoters, which are preferred to drive gRNAs in plants, are A and G, respectively, which requires the selection of genomic targets having the same nucleobases at their 5′-end [38]. In human cell lines and zebrafish, it has been shown that the length of the target-specific part of the gRNA can be shortened to 18 nucleotides without a noticeable effect on mutation efficiency. Surprisingly, such shorter gRNAs even had increased specificity for the on-target sequence as compared with off-targets [52,54,55]. Guide-RNAs with less than 20 target-specific nucleobases also showed high performance in *Arabidopsis* and barley [56,57]. On the other hand, an extension of the gRNA 5′-part beyond 20 nucleotides was reported to cause reduced cleavage efficiency [58].

Quite a number of online platforms have been developed for the selection of target motifs and corresponding gRNAs. For instance, CRISPR-Plant was especially developed for plants [59,60]. While a whole array of plant reference genomes can be directly screened in this platform, the barley and wheat genomes are not implemented. However, for the selection of gRNAs in cereals, DeskGen is a highly instrumental alternative, as these cereal genomes are available for target validation [61,62]. This platform offers a comparatively large scope of options, including various Cas endonucleases and a useful range of lengths of the target-specific gRNA 5′-part. Also, the CRISPOR online tool can be specifically used for barley and wheat. It comprises a comprehensive choice of different Cas enzymes [63,64]. In addition to the selection of gRNAs predicted to be well-performing at their target motif, all above online tools are also capable of indicating potential off-targets in the chosen reference genome. WU-CRISPR is another useful platform, as its algorithm is comparatively strict in selecting particularly useful target motifs. However, this tool is confined to SpCas9 and does not offer off-target screens in reference genomes [65]. A general disadvantage of all these platforms is that their algorithms have been established using data from other organisms than plants, which is thought to be one of the reasons that the reliability of their results is still fairly limited.

Liang et al. [66] have focused on the preservation of the gRNA secondary structure. They determined that three of the stem loops frequently occurring in the gRNA 3′-part are essential to appropriately bind to the Cas9 protein, and hence are also of decisive importance to the overall functionality of the gRNA/Cas complex. These authors also found out that 98 percent of well-performing gRNAs have no more than a total of 12 base pairs formed between their target-specific and 3′-parts and no more than 7 target-specific nucleobases consecutively involved in intra-molecular base pairs. In addition, no more than 6 base pairs should be formed within the target-specific part of the gRNA. In order to ensure these gRNA features, it is recommended to thoroughly investigate the secondary structure of candidate gRNAs. For the prediction of RNA secondary structures, online platforms such as mfold [67–69] or RNAfold [70,71] are available.

In cases where genetic modifications are intended to be performed in genotypes other than those with reference genomes, it is recommended to countercheck pre-selected target motifs for their presence and integrity, because even single nucleotide polymorphisms typically result in a dramatic drop in gRNA/Cas efficiency.

2.3. Delivery of Cas Endonucleases and Associated Reagents into Plant Cells

A number of methods are available for the transfer of DNA, RNA, and proteins into plant cells, which in principle can also be used for Cas endonucleases, customized gRNAs, DNA repair templates, and any further reagents that may be used to support site-directed genome modification. Nowadays, the most widely used approach is based on the genomic integration of expression units with gRNA and Cas-encoding DNA sequences. However, the production of stably transgenic plants is a particular challenge for cereals, because agrobacteria, which are mostly used for plant transformation, have a very limited compatibility with these non-host plants. Moreover, the formation of adventitious shoots from leaf or shoot explants, which is readily achieved in the context of DNA transfer methods applicable to most dicotyledonous species, has only been successful in exceptional, hardly reproducible cases in cereals [72]. On the basis of special methodological approaches, routine genetic transformation of barley and wheat is nevertheless possible, at least by using selected accessions that are comparatively amenable to methods of DNA transfer. Such experimental model genotypes are, for example, the barley cultivar Golden Promise and the Mexican wheat breeding line Bobwhite SH9826 [11]. For stable DNA transfer in cereal crops, immature zygotic embryos are most widely used. Importantly, such embryos must be of a developmental stage at which highly totipotent cells are still abundantly present. In immature embryos of barley and wheat, such cells are preferentially residing in the area that surrounds the shoot apical meristem and the scutellum, respectively. Stable transgenic wheat and barley plants were already produced by ballistic transfer of plasmid-coated gold particles to immature embryos in the early 1990s [73,74]. Using hypervirulent bacterial strains, which had initially been used for the transformation of rice [75], thereafter it was soon also possible to generate transgenic barley and wheat plants using this principle [76,77]. Today, it is routinely possible to generate stable transgenic plants from over 10% of inoculated immature embryos of barley [78], while in wheat, efficiencies of about 5 percent have been achieved both via *Agrobacterium*-mediated and ballistic DNA transfer [79,80]. However, a largely improved protocol was more recently demonstrated to allow for much improved transformation efficiency in wheat ([81,82]. The comparative examination of a variety of genotypes has shown that although some can indeed be transformed in addition to the model lines, the efficiency, however, is much reduced [83,84]. Previous studies on Cas9-induced mutagenesis of barley have been based without exception on *Agrobacterium*-mediated DNA transfer. In contrast to this, site-directed mutagenesis in wheat has mostly relied on ballistic gene transfer [85].

Another particularly interesting principle of DNA transfer into cereal cells is the use of immature, single-celled pollen (microspores), which can undergo cell proliferation and embryogenic development under suitable culture conditions [86]. Since pollen consists of haploid cells, it is possible that homozygous transgenic plants can be directly produced via this developmental pathway in association with DNA transfer and whole genome duplication. *Agrobacterium*-mediated DNA transfer in embryogenic pollen cultures of barley has the further advantage that a winter barley variety can be used, which is comparatively difficult via DNA transfer to immature embryos [87,88]. A similar transformation method has been reported also for wheat, which is, however, based on ballistic DNA transfer [89]. In the authors' laboratory, barley mutants have already been produced via the pollen embryogenesis pathway using TALENs [90], as well as by Cas endonucleases. Furthermore, the production of stable transgenic barley and wheat by ballistic DNA transfer into meristematic tissue from the shoot apex of embryos prepared from mature grains was also successful [91,92]. This method has potential because of particularly low genotype dependence, since plant regeneration is based on conventional germination of the embryos, and therefore no formation of adventitious shoots is necessary. More recently, Hamada et al. [93] have also demonstrated that the production of plants with site-directed modifications using gRNA/cas9 constructs is possible using this method.

After alteration of the genomic target motif, the presence of transgenes coding for gRNA and endonuclease is not only unnecessary but also undesirable, as off-targets can still be mutated, even if they are not identical with the on-target. In this context, an advantage of the application of Cas endonucleases compared to conventional genetic engineering is that the integration site of the transgenes is mostly not coupled with the site of the desired genetic modification. Thus, the elimination of these transgenes by genetic segregation of progeny is comparatively straightforward while maintaining the desired modification. Endonuclease-triggered genetic alterations in primary transgenic plants are often heterozygous and restricted to tissue sectors. Therefore, progeny needs to be screened in order to select homozygously mutated individuals. Consequently, no extra time is required to obtain transgene-free mutants. Haploid technology is a particularly elegant solution for the separation of achieved genetic modifications from unnecessary transgene insertions or for the segregation of several mutations present in chimeric plants of the M1 generation. With this technology, populations of completely homozygous recombinants can be obtained, in which desired genotypes are present at a much higher frequency than among offspring derived from conventional selfing [94,95].

In addition to the various possibilities of genomic integration of gRNA and cas endonuclease genes, methods for transient expression have also been established, which is of considerable value for the preliminary validation of target-specific gRNAs, Cas endonuclease variants, and any further components used in this context. The most widely used transient expression method is based on the transfection of isolated mesophyll protoplasts, whose plasma membrane is rendered porous by application of polyethylene glycol. This enables transgenic plasmid DNA, in vitro transcribed RNA, bacterially pre-produced Cas protein, or ribonucleoprotein complexes to be taken up by the protoplasts. This has been exemplified, amongst others, in wheat and barley [39,57,96]. The activity of the transferred components can then be checked after amplification of the genomic target regions using the T7E1 assay, by Sanger or deep sequencing, as is described in more detail further down below. Another method for the functional validation of gRNA/cas constructs by transient expression is based on the ballistic transformation of leaf epidermis. Epidermal cells are comparatively well suited to be screened microscopically, for instance in regards to Cas-induced restoration of a reporter gene construct [97]. Barley leaf tissue has been used to demonstrate precise editing via homology-based DNA repair using a synthetic DNA repair template [98].

As an alternative to the expression of gRNA- and Cas-encoding DNA, in vitro transcribed RNA or Cas protein can be transferred to plant cells in order to make specific modifications to the genome. Guide RNA and Cas endonuclease can also be pre-assembled to form gRNA/Cas9 ribonucleoprotein complexes prior to their transfer into plant cells. The use of pre-produced RNA and protein molecules or complexes thereof has the advantage over DNA that their effective amount is independent of the expression profile and strength of plant promoters. In addition, the activity of these components is limited in time, as they are not continuously delivered by gene expression but are subject to cellular degradation. Accordingly, fewer mutations will occur in off-target motifs during the subsequent development of the plants. A further advantage is that mutated progeny do not have to be examined for the loss of integrated gRNA- and Cas9-encoding DNA sequences, as all mutated individuals can be used without restriction due to their transgene-free nature. In wheat, the GW2 gene was used as an example to show that the transfer of in vitro transcribed RNA coding for both gRNA and Cas9 leads to mutations in one percent of ballistically transformed cells [99]. About one third of the resulting plants were mutated in all six copies of the hexaploid genome. Compared to the transfer of gRNA and cas9 transgenes, however, this method had a mutation rate that was about 60 percent lower. Liang et al. [96] achieved an improvement in mutation efficiency by assembling gRNA and Cas9 into ribonucleoprotein complexes before ballistic transfer into immature embryos. After selection-free plant regeneration, mutants were identified with a frequency of more than four percent, which is on a par with the efficiency of conventional wheat transformation.

2.4. From Site-Directed Mutagenesis to Precise Genome Editing

Current utilization of Cas endonuclease technology is still largely limited to random alterations of the DNA sequence at the user-defined genomic sites. This comparatively simple approach of site-directed mutagenesis relies on the error rate of the DNA repair mechanism, called non-homologous end-joining (NHEJ), which is predominant in plant cells. In this process, the two DNA ends resulting from a double-strand break are recognized and relegated irrespective of their nucleobase sequence.

Increased predictability of site-directed modifications was shown to result, for example, from two simultaneously induced DNA breaks, which can lead to the precise deletion of the interjacent fragment [22]. In addition, microhomology-based DNA repair produces predictable deletions in a comparatively simple way, provided that identical sequence repeats are present at both DNA ends that are to be relegated [100].

A more ambitious approach is the so-called base editing, in which a single nucleotide is specifically converted into another so that no more than one amino acid of the encoded protein is altered at a time [101]. For this purpose, a Cas-based nickase, that is, a Cas endonuclease derivative that cuts only one strand of DNA due to the mutative alteration of one of its two nucleolytic domains, is bound to a natural or artificial nucleobase deaminase enzyme [102]. Cytidine deaminases can convert C/G basepairs to T/A in the target region, whereas adenosine deaminases induce A/T to G/C conversions [102,103]. The functionality of cytidine and adenosine deaminases has already been demonstrated in several plant species, including wheat [104,105]. The effective base editing area ranges in dependence of the base editor used, e.g., the cytidine deaminase used by Zong et al. [104] can convert positions 1 to 17 within the target site and the adenosine deaminase used by Li et al. [105] is capable of converting positions 4 to 8.

In addition to the aforementioned error-prone non-homologous end-joining (NHEJ) DNA repair mechanism that predominates in plant cells, homology-directed repair (HDR) can also be used to generate precise genome alterations, albeit this process is much less active in somatic cells. Homology-directed repair naturally involves the sister chromatid of the same chromosome or the homologous chromosome as the correct sequence template, allowing the original sequence to be restored, even in cases of comparatively severe DNA damage. By using artificial DNA repair templates that are partially complementary to the site of the plant genome to be modified, it is possible to make even fairly large modifications precisely, as specified by the experimenter [106]. However, due to the methodological challenges, there are very few examples of this precise editing in plants published thus far. A first experimental approach involved the stable integration of the repair template together with the gRNA- and Cas-encoding expression units into the plant genome. The integrated repair template is flanked by the target sequences of the gRNA-mediated endonuclease used, so that it is cut out by the Cas restriction enzyme. This principle has proven to be sufficient in *Arabidopsis* [107]. The use of paired Cas9 nickases, which generate two single-strand breaks on the two opposite DNA strands, has increased the efficiency of homology-directed DNA repair [50,108,109]. With geminivirus replicons as carriers of the repair template, the dose per cell is increased. In this case, a strain of the Bean Yellow Dwarf Virus was modified so that only the elements essential for the method remained and the artificial repair template was amplified with high frequency [110]. In barley and wheat, precise editing has been achieved either only at the cellular level [98] or has the limiting prerequisite that the obtained genetic modification leads to an in vitro selectable trait, e.g., an herbicide resistance [111].

2.5. Identification and Characterization of Site-Specifically Modified Plants

The presence of gRNA- and Cas-encoding expression units is usually tested by PCR amplification [112,113]. In some studies, the expression of gRNA and cas9 was additionally assessed by RT-PCR, which can be very informative, especially in the process of method establishment [37]. For the detection of site-directed modifications, PCR products derived from the genomic target region can be analyzed using various methods. Digestion of the PCR product using conventional restriction enzymes is the most straightforward approach. However, the prerequisite for this is that the restriction site of the Cas endonuclease lies within a recognition sequence of the conventional restriction enzyme that then can be used for the test. In contrast to the wild-type sequence, PCR products of mutated target motifs cannot be digested by this enzyme [34]. A disadvantage of this method is that the choice of target motifs is substantially limited. This can be circumvented using the T7E1 assay, in which a mixture of PCR products of mutant and wild-type alleles is digested by the mismatch-sensitive T7 endonuclease I [39,114]. Another disadvantage these methods have in common is that their sensitivity is low, i.e., if mutated alleles of a plant are limited to small tissue sectors, detection may be not possible. This leads to a certain number of mutants not being identified.

To avoid this problem, Liang et al. [115] used target-specific synthetic gRNA and Cas9 protein for in vitro digestion of PCR products from genomic targets. This method is independent of conventional restriction sites in the target motif and was reported to be extremely sensitive. However, a large-scale application of this approach is hampered by the fact that recombinant Cas protein is fairly expensive.

As compared with the aforementioned restriction-based assays, sequencing of PCR product derived from the genomic target region is more informative. In heterozygous or chimeric mutated plants, the sequencing chromatogram shows double and multiple peaks at the individual base positions, which usually starts at the restriction site of the Cas endonuclease and continues in the same direction as the sequencing was conducted. Various bioinformatic tools, such as Tracking of Indels by DEcomposition (TIDE) [116] and DSDecode [117], provide the opportunity to read out such peaks, thereby determining the different allele variants present in the analyzed plant sample. A more reliable approach involves the analysis of several single clones via Sanger sequencing, as this provides a detailed breakdown of the mutations present in the analyzed sample. In addition, more conclusive information on the proportions of the mutated alleles can be obtained. In this context, it is important to consider that Cas endonuclease-triggered mutations are usually heterozygous and often limited to sectors of the plant. Moreover, leaf samples are not necessarily representative for the whole plant. This phenomenon is reflected by the observation that not all mutations detected in primary mutants are inherited by subsequent generations. In addition to Sanger sequencing, deep sequencing methods can be used to analyze mutations in individual plants, which is particularly informative owing to the obtained amount of data [48,118]. In practice, it is typically necessary to find a reasonable compromise between workload, conclusiveness of assays, and costs. Therefore, a combination of different methods is often the most suitable solution [40,42].

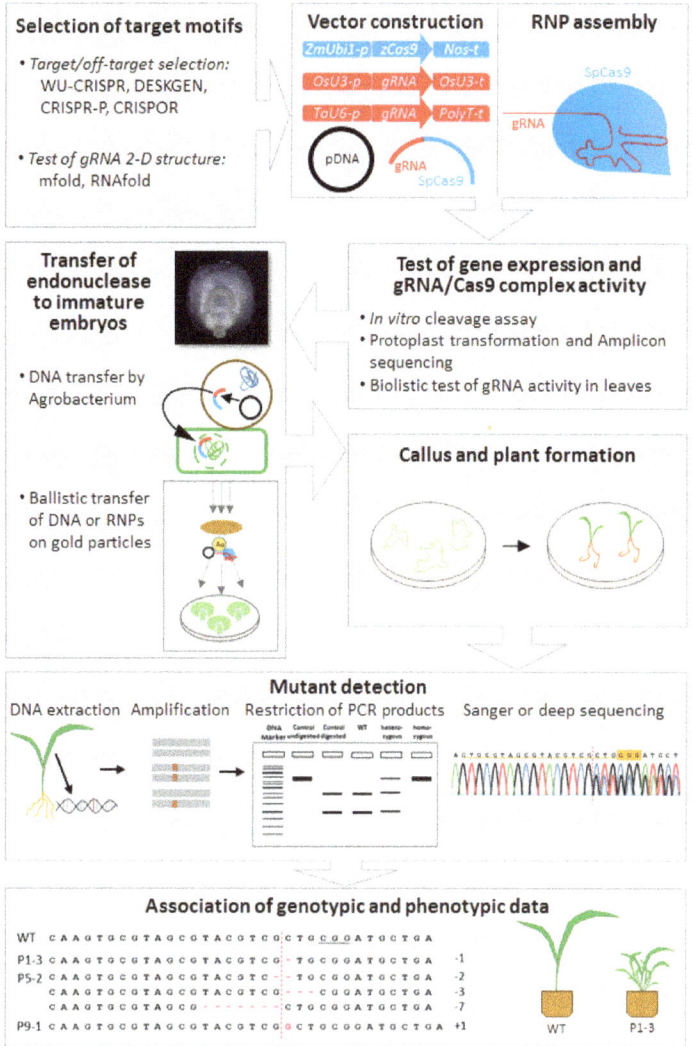

Figure 1. General workflow of targeted genome modification in barley and wheat, including target motif selection, vector construction, vector tests, gRNA and Cas9 delivery, plant regeneration, mutant detection, and genotypic and phenotypic analyses.

The molecular analysis of candidate plants can also address accidental modification of so-called off-target sites, which can cause undesired side effects and also influence the efficiency of on-target mutagenesis. In plants, the proportion of off-target mutations is typically below 1% [119]. In contrast to human cells, this does not pose a serious problem. These mutations are often confined to small sectors that are usually not passed on to offspring. Furthermore, due to the high specificity of the gRNA/Cas complex, possible off-target sites are well predictable so that progeny can be readily tested to eliminate segregants with unwanted mutations.

Since primary mutants are often chimeric or heterozygous and integrated cas9 and gRNA transgenes can be inherited, the analysis of progeny is essential to identify homozygous mutants that are null segregants in regards to the transgenes [35,112].

It is not recommended to spend too much effort for phenotypic analyses of primary mutants, because in addition to their uncertain genetic homogeneity and zygosity, the general fitness of these individuals is typically strongly affected owing to the manipulation and culture procedure they have gone through [120]. Importantly, it is advisable to use non-transgenic, non-mutant segregants as wild-type controls for the analysis of plant traits, because the comparison with other plants bears a fairly high risk of obtaining misleading results. A survey of the general workflow of targeted genome modification in barley and wheat is depicted in Figure 1.

3. Applications

While in this section only some representative examples for applications of Cas endonuclease technology in barley and wheat are presented in more detail, Table A1 provides a comprehensive overview of studies hitherto published on these crops.

In 2013, Upadhyay et al. [37] were the first to show that Cas endonucleases are in principle applicable in a *Triticeae* species and especially in the large and complex genome of wheat. However, this study was still confined to site-directed mutagenesis in a cell suspension, from which no plants can be regenerated. Mutations were detected in 18 to 22 percent of the sequenced amplicons derived from target regions of the INOSITOL OXYGENASE (TaINOX), and PHYTOENE DESATURASE (TaPDS) genes. The first plants carrying Cas endonuclease-induced mutations were presented by Wang et al. [39]. While heritable loss-of-function mutations were induced in all three homoeologues of the MILDEW-RESISTANCE LOCUS O (TaMLO) gene by the use of TALENs, Cas9-triggered mutations were still confined to the A subgenome. Using two wheat backgrounds, the authors produced a total of 687 gRNA/cas9 primary transgenic plants, of which about 5 percent proved to carry mutations in the target motif. However, in contrast to the TALEN-induced ones, the heritability of these mutations was not shown.

The first published use of Cas endonucleases in barley aimed to induce mutations in HvPM19 [35], which encodes an ABA-induced plasma membrane protein previously described in wheat as a positive regulator of dormancy. Four copies of this gene are present in barley. For two of those, knockout mutants were generated. Amongst 13 primary transgenic plants, three carried mutations in HvPM19-1, whereas one HvPM19-3 mutant was found amongst ten T0 plants. While this study was the first to provide evidence for the heritability of Cas9-induced mutations in a *Triticeae* crop, a resultant phenotype was not described.

Once the applicability of Cas endonuclease technology to achieve heritable mutations was shown for barley and wheat, this principle was used for many further approaches. Holme et al. [40] used Cas endonuclease-induced mutations to investigate the function of the PHYTASE GENE A of barley. HvPAPhy_a acts as the main regulator of phytase content in the barley grain. Phytases are important sources of bioavailable phosphorus, and therefore particularly important for germination. To reduce the activity of HvPAPhy_a independently of HvPAPhy_b, which is very similar in its gene sequence, Holme et al. specifically mutated the promoter of HvPAPhy_a. While small insertions and deletions in homozygous mutant progeny had only little effect, larger deletions, which also affected exon 1 of the target gene, caused a significant reduction of phytase activity.

Taking a Cas9-mediated knockout approach, Wang et al. [118] investigated the function of the GW2 gene of wheat. The rice orthologue of TaGW2 had been previously described as a negative regulator of grain size and thousand-grain weight. Wang et al. produced knockout mutants for all three subgenomes of wheat by RNA-guided Cas9 and generated all possible combinations of mutated homeoalleles by crossing. The knockout of individual TaGW2 homeologues in any of the subgemomes increased the size of the wheat grain, as well as the thousand-grain weight. Consequently, the authors concluded an additive effect of all three TaGW2 homoeologues [118].

Gluten in the wheat grain is relevant for the baking properties of the flour. This protein fraction is encoded by about 100 genes. Gluten intolerance is mainly based on the immune response to a particular peptide within the α-gliadins, which are encoded by 45 genes. Sánchez-León et al. [113] designed two guide-RNAs specific to the gene region encoding the hyperallergenic epitopes and produced bread and durum wheat lines with knockout mutations in up to 33 of the 45 α-gliadin genes. Thus, immunoreactivity was reduced by as much as 85 percent. The inheritance of the mutations and their effect on the α-gliadin content were followed over three generations. In order to further reduce the immunogenic α-gliadins in wheat, further work is still necessary [113].

In barley and wheat, Cas endonucleases have been mainly used for the investigation of gene functions by knockout, for the modification of metabolite contents and to increase resistance to fungal pathogens (Table A1). These studies show the great potential of this technology for research and breeding. In addition to the generation of random sequences at the target site, more recent work has demonstrated predictable nucleobase exchanges in wheat by the use of a chimeric Cas9 derivative, which consists of a nickase and an adenosine deaminase [105]. This base editing approach opens up further possibilities of generating functional gene variants rather than knockout alleles.

4. Regulation

Parallel to the development of the Cas endonuclease technology, science organizations, politicians, and society are dealing with the effects of the technology. The following section will briefly address some related aspects, in particular the regulation of plants that carry site-directed mutations. The focus here is on the USA and Europe, as these two have very contrasting approaches.

For the United States, Wolt and Wolf [121] summarized the regulation of the so-called green biotechnology, and in particular the use of customizable endonucleases. Commencing at the Asilomar conference on biosafety in 1975, a regulatory framework for plants with recombinant DNA has been developed in the USA since the mid-1980s. Over the years, this framework became more and more complex due to more accompanying research and the increasingly lengthy periods needed for decision-making. Although the focus of regulation in the United States is on the product, the procedure also includes process-based facets. For example, genetically engineered plants have to undergo special approval procedures. However, these differ between plants that have been transformed by use of the plant pathogen *Agrobacterium* and those that have been ballistically transformed. If plants with an identical trait are produced without genetic engineering, no special regulation takes place. For plants bred by use of customized endonuclease technology, the portal "Am I regulated?" has been established, in which the USDA made case-by-case statements at short notice as to whether a plant needs to be regulated or not. In 2017, work began on revising the regulations on biotechnology. The major trigger for regulation was shifted towards the process by a redefinition of "genetic engineering" as "mean techniques that use recombinant or synthetic nucleic acids with the intent to create or alter a genome". This is not applied to processes of targeted genome modification that cause deletions or base edits or result in the targeted insertion of a DNA fragment, which would also be possible by conventional breeding, conventionally induced mutagenesis or even without any human intervention [121].

In a detailed report in 2017 [122], the European Academies' Science Advisory Council (EASAC) stressed the possibilities that targeted genome modification offers. In plant breeding, improved precision compared to undirected mutagenesis methods represents a significant advance. At the same time, some representatives of non-governmental organizations and political parties are viewing biotechnological breeding methods critically. According to EASAC, this gives science a special responsibility to explain its work. In regards to plant breeding, EASAC recommends to politicians not to regulate transgene-free plants carrying site-directed modifications as being "genetically engineered" in the sense of the EU Directive 2001/18/EC on the deliberate release into the environment of genetically modified organisms. Additionally, it advises a revision of the European genetic engineering legislation with a stronger focus on the product instead of the breeding process. The need for international compatibility of regulation is emphasized and reference is also made to best practices from other countries, such as the United States, where site-directed mutagenesis is not regulated as genetic engineering [122].

While some countries, including the United States, Argentina, Brazil, Chile, and more recently Australia, have already made decisions to regulate targeted mutagenesis on a case-by-case or general basis not differently than conventional methods of plant breeding, in Europe, those varieties are the subject of the Directive 2001/18/EC. According to that, field trials with plants carrying targeted mutations need special permission on application, and placing respective varieties on the market needs tedious and very costly approval procedures. The European plant research community has responded to this with demands for a revision of European genetic engineering legislation. For more details see Sprink et al. [123] or the position paper of 95 European plant research institutions (http://www.vib.be/en/news/Documents/Position%20paper%20on%20the%20ECJ%20ruling%20on%20CRISPR%2012%20Nov%202018.pdf).

In many countries in the world, decisions on how to regulate plants generated using customized endonucleases are currently pending. They will play a key role in deciding how and by whom these technologies and their possibilities can be used.

5. Perspective

The new possibility of utilizing customized endonucleases in barley and wheat opens up a broad spectrum of opportunities for applications. A generally limiting aspect, however, is that the genetic transformation of cereals poses a particular challenge. In the context of the question for new or improved methods of transferring DNA into regenerable cells of cereals, it has to be considered that a number of methods have already been developed, but have not yet been utilized for the application of Cas endonucleases. A remarkable example of this is the use of isolated microspores or embryogenic pollen cultures generated from them for DNA transfer using *Agrobacterium* [87], ballistics [89] or electroporation [124]. When it comes to the development of new methods, particular attention must also be paid to the fact that they should be applicable to a broad spectrum of genetic backgrounds, which unfortunately does not apply to any of the methods used to date. On the other hand, the particular advantage of Cas endonuclease technology will only be fully exploited if it is possible to directly modify any plant genotype of choice. This is the only way to completely overcome the cumbersome introgression of advantageous gene variants into elite germplasms, and above all, the genetic coupling of desired alleles with undesired ones. In this context, the dissection and culture of ovules is another remarkable possibility for the in vitro regeneration of wheat and barley [125,126]. For the *Agrobacterium*-mediated DNA transfer into cultivated ovules, in which the T-DNA is transferred into few-celled proembryos, it was shown that this method has comparatively low genotype dependence [127].

In all studies published to date in which heritable site-directed modifications have been achieved in the barley and wheat genomes using Cas endonuclease technology, the POLYUBIQUITIN1 promoter from maize was used to express the cas9 gene. For the stable expression of protein-encoding transgenes in the *Triticeae*, however, a vast number of further promoters with different activity profiles are available, some of which may also be used for Cas endonucleases [128–130]. In addition, there is, thus far, unexploited potential for the expression of gRNAs in barley and wheat. For instance in rice, a t-RNA processing-based system was established, in which multiple gRNAs are released from a complex transcript by endogenous RNases [50,131].

To keep pace with the frequent emergence of novel Cas and gRNA formats, as well as regulatory and further functional elements, several modular vector systems have been developed that allow for a rapid and versatile assembly of any combination of components of choice. Some of these systems may also be useful for the application in the *Triticeae* cereals [45,50].

Whereas site-directed mutagenesis represents the currently well-established state of the art of Cas endonuclease technology, more precise techniques of targeted genome modification have unfortunately not yet become routine, and are at best exemplary in plants. A portfolio of chimeric Cas derivatives is now available for the so-called base editing using nucleoside deaminases, which allows conversions from G to A, C to T, T to C, and A to G. Thus, the vast majority of spontaneous point mutations occurring in nature can be specifically generated, or if desired, the corresponding wild-type alleles can be restored as well [103]. A broad application of precise genome editing in association with DNA repair templates, on the other hand, still poses particularly great methodological challenges. Homology-based repair, for example, is comparatively rarely occurring, which entails a correspondingly low efficiency of conventional selection marker genes whose genomic integration and expression is too poorly associated with template-mediated repair events. In addition, the artificial repair templates must compete with the naturally recruited homologous sequences present in the cell on sister chromatid and homologous chromosome(s), which further reduces the efficiency of targeted modification.

A further limitation of Cas endonuclease technology is that the binding site of the gRNA/Cas complex cannot be predefined entirely at will. The most commonly used enzyme for target sequence-specific genome modification is Cas9 from *Streptococcus pyogenes*. This enzyme requires the aforementioned protospacer-adjacent motif (PAM) NGG for binding to the genomic target. However, there are several other Cas9 variants that require other cognate DNA motifs. Such Cas9-orthologous enzymes are known from various representatives of the genera *Streptococcus*, *Staphylococcus*, *Campylobacter*, and *Neisseria* [132–134]. In *Arabidopsis*, for example, the *Staphylococcus aureus* and *Streptococcus thermophilus* Cas9 variants have already been used in addition to the standard SpCas9 [135]. Furthermore, in order to extend the flexibility in the choice of targets beyond the naturally occurring PAMs, Hu et al. [136] developed an artificial Cas9 variant. The advantage of this "expanded" Cas9 (xCas9) is that almost any nucleobase triplet is accepted for its binding to the genomic target. The synthetic Cas9-NG, which was developed in a different way and published shortly thereafter, proved to be even more efficient than xCas9 in many PAM variants [137].

While Cas9 endonucleases are widely employed in plants, Zetsche et al. [138] described another Cas endonuclease, namely Cas12a (Cpf1) from *Francisella novicida*, that can also be used to induce targeted alterations in genomes. Later, Cas12a-orthologous enzymes from other bacteria were also successfully used [139]. Besides some other peculiarities, Cas12a, in contrast to Cas9, produces 5'- overhangs rather than blunt DNA ends. This provides a particularly attractive option for the establishment of robust methods of precise genome editing using homology-directed DNA repair.

Cas endonucleases have also been modified in such a way that they induce single-strand breaks in double-stranded DNA. Among such nickases, there are specific variants for each of the two DNA strands [12,140]. In *Arabidopsis*, paired nickases derived from Cas9 were used to demonstrate that both the specificity of the binding to the genomic DNA and the efficiency of homology-directed DNA repair can be increased [50,108,109].

Whereas knockouts are the standard outcome of site-directed mutagenesis, the possibility to achieve quantitative changes in the functionality of target genes has not yet been exploited in cereals. Using tobacco as experimental model, Schedel et al. [141] have shown that in-frame mutations can maintain gene functionality to a reduced extent. This can be of particular value for the targeted modification of genes that are essential for the plant or the generation of new allelic diversity for the use in plant breeding. In a particularly elegant approach, Rodriguez-Leal et al. [142] produced tomato lines with various CLAVATA3 alleles by simultaneous expression of several gRNAs specified for different target motifs in the promoter of this gene. The resultant gene variants led to a whole series of lines with different fruit sizes according to the individual strength of CLV3 expression.

In other crop plants, there are various application examples of customizable endonucleases, which may also be considered for breeding programs in the *Triticeae* cereals. Since every characteristic of plants is largely genetically determined, it must in fact be possible to improve every trait that is relevant for the utility value of crops by employing Cas endonuclease technology. Due to these extensive possibilities, only a few of the promising options can be mentioned here as examples. A potential field of application for Cas endonucleases is the establishment of resistance to various potyviruses, which has already been achieved, for example, in cucumber. The established resistance relies on the knockout of the host plant's EIF4E gene, which is essential for the translation of viral mRNAs [143]. Resistance-conferring alleles are also known from a related barley gene [144,145]. In addition, there are a number of susceptibility genes for diseases caused by other pathogens, which is similarly suitable for resistance breeding.

The ability of Cas endonucleases to completely eliminate the function of a gene seems particularly attractive when it comes to freeing crops from toxic, carcinogenic, allergenic, or bad-tasting gene products or metabolites. The aforementioned study by Sanchez-Léon et al. [113] on the production of wheat with gluten, whose fraction of immunodominant gliadins is drastically reduced, is a promising example in this respect. Another potential use of Cas endonuclease technology results from the fact that most modern durum wheat varieties carry a mutation in a heavy metal transporter gene, which entails a reduced sequestration of cadmium in the roots so that this toxic element accumulates in the grain to an extent that is of concern in terms of human health [146]. The restoration of the corresponding wild-type allele would, therefore, be a particularly promising way of improving the quality of pasta, which is widely consumed across the entire globe.

As has already been demonstrated in potato [147] and rice [148], it is also possible to modify the starch quality by specifically modulating the biosynthesis of the major starch components amylose and amylopectin. While amylose-free starch is of particularly high value for the paper and chemical industries, starch with reduced amyplopectin content is referred to as resistant starch, which, due to its fiber-like properties, has the potential to counteract type-2 diabetes, a civilization disease that is now increasingly spreading even in developing countries.

For future approaches, it is also conceivable to modify many different genes in one step, for instance to influence complex processes such as photosynthesis [149]. Cas endonuclease technology offers unprecedented prerequisites for this due to the option of simultaneous use of numerous gRNAs.

In summary, it can be stated that the establishment of Cas endonuclease technology represents a quantum leap for plant research and breeding. Even with regard to major methodological challenges, the current pace of research and development in this field means that further very useful solutions are highly probable in the near future. The implementation of this technology, especially in the agronomically important cereals of the *Triticeae* cereals, will likely continue to gain momentum, and is thus expected to increasingly contribute to the effective production of sufficient and high-quality food, feed, and industrial raw materials, while ensuring largely improved environmental compatibility of agriculture.

Author Contributions: Conceptualization, J.K.; collection of information, I.K., C.H., R.H., J.K.; writing—original draft preparation, I.K., C.H., R.H., J.K.; writing—review and editing, I.K., C.H., R.H., J.K.; supervision, J.K.; funding acquisition, J.K.

Funding: This research was funded by the German Federal Ministry of Education and Research, project grants OSIRIS (FKZ 031B0201) and IDEMODERESBAR (FKZ 031B0199).

Conflicts of Interest: The authors declare no conflict of interest. The funders had no role in the design of the study; in the collection, analyses, or interpretation of data; in the writing of the manuscript, or in the decision to publish the results.

Appendix A

Table A1. Survey of studies employing Cas endonuclease technology in wheat and barley. If not stated otherwise, gRNA and Cas9 were delivered as transgenes, the outcome was site-directed mutagenesis (i.e., random InDels at the target sites), and genetic modifications were shown to be heritable. IEs: immature embryos, RNP: gRNA/Cas9 ribonucleoprotein.

Target Gene	Reagent Delivery, Recipient Cells, Genotype, Notes	Aim of the Study	Efficiency of Targeted Mutagenesis	Observed Phenotype	Reference
Wheat					
INOSITOL OXYGENASE (INOX), PHYTOENE DESATURASE (PDS)	Agrobacterium, callus-derived cell suspension, genotype not reported	method establishment	mutations detected in cell suspension, no plants generated	not reported	[37]
MILDEW RESISTANCE LOCUS O (MLO)	ballistic, IEs of "Kenong199" and "Bobwhite"	resistance to powdery mildew	4 out of 72 T0 plants mutated in A-genome, heritability not shown	not reported	[39]
GRAIN WIDTH 2 (GW2)	ballistic transfer of RNP, IEs of "Kenong199", selection-free regeneration	method establishment	28 primary mutants from 640 IEs, heritability not shown	not reported	[96]
α-GLIADIN gene family	ballistic, IEs of "BW208", "THA53", "Don Pedro"	reduced α-gliadin content in grains	Mutation rate in T0 not reported, 17 out of 17 analyzed T1 plants mutated, heritability of genotype and phenotype shown up to T3	reduced α-gliadin content in grains	[113]
ENHANCED DISEASE RESISTANCE 1 (EDR1)	ballistic, IEs of "KN199"	reduced susceptibility to Blumeria graminis	5 mutant T0 plants identified, 5 out of 207 analyzed T1 plants homozygously mutated in all 3 copies	resistance shown in homozygous T1 mutants	[150]
PHYTOENE DESATURASE (PDS)	Agrobacterium, IEs of "Fielder"	method establishment	up to 13 out of 73 T0 plants mutated, homozygous mutations in individual subgenomes shown in T2	not reported (no homozygous triple mutants)	[85]
DEP1, GRAIN WIDTH 2 (GW2)	PEG-mediated transfection, protoplasts of "Kenong199", ballistic, IEs of "Kenong199"	establishment of base editing using adenosine deaminase fusion to nCas9	up to 7.5% of NGS reads from target amplicons showed A -> G conversion, 5 (DEP1) and 2 (GW2) regenerated T0 plants showed heterozygous mutations, no heritability reported	not reported	[105]
ZIP4-B2 (homoeolog of B genome only)	Agrobacterium, IEs of "Fielder"	increase of crossover frequency	4 out of 81 analyzed T0 plants mutated, 24 T1 plants (progeny of 1 T0 plant) analyzed	increase of homologous crossovers in KO-mutants under presence of magnesium	[151]
GRAIN WIDTH 2 (GW2-ABD), 9-LIPOXYGENASE (LPX-1-BD), MILDEW RESISTANCE LOCUS O (MLO)	Ballistic, IEs of "Bobwhite"	GW2: increased grain size and weight, LPX-1 and TaMLO: resistance to fungi	20 out of 102 T0 plants with triple gRNAs mutated, 6 out of 61 T0 plants with GW2-gRNA mutated, heritability analyzed by amplicon NGS in progeny of 2 T0 plants, several T1 plants heterozygous for GW2 or MLO, homozygous T1 for GW2	GW2-KO: increased thousand-grain weight by 27%	[48]

Table A1. Cont.

Target Gene	Reagent Delivery, Recipient Cells, Genotype, Notes	Aim of the Study	Efficiency of Targeted Mutagenesis	Observed Phenotype	Reference
MALE STERILITY 45 (MS45)	Agrobacterium, IEs of "Fielder" and 'SBC0456D'	male sterility	25 out of 181 analyzed T1 plants mutated in different combinations (single, double, triple KOs)	triple mutants (ABD-KO) proved to be male sterile	[152]
GRAIN WIDTH 2 (GW2)	ballistic, IEs of "Bobwhite"	increased grain size and weight	T1, T2, T3 mutants derived from Wang et al. (2018a) used to generate single, double, and triple KO-mutants	thousand-grain weight increased, dosage effect of homeoalleles shown	[118]
PUROINDOLINE b (PINb), GRANULE-BOUND STARCH SYNTHASE (WAXY), DA1	Agrobacterium, IEs of "Fielder" (DA1 only)	PINb: grain hardness, WAXY: starch composition, DA1: grain size	13 out of 24 T0 plants DA1-target mutated in A or B genome, heritability not shown	not reported	[153]
Barley					
PM19-1 and PM19-3	Agrobacterium, IEs of 'Golden Promise'	seed dormancy	3 out of 13 analyzed T0 plants mutated (PM19-1), 1 out of 10 analyzed T0 plants mutated (PM19-3), heritability of mutations shown in progeny of 4 identified mutants in T2, off-target mutations found in other copies of PM-19	not reported	[35]
Promoter of PHYTASE GENE A (PAPhy_a)	Agrobacterium, IEs of "Golden Promise"	reduced phytase content in the grain	28 out of 64 T0 plants 28 mutated, heritability shown in T2	homozygous T1 mutants showed reduced phytase activity	40]
ENDO-N-ACETYL-β-D-GLUCOSAMINIDASE (ENGase)	ballistic and Agrobacterium, IEs of "Golden Promise", co-transformation for fragment deletion	N-glycan modification in the grain	ballistic: 7 out of 8 T0 plants mutated, Agrobacterium: 15 out of 23 T0 plants mutated, 6 plants with targeted deletion between two targeted positions, heritability shown in T1, homozygous mutants	N-glycan modification in the grain	[112]
CYTOKININ DEHYDROGENASE (CKX)	Agrobacterium, IEs of "Golden Promise"	gene function in cytokinin metabolism	15 out of 23 T0 plants mutated, heritability shown in T1, 4 homozygous KO-mutants selected for further analysis	KO-mutants showed higher cytokinin content and slightly reduced root growth	[154]
PHYTOENE DESATURASE (PDS)	Agrobacterium, IEs of "Golden Promise"	photo-bleaching	6 out of 40 T0 plants mutated, heritability not reported	chimeric photobleaching in T0 mutants	[85]
Member of MICRORCHIDIA GHKL ATPASE subfamily (MORC1)	Agrobacterium, IEs of "Golden Promise"	resistance to fungal pathogens	callus screened for mutations, progeny of randomly selected T0 plants screened for mutations, 12 and 10 T1 families analyzed for two constructs, mutations in 19 of the 22 families, proportion of mutated plants not reported	homozygous morc1-KO T1 plants were less susceptible to fungal pathogens and showed higher expression of transposons	[45]
PROTEIN TARGETING TO STARCH 1 (PTST1), GRANULE-BOUND STARCH SYNTHASE (GBSS1a)	Agrobacterium, IEs of "Golden Promise"	starch accumulation in the grain	6 out of 8 (PTST1) and 5 out of 9 (GBSS1a) T0 plants mutated, heritability shown in T1	ptst1-KO: starch-free grains, no germination, gbss1a-KO: amylose-free grains	[155]

References

1. Baulcombe, D.; Crute, I.; Davies, B.; Dunwell, J.; Gale, M.; Jones, J.; Pretty, J.; Sutherland, W.; Toulmin, C. *Reaping the Benefits: Science and the Sustainable Intensification of Global Agriculture;* The Royal Society: London, UK, 2009; p. 72.
2. Ronald, P. Plant genetics, sustainable agriculture and global food security. *Genetics* **2011**, *188*, 11–20. [CrossRef]
3. *Global Warming of 1.5 °C;* IPCC Special Report; IPCC: Geneva, Switzerland, 2018. Available online: https://www.ipcc.ch/sr15/ (accessed on 28 May 2019).
4. Modrzejewski, D.; Hartung, F.; Sprink, T.; Krause, D.; Kohl, C.; Schiemann, J.; Wilhelm, R. What is the available evidence for the application of genome editing as a new tool for plant trait modification and the potential occurrence of associated off-target effects: A systematic map protocol. *Environ. Evid.* **2018**, *7*, 11. [CrossRef]
5. Parry, M.A.J.; Madgwick, P.J.; Bayon, C.; Tearall, K.; Hernandez-Lopez, A.; Baudo, M.; Rakszegi, M.; Hamada, W.; Al-Yassin, A.; Ouabbou, H.; et al. Mutation discovery for crop improvement. *J. Exp. Bot.* **2009**, *60*, 2817–2825. [CrossRef] [PubMed]
6. Reynolds, M.; Foulkes, M.J.; Slafer, G.A.; Berry, P.; Parry, M.A.J.; Snape, J.W.; Angus, W.J. Raising yield potential in wheat. *J. Exp. Bot.* **2009**, *60*, 1899–1918. [CrossRef] [PubMed]
7. Sovová, T.; Kerins, G.; Demnerová, K.; Ovesná, J. Genome Editing with Engineered Nucleases in Economically Important Animals and Plants: State of the Art in the Research Pipeline. *Curr. Issues Mol. Biol.* **2017**, *21*, 41–62.
8. McCallum, C.M.; Comai, L.; Greene, E.A.; Henikoff, S. Targeting Induced LocalLesions IN Genomes (TILLING) for Plant Functional Genomics. *Plant Physiol.* **2000**, *123*, 439–442. [CrossRef] [PubMed]
9. Comai, L.; Young, K.; Till, B.J.; Reynolds, S.H.; Greene, E.A.; Codomo, C.A.; Enns, L.C.; Johnson, J.E.; Burtner, C.; Odden, A.R.; et al. Efficient discovery of DNA polymorphisms in natural populations by Ecotilling. *Plant J.* **2004**, *37*, 778–786. [CrossRef] [PubMed]
10. Herrera-Estrella, L.; Depicker, A.; van Montagu, M.; Schell, J. Expression of chimaeric genes transferred into plant cells using a Ti-plasmid-derived vector. *Nature* **1983**, *303*, 209–213. [CrossRef]
11. Kumlehn, J.; Hensel, G. Genetic transformation technology in the Triticeae. *Breed. Sci.* **2009**, *59*, 553–560. [CrossRef]
12. Jinek, M.; Chylinski, K.; Fonfara, I.; Hauer, M.; Doudna, J.A.; Charpentier, E. A programmable dual-RNA-guided DNA endonuclease in adaptive bacterial immunity. *Science* **2012**, *337*, 816–821. [CrossRef]
13. Ramírez-González, R.H.; Borrill, P.; Lang, D.; Harrington, S.A.; Brinton, J.; Venturini, L.; Davey, M.; Jacobs, J.; van Ex, F.; Pasha, A.; et al. The transcriptional landscape of polyploid wheat. *Science* **2018**, *361*. [CrossRef] [PubMed]
14. Brenchley, R.; Spannagl, M.; Pfeifer, M.; Barker, G.L.A.; D'Amore, R.; Allen, A.M.; McKenzie, N.; Kramer, M.; Kerhornou, A.; Bolser, D.; et al. Analysis of the bread wheat genome using whole-genome shotgun sequencing. *Nature* **2012**, *491*, 705–710. [CrossRef]
15. International Barley Genome Sequencing Consortium; Mayer, K.F.; Waugh, R.; Brown, J.W.; Schulman, A.; Langridge, P.; Platzer, M.; Fincher, G.B.; Muehlbauer, G.J.; Sato, K.; et al. A physical, genetic and functional sequence assembly of the barley genome. *Nature* **2012**, *491*, 711–716. [CrossRef] [PubMed]
16. Mascher, M.; Gundlach, H.; Himmelbach, A.; Beier, S.; Twardziok, S.O.; Wicker, T.; Radchuk, V.; Dockter, C.; Hedley, P.E.; Russell, J.; et al. A chromosome conformation capture ordered sequence of the barley genome. *Nature* **2017**, *544*, 427–433. [CrossRef] [PubMed]
17. Milner, S.G.; Jost, M.; Taketa, S.; Mazón, E.R.; Himmelbach, A.; Oppermann, M.; Weise, S.; Knüpffer, H.; Basterrechea, M.; König, P.; et al. Genebank genomics highlights the diversity of a global barley collection. *Nat. Genet.* **2019**, *51*, 319–326. [CrossRef]
18. Plessis, A.; Perrin, A.; Haber, J.E.; Dujon, B. Site-specific recombination determined by I-SceI, a mitochondrial group I intron-encoded endonuclease expressed in the yeast nucleus. *Genetics* **1992**, *130*, 451–460.
19. Pauwels, K.; Podevin, N.; Breyer, D.; Carroll, D.; Herman, P. Engineering nucleases for gene targeting: Safety and regulatory considerations. *New Biotechnol.* **2014**, *31*, 18–27. [CrossRef]

20. Puchta, H.; Dujon, B.; Hohn, B. Homologous recombination in plant cells is enhanced by in vivo induction of double strand breaks into DNA by a site-specific endonuclease. *Nucleic Acids Res.* **1993**, *21*, 5034–5040. [CrossRef]

21. Gao, H.; Smith, J.; Yang, M.; Jones, S.; Djukanovic, V.; Nicholson, M.G.; West, A.; Bidney, D.; Falco, S.C.; Jantz, D.; et al. Heritable targeted mutagenesis in maize using a designed endonuclease. *Plant J.* **2010**, *61*, 176–187. [CrossRef] [PubMed]

22. Puchta, H.; Fauser, F. Synthetic nucleases for genome engineering in plants: Prospects for a bright future. *Plant J.* **2014**, *78*, 727–741. [CrossRef] [PubMed]

23. Kouranova, E.; Forbes, K.; Zhao, G.; Warren, J.; Bartels, A.; Wu, Y.; Cui, X. CRISPRs for Optimal Targeting: Delivery of CRISPR Components as DNA, RNA, and Protein into Cultured Cells and Single-Cell Embryos. *Hum. Gene Ther.* **2016**, *27*, 464–475. [CrossRef] [PubMed]

24. Vu, G.T.H.; Cao, H.X.; Watanabe, K.; Hensel, G.; Blattner, F.R.; Kumlehn, J.; Schubert, I. Repair of Site-Specific DNA Double-Strand Breaks in Barley Occurs via Diverse Pathways Primarily Involving the Sister Chromatid. *Plant Cell* **2014**, *26*, 2156–2167. [CrossRef]

25. Kim, Y.G.; Cha, J.; Chandrasegaran, S. Hybrid restriction enzymes: Zinc finger fusions to Fok I cleavage domain. *Proc. Natl. Acad. Sci. USA* **1996**, *93*, 1156–1160. [CrossRef]

26. Voytas, D.F. Plant genome engineering with sequence-specific nucleases. *Annu. Rev. Plant Biol.* **2013**, *64*, 327–350. [CrossRef]

27. Smith, J.; Berg, J.M.; Chandrasegaran, S. A detailed study of the substrate specificity of a chimeric restriction enzyme. *Nucleic Acids Res.* **1999**, *27*, 674–681. [CrossRef]

28. Doyon, Y.; McCammon, J.M.; Miller, J.C.; Faraji, F.; Ngo, C.; Katibah, G.E.; Amora, R.; Hocking, T.D.; Zhang, L.; Rebar, E.J.; et al. Heritable targeted gene disruption in zebrafish using designed zinc-finger nucleases. *Nat. Biotechnol.* **2008**, *26*, 702–708. [CrossRef]

29. Bonas, U.; Stall, R.E.; Staskawicz, B. Genetic and structural characterization of the avirulence gene avrBs3 from Xanthomonas campestris pv. vesicatoria. *Mol. Gen. Genet. MGG* **1989**, *218*, 127–136. [CrossRef]

30. Christian, M.; Cermak, T.; Doyle, E.L.; Schmidt, C.; Zhang, F.; Hummel, A.; Bogdanove, A.J.; Voytas, D.F. Targeting DNA double-strand breaks with TAL effector nucleases. *Genetics* **2010**, *186*, 757–761. [CrossRef]

31. Boch, J.; Scholze, H.; Schornack, S.; Landgraf, A.; Hahn, S.; Kay, S.; Lahaye, T.; Nickstadt, A.; Bonas, U. Breaking the code of DNA binding specificity of TAL-type III effectors. *Science* **2009**, *326*, 1509–1512. [CrossRef]

32. Doudna, J.A.; Charpentier, E. Genome editing. The new frontier of genome engineering with CRISPR-Cas9. *Science* **2014**, *346*, 1258096. [CrossRef] [PubMed]

33. Kumlehn, J.; Pietralla., J.; Hensel, G.; Pacher, M.; Puchta, H. The CRISPR/Cas revolution continues: From efficient gene editing for crop breeding to plant synthetic biology. *J. Integr. Plant Biol.* **2018**, *60*, 1127–1154. [CrossRef]

34. Shan, Q.; Wang, Y.; Li, J.; Zhang, Y.; Chen, K.; Liang, Z.; Zhang, K.; Liu, J.; Xi, J.J.; Qiu, J.-L.; et al. Targeted genome modification of crop plants using a CRISPR-Cas system. *Nat. Biotechnol.* **2013**, *31*, 686–688. [CrossRef] [PubMed]

35. Lawrenson, T.; Shorinola, O.; Stacey, N.; Li, C.; Østergaard, L.; Patron, N.; Uauy, C.; Harwood, W. Induction of targeted, heritable mutations in barley and Brassica oleracea using RNA-guided Cas9 nuclease. *Genome Biol.* **2015**, *16*, 258. [CrossRef] [PubMed]

36. Liu, X.; Wu, S.; Xu, J.; Sui, C.; Wei, J. Application of CRISPR/Cas9 in plant biology. *Acta Pharm. Sin. B* **2017**, *7*, 292–302. [CrossRef] [PubMed]

37. Upadhyay, S.K.; Kumar, J.; Alok, A.; Tuli, R. RNA-guided genome editing for target gene mutations in wheat. *G3 Genes Genome Genet.* **2013**, *3*, 2233–2238. [CrossRef] [PubMed]

38. Shan, Q.; Wang, Y.; Li, J.; Gao, C. Genome editing in rice and wheat using the CRISPR/Cas system. *Nat. Protoc.* **2014**, *9*, 2395–2410. [CrossRef]

39. Wang, Y.; Cheng, X.; Shan, Q.; Zhang, Y.; Liu, J.; Gao, C.; Qiu, J.-L. Simultaneous editing of three homoeoalleles in hexaploid bread wheat confers heritable resistance to powdery mildew. *Nat. Biotechnol.* **2014**, *32*, 947–951. [CrossRef]

40. Holme, I.B.; Wendt, T.; Gil-Humanes, J.; Deleuran, L.C.; Starker, C.G.; Voytas, D.F.; Brinch-Pedersen, H. Evaluation of the mature grain phytase candidate HvPAPhy_a gene in barley (*Hordeum vulgare* L.) using CRISPR/Cas9 and TALENs. *Plant Mol. Biol.* **2017**, *95*, 111–121. [CrossRef]

41. Čermák, T.; Baltes, N.J.; Čegan, R.; Zhang, Y.; Voytas, D.F. High-frequency, precise modification of the tomato genome. *Genome Biol.* **2015**, *16*, 232. [CrossRef]

42. Gil-Humanes, J.; Wang, Y.; Liang, Z.; Shan, Q.; Ozuna, C.V.; Sánchez-León, S.; Baltes, N.J.; Starker, C.; Barro, F.; Gao, C.; et al. High-efficiency gene targeting in hexaploid wheat using DNA replicons and CRISPR/Cas9. *Plant J.* **2017**, *89*, 1251–1262. [CrossRef]

43. Gao, Y.; Zhao, Y. Self-processing of ribozyme-flanked RNAs into guide RNAs in vitro and in vivo for CRISPR-mediated genome editing. *J. Integr. Plant Biol.* **2014**, *56*, 343–349. [CrossRef] [PubMed]

44. Xing, H.-L.; Dong, L.; Wang, Z.-P.; Zhang, H.-Y.; Han, C.-Y.; Liu, B.; Wang, X.-C.; Chen, Q.-J. A CRISPR/Cas9 toolkit for multiplex genome editing in plants. *BMC Plant Biol.* **2014**, *14*, 327. [CrossRef]

45. Kumar, N.; Galli, M.; Ordon, J.; Stuttmann, J.; Kogel, K.-H.; Imani, J. Further analysis of barley MORC1 using a highly efficient RNA-guided Cas9 gene-editing system. *Plant Biotechnol. J.* **2018**, *16*, 1892–1903. [CrossRef] [PubMed]

46. Ma, X.; Zhang, Q.; Zhu, Q.; Liu, W.; Chen, Y.; Qiu, R.; Wang, B.; Yang, Z.; Li, H.; Lin, Y.; et al. A Robust CRISPR/Cas9 System for Convenient, High-Efficiency Multiplex Genome Editing in Monocot and Dicot Plants. *Mol. Plant* **2015**, *8*, 1274–1284. [CrossRef]

47. Lowder, L.G.; Zhang, D.; Baltes, N.J.; Paul, J.W.; Tang, X.; Zheng, X.; Voytas, D.F.; Hsieh, T.-F.; Zhang, Y.; Qi, Y. A CRISPR/Cas9 Toolbox for Multiplexed Plant Genome Editing and Transcriptional Regulation. *Plant Physiol.* **2015**, *169*, 971–985. [CrossRef] [PubMed]

48. Wang, W.; Pan, Q.; He, F.; Akhunova, A.; Chao, S.; Trick, H.; Akhunov, E. Transgenerational CRISPR-Cas9 Activity Facilitates Multiplex Gene Editing in Allopolyploid Wheat. *Cris. J.* **2018**, *1*, 65–74. [CrossRef]

49. Tang, X.; Zheng, X.; Qi, Y.; Zhang, D.; Cheng, Y.; Tang, A.; Voytas, D.F.; Zhang, Y. A Single Transcript CRISPR-Cas9 System for Efficient Genome Editing in Plants. *Mol. Plant* **2016**, *9*, 1088–1091. [CrossRef] [PubMed]

50. Čermák, T.; Curtin, S.J.; Gil-Humanes, J.; Čegan, R.; Kono, T.J.Y.; Konečná, E.; Belanto, J.J.; Starker, C.G.; Mathre, J.W.; Greenstein, R.L.; et al. A Multipurpose Toolkit to Enable Advanced Genome Engineering in Plants. *Plant Cell* **2017**, *29*, 1196–1217. [CrossRef]

51. Ren, X.; Yang, Z.; Xu, J.; Sun, J.; Mao, D.; Hu, Y.; Yang, S.J.; Qiao, H.H.; Wang, X.; Hu, Q.; et al. Enhanced specificity and efficiency of the CRISPR/Cas9 system with optimized sgRNA parameters in Drosophila. *Cell Rep.* **2014**, *9*, 1151–1162. [CrossRef]

52. Moreno-Mateos, M.A.; Vejnar, C.E.; Beaudoin, J.D.; Fernandez, J.P.; Mis, E.K.; Khokha, M.K.; Giraldez, A.J. CRISPRscan: Designing highly efficient sgRNAs for CRISPR-Cas9 targeting in vivo. *Nat. Methods* **2015**, *12*, 982–988. [CrossRef]

53. Cong, L.; Ran, F.A.; Cox, D.; Lin, S.; Barretto, R.; Habib, N.; Hsu, P.D.; Wu, X.; Jiang, W.; Marraffini, L.A.; et al. Multiplex genome engineering using CRISPR/Cas systems. *Science* **2013**, *339*, 819–823. [CrossRef] [PubMed]

54. Fu, Y.; Sander, J.D.; Reyon, D.; Cascio, V.M.; Joung, J.K. Improving CRISPR-Cas nuclease specificity using truncated guide RNAs. *Nat. Biotechnol.* **2014**, *32*, 279–284. [CrossRef]

55. Zhang, J.P.; Li, X.L.; Neises, A.; Chen, W.; Hu, L.P.; Ji, G.Z.; Yu, J.Y.; Xu, J.; Yuan, W.P.; Cheng, T.; et al. Different Effects of sgRNA Length on CRISPR-mediated Gene Knockout Efficiency. *Sci. Rep.* **2016**, *6*, 28566. [CrossRef] [PubMed]

56. Osakabe, Y.; Watanabe, T.; Sugano, S.S.; Ueta, R.; Ishihara, R.; Shinozaki, K.; Osakabe, K. Optimization of CRISPR/Cas9 genome editing to modify abiotic stress responses in plants. *Sci. Rep.* **2016**, *6*, 26685. [CrossRef] [PubMed]

57. Gerasimova, S.V.; Korotkova, A.M.; Hertig, C.; Hiekel, S.; Hoffie, R.; Budhagatapalli, N.; Otto, I.; Hensel, G.; Shumny, V.K.; Kochetov, A.V.; et al. Targeted genome modifcation in protoplasts of a highly regenerable Siberian barley cultivar using RNA-guided Cas9 endonuclease. *Vavilovskii Zhurnal Genet. I Selektsii* **2019**, *22*, 1033–1039. [CrossRef]

58. Cho, S.W.; Kim, S.; Kim, Y.; Kweon, J.; Kim, H.S.; Bae, S.; Kim, J.S. Analysis of off-target effects of CRISPR/Cas-derived RNA-guided endonucleases and nickases. *Genome Res.* **2014**, *24*, 132–141. [CrossRef]

59. Lei, Y.; Lu, L.; Liu, H.Y.; Li, S.; Xing, F.; Chen, L.L. CRISPR-P: A web tool for synthetic single-guide RNA design of CRISPR-system in plants. *Mol. Plant* **2014**, *7*, 1494–1496. [CrossRef] [PubMed]

60. Liu, H.; Ding, Y.; Zhou, Y.; Jin, W.; Xie, K.; Chen, L.L. CRISPR-P 2.0: An Improved CRISPR-Cas9 Tool for Genome Editing in Plants. *Mol. Plant* **2017**, *10*, 530–532. [CrossRef] [PubMed]

61. Doench, J.G.; Hartenian, E.; Graham, D.B.; Tothova, Z.; Hegde, M.; Smith, I.; Sullender, M.; Ebert, B.L.; Xavier, R.J.; Root, D.E. Rational design of highly active sgRNAs for CRISPR-Cas9-mediated gene inactivation. *Nat. Biotechnol.* **2014**, *32*, 1262–1267. [CrossRef]

62. Doench, J.G.; Fusi, N.; Sullender, M.; Hegde, M.; Vaimberg, E.W.; Donovan, K.F.; Smith, I.; Tothova, Z.; Wilen, C.; Orchard, R.; et al. Optimized sgRNA design to maximize activity and minimize off-target effects of CRISPR-Cas9. *Nat. Biotechnol.* **2016**, *34*, 184–191. [CrossRef] [PubMed]

63. Haeussler, M.; Schonig, K.; Eckert, H.; Eschstruth, A.; Mianne, J.; Renaud, J.B.; Schneider-Maunoury, S.; Shkumatava, A.; Teboul, L.; Kent, J.; et al. Evaluation of off-target and on-target scoring algorithms and integration into the guide RNA selection tool CRISPOR. *Genome Biol.* **2016**, *17*, 148. [CrossRef]

64. Concordet, J.P.; Haeussler, M. CRISPOR: Intuitive guide selection for CRISPR/Cas9 genome editing experiments and screens. *Nucleic Acids Res.* **2018**, *46*, W242–W245. [CrossRef] [PubMed]

65. Wong, N.; Liu, W.; Wang, X. WU-CRISPR: Characteristics of functional guide RNAs for the CRISPR/Cas9 system. *Genome Biol.* **2015**, *16*, 218. [CrossRef]

66. Liang, G.; Zhang, H.; Lou, D.; Yu, D. Selection of highly efficient sgRNAs for CRISPR/Cas9-based plant genome editing. *Sci. Rep.* **2016**, *6*, 21451. [CrossRef]

67. Zuker, M.; Jacobson, A.B. Using reliability information to annotate RNA secondary structures. *RNA* **1998**, *4*, 669–679. [CrossRef]

68. Waugh, A.; Gendron, P.; Altman, R.; Brown, J.W.; Case, D.; Gautheret, D.; Harvey, S.C.; Leontis, N.; Westbrook, J.; Westhof, E.; et al. RNAML: A standard syntax for exchanging RNA information. *RNA* **2002**, *8*, 707–717. [CrossRef]

69. Zuker, M. Mfold web server for nucleic acid folding and hybridization prediction. *Nucleic Acids Res.* **2003**, *31*, 3406–3415. [CrossRef]

70. Gruber, A.R.; Lorenz, R.; Bernhart, S.H.; Neubock, R.; Hofacker, I.L. The Vienna RNA websuite. *Nucleic Acids Res.* **2008**, *36*, W70–W74. [CrossRef] [PubMed]

71. Lorenz, R.; Bernhart, S.H.; Honer Zu Siederdissen, C.; Tafer, H.; Flamm, C.; Stadler, P.F.; Hofacker, I.L. ViennaRNA Package 2.0. *Algorithms Mol. Biol.* **2011**, *6*, 26. [CrossRef] [PubMed]

72. Pasternak, T.P.; Rudas, V.A.; Lörz, H.; Kumlehn, J. Embryogenic Callus Formation and Plant Regeneration from Leaf Base Segments of Barley (*Hordeum vulgare* L.). *J. Plant Physiol.* **1999**, *155*, 371–375. [CrossRef]

73. Vasil, V.; Castillo, A.M.; Fromm, M.E.; Vasil, I.K. Herbicide Resistant Fertile Transgenic Wheat Plants Obtained by Microprojectile Bombardment of Regenerable Embryogenic Callus. *Bio-Technol.* **1992**, *10*, 667–674. [CrossRef]

74. Wan, Y.; Lemaux, P.G. Generation of Large Numbers of Independently Transformed Fertile Barley Plants. *Plant Physiol.* **1994**, *104*, 37–48. [CrossRef]

75. Hiei, Y.; Ohta, S.; Komari, T.; Kumashiro, T. Efficient transformation of rice (*Oryza sativa* L.) mediated by Agrobacterium and sequence analysis of the boundaries of the T-DNA. *Plant J.* **1994**, *6*, 271–282. [CrossRef] [PubMed]

76. Tingay, S.; McElroy, D.; Kalla, R.; Fieg, S.; Wang, M.; Thornton, S.; Brettell, R. Agrobacterium tumefaciens-mediated barley transformation. *Plant J.* **1997**, *11*, 1369–1376. [CrossRef]

77. Cheng, M.; Fry, J.E.; Pang, S.; Zhou, H.; Hironaka, C.M.; Duncan, D.R.; Conner, T.W.; Wan, Y. Genetic Transformation of Wheat Mediated by Agrobacterium tumefaciens. *Plant Physiol.* **1997**, *115*, 971–980. [CrossRef]

78. Hensel, G.; Kastner, C.; Oleszczuk, S.; Riechen, J.; Kumlehn, J. Agrobacterium-mediated gene transfer to cereal crop plants: Current protocols for barley, wheat, triticale, and maize. *Int. J. Plant Genom.* **2009**, *2009*, 835608. [CrossRef]

79. Rasco-Gaunt, S.; Riley, A.; Cannell, M.; Barcelo, P.; Lazzeri, P.A. Procedures allowing the transformation of a range of European elite wheat (*Triticum aestivum* L.) varieties via particle bombardment. *J. Exp. Bot.* **2001**, *52*, 865–874. [CrossRef] [PubMed]

80. Wu, H.; Sparks, C.; Amoah, B.; Jones, H.D. Factors influencing successful Agrobacterium-mediated genetic transformation of wheat. *Plant Cell Rep.* **2003**, *21*, 659–668. [CrossRef]

81. Richardson, T.; Thistleton, J.; Higgins, T.J.; Howitt, C.; Ayliffe, M. Efficient Agrobacterium transformation of elite wheat germplasm without selection. *Plant Cell Tissue Organ Cult.* **2014**, *119*, 647–659. [CrossRef]

82. Ishida, Y.; Tsunashima, M.; Hiei, Y.; Komari, T. Wheat (*Triticum aestivum* L.) transformation using immature embryos. *Methods Mol. Biol.* **2015**, *1223*, 189–198. [CrossRef] [PubMed]

83. Hensel, G.; Valkov, V.; Middlefell-Williams, J.; Kumlehn, J. Efficient generation of transgenic barley: The way forward to modulate plant-microbe interactions. *J. Plant Physiol.* **2008**, *165*, 71–82. [CrossRef]

84. Yeo, F.K.S.; Hensel, G.; Vozábová, T.; Martin-Sanz, A.; Marcel, T.C.; Kumlehn, J.; Niks, R.E. Golden SusPtrit—A genetically well transformable barley line for studies on the resistance to rust fungi. *Theor. Appl. Genet.* **2014**, *127*, 325–337. [CrossRef] [PubMed]

85. Howells, R.M.; Craze, M.; Bowden, S.; Wallington, E.J. Efficient generation of stable, heritable gene edits in wheat using CRISPR/Cas9. *BMC Plant Biol.* **2018**, *18*, 215. [CrossRef] [PubMed]

86. Daghma, D.E.S.; Hensel, G.; Rutten, T.; Melzer, M.; Kumlehn, J. Cellular dynamics during early barley pollen embryogenesis revealed by time-lapse imaging. *Front. Plant Sci.* **2014**, *5*, 675. [CrossRef]

87. Kumlehn, J.; Serazetdinova, L.; Hensel, G.; Becker, D.; Loerz, H. Genetic transformation of barley (*Hordeum vulgare* L.) via infection of androgenetic pollen cultures with Agrobacterium tumefaciens. *Plant Biotechnol. J.* **2006**, *4*, 251–261. [CrossRef]

88. Otto, I.; Muller, A.; Kumlehn, J. Barley (*Hordeum vulgare* L.) transformation using embryogenic pollen cultures. *Methods Mol. Biol.* **2015**, *1223*, 85–99. [CrossRef] [PubMed]

89. Shim, Y.S.; Pauls, K.P.; Kasha, K.J. Transformation of isolated barley (*Hordeum vulgare* L.) microspores: II. Timing of pretreatment and temperatures relative to results of bombardment. *Genome* **2009**, *52*, 175–190. [CrossRef]

90. Gurushidze, M.; Hensel, G.; Hiekel, S.; Schedel, S.; Valkov, V.; Kumlehn, J. True-breeding targeted gene knock-out in barley using designer TALE-nuclease in haploid cells. *PLoS ONE* **2014**, *9*, e92046. [CrossRef] [PubMed]

91. Zhang, S.; Cho, M.J.; Koprek, T.; Yun, R.; Bregitzer, P.; Lemaux, P.G. Genetic transformation of commercial cultivars of oat (*Avena sativa* L.) and barley (*Hordeum vulgare* L.) using in vitro shoot meristematic cultures derived from germinated seedlings. *Plant Cell Rep.* **1999**, *18*, 959–966. [CrossRef]

92. Hamada, H.; Linghu, Q.; Nagira, Y.; Miki, R.; Taoka, N.; Imai, R. An in planta biolistic method for stable wheat transformation. *Sci. Rep.* **2017**, *7*, 11443. [CrossRef]

93. Hamada, H.; Liu, Y.; Nagira, Y.; Miki, R.; Taoka, N.; Imai, R. Biolistic-delivery-based transient CRISPR/Cas9 expression enables in planta genome editing in wheat. *Sci. Rep.* **2018**, *8*, 14422. [CrossRef]

94. Kapusi, E.; Hensel, G.; Coronado, M.J.; Broeders, S.; Marthe, C.; Otto, I.; Kumlehn, J. The elimination of a selectable marker gene in the doubled haploid progeny of co-transformed barley plants. *Plant Mol. Biol.* **2013**, *81*, 149–160. [CrossRef]

95. Gurushidze, M.; Trautwein, H.; Hoffmeister, P.; Otto, I.; Müller, A.; Kumlehn, J. Doubled Haploidy as a Tool for Chimaera Dissolution of TALEN-Induced Mutations in Barley. In *Biotechnologies for Plant Mutation Breeding*; Springer: Cham, Switzerland, 2017; Chapter 8; pp. 129–141. [CrossRef]

96. Liang, Z.; Chen, K.; Li, T.; Zhang, Y.; Wang, Y.; Zhao, Q.; Liu, J.; Zhang, H.; Liu, C.; Ran, Y.; et al. Efficient DNA-free genome editing of bread wheat using CRISPR/Cas9 ribonucleoprotein complexes. *Nat. Commun.* **2017**, *8*, 14261. [CrossRef] [PubMed]

97. Budhagatapalli, N.; Schedel, S.; Gurushidze, M.; Pencs, S.; Hiekel, S.; Rutten, T.; Kusch, S.; Morbitzer, R.; Lahaye, T.; Panstruga, R.; et al. A simple test for the cleavage activity of customized endonucleases in plants. *Plant Methods* **2016**, *12*, 18. [CrossRef]

98. Budhagatapalli, N.; Rutten, T.; Gurushidze, M.; Kumlehn, J.; Hensel, G. Targeted Modification of Gene Function Exploiting Homology-Directed Repair of TALEN-Mediated Double-Strand Breaks in Barley. *G3 Genes Genome Genet.* **2015**, *5*, 1857–1863. [CrossRef]

99. Zhang, Y.; Liang, Z.; Zong, Y.; Wang, Y.; Liu, J.; Chen, K.; Qiu, J.L.; Gao, C. Efficient and transgene-free genome editing in wheat through transient expression of CRISPR/Cas9 DNA or RNA. *Nat. Commun.* **2016**, *7*, 12617. [CrossRef] [PubMed]

100. Bae, S.; Kweon, J.; Kim, H.S.; Kim, J.-S. Microhomology-based choice of Cas9 nuclease target sites. *Nat. Methods* **2014**, *11*, 705–706. [CrossRef]

101. Zong, Y.; Wang, Y.; Li, C.; Zhang, R.; Chen, K.; Ran, Y.; Qiu, J.-L.; Wang, D.; Gao, C. Precise base editing in rice, wheat and maize with a Cas9-cytidine deaminase fusion. *Nat. Biotechnol.* **2017**, *35*, 438–440. [CrossRef]

102. Komor, A.C.; Kim, Y.B.; Packer, M.S.; Zuris, J.A.; Liu, D.R. Programmable editing of a target base in genomic DNA without double-stranded DNA cleavage. *Nature* **2016**, *533*, 420–424. [CrossRef] [PubMed]

103. Gaudelli, N.M.; Komor, A.C.; Rees, H.A.; Packer, M.S.; Badran, A.H.; Bryson, D.I.; Liu, D.R. Programmable base editing of A•T to G•C in genomic DNA without DNA cleavage. *Nature* **2017**, *551*, 464–471. [CrossRef] [PubMed]

104. Zong, Y.; Song, Q.; Li, C.; Jin, S.; Zhang, D.; Wang, Y.; Qiu, J.-L.; Gao, C. Efficient C-to-T base editing in plants using a fusion of nCas9 and human APOBEC3A. *Nat. Biotechnol.* **2018**, *36*, 950–953. [CrossRef]

105. Li, C.; Zong, Y.; Wang, Y.; Jin, S.; Zhang, D.; Song, Q.; Zhang, R.; Gao, C. Expanded base editing in rice and wheat using a Cas9-adenosine deaminase fusion. *Genome Biol.* **2018**, *19*, 59. [CrossRef]

106. Puchta, H.; Dujon, B.; Hohn, B. Two different but related mechanisms are used in plants for the repair of genomic double-strand breaks by homologous recombination. *Proc. Natl. Acad. Sci. USA* **1996**, *93*, 5055–5060. [CrossRef]

107. Fauser, F.; Roth, N.; Pacher, M.; Ilg, G.; Sánchez-Fernández, R.; Biesgen, C.; Puchta, H. In planta gene targeting. *Proc. Natl. Acad. Sci. USA* **2012**, *109*, 7535–7540. [CrossRef]

108. Ran, F.A.; Hsu, P.D.; Lin, C.-Y.; Gootenberg, J.S.; Konermann, S.; Trevino, A.E.; Scott, D.A.; Inoue, A.; Matoba, S.; Zhang, Y.; et al. Double nicking by RNA-guided CRISPR Cas9 for enhanced genome editing specificity. *Cell* **2013**, *154*, 1380–1389. [CrossRef]

109. Mikami, M.; Toki, S.; Endo, M. Precision Targeted Mutagenesis via Cas9 Paired Nickases in Rice. *Plant Cell Physiol.* **2016**, *57*, 1058–1068. [CrossRef] [PubMed]

110. Baltes, N.J.; Gil-Humanes, J.; Cermak, T.; Atkins, P.A.; Voytas, D.F. DNA replicons for plant genome engineering. *Plant Cell* **2014**, *26*, 151–163. [CrossRef]

111. Ran, Y.; Patron, N.; Kay, P.; Wong, D.; Buchanan, M.; Cao, Y.Y.; Sawbridge, T.; Davies, J.P.; Mason, J.; Webb, S.R.; et al. Zinc finger nuclease-mediated precision genome editing of an endogenous gene in hexaploid bread wheat (Triticum aestivum) using a DNA repair template. *Plant Biotechnol. J.* **2018**, *16*, 2088–2101. [CrossRef] [PubMed]

112. Kapusi, E.; Corcuera-Gomez, M.; Melnik, S.; Stoger, E. Heritable Genomic Fragment Deletions and Small Indels in the Putative ENGase Gene Induced by CRISPR/Cas9 in Barley. *Front. Plant Sci.* **2017**, *8*, 540. [CrossRef] [PubMed]

113. Sánchez-León, S.; Gil-Humanes, J.; Ozuna, C.V.; Gimenez, M.J.; Sousa, C.; Voytas, D.F.; Barro, F. Low-gluten, nontransgenic wheat engineered with CRISPR/Cas9. *Plant Biotechnol. J.* **2018**, *16*, 902–910. [CrossRef]

114. Xie, K.; Yang, Y. RNA-guided genome editing in plants using a CRISPR-Cas system. *Mol. Plant* **2013**, *6*, 1975–1983. [CrossRef]

115. Liang, Z.; Chen, K.; Yan, Y.; Zhang, Y.; Gao, C. Genotyping genome-edited mutations in plants using CRISPR ribonucleoprotein complexes. *Plant Biotechnol. J.* **2018**, *16*, 2053–2062. [CrossRef]

116. Brinkman, E.K.; Chen, T.; Amendola, M.; van Steensel, B. Easy quantitative assessment of genome editing by sequence trace decomposition. *Nucleic Acids Res.* **2014**, *42*, e168. [CrossRef]

117. Liu, W.; Xie, X.; Ma, X.; Li, J.; Chen, J.; Liu, Y.G. DSDecode: A Web-Based Tool for Decoding of Sequencing Chromatograms for Genotyping of Targeted Mutations. *Mol. Plant* **2015**, *8*, 1431–1433. [CrossRef]

118. Wang, W.; Simmonds, J.; Pan, Q.; Davidson, D.; He, F.; Battal, A.; Akhunova, A.; Trick, H.N.; Uauy, C.; Akhunov, E. Gene editing and mutagenesis reveal inter-cultivar differences and additivity in the contribution of TaGW2 homoeologues to grain size and weight in wheat. *Theor. Appl. Genet.* **2018**, *131*, 2463–2475. [CrossRef]

119. Kim, D.; Alptekin, B.; Budak, H. CRISPR/Cas9 genome editing in wheat. *Funct. Integr. Genome* **2018**, *18*, 31–41. [CrossRef]

120. Hensel, G.; Marthe, C.; Kumlehn, J. Agrobacterium-Mediated Transformation of Wheat Using Immature Embryos. *Methods Mol. Biol.* **2017**, *1679*, 129–139. [CrossRef] [PubMed]

121. Wolt, J.D.; Wolf, C. Policy and Governance Perspectives for Regulation of Genome Edited Crops in the United States. *Front. Plant Sci.* **2018**, *9*, 1606. [CrossRef]

122. E.A.S.A. *Council Genome Editing: Scientific Opportunities, Public Interests and Policy Options in the European Union*; EASAC Secretariat Deutsche Akademie der Naturforscher Leopoldina German National Academy of Sciences: Halle, Germany, 2017; p. 34.

123. Sprink, T.; Eriksson, D.; Schiemann, J.; Hartung, F. Regulatory hurdles for genome editing: Process-vs. product-based approaches in different regulatory contexts. *Plant Cell Rep.* **2016**, *35*, 1493–1506. [CrossRef]

124. Bhowmik, P.; Ellison, E.; Polley, B.; Bollina, V.; Kulkarni, M.; Ghanbarnia, K.; Song, H.; Gao, C.; Voytas, D.F.; Kagale, S. Targeted mutagenesis in wheat microspores using CRISPR/Cas9. *Sci. Rep.* **2018**, *8*, 6502. [CrossRef]

125. Kumlehn, J.; Schieder, O.; Lorz, H. In vitro development of wheat (*Triticum aestivum* L.) from zygote to plant via ovule culture. *Plant Cell Rep.* **1997**, *16*, 663–667. [CrossRef]

126. Holme, I.B.; Brinch-Pedersen, H.; Lange, M.; Holm, P.B. Transformation of barley (*Hordeum vulgare* L.) by Agrobacterium tumefaciens infection of in vitro cultured ovules. *Plant Cell Rep.* **2006**, *25*, 1325–1335. [CrossRef] [PubMed]

127. Holme, I.B.; Brinch-Pedersen, H.; Lange, M.; Holm, P.B. Transformation of different barley (*Hordeum vulgare* L.) cultivars by Agrobacterium tumefaciens infection of in vitro cultured ovules. *Plant Cell Rep.* **2008**, *27*, 1833–1840. [CrossRef]

128. Hensel, G.; Himmelbach, A.; Chen, W.; Douchkov, D.K.; Kumlehn, J. Transgene expression systems in the Triticeae cereals. *J. Plant Physiol.* **2011**, *168*, 30–44. [CrossRef]

129. Freeman, J.; Sparks, C.A.; West, J.; Shewry, P.R.; Jones, H.D. Temporal and spatial control of transgene expression using a heat-inducible promoter in transgenic wheat. *Plant Biotechnol. J.* **2011**, *9*, 788–796. [CrossRef] [PubMed]

130. Jöst, M.; Hensel, G.; Kappel, C.; Druka, A.; Sicard, A.; Hohmann, U.; Beier, S.; Himmelbach, A.; Waugh, R.; Kumlehn, J.; et al. The INDETERMINATE DOMAIN Protein BROAD LEAF1 Limits Barley Leaf Width by Restricting Lateral Proliferation. *Curr. Biol.* **2016**, *26*, 903–909. [CrossRef]

131. Xie, K.; Minkenberg, B.; Yang, Y. Boosting CRISPR/Cas9 multiplex editing capability with the endogenous tRNA-processing system. *Proc. Natl. Acad. Sci. USA* **2015**, *112*, 3570–3575. [CrossRef] [PubMed]

132. Esvelt, K.M.; Mali, P.; Braff, J.L.; Moosburner, M.; Yaung, S.J.; Church, G.M. Orthogonal Cas9 proteins for RNA-guided gene regulation and editing. *Nat. Methods* **2013**, *10*, 1116–1121. [CrossRef] [PubMed]

133. Ran, F.A.; Cong, L.; Yan, W.X.; Scott, D.A.; Gootenberg, J.S.; Kriz, A.J.; Zetsche, B.; Shalem, O.; Wu, X.; Makarova, K.S.; et al. In vivo genome editing using Staphylococcus aureus Cas9. *Nature* **2015**, *520*, 186–191. [CrossRef]

134. Kim, H.; Kim, S.T.; Ryu, J.; Kang, B.C.; Kim, J.S.; Kim, S.G. CRISPR/Cpf1-mediated DNA-free plant genome editing. *Nat. Commun.* **2017**, *8*, 14406. [CrossRef]

135. Steinert, J.; Schiml, S.; Fauser, F.; Puchta, H. Highly efficient heritable plant genome engineering using Cas9 orthologues from Streptococcus thermophilus and Staphylococcus aureus. *Plant J.* **2015**, *84*, 1295–1305. [CrossRef] [PubMed]

136. Hu, J.H.; Miller, S.M.; Geurts, M.H.; Tang, W.; Chen, L.; Sun, N.; Zeina, C.M.; Gao, X.; Rees, H.A.; Lin, Z.; et al. Evolved Cas9 variants with broad PAM compatibility and high DNA specificity. *Nature* **2018**, *556*, 57–63. [CrossRef]

137. Nishimasu, H.; Shi, X.; Ishiguro, S.; Gao, L.; Hirano, S.; Okazaki, S.; Noda, T.; Abudayyeh, O.O.; Gootenberg, J.S.; Mori, H.; et al. Engineered CRISPR-Cas9 nuclease with expanded targeting space. *Science* **2018**, *361*, 1259–1262. [CrossRef] [PubMed]

138. Zetsche, B.; Gootenberg, J.S.; Abudayyeh, O.O.; Slaymaker, I.M.; Makarova, K.S.; Essletzbichler, P.; Volz, S.E.; Joung, J.; van der Oost, J.; Regev, A.; et al. Cpf1 is a single RNA-guided endonuclease of a class 2 CRISPR-Cas system. *Cell* **2015**, *163*, 759–771. [CrossRef] [PubMed]

139. Tang, X.; Lowder, LG.; Zhang, T.; Malzahn, AA.; Zheng, X.; Voytas, DF.; Zhong, Z.; Chen, Y.; Ren, Q.; Li, Q.; Kirkland, ER.; Zhang, Y.; Qi, Y. A CRISPR-Cpf1 system for efficient genome editing and transcriptional repression in plants. *Nat. Plants* **2017**, *3*, 17018. [CrossRef] [PubMed]

140. Fauser, F.; Schiml, S.; Puchta, H. Both CRISPR/Cas-based nucleases and nickases can be used efficiently for genome engineering in Arabidopsis thaliana. *Plant J.* **2014**, *79*, 348–359. [CrossRef]

141. Schedel, S.; Pencs, S.; Hensel, G.; Muller, A.; Rutten, T.; Kumlehn, J. RNA-Guided Cas9-Induced Mutagenesis in Tobacco Followed by Efficient Genetic Fixation in Doubled Haploid Plants. *Front. Plant Sci.* **2017**, *7*, 1995. [CrossRef]

142. Rodriguez-Leal, D.; Lemmon, Z.H.; Man, J.; Bartlett, M.E.; Lippman, Z.B. Engineering Quantitative Trait Variation for Crop Improvement by Genome Editing. *Cell* **2017**, *171*, 470–480. [CrossRef] [PubMed]

143. Chandrasekaran, J.; Brumin, M.; Wolf, D.; Leibman, D.; Klap, C.; Pearlsman, M.; Sherman, A.; Arazi, T.; Gal-On, A. Development of broad virus resistance in non-transgenic cucumber using CRISPR/Cas9 technology. *Mol. Plant Pathol.* **2016**, *17*, 1140–1153. [CrossRef] [PubMed]

144. Stein, N.; Perovic, D.; Kumlehn, J.; Pellio, B.; Stracke, S.; Streng, S.; Ordon, F.; Graner, A. The eukaryotic translation initiation factor 4E confers multiallelic recessive Bymovirus resistance in *Hordeum vulgare* (L.). *Plant J.* **2005**, *42*, 912–922. [CrossRef]

145. Yang, P.; Lupken, T.; Habekuss, A.; Hensel, G.; Steuernagel, B.; Kilian, B.; Ariyadasa, R.; Himmelbach, A.; Kumlehn, J.; Scholz, U.; et al. PROTEIN DISULFIDE ISOMERASE LIKE 5-1 is a susceptibility factor to plant viruses. *Proc. Natl. Acad. Sci. USA* **2014**, *111*, 2104–2109. [CrossRef]

146. Maccaferri, M.; Harris, N.S.; Twardziok, S.O.; Pasam, R.K.; Gundlach, H.; Spannagl, M.; Ormanbekova, D.; Lux, T.; Prade, V.M.; Milner, S.G.; et al. Durum wheat genome highlights past domestication signatures and future improvement targets. *Nat. Genet.* **2019**, *51*, 885–895. [CrossRef]

147. Andersson, M.; Turesson, H.; Nicolia, A.; Falt, A.S.; Samuelsson, M.; Hofvander, P. Efficient targeted multiallelic mutagenesis in tetraploid potato (*Solanum tuberosum*) by transient CRISPR-Cas9 expression in protoplasts. *Plant Cell Rep.* **2017**, *36*, 117–128. [CrossRef] [PubMed]

148. Sun, Y.; Jiao, G.; Liu, Z.; Zhang, X.; Li, J.; Guo, X.; Du, W.; Du, J.; Francis, F.; Zhao, Y.; et al. Generation of High-Amylose Rice through CRISPR/Cas9-Mediated Targeted Mutagenesis of Starch Branching Enzymes. *Front. Plant Sci.* **2017**, *8*, 298. [CrossRef]

149. Simkin, A.J.; Lopez-Calcagno, P.E.; Raines, C.A. Feeding the world: Improving photosynthetic efficiency for sustainable crop production. *J. Exp. Bot.* **2019**, *70*, 1119–1140. [CrossRef]

150. Zhang, Y.; Bai, Y.; Wu, G.; Zou, S.; Chen, Y.; Gao, C.; Tang, D. Simultaneous modification of three homoeologs of TaEDR1 by genome editing enhances powdery mildew resistance in wheat. *Plant J.* **2017**, *91*, 714–724. [CrossRef] [PubMed]

151. Rey, M.-D.; Martín, A.C.; Smedley, M.; Hayta, S.; Harwood, W.; Shaw, P.; Moore, G. Magnesium Increases Homoeologous Crossover Frequency During Meiosis in ZIP4 (Ph1 Gene) Mutant Wheat-Wild Relative Hybrids. *Front. Plant Sci.* **2018**, *9*, 509. [CrossRef] [PubMed]

152. Singh, M.; Kumar, M.; Albertsen, M.C.; Young, J.K.; Cigan, A.M. Concurrent modifications in the three homeologs of Ms45 gene with CRISPR-Cas9 lead to rapid generation of male sterile bread wheat (*Triticum aestivum* L.). *Plant Mol. Biol.* **2018**, *97*, 371–383. [CrossRef]

153. Zhang, S.; Zhang, R.; Song, G.; Gao, J.; Li, W.; Han, X.; Chen, M.; Li, Y.; Li, G. Targeted mutagenesis using the Agrobacterium tumefaciens-mediated CRISPR-Cas9 system in common wheat. *BMC Plant Biol.* **2018**, *18*, 302. [CrossRef] [PubMed]

154. Holubová, K.; Hensel, G.; Vojta, P.; Tarkowski, P.; Bergougnoux, V.; Galuszka, P. Modification of Barley Plant Productivity Through Regulation of Cytokinin Content by Reverse-Genetics Approaches. *Front. Plant Sci.* **2018**, *9*, 1676. [CrossRef]

155. Zhong, Y.; Blennow, A.; Kofoed-Enevoldsen, O.; Jiang, D.; Hebelstrup, K.H. Protein Targeting to Starch 1 is essential for starchy endosperm development in barley. *J. Exp. Bot.* **2019**, *70*, 485–496. [CrossRef] [PubMed]

© 2019 by the authors. Licensee MDPI, Basel, Switzerland. This article is an open access article distributed under the terms and conditions of the Creative Commons Attribution (CC BY) license (http://creativecommons.org/licenses/by/4.0/).

International Journal of
Molecular Sciences

MDPI

Review

Resistance to Cereal Cyst Nematodes in Wheat and Barley: An Emphasis on Classical and Modern Approaches

Muhammad Amjad Ali [1,*], **Mahpara Shahzadi** [1], **Adil Zahoor** [1], **Abdelfattah A. Dababat** [2], **Halil Toktay** [3], **Allah Bakhsh** [4], **Muhammad Azher Nawaz** [5] and **Hongjie Li** [6,*]

[1]   Department of Plant Pathology, Faculty of Agriculture, University of Agriculture, Faisalabad 38040, Pakistan; mahpara015@gmail.com (M.S.); adilzahoor3253@gmail.com (A.Z.)
[2]   International Maize and Wheat Improvement Center (CIMMYT), Ankara 06511, Turkey; A.Dababat@cgiar.org
[3]   Department of Plant Production and Technologies, Faculty of Agricultural Sciences and Technologies, Nigde Omer Halisdemir University, Nigde 51240, Turkey; toktay@yahoo.com
[4]   Department of Agricultural Genetic Engineering, Faculty of Agricultural Sciences and Technologies, Nigde Omer Halisdemir University, Nigde 51240, Turkey; abthebest@gmail.com
[5]   College of Agriculture, University of Sargodha, Sargodha 40100, Pakistan; azher490@hotmail.com
[6]   National Engineering Laboratory for Crop Molecular Breeding, Institute of Crop Sciences, Chinese Academy of Agricultural Sciences, Beijing 100081, China
*   Correspondence: amjad.ali@uaf.edu.pk (M.A.A.); lihongjie@caas.cn (H.L.); Tel.: +92-345-788-6980 (M.A.A.); +86-10-8210-5321 (H.L.)

Received: 7 December 2018; Accepted: 15 January 2019; Published: 19 January 2019

Abstract: Cereal cyst nematodes (CCNs) are among the most important nematode pests that limit production of small grain cereals like wheat and barley. These nematodes alone are estimated to reduce production of crops by 10% globally. This necessitates a huge enhancement of nematode resistance in cereal crops against CCNs. Nematode resistance in wheat and barley in combination with higher grain yields has been a preferential research area for cereal nematologists. This usually involved the targeted genetic exploitations through natural means of classical selection breeding of resistant genotypes and finding quantitative trait luci (QTLs) associated with resistance genes. These improvements were based on available genetic diversity among the crop plants. Recently, genome-wide association studies have widely been exploited to associate nematode resistance or susceptibility with particular regions of the genome. Use of biotechnological tools through the application of various transgenic strategies for enhancement of nematode resistance in various crop plants including wheat and barley had also been an important area of research. These modern approaches primarily include the use of gene silencing, exploitation of nematode effector genes, proteinase inhibitors, chemodisruptive peptides and a combination of one or more of these approaches. Furthermore, the perspective genome editing technologies including CRISPR-Cas9 could also be helpful for improving CCN resistance in wheat and barley. The information provided in this review will be helpful to enhance resistance against CCNs and will attract the attention of the scientific community towards this neglected area.

Keywords: cereal cyst nematodes; wheat; barely; breeding; biotechnology; resistance

---

1. Introduction

Small grain cereals have served as the basis for staple foods, beverages, and animal feed for thousands of years [1–3]. Wheat (*Ttriticum aestivum* L.), barley (*Hordeum vulgare* L.), oats (*Avena sativa* L.), rye (*Secale cereale* L.), triticale (*X Triticicosecale* Wittm.), rice (*Oryza sativa* L.), and some other cereals are rich in calories, proteins, carbohydrates, vitamins, and minerals. These cereals supply around 20% of the calories consumed by people worldwide and are therefore a primary source of energy for humans. Global production of small grains increased exponentially from 1960 to 2005, and then began to decline [4]. Further decline in production is projected to continue through 2050 [5], while global demand for these grains is projected to increase by 1% per annum [4].

Currently, wheat, barley, and oat production exceeds consumption in developed countries, while in developing countries the consumption rate is higher than production [5]. Current production levels and trends will not be sufficient to fulfill the projected global demand spawned by the increasing population pressure. For wheat, global production will need to be increased by 60% to fulfill the estimated demand in 2050 [6]. Until recently, global wheat production increased mostly in response to development of improved cultivars and farming practices and technologies. Production is now limited by biotic and abiotic constraints, including diseases, nematodes, insect pests, weeds, and climate. Among these constraints, plant-parasitic nematodes (PPNs) alone are estimated to reduce production of all world crops by 10% [7]. Nematodes are the second biggest group of animals after insects and are present everywhere on the earth ranging from the Polar Regions to the bottom of the oceans [8]. They are present in the ecosystem as free living and as saprophytes, bacteriovores, fungivores, algaevores and as parasites of human, animals and plants. PPNs are 7% of total species of the phylum Nematoda belonging to 4300 species and 197 genera and infect a huge range of economically important crop plants, including wheat and barley [9,10]. The most dangerous species of PPNs belong to Heteroderidae, which also exhibits the cereal cyst nematodes (CCNs). These nematodes are obligate sedentary endo-parasites and are among the important pests that limit the production of small grain cereals. Heavily invaded young cereal plants are stunted and their lower leaves are often chlorotic, forming pale green patches in the field. Mature plants are also stunted and have a reduced number of tillers, and the roots are shallow and abbreviated and have a "bushy-knotted" appearance [11,12]. CCNs comprise a number of closely-related species and are found in most regions where cereals are produced [13–17]. Several efforts have been made to enhance resistance against CCNs in wheat and barley. A major proportion of these efforts are screening and selection of suitable parents for breeding programs followed by use of molecular markers associated with this resistance. Similarly, some attempts have also been made to silence certain nematode parasitism genes and to develop plants with increased resistance using biotechnological approaches. This review is an update on the use of various classical and modern approaches to induce resistance against CCNs in wheat and barley.

2. Economic Impact of CCNs in Wheat and Barely

CCNs have reduced yields in individual research trials or fields by as much as 20% in Pakistan, 50% in Australia, 50% in Turkey, and 90% in Saudi Arabia [14,18]. More than half the fields are reported to be infested by CCNs in selected cereal-producing regions of Turkey [19,20], Iran [21], the U.S.A. [22], and Europe [23]. Wheat fields are infested by CCNs in at least 16 provinces of China [24]. A bibliography of 123 CCN publications relating to all aspects of CCN biology and management in China, from 1991 to 2014, was published by Riley and Qi [25]. Reports of crop losses at the magnitudes shown above do not accurately depict the magnitude of economic losses at the regional or national level because documentation was based mostly on research plots located in infested areas of fields. Since the nematode density varies greatly across most fields, published estimates nearly always fail to represent field-wide yield reductions, which are rarely documented. A further complication is that some reports initially attributed to yield reduction by *H. avenae* are now known or assumed to have been attributable to species recently reclassified as *H. australis*, *H. filipjevi*, *H. latipons*, or *H. sturhani*. Nevertheless, several reports of regional or national crop losses caused by CCNs are available. In Australia, annual yield losses due to Australian populations of *H. avenae* were estimated at 300 thousand tonnes [26]. Losses in Australia were at one time much higher but have been reduced greatly by deploying resistant varieties because only one biotype is present [18,27]. Yield losses in three provinces of China, caused by *H. avenae*, were estimated at 1.2 Mt, assuming that 22% of the production area was infested and that the overall yield reduction was 10% in those areas [24]. National production of cereals in Norway was estimated to be reduced by between 1% and 5% by multiple CCN species [28]. Losses from *H. avenae* and *H. filipjevi* in four northwestern states of the U.S.A. are estimated at 22 thousand tons, assuming that 0.04% of the wheat and barley fields are infested and the average field-wide yield reduction in infested fields is 10% [29]. Economic losses in other infested regions of the western U.S.A. and globally are poorly documented [14,30,31].

3. Factors Affecting Yield Losses in CCN-Disease

Nematode population has a strong correlation with yield reduction and this relationship is primarily influenced by crop variety, prevailing climatic conditions, cultural techniques, soil quality, and the distribution and density of nematode at a particular location. Similarly, improper use of fertilizer, drought conditions, weakness of root proliferation in soil and unfavorable temperature affected plants are also important determinants of nematode density at the planting time. The varieties have the capability to replace the damaged roots, and thus it is considered to have a better potential among varieties. Nevertheless, associations have been made throughout the history of CCN research to demonstrate a generally linear relationship between initial population density and potential for reduced grain yield, and reductions of other growth and yield components. As an example of the relationship between CCN density and grain yield, *H. filipjevi* in Iran reduced the yield of rainfed winter wheat at all densities ranging from 2.5 to 20 eggs plus J2s/g of soil, with the lowest and highest densities causing 11 and 48% reductions in yield, respectively [32]. Andersen [33] reported that the numbers of *H. avenae* cysts produced on barley were 12, 24, 26, and 33 cysts/plant at initial densities of 1, 2.5, 5, and 10 eggs plus J2s/g of soil. In the northwestern U.S.A., rainfed wheat yields are generally reduced when the number of *H. avenae* eggs plus J2s from extracted cysts, present within the soil, exceeds 3/g of soil [29]. Densities of five *H. avenae* or *H. filipjevi* J2s/g of soil are capable of causing economic damage to irrigated wheat in India [34]. This demonstrates that the initial population pressure is a key player involved in the reduction of grain yield in addition to various other biotic, abiotic and edaphic factors associated with the plants and nematodes.

4. Resistance and Tolerance Responses against CCNs

Use of host-plant resistance and tolerance against CCNs is one of the most cost effective and prominent approaches to minimize crop losses below the threshold level [35]. Continuous cultivation of resistant wheat or barley crops has successfully reduced CCN densities to negligible levels in several countries around the globe (Reviewed by Smiley et al. [29]). Cook and Evans [36] and Cui et al. [37] documented the suppression of nematode reproduction due to the plantation of resistant cultivars. It is established that most of the resistances reported against CCNs in commercial cultivars have been based on introgressions of single dominant genes [12,15,38–41]. However, the resistance response must be combined with tolerance to attain yield stability [36,42]. Similarly, Cook and Evans [36] reported that the ability to get better crop yield from tolerant cultivars is possible, which is normally coupled with a substantial degree of nematode control. However, generally tolerance is tested under field conditions by comparing the yield of a control treated with a nematocide, i.e., aldicarb, and untreated infested soil that is also used to assess the effect of pre-planting nematode population density [16,42,43].

Roots of resistant and susceptible cultivars are initially attacked by J2s, which may result in an intolerant reaction prior to the expression of resistance in a resistant cultivar [44–48]. The tolerance character is primarily attributed to particular characteristics of root growth and physiological response of the plants to nematode invasion [49,50]. During the course of establishment of syncytia, root development is highly compromised due to root abbreviation and sometimes proliferated adventitious roots. The growth of infected roots, in the increasing depth of the soil, mostly failed. Nematode resistance is mostly negatively associated with grain yield and susceptible cultivars without nematode infestation showed a higher grain yield as compared to that of resistant cultivars [51]. Due to this reason, farmers mostly show unwillingness to use resistant cultivars, which presents a lower yield than susceptible in non-infested soils [23].

Andersson [44] and O'Brien and Fisher [45] concluded that barley normally shows more tolerance against *H. avenae* as compared to oats or wheat. Andersen [33] reported comparative damage thresholds of 5, 1, and 0.2 eggs plus J2s per gram of soil for these crops, respectively. Those results were further confirmed in selected cultivars of barley and wheat in the U.S.A. [43]. The mechanism behind this enhanced tolerance in barley is generally associated with earlier development of crown roots in barley seedlings than in wheat seedlings, which enables the crown roots to compensate more rapidly for early damage on seminal roots [52].

On the other hand, a combination of resistance and tolerance responses in wheat and barley cultivars leads to improved profitability and production efficiency [29]. However, most wheat breeding programs in the U.S.A. and other regions with established populations of CCNs are not yet breeding for resistance or selecting for tolerance against CCNs. Less financial support, technical and institutional support, uniformity of infestation and a lack of field testing sites could be the possible reasons behind this. Nonetheless, an understanding of accurate resistance mechanisms would be the key to enhance nematode resistant and/or tolerant wheat and barley cultivars [53–56]. Various isolates of *Heterodera* from different regions of the world differ in their behavior against the resistance response in different cereal crops. The resistant cultivars of wheat, barley, or oat to the populations of *Heterodera* in one region are susceptible to the CCN populations in the other region [29]. Oat cultivars exhibiting resistance and tolerance to *H. avenae* in Australia were susceptible to *H. avenae* in Britain [57].

5. Marker-Assisted Breeding and QTL Mapping for Nematode Resistance

The first resistance gene against *H. avenae* was discovered in barley during 1920 in Sweden; however, it was not characterized until 1961 [33]. After that, a number of scientists in several countries with five decades focused on the improvement of resistance in barley against CCN (Reviewed by Smiley et al. [29]). The source of resistant to *H. avenae* was found on chromosome 2H of barley, which was mapped as the *Ha2* locus [58] through restriction fragment length polymorphism (RFLP) molecular markers. Similarly, Barr et al. [59] mapped *Ha4* locus on chromosome 5H in barley using RFLP.

It is well established that most of the resistant sources against various diseases in common wheat were obtained from wild wheat relatives through breeding programs [46]. Barloy et al. [60] reported that 9 resistance genes were transferred into common wheat from its wild relatives like *Aegilops* and other *Triticum* spp. to enhance resistance against *H. avenae*. These genes include *Cre1* and *Cre8* from *T. aestivum*; *Cre3* and *Cre4* from *Ae. tauschii* Coss.; *Cre2*, *Cre5* and *Cre6* from *Ae. ventricosa* (Zhuk.); *Cre7* from *Ae. triuncialis* L.; *CreR* from rye and *CreV* from *Dasypium villosum* L. Can. [61]. Recently, Baloch et al. [62] have reviewed and given detailed information regarding the transfer of these genes and their associated QTL in common wheat from its wild relatives.

Furthermore, some other resistance loci, i.e., *CreX* and *CreY* from *Ae. variabilis* Eig. [60], were documented, while their location on the chromosomes and mode of inheritance is still uncharacterized. Similarly, in the hexaploid wheat, many resistance genes are introgressed from its progenitors. Mokabli et al. [63] and Rivoal et al. [41] reported that *Cre1* gene is highly responsive against *H. avenae* population in north Africa, Europe, and North America, while in Asia and Australia it is less effective or ineffective against the population of CCNs [41,63]. However, the virulence to *Cre1* gene, compared with *H. avenae*, differs for the population of *H. latipons* in Syria and in India for *H. filipjevi* [63]. Gene *Cre1* was effective against *H. filipjevi* but *Cre3* was not effective against the Turkish population of *H. filipjevi* [29]. *Cre3* is mostly effective against the population of *H. avenae* in Australia [64], while not being effective against the European populations of *H. avenae* [65,66]. Nicol et al. [67] reported that *Cre2* and *Cre4* genes from *Aegilops* spp. along with an unidentified gene from wheat line AUS4930 have broad spectrum resistance against several species of genus *Heterodera* and its pathotypes. The coordination of CIMMYT established these loci in wheat in many regions of the world, which included *Cre1* to *Cre7* with a substantial degree of resistance against CCNs. The CCN resistance QTLs are mapped on chromosomes 1A, 1D, 4D, 5A, 5B, 5D, 6A, 6B, 7A, and 7D of hexaploid wheat [68,69]. Additionally, 11 DArT markers associated with resistance against CCNs have also been reported in wheat and could lead to the identification of new resistance loci and tools that may become useful in wheat breeding programs [68]. Smiley et al. [29] reviewed the findings of previous scientists that a single resistance gene to *H. avenae*, which has been used for the identification of new pathotypes for a long period of time in wheat, oat, and barley.

Some researchers have developed the molecular markers linked to the resistance against *H. avenae* in wheat and barley (Reviewed by Smiley et al. [29]). Marker-assisted selection (MAS) and backcrossing to improve genetic resistance is being applied, but effective resistance genes for CCNs are not yet available for all crops and are not effective against all pathotypes. Large scale tube, pot, or field test screening to identify lines of wheat, barley, oats, and triticale resistant to the Australian populations of *H. avenae* have been undertaken in Australia for more than 30 years [27]. Initially, the pot test was the method of choice, with resistance determined by white cysts counted on the surface of root balls enabling up to 600 pots to be evaluated each day. Up to 130,000 plants per annum have been screened in this way, resulting in the release of many cultivars resistant or moderately resistant to the Australian populations [27]. However, this approach is labor intensive and time consuming, taking a full growing season to complete. With the development and validation of codominant molecular markers linked to resistance to the Australian populations of *H. avenae*, selection can be applied to leaf samples from small seedlings, and the tests can be automated to determine the presence of resistance genes in 1 to 2 days, with substantial savings in costs and time. As a result, MAS for resistance to CCNs in wheat is now used routinely in Australia to identify resistant germplasm in breeding programs [29].

The combination of pot tests and MAS has been used very successfully to reduce infestation levels and losses caused by the CCN in Australia [70]. The strategy followed for MAS involves two phases: prebreeding, to identify and characterize resistant sources and the development of linked markers, followed by their incorporation by backcrossing into advanced breeding lines. This includes pyramiding of resistance genes from different sources (e.g., *Cre1* on chromosome arm 2BL, *Cre3* on chromosome arm 2DL, and *Cre8* on chromosome arm 6BL), using specific linked PCR-based molecular markers to follow each gene [70]. Deployment of resistant cultivars, starting in about 1975, was also responsible for a strong decrease in damage caused by *H. avenae* in Sweden [44].

Transgenic expression of resistance genes results in the induction of a variety of defense responses, including the up-regulation of pathogenesis related (PR) proteins to establish nematode resistance in plants [71]. Uehara et al. [72] reported that induction of PR-1(P4) involved in the regulation of resistance conferred by *HeroA* R gene against potato cyst nematodes [PCNs, *Globodera rostochiensis* (Wollenweber, 1923) and *G. pallida* (Stone, 1973)] through salicylic acid (SA) signaling and nematode infection resulted in the inhibition of the SA signaling pathway in the susceptible cultivars. Similar effects were found in resistant line of hexaploid wheat carrying *Cre2* gene, which showed upregulation of ascorbate peroxidase coding gene in response to cereal cyst nematode (*H. avenae*) when compared with the expression in the susceptible lines [73].

6. Genome-Wide Association Studies for CCN Resistance

Different molecular markers linked to genes for nematode resistance have been identified and reported by developing and utilizing bi-parental populations acquired by crossing contrasting parents, i.e., resistant and susceptible to particular isolates of CCNs. This provides the basis of QTL mapping for nematode resistance which normally involves at least a couple years to develop mapping populations and gene discovery based on the genetic background of only two parental genotypes. Recently, 3 QTL linked to the genomic regions associated with resistance against *Pratylechus thorni* have been identified on chromosomes 2B and 6D in wheat, validating the robustness of these QTL as useful sources of resistance in different genetic backgrounds [74]. Wheat genotypes resistant to *H. avenae* in one region of a same country could be highly susceptible to isolates of *H. avenae* from other regions. This was confirmed by Imren et al. [75] who assessed various landraces and national wheat cultivars from Turkey that wheat genotypes and isolates from different regions in the same country behaved differently. Moreover, QTL recognized from a particular genetic background and working in a particular environment may not be similarly responsive in the other. This shows the need for the development and use of other alternative strategies like association mapping (AM) for the identification of molecular markers which are tightly associated with nematode resistance in crop plants.

Association mapping makes the use of natural plant populations instead of developing the bi-parental mapping populations. However, substantial diversity is required in AM for the detection of QTL with enhanced resolution because AM employs linkage disequilibrium (LD) between alleles within diverse populations to identify markers associated with particular traits. This technique primarily involves genotyping of crop populations, i.e., landraces, cultivars, and advanced lines etc., using particular molecular markers like simple sequence repeats (SSRs) and single-nucleotide polymorphisms (SNPs) and simultaneously phenotying this particular collection of germplasm against a specific biotic stress. For instance, Yang et al. [76] used this approach to map genomic regions associated with stripe rust resistance in wheat. Likewise, recent approaches, such as genome-wide association mapping, utilize the adapted germplasm for identification of QTL associated with a particular plant trait. Moreover, comparing with bi-parental mapping populations, in addition to higher mapping resolution, genome-wide association mapping leads to an increased number of identified alleles and immediate application of the mapping results in a breeding program [62].

Recently, genome-wide association studies (GWAS) have been largely exploited to map the genomic locations associated with a specific plant trait. This approach comprises the high-throughput genotyping of germplasm accessions by sequencing (GBS) through PCR based and next generation sequencing technologies followed by the phenotying. This genotyping results in high-density SNP data that could be utilized to detect marker-character associations in the mapping experiments [77–79]. Advancement in next-generation sequencing technologies has considerably enabled the discovery of SNPs by whole genome sequencing [80–82], transcriptome sequencing [83,84], or reduced-representation sequencing in diverse populations of individuals [85]. GWAS could also be an effective way to identity resistance QTL associated with nematode resistance. For instance, very recently, 13 SNPs associated with resistance to soybean cyst nematode (*Heterodera glycines* Ichinohe, 1952) were recognized, out of which 3 SNPs were associated with already reported QTLs *Rhg1* and *Rhg4* [86].

Similarly, Pariyar et al. [87] performed GWAS of 161 winter wheat accessions using 90K iSelect SNP chip and identified 11 QTL on chromosomes 1AL, 2AS, 2BL, 3AL, 3BL, 4AS, 4AL, 5BL, and 7BL associated with resistance against *H. filipjevi*. They further reported that 8 of these 11 QTL present on chromosome arms 1AL, 2AS, 2BL, 3AL, 4AL, and 5BL were tightly associated with putative genes important for resistant response during plant-pathogen interactions. For instance, methyl transferase 1-associated protein 1 (DMAP1) gene and a putative a RING/FYVE/PHD-type Zinc finger gene linked to chromosome arms 1AL and 2AS are involved in programmed cell death and resistant response against pests. According to the authors, it was the first report of GWAS to map the resistance against CCNs in wheat.

GWAS was recently used to identify Diversity Array Technology (DArT) markers associated with resistance against *H. avenae*, *P. neglectus*, and *P. thornei* from 126 CIMMYT advanced lines of spring wheat adapted to semi-arid conditions [68]. The results demonstrated association of 11 markers with resistance against Ha21 pathotype of *H. avenae*, 25 markers with resistance against *P. neglectus*, and 9 significant markers linked to resistance against *P. thornei*. In that work, chromosome 4A (~90–105 cM) proved to be a source of resistance to *P. thornei*, as has been recently reported. All the QTL associated with resistance against *H. avenae* were mapped on chromosomes 5A, 6A, and 7A of wheat.

7. Nematode Resistance through Modern Approaches

Conventional breeding for pest resistance requires either large scale screening of genotypes to identify suitable resistance genes and multiple generations of plants to incorporate the resistance, or more recently, the application of MAS to combine validated resistance genes. Transgenic plants provide an alternative strategy to develop new forms of resistance by homologous or heterologous transfer of functional resistance genes directly, or by developing 'synthetic' resistance genes that interfere with vital processes in the nematode-plant interaction. We have recently reviewed various transgenic strategies to enhance nematode resistance in plants [10]. Here, some of these important approaches are discussed to develop plants that are resistant against CCNs.

7.1. Gene Silencing for the Enhancement of Nematode Resistance

PPNs can take up macromolecules from the cytoplasm of plant cell during the feeding process. Initially it was thought that J2s would not ingest external solution when outside a host plant. It is now evident that J2s can take up external solutions containing dsRNA when they are 'soaked' or incubated in a solution containing dsRNA [88], with some studies showing that the addition of neurostimulants like octopamine to the soaking solution can enhance the uptake of solution and thus dsRNA by PPNs. This finding enabled the study of gene silencing through in vitro soaking or feeding of dsRNA to the infective J2s. As a result, many effectors and other genes involved in biochemical and developmental processes have been targeted in functional studies to determine how their silencing can affect nematode parasitism, reproduction, viability and ability to establish feeding sites [10]. In general, in vitro studies involving the soaking of J2s in dsRNA solutions for up to 24 h have been used to assess the phenotypic effects (e.g., reduced motility or rigidity, aberrant behavior, reduced attraction to roots, reduction in migration in roots), and subsequent quantification of changes in the expression of the targeted gene [89,90]. Such studies target genes involved in nematode parasitism, development, locomotion, invasion and important biological pathways. To date, there have been relatively few studies on gene silencing for CCNs, partly because these species are relatively difficult to handle and culture, and because genomic or transcriptomic data were rather limited. Recently, Gantasala et al. [91] used in vitro silencing of 4 genes from *H. avenae* viz., nuclear hormone receptor, ployadenalyte binding protein, intron binding protein and epsin, through siRNA soaking. They reported that silencing of these genes resulted in a 71%, 26%, and 60% reduction in females and eggs due to the silencing of epsin, intron binding protein and ployadenalyte binding protein, respectively. Conversely, a 25% increase in females and eggs was reported due to the suppression of the nuclear hormone receptor [91].

Host-induced gene silencing (HIGS) is another transgenic approach that has been widely used by the scientists to silence nematode genes involved in the invasion, virulence and establishment of feeding sites. This technique involves the cloning and transformation of dsRNA of nematode gene into the host plants from where it is taken up by the nematodes through their sylet in the form of siRNAs. These siRNAs interfere with expression of nematode genes at the transcript level to induce silencing of that particular gene. A diagrammatic representation of HIGS in provided in Figure 1. We have recently reviewed gene silencing studies that have been undertaken for cyst nematodes [10,29,92].

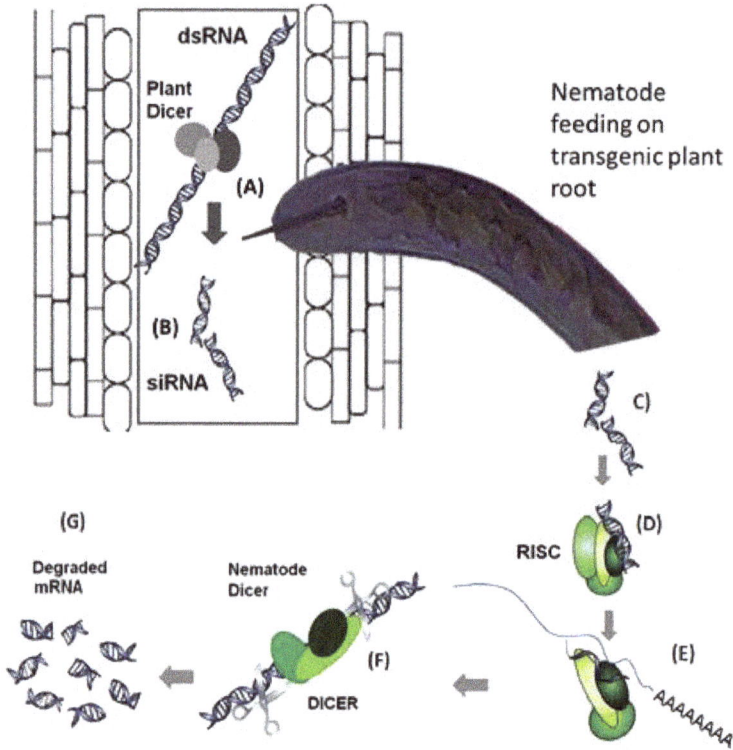

Figure 1. Host induced gene silencing for nematode resistance in plants. (**A**) and (**B**): Double stranded RNA (dsRNA) sequence of the target nematode effector gene is transformed into the plant which is cut by the plant DICER enzyme into small interfering RNAs (siRNAs). (**C**), (**D**) and (**E**): These siRNAs are taken up by the nematodes through their stylets which are detected by the RISC complex that binds to mRNA (complementary to the siRNA sequences) of the target effector gene. (**F**) and (**G**): This is followed by the activation of nematode DICER to cut the double stranded RNA to degrade the mRNA of the target nematode effector gene. Reproduced from Ali et al. [10].

In most of the HIGS studies, different goals set by the researchers have been achieved by knocking down the vital genes from nematodes to reduce their ability to parasitize the plant, or by interfering with chemotaxis towards and invasion into host roots, migration into the root tissues, development of nematode feeding structures, or reproduction (reviewed by references [8,10,93]). For example, some data indicates that more than a 90% reduction can be achieved in cyst development for *H. schachtii* on *Arabidopsis*, and a reduction in cyst formation of up to 94% was found after silencing the synaptobrevin (*snb-1*) gene in soybean cyst nematode through HIGS [94]. It is important to perform bioinformatic analyses of the target gene sequences before using them in the gene silencing studies to get rid of off-target effects. In addition, this approach can be used to silence genes that are induced in the host plants and interact with nematode-secreted effector proteins for compatible plant-nematode interactions. For example, the CLE-like nematode effectors are recognized by plant CLE receptors that are required for syncytium development in plants [95,96]. Knockdown of these receptors in soybean roots resulted in reduced soybean cyst nematode infestations [97]. However, reducing expression of plant genes involved in the nematode-plant interaction is likely to be detrimental to plant crops, since the plant genes will have a functional role in the plant, and interfering with this could reduce field performance. To overcome this problem, we have recently reported the site specific delivery of nematode resistance genes and miRNAs of plant genes involved in the establishment of nematode feeding sites using syncytia specific promoters [98,99].

Some of the nematode effector proteins are localized in the nucleus of the host plant cells, which act as transcription factors to regulate the expression of plant genes [92]. It has been shown recently that annexin like protein (Ha-annexin) from *H. avenae* is localized in the nuclei of host plant cells and is involved in the suppression of basal plant defense responses [100]. A transgenic wheat line containing a HIGS construct of Ha-annexin revealed compromised nematode establishment on the plants. Similarly, the transient expression of Ha-annexin led to the down-regulation of the host hypersensitive (HR) response induced by BAX protein and different pathogen associated molecular patterns (PAMPs) such as flagellin [100]. Very recently two venom allergen-like effector proteins (HaVAP1 and HaVAP2) from *H. avenae* have been characterized through gene silencing [101]. The results indicated that both of them are involved in the suppression of programmed cell death induced by BAX in *Nicotiana benthamiana* leaves. Kumar et al. [102] have reported the de novo transcriptome of *H. avenae* which may lead to the identification of various candidate target genes for HIGS to enhance CCN resistance in wheat.

However, most of the times, HIGS using RNAi does not confer 100% resistance to nematodes. There are a number of possible explanations for this. For instance, the first reason could be the specific target gene chosen with several questions like, is its expression vital, is it unique, or does it belong to a multigene family? Similarly, other factors that affect the extent of gene silencing include where and when the target gene is expressed, the specific dsRNA sequence chosen, the stability of target mRNA or encoded protein, the presence of a 'recovery' phenomenon, and experimental variables such as the vigor of the nematodes treated, and differences in the RNAi machinery (e.g., for systemic spread of the silencing signal) between genera or species [103]. This shows the need for other appropriate biotechnologies to increase nematode resistance in plants.

7.2. Utilization of Proteinase Inhibitors and Chemosensory Disruptive Peptides

RNAi is not the only transgenic approach to confer resistance to nematode pests. Quite a number of different protease inhibitors (PIs) such as cystatins, cowpea trypsin inhibitor (CpTI) and serine proteinase inhibitors have been documented to be successful in producing nematode resistant plants [104]. Heterologous expression of a serine proteinase inhibitor, PIN2 from potato (*Solanum tuberosum* L.) into durum wheat (*Triticum durum* Desf.) led to increased resistance in wheat against *H. avenae* [105]. The main digestive enzymes of many nematodes are cysteine proteinases which have been tackled with the transgenic expression of small proteinase inhibitors, i.e., cystatins, in several plants species to enhance nematode resistance. This strategy has been successfully utilized in a number of crop plants like tomato (*Lycopersicon esculentum* Mill.), rice (*Oryza sativa* L.), potato, banana (*Musa sapientum* L.), and plantain [10]. The human diet normal part is cystatin and is quickly sullied by gastric juice and also it is not allergenic, indicating there is no biosafety issue: the peptide is too small to be allergenic and is degraded in the human small intestine. Similarly, there is no evidence for environmental safety concerns.

In addition to utilization of anti-feedant cysteine proteinase inhibitors to interfere with nematode digestion, anti-root-invasion non-lethal synthetic peptides have been used to hinder the invasion of the nematodes in host plants. These peptides are able to disrupt the chemosensory ability of the nematodes, which make them unable to sense the presence of the plants and the process of nematode chemotaxis towards plants is greatly affected. Winter et al. [106] first reported that synthetic peptides can interfere with the chemoreceptive ability of nematodes to chemically signal in very small concentrations. It was reported that expression of a synthetic peptide nAChRbp in potato resulted in the inhibition of nematode acetylcholinesterase (AChE) gene, leading to disorientation of invading J2s of PCN (*G. pallida*) that led to a 52% reduction of female nematodes established on potato roots [107]. Moreover, Costa et al. [108] revealed that *AChE* is highly expressed in chemo- and mechano-sensory neurons of *C. elegan*, further supporting the hypothesis that inhibition of this gene results in chemodisrution of the invading J2s of PPNs [109]. These reports suggest a potential use of chemoreceptive nematode repellent peptides to induce transgenic resistance in cereal crops against various species of CCNs.

7.3. Coupling of Various Resistance Strategies to Augment Nematode Resistance

Sometimes, if the HIGS does not work efficiently, it could be coupled with some other transgenic approaches to augment nematode resistance in plants. One way to improve transgenic resistance to nematodes is to combine two different modes of resistance, such as a cystatin and anti-invasion peptide [110]. The coupling of these approaches in potato has resulted in high degree of resistance against PCN without effecting soil quality [111]. The same combination approach has been utilized by the scientists at International Institute of Tropical Agriculture (IITA, Ibadan, Nigeria), in collaboration with researchers at University of Leeds, UK using maize (*Zea mays* L.) cystatin and synthetic nematode repellent peptide, nAChRbp to developed transgenic plantain for resistance against three nematode species, i.e., *R. similis*, *H. multicinctus* and *Meloidogyne* spp. [112,113]. Furthermore, pyramiding of cystatins and chemodisruptive peptide into different crop plants has shown a high degree of nematode resistance and enhanced crop yields [114–116]. These approaches could be further combined with RNAi to give more effective and more durable transgenic resistance. Durability of resistance to pests and diseases is an important consideration, and where RNAi-based traits have already been deployed commercially, the expression of the trait seems to be consistent in the following generations [117,118]. This emphasizes the potential of combining all these approaches to multiply the degree of resistance in cereal crops against various CCN species.

8. Genome Editing Technologies: A Potential Perspective for Nematode Resistance in Plants

To date, conventional breeding practices, due to their laborious and time consuming nature, are being replaced with genome editing and other advanced molecular techniques. However, genome editing is being integrated with plant breeding for the development of crop cultivars with improved resistance against pests and other diseases. Until now, different genome editing technologies have been practiced to enhance disease resistance in different crop plants. However, one of the recent and breakthrough in genome editing has been accomplished by the CRISPR (Clustered Regular Interspaced Palindromic Repeats) Cas9 (CRISPR-associated protein) system that is the key player of bacterial immune system.

CRISPR/Cas9 is RNA-guided machinery for effective and precise editing of genomes as compared to any other gene editing technique like Transcription Activator Like Effector Nucleases (TALENS) and Zinc Finger Nucleases (ZFNs) [119,120]. Being precise, highly specific, a multi-gene editor and highly efficient, CRISPR/Cas9 is regarded as promising gene editing system for crop plants [119]. Because CRISPR/Cas9 technology is sequence specific nuclease, it could be employed to exploit defense related mechanisms in plants against invading pathogens. For instance, Ali et al. [119] and Baltes et al. [120] described the implementation of the CRISPR/Cas9 technique to develop resistant plants against geminiviruses and exhibit great potential for enhancement of CCN resistance in cereal crops.

Genome editing using this system is an exciting and powerful alternative to RNAi for gene silencing. It is essentially targeted mutagenesis, in which mutations can be induced in target genes in a nontransgenic manner termed nonhomologous end joining (NHEJ). Alternatively, by insertion of oligonucleotide sequences with ends homologous to each side of the cut site, specific additions to a sequence can be made, known as homologous end joining (HEJ). The technology has passed through a number of iterations, with the use of a 'guide' RNA sequence directing a dsDNAse enzyme (CRISPR-Cas9) to cut a target sequence at a specific site [121]. In NHEJ, the cell repair enzymes frequently make a mistake in joining the ends, resulting in a targeted mutation or total inactivation of the gene. When a cassette consisting of a selectable marker gene with CRISPR and Cas 9 is used to select edited cells for regeneration to plants, the site of gene editing will be elsewhere in the genome from the editing cassette. Hence, for cereals, it is possible to generate edited genotypes that no longer contain an editing cassette. This can be achieved by making a cross with the original (or another) genotype, and identifying genotypes with the edited gene but without the introduced cassette. A plant with a targeted mutation but lacking any introduced DNA may well be regulated as non-Genetically Modified (Non-GM) [29]. However, since the silencing trigger is delivered from the plant to the pest in HIGS, it is not possible to deliver a (nontransgenic) genome editing signal in this way. Application of this technology for nematode management necessitates the identification of a nonvital host plant gene whose expression is very much needed for nematode parasitism to edit it to be nonfunctional.

Recently, genome editing using CRISPR/Cas 9 system has been established in free living nematode, *Caenorhabditis elegans* [122,123]. This development would lead to the characterization of several important genes involved in different physiological processes of nematodes. Nonetheless, there are very few reports available on the application of CRISPR/Cas 9 system to study the resistance responses of the plants against nematodes. Kang [124] recently used soybean hairy roots to study the resistance response knockouts of two serine hydroxymethyltransferase genes, *GmSHMT08* and *GmSHMT05*, generated through this system in soybean-*H. glycines* model organisms. However, to date no report is available regarding the use of this technology in CCN-plant interactions. The recent availability of genome sequences for hosts of CCNs could well lead to the development of nontransgenic cereals genome edited for resistance [125,126].

9. Conclusions and Future Perspectives

Current production levels and trends of wheat and barley will not be sufficient to fulfill the projected global demand created by increased populations. For wheat, global production will need to be increased by 60% to fulfill the estimated demand in 2050. Recently, the global wheat production has increased mostly in response to the development of improved cultivars and farming practices and technologies. However, its production is still limited by biotic and abiotic constraints, including diseases, nematodes, insect pests, weeds, and changing climate. Since domestication, improvements in crop plants regarding pest resistance and higher yields are preferential research area of all times in plant sciences, and usually involve targeted genetic exploitations through natural means as well as transgenic means.

Genotyping by sequencing (GBS) followed by association mapping of resistance genes in crop plants could be an important strategy. Similarly, whole genome sequence availability of various CCNs could provide a great platform regarding the vital effector genes that could be manipulated via different transgenic technologies [126,127]. Moreover, full sequence and annotation of wheat genome has been carried out recently containing huge data that could be helpful in understanding nematode-wheat interactions at the molecular level, which could be used for the enhancement of resistance [127]. Furthermore, wheat and barley transcriptome studies in response to CCN infection could be used to manipulate different up- and down-regulated genes. The genes involved in nematode establishment could be down-regulated and suppressed defense genes in syncytia could be overexpressed using syncytia specific genes, as has been demonstrated in Arabidopsis [10,97,98]. Recently, the role of silicon derived resistance has been described in plants against a variety of phytopathogens [128]. This silicon based resistance could be employed for managing CCNs.

Although about 10% of the world's crops are transgenic at present, the costs and issues that must be overcome to deploy any form of transgenic crop resistance to nematodes, and to CCN in particular, are not insubstantial. Hence, there is current research aimed at delivering dsRNA in spray form (ectopic delivery) rather than by transgenic plants. This strategy requires low cost production of dsRNA sequences, methods to stabilize them for field delivery, uptake of dsRNA by leaves, its systemic basipetal movement through plants to roots, and uptake by nematode on feeding [129,130]. If the technical aspects of ectopic delivery of dsRNA can be overcome in a cost-effective manner, this could bypass the issues of RNAi-based transgenic nematode control. In addition to ectopic delivery of RNAi, gene silencing technology can be used to determine targets for new nematicides. This process involves genome-enabled novel chemical nematicides [129]. Very recently, it has been reported that the tryptophan decarboxylase 1 (*AeVTDC1*) gene from a wide relative of wheat *Aegilops variabilis* regulates the resistance against *H. avenae* by altering the downstream secondary metabolite contents rather than auxin synthesis [131]. This shows the potential use of genetic resources from wide relatives for the enhancement of CCN resistance in cereal crops.

Funding: The financial supports provided by the National Key Research and Development Program of China (2017YFD0101000) and the International Cooperation Project in the Innovative Engineering of Chinese Academy of Agricultural Sciences (CAAS-XTCX2018-020-2) are gratefully appreciated.

Conflicts of Interest: The authors declare no conflict of interest.

References

1. Breiman, A.; Graur, D. Wheat evolution. *Israel J. Plant Sci.* **1995**, *43*, 85–98. [CrossRef]
2. Gustafson, P.; Raskina, O.; Ma, X.-F.; Nevo, E. Wheat evolution, domestication, and improvement. In *Wheat Science and Trade*; Carver, B., Ed.; Wiley-Blackwell: Ames, IA, USA, 2009; pp. 5–30.
3. Newman, R.K.; Newman, C.W. Barley history: Relationship of humans and barley through the ages. In *Barley for Food and Health: Science, Technology, and Products*; Newman, R.K., Walter Newman, C., Eds.; John Wiley and Sons: Hoboken, NJ, USA, 2008; pp. 1–17.
4. Alexandratos, N.; Bruinsma, J. *World Agriculture towards 2030/2050: The 2012 Revision*; Division, A.D.E., Ed.; Food and Agriculture Organization of the United Nations: New York, NY, USA, 2012.
5. OECD-FAO. Agricultural Outlook Organisation for Economic Cooperation and Development (OECD) and Food and Agricultural Organization of the United Nations (FAO). OECD Agricultural Statistics (Database). Available online: http://dx.doi.org/10.1787/agr-data-en (accessed on 7 December 2018).
6. Ackerman, F.; Stanton, E.A. Can climate change save lives? A comment on economy-wide estimates of the implications of climate change: Human health. *Ecol. Econ.* **2008**, *66*, 8–13. [CrossRef]
7. Whitehead, A. *Plant Nematode Control*; CAB International: Oxon, UK; New York, NY, USA, 1998; Volume VIII.
8. Ali, M.A.; Abbas, A.; Azeem, F.; Javed, N.; Bohlmann, H. Plant-nematode Interactions: From genomics to metabolomics. *Int. J. Agric. Biol.* **2015**, *17*, 1071–1082.
9. Decraemer, W.; Hunt, D.J. Structure and classification. In *Plant Nematology*; Perry, R.N., Moens, M., Eds.; CABI1: Oxfordshire, UK, 2006; pp. 4–32.
10. Ali, M.A.; Azeem, F.; Abbas, A.; Joyia, F.A.; Li, H.J.; Dababat, A.A. Transgenic strategies for enhancement of nematode resistance in plants. *Front. Plant Sci.* **2017**, *8*, 750. [CrossRef]
11. Nicol, J.M.; Turner, S.J.; Coyne, D.L.; den Nijs, L.; Hockland, S.; Tahna Maafi, Z. Current nematode threats to world agriculture. In *Genomics and Molecular Genetics of Plant-Nematode Interactions*; Jones, J., Gheysen, G., Fenoll, C., Eds.; Springer: Dordrecht, The Netherlands, 2011; pp. 21–43.
12. Smiley, R.W.; Nicol, J.M. Nematodes which challenge global wheat production. In *Wheat Science and Trade*; Carver, B.F., Ed.; Wiley-Blackwell: Ames, IA, USA, 2009; pp. 171–187.
13. Dababat, A.A.; Pariyar, S.; Nicol, J.; Erginbasx-Orakçi, G.; Wartin, C.; Klix, M.; Bolat, N.; Braun, H.; Sikora, R. Influence of fungicide seed treatment on the integrated control of *Heterodera filipjevi* on six wheat germplasm with different levels of genetic resistance under controlled conditions. *Nematropica* **2014**, *44*, 25–30.
14. Dababat, A.A.; Muminjanov, H.; Smiley, R.W. *Nematodes of Small Grain Cereals: Current Status and Research*; FAO: Ankara, Turkey, 2015.
15. Nicol, J.M.; Rivoal, R. Global knowledge and its application for the integrated control and management of nematodes on wheat. In *Integrated Management and Biocontrol of Vegetable and Grain Crops Nematodes*; Ciancio, A., Mukerji, K.G., Eds.; Springer: Dordrecht, The Netherlands, 2008; pp. 243–287.
16. Smiley, R.W. Occurrence, distribution and control of *Heterodera avenae* and *H. filipjevi* in the western USA. In *Cereal Cyst Nematodes: Status, Research and Outlook*; Riley, I.T., Nicol, J.M., Dababat, A.A., Eds.; CIMMYT: Ankara, Turkey, 2009; pp. 35–40.
17. Subbotin, S.A.; Sergei, A.; Baldwin, J. *Systematics of Cyst Nematodes (Nematoda: Heteroderinae)*; Brill: Leiden, The Netherlands, 2010; Volume 8A.
18. Riley, I.T.; McKay, A. Cereal cyst nematode in Australia: Biography of a biological invader. In *Cereal Cyst Nematodes: Status, Research and Outlook*; Riley, I.T., Nicol, J.M., Dababat, A.A., Eds.; CIMMYT: Ankara, Turkey, 2009; pp. 23–28.

19. Abidou, H.; El-Ahmed, A.; Nicol, J.M.; Bolat, N.; Rivoal, R.; Yahyaoui, A. Occurrence and distribution of species of the *Heterodera avenae* group in Syria and Turkey. *Nematol. Mediterr.* **2005**, *33*, 195–201.

20. Elekçioğlu, İ.H.; Nicol, J.; Bolat, N.; Sahin, E.; Yorgancılar, A.; Braun, H.; Yorgancılar, O.; Yılddırım, A.; Kılınç, A.; Toktay, H.; Çalisxkan, M. Longterm studies on the cereal cyst nematode *Heterodera filipjevi* in Turkey: International collaboration with regional implications. In *Cereal Cyst Nematodes: Status, Research and Outlook*; Riley, I.T., Nicol, J.M., Dababat, A.A., Eds.; CIMMYT: Ankara, Turkey, 2009; pp. 11–16.

21. Tanha Maafi, Z.; Nicol, J.M.; Kazemi, H.; Ebrahimi, N.; Gitty, M.; Ghalandar, M.; Mohammadi-Pour, M.; Khoshkhabar, Z. Cereal cyst nematodes, root rot pathogens and root lesion nematodes affecting cereal production in Iran. In *Cereal Cyst Nematodes: Status, Research and Outlook*; Riley, I.T., Nicol, J.M., Dababat, A.A., Eds.; CIMMYT: Ankara, Turkey, 2009; pp. 51–55.

22. Smiley, R.W.; Ingham, R.E.; Uddin, W.; Cook, G.H. Crop sequences for managing cereal cyst nematode and fungal populations of winter wheat. *Plant Dis.* **1994**, *78*, 1142–1149. [CrossRef]

23. Rivoal, R.; Cook, R. Nematode pests of cereals. In *Plant Parasitic Nematodes in Temperate Agriculture*; Evans, K., Trudgill, D.L., Webster, J.M., Eds.; CAB International: Wallingford, UK, 1993; pp. 259–303.

24. Peng, D.L.; Nicol, J.; Li, H.M.; Hou, S.Y.; Li, H.X.; Chen, S.L.; Ma, P.; Li, H.L.; Riley, I. Current knowledge of cereal cyst nematode (*Heterodera avenae*) on wheat in China. In *Cereal Cyst Nematodes: Status, Research and Outlook*; Riley, I.T., Nicol, J.M., Dababat, A.A., Eds.; CIMMYT: Ankara, Turkey, 2009; pp. 29–34.

25. Riley, I.T.; Qi, R.D. Annotated bibliography of cereal cyst nematodes (*Heterodera avenae* and *H. filipjevi*) in China, 1991 to 2014. *Australas. Nematol. Newsl.* **2015**, *26*, 1–46.

26. Murray, G.; Brennan, J. Estimating disease losses to the Australian wheat industry. *Australas. Plant Pathol.* **2009**, *38*, 558–570. [CrossRef]

27. Lewis, J.M.; Matic, M.; McKay, A.C. Success of cereal cyst nematode resistance in Australia: History and status of resistance screening systems. In *Cereal Cyst Nematodes: Status, Research and Outlook*; Riley, I.T., Nicol, J.M., Dababat, A.A., Eds.; CIMMYT: Ankara, Turkey, 2009; pp. 137–142.

28. Holgado, R.; Støen, M.; Magnusson, C.; Hammeraas, B. The occurrence and hosts of cereal cyst nematodes (*Heterodera* spp.) in Norway. *Int. J. Nematol. Mediterr.* **2003**, *13*, 1–19.

29. Smiley, R.W.; Dababat, A.A.; Iqbal, S.; Jones, M.G.K.; Maafi, Z.T.; Peng, D.L.; Subbotin, S.A.; Waeyenberge, L. Cereal cyst nematodes: A complex and destructive group of *Heterodera* species. *Plant Dis.* **2017**, *101*, 1692–1720. [CrossRef]

30. Riley, I.T.; Nicol, J.M.; Dababat, A.A. *Cereal Cyst Nematodes: Status, Research and Outlook*; CIMMYT: Ankara, Turkey, 2009.

31. Dababat, A.; Mokrini, F.; Smiley, R.W. *Proceedings of the Sixth International Cereal Nematodes Symposium (Agadir, Morocco)*; CIMMYT: Ankara, Turkey, 2017.

32. Hajihasani, A.; Tanha Maafi, Z.; Nicol, J.M.; Rezaee, S. Effect of the cereal cyst nematode, *Heterodera filipjevi*, on wheat in microplot trials. *Nematology* **2010**, *12*, 357–363.

33. Andersen, S. *Resistens mod Havreål Heterodera avenae*; I Kommission hos Dansk Videnskabs Forlag: Copenhagen, Denmark, 1961.

34. Singh, A.; Sharma, A.K.; Shoran, J. Heterodera avenae and its management in India. In *Cereal Cyst Nematodes: Status, Research and Outlook*; Riley, I.T., Nicol, J.M., Dababat, A.A., Eds.; CIMMYT: Ankara, Turkey, 2009; pp. 17–22.

35. Trudgill, D.L. Resistance to and tolerance of plant parasitic nematodes in plants. *Annu. Rev. Phytopathol.* **1991**, *29*, 167–192. [CrossRef]

36. Cook, R.; Evans, K. Resistance and tolerance. In *Principles and Practice of Nematode Control in Crops*; Brown, R.H., Kerry, B.R., Eds.; Academic Press: Sydney, Australia, 1987; pp. 179–231.

37. Cui, L.; Sun, L.; Gao, X.; Song, W.; Wang, X.M.; Li, H.L.; Liu, Z.Y.; Tang, W.H.; Li, H.J. The impact of resistant and susceptible wheat cultivars on the multiplication of *Heterodera filipjevi* and *H. avenae* in parasite-infected soil. *Plant Pathol.* **2016**, *65*, 1192–1199. [CrossRef]

38. McDonald, A.; Nicol, J. Nematode parasites of cereals. In *Plant Parasitic Nematodes in Subtropical and Tropical Agriculture*; Luc, M., Sikora, R.A., Bridge, J., Eds.; CAB Int.: Wallingford, UK, 2005; pp. 131–191.

39. Nicol, J.M. Important nematode pests of cereals. In *Bread Wheat: Improvement and Production*; Curtis, B.C., Rajaram, S., Gómez, M., Eds.; FAO: Rome, Italy, 2002; pp. 345–366.

40. Nicol, J.M.; Rivoal, R.; Taylor, S.; Zaharieva, M. Global importance of cyst (*Heterodera* spp.) and lesion nematodes (*Pratylenchus* spp.) on cereals: Distribution, yield loss, use of host resistance and integration of molecular tools. *Nematol. Monogr. Perspect.* **2003**, *2*, 1–19.

41. Rivoal, R.; Bekal, S.; Valette, S.; Gauthier, J.P.; Fradj, M.B.; Mokabli, A.; Jahier, J.; Nicol, J.M.; Yahyaoui, A. Variation in reproductive capacity and virulence on different genotypes and resistance genes of Triticeae, in the cereal cyst nematode species complex. *Nematology* **2001**, *3*, 581–590.

42. Brown, R.H. Control strategies in low-value crops. In *Principles and Practice of Nematode Control in Crops*; Brown, R.H., Kerry, B.R., Eds.; Academic Press: Sydney, Australia, 1987; pp. 351–387.

43. Smiley, R.W.; Marshall, J.M. Detection of dual *Heterodera avenae* resistance plus tolerance traits in spring wheat. *Plant Dis.* **2016**, *100*, 1677–1685. [CrossRef]

44. Andersson, S. Population dynamics and control of *Heterodera avenae*—A review with some original results. *EPPO Bull.* **1982**, *12*, 463–475. [CrossRef]

45. O'Brien, P.C.; Fisher, J.M. Development of *Heterodera avenae* on resistant wheat and barley cultivars. *Nematologica* **1977**, *23*, 390–397. [CrossRef]

46. Ogbonnaya, F.C.; Seah, S.; Delibes, A.; Jahier, J.; Lopez-Braña, I.; Eastwood, R.F.; Lagudah, E.S. Molecular genetic characterisation of a new nematode resistance gene in wheat. *Theor. Appl. Genet.* **2001**, *102*, 623–629. [CrossRef]

47. Oka, Y.; Chet, I.; Speigel, Y. Accumulation of lectins in cereal roots invaded by the cereal cyst nematode *Heterodera avenae*. *Physiol. Mol. Plant Pathol.* **1997**, *51*, 333–345. [CrossRef]

48. Cui, L.; Gao, X.; Wang, X.M.; Jian, H.; Tang, W.H.; Li, H.L.; Li, H.J. Characterization of interaction between wheat roots with different resistance and *Heterodera filipjevi*. *Acta Agron. Sin.* **2012**, *38*, 1009–1017. [CrossRef]

49. Stanton, J.M.; Fisher, J.M. Factors of early growth associated with tolerance of wheat to *Heterodera avenae*. *Nematologica* **1988**, *34*, 188–197. [CrossRef]

50. Volkmar, K.M. The cereal cyst nematode (*Heterodera avenae*) on oats. I. Identification of attributes useful in early screening for tolerance to *H. avenae*. *Aust. J. Agric. Res.* **1990**, *41*, 39–49. [CrossRef]

51. Wilson, R.E.; Hollamby, G.J.; Bayraktar, A. Selecting for high yield potential in wheat with tolerance to cereal cyst nematode. *Aust. Field Crops Newsl.* **1983**, *18*, 21–25.

52. Williams, T.D.; Salt, G.A. The effects of soil sterilants on the cereal cyst-nematode (*Heterodera avenae* Woll.), take-all (*Ophiobolus graminis* Sacc.) and yields of spring wheat and barley. *Ann. Appl. Biol.* **1970**, *66*, 329–338. [CrossRef]

53. Andres, M.F.; Melillo, M.T.; Delibes, A.; Romero, M.D.; Bleve-Zacheo, T. Changes in wheat root enzymes correlated with resistance to cereal cyst nematodes. *New Phytol.* **2001**, *152*, 343–354. [CrossRef]

54. Montes, M.J.; López-Braña, I.; Delibes, A. Root enzyme activities associated with resistance to *Heterodera avenae* conferred by gene *Cre7* in a wheat/*Aegilops triuncialis* introgression line. *J. Plant Physiol.* **2004**, *161*, 493–495. [CrossRef]

55. Montes, M.J.; López-Braña, I.; Romero, M.D.; Sin, E.; Andrés, M.F.; Martín-Sánchez, J.; Delibes, A. Biochemical and genetic studies of two *Heterodera avenae* resistance genes transferred from *Aegilops ventricosa* to wheat. *Theor. Appl. Genet.* **2003**, *107*, 611–618. [CrossRef]

56. Seah, S.; Miller, C.; Sivasithamparam, K.; Lagudah, E.S. Root responses to cereal cyst nematode (*Heterodera avenae*) in hosts with different resistance genes. *New Phytol.* **2000**, *146*, 527–533. [CrossRef]

57. Cook, R.; York, P.A. Reaction of some European and Australian oat cultivars to cereal cyst nematode. *Ann. Appl. Biol.* **1988**, *122*, 84–85.

58. Kretschmer, J.M.; Chalmers, K.J.; Manning, S.; Karakousis, A.; Barr, A.R.; Islam, A.K.M.R.; Logue, S.J.; Choe, Y.W.; Barker, S.J.; Lance, R.C.M.; et al. RFLP mapping of the *Ha2* cereal cyst nematode resistance gene in barley. *Theor. Appl. Genet.* **1997**, *94*, 1060–1064. [CrossRef]

59. Barr, A.R.; Chalmers, K.J.; Karakousis, A.; Kretschmer, J.M.; Manning, S.; Lance, R.C.M.; Lewis, J.; Jeffries, S.P.; Langridge, P. RFLP mapping of a new cereal cyst nematode resistance locus in barley. *Plant Breed* **1998**, *117*, 185–187. [CrossRef]

60. Barloy, D.; Lemoine, J.; Abelard, P.; Tanguy, A.M.; Rivoal, R.; Jahier, J. Marker-assisted pyramiding of two cereal cyst nematode resistance genes from *Aegilops variabilis* in wheat. *Mol. Breed* **2007**, *20*, 31–40. [CrossRef]

61. Zhang, R.Q.; Feng, Y.X.; Li, H.F.; Yuan, H.X.; Dai, J.L.; Cao, A.Z.; Xing, X.P.; Li, H.L. Cereal cyst nematode resistance gene *CreV* effective against *Heterodera filipjevi* transferred from chromosome 6VL of *Dasypyrum villosum* to bread wheat. *Mol. Breed.* **2016**, *36*, 122. [CrossRef]

62. Baloch, F.S.; Cömertbay, G.; Özkan, H. DNA molecular markers for disease resistance in plant breeding with example in wheat. In *Nematodes of Small Grain Cereals: Current Status and Research*; Dababat, A., Muminjanov, H., Smiley, R.W., Eds.; FAO: Ankara, Turkey, 2015; pp. 159–166.

63. Mokabli, A.; Rivoal, R.; Gauthier, J.P.; Valette, S. Variation in virulence of cereal cyst nematode populations from North Africa and Asia. *Nematology* **2002**, *4*, 521–525.

64. Vanstone, V.A.; Hollaway, G.J.; Stirling, G.R. Managing nematode pests in the southern and western regions of the Australian cereal industry: Continuing progress in a challenging environment. *Australas. Plant Pathol.* **2008**, *37*, 220–234. [CrossRef]

65. De Majnik, J.; Ogbonnaya, F.C.; Moullet, O.; Lagudah, E.S. The *Cre1* and *Cre3* nematode resistance genes are located at homeologous loci in the wheat genome. *Mol. Plant-Microbe Interact.* **2003**, *16*, 1129–1134. [CrossRef] [PubMed]

66. Safari, E.; Gororo, N.N.; Eastwood, R.F.; Lewis, J.; Eagles, H.A.; Ogbonnaya, F.C. Impact of *Cre1*, *Cre8* and *Cre3* genes on cereal cyst nematode resistance in wheat. *Theor. Appl. Genet.* **2005**, *110*, 567–572. [CrossRef] [PubMed]

67. Nicol, J.M.; Rivoal, R.; Trethowan, R.M.; van Ginkel, M.; Mergoum, M.; Singh, R.P. CIMMYT's approach to identify and use resistance to nematodes and soil-borne fungi, in developing superior wheat germplasm. In *Wheat in a Global Environment*; Bedo, Z., Lang, L., Eds.; Kluwer Academic Publishers: Dordrecht, The Netherlands, 2001; pp. 381–389.

68. Dababat, A.A.; Ferney, G.-B.H.; Erginbas-Orakci, G.; Dreisigacker, S.; Imren, M.; Toktay, H.; Elekçioğlu, İ.H.; Mekete, T.; Nicol, J.M.; Ansari, O.; et al. Association analysis of resistance to cereal cyst nematodes (*Heterodera avenae*) and root lesion nematodes (*Pratylenchus neglectus* and *P. thornei*) in CIMMYT advanced spring wheat lines for semi-arid conditions. *Breed. Sci.* **2016**, *66*, 692–702. [CrossRef] [PubMed]

69. Mulki, M.A.; Jighly, A.; Ye, G.; Emebiri, L.C.; Moody, D.; Ansari, O.; Ogbonnaya, F.C. Association mapping for soilborne pathogen resistance in synthetic hexaploid wheat. *Mol. Breed.* **2013**, *31*, 299–311. [CrossRef]

70. Ogbonnaya, F.C.; Eastwood, R.F.; Lagudah, E. Identification and utilisation of genes for cereal cyst nematode (*Heterodera avenae*) resistance in wheat: The Australian experience. In *Cereal Cyst Nematodes: Status, Research and Outlook*; Riley, I.T., Nicol, M.J., Dababat, A.A., Eds.; CIMMYT: Ankara, Turkey, 2009; pp. 166–171.

71. Ali, M.; Anjam, M.; Nawaz, M.; Lam, H.M.; Chung, G. Signal transduction in plant–nematode interactions. *Int. J. Mol. Sci.* **2018**, *19*, 1648. [CrossRef]

72. Uehara, T.; Sugiyama, S.; Matsuura, H.; Arie, T.; Masuta, C. Resistant and susceptible responses in tomato to cyst nematode are differentially regulated by salicylic acid. *Plant Cell Physiol.* **2010**, *51*, 1524–1536. [CrossRef]

73. Simonetti, E.; Alba, E.; Montes, M.J.; Delibes, A.; López-Braña, I. Analysis of ascorbate peroxidase genes expressed in resistant and susceptible wheat lines infected by the cereal cyst nematode, *Heterodera avenae*. *Plant Cell Rep.* **2010**, *29*, 1169–1178. [CrossRef]

74. Linsell, K.J.; Rahman, M.S.; Taylor, J.D.; Davey, R.S.; Gogel, B.J.; Wallwork, H.; Forrest, K.L.; Hayden, M.J.; Taylor, S.P.; Oldach, K.H. QTL for resistance to root lesion nematode (*Pratylenchus thornei*) from a synthetic hexaploid wheat source. *Theor. Appl. Genet.* **2014**, *127*, 1409–1421. [CrossRef]

75. Imren, M.; Toktay, H.; Bozbuğa, R.; Dababat, A.; Elekçioğlu, İ.H. Pathotype determination of the cereal cyst nematode, *Heterodera avenae* (Wollenweber, 1924) in the Eastern Mediterranean Region in Turkey. *Türk. J. Entomol.* **2013**, *37*, 13–19.

76. Yang, T.M.; Xie, C.J.; Sun, Q.X. Situation of the sources of stripe rust resistance of wheat in the post-CYR32 era in China. *Acta Agron. Sin.* **2003**, *29*, 161–168. [CrossRef]

77. Jia, J.Z.; Zhao, S.C.; Kong, X.Y.; Li, Y.R.; Zhao, G.Y.; He, W.M.; Appels, R.; Pfeifer, M.; Tao, Y.; Zhang, X.Y.; et al. International Wheat Genome Sequencing Consortium, Yang, H.M.; Liu, X.; He, Z.H.; Mao, L.; Wang, J. *Aegilops tauschii* draft genome sequence reveals a gene repertoire for wheat adaptation. *Nature* **2013**, *496*, 91–95. [CrossRef]

78. Tian, F.; Bradbury, P.J.; Brown, P.J.; Hung, H.; Sun, Q.; Flint-Garcia, S.; Rocheford, T.R.; McMullen, M.D.; Holland, J.B.; Buckler, E.S. Genome-wide association study of leaf architecture in the maize nested association mapping population. *Nat. Genet.* **2011**, *43*, 159–162. [CrossRef]

79. Zhao, K.; Tung, C.-W.; Eizenga, G.C.; Wright, M.H.; Ali, M.L.; Price, A.H.; Norton, G.J.; Islam, M.R.; Reynolds, A.; Mezey, J.; et al. Genome-wide association mapping reveals a rich genetic architecture of complex traits in *Oryza sativa*. *Nat. Commun.* **2011**, *2*, 467. [CrossRef]

80. Berkman, P.J.; Lai, K.; Lorenc, M.T.; Edwards, D. Next-generation sequencing applications for wheat crop improvement. *Am. J. Bot.* **2012**, *99*, 365–371. [CrossRef]

81. Chia, J.-M.; Song, C.; Bradbury, P.J.; Costich, D.; de Leon, N.; Doebley, J.; Elshire, R.J.; Gaut, B.; Geller, L.; Glaubitz, J.C.; et al. Maize HapMap2 identifies extant variation from a genome in flux. *Nat. Genet.* **2012**, *44*, 803–807. [CrossRef]

82. Xu, X.; Liu, X.; Ge, S.; Jensen, J.D.; Hu, F.Y.; Li, X.; Dong, Y.; Gutenkunst, R.N.; Fang, L.; Huang, L.; et al. Resequencing 50 accessions of cultivated and wild rice yields markers for identifying agronomically important genes. *Nat. Biotechnol.* **2011**, *30*, 105–111. [CrossRef]

83. Allen, A.M.; Barker, G.L.A.; Berry, S.T.; Coghill, J.A.; Gwilliam, R.; Kirby, S.; Robinson, P.; Brenchley, R.C.; D'Amore, R.; McKenzie, N.; et al. Transcript-specific, single-nucleotide polymorphism discovery and linkage analysis in hexaploid bread wheat (*Triticum aestivum* L.). *Plant Biotechnol. J.* **2011**, *9*, 1086–1099. [CrossRef]

84. Cavanagh, C.R.; Chao, S.; Wang, S.; Huang, B.E.; Stephen, S.; Kiani, S.; Forrest, K.; Saintenac, C.; Brown-Guedira, G.L.; Akhunova, A.; et al. Genome-wide comparative diversity uncovers multiple targets of selection for improvement in hexaploid wheat landraces and cultivars. *Proc. Natl. Acad. Sci. USA* **2013**, *110*, 8057–8062. [CrossRef]

85. Elshire, R.J.; Glaubitz, J.C.; Sun, Q.; Poland, J.A.; Kawamoto, K.; Buckler, E.S.; Mitchell, S.E. A robust, simple genotyping-by-sequencing (GBS) approach for high diversity species. *PLoS ONE* **2011**, *6*, e19379. [CrossRef]

86. Zhang, J.; Wen, Z.X.; Li, W.; Zhang, Y.W.; Zhang, L.F.; Dai, H.Y.; Wang, D.C.; Xu, R. Genome-wide association study for soybean cyst nematode resistance in Chinese elite soybean cultivars. *Mol. Breed.* **2017**, *37*, 60. [CrossRef]

87. Pariyar, S.R.; Dababat, A.A.; Sannemann, W.; Erginbas-Orakci, G.; Elashry, A.; Siddique, S.; Morgounov, A.; Leon, J.; Grundler, F.M.W. Genome-wide association study in wheat identifies resistance to the cereal cyst nematode *Heterodera filipjevi*. *Phytopathology* **2016**, *106*, 1128–1138. [CrossRef]

88. Urwin, P.E.; Lilley, C.J.; Atkinson, H.J. Ingestion of double-stranded RNA by preparasitic juvenile cyst nematodes leads to RNA interference. *Mol. Plant-Microbe Interact.* **2002**, *15*, 747–752. [CrossRef]

89. Fosu-Nyarko, J.; Jones, M.G.K. Advances in understanding the molecular mechanisms of root lesion nematode host interactions. *Annu. Rev. Phytopathol.* **2016**, *54*, 253–278. [CrossRef]

90. Tan, J.A.C.H.; Jones, M.G.K.; Fosu-Nyarko, J. Gene silencing in root lesion nematodes (*Pratylenchus* spp.) significantly reduces reproduction in a plant host. *Exp. Parasitol.* **2013**, *133*, 166–178. [CrossRef]

91. Gantasala, N.P.; Kumar, M.; Banakar, P.; Thakur, P.K.; Rao, U. Functional validation of genes in cereal cyst nematode, *Heterodera avenae*, using siRNA gene silencing. In *Nematodes of Small Grain Cereals: Current Status and Research*; Dababat, A., Muminjanov, H., Smiley, R.W., Eds.; FAO: Ankara, Turkey, 2015; pp. 353–356.

92. Ali, M.A.; Azeem, F.; Li, H.J.; Bohlmann, H. Smart parasitic nematodes use multifaceted strategies to parasitize plants. *Front. Plant Sci.* **2017**, *8*, 1699. [CrossRef]

93. Dutta, T.K.; Banakar, P.; Rao, U. The status of RNAi-based transgenic research in plant nematology. *Front. Microbiol.* **2015**, *5*, 760. [CrossRef]

94. Klink, V.P.; Kim, K.-H.; Martins, V.; MacDonald, M.H.; Beard, H.S.; Alkharouf, N.W.; Lee, S.-K.; Park, S.-C.; Matthews, B.F. A correlation between host-mediated expression of parasite genes as tandem inverted repeats and abrogation of development of female *Heterodera glycines* cyst formation during infection of *Glycine max*. *Planta* **2009**, *230*, 53–71. [CrossRef]

95. Replogle, A.; Wang, J.; Bleckmann, A.; Hussey, R.S.; Baum, T.J.; Sawa, S.; Davis, E.L.; Wang, X.; Simon, R.; Mitchum, M.G. Nematode CLE signaling in Arabidopsis requires CLAVATA2 and CORYNE. *Plant J.* **2011**, *65*, 430–440. [CrossRef]

96. Replogle, A.; Wang, J.; Paolillo, V.; Smeda, J.; Kinoshita, A.; Durbak, A.; Tax, F.E.; Wang, X.; Sawa, S.; Mitchum, M.G. Synergistic interaction of CLAVATA1, CLAVATA2, and RECEPTOR-LIKE PROTEIN KINASE 2 in cyst nematode parasitism of Arabidopsis. *Mol. Plant-Microbe Interact.* **2013**, *26*, 87–96. [CrossRef]

97. Guo, X.; Chronis, D.; De La Torre, C.M.; Smeda, J.; Wang, X.; Mitchum, M.G. Enhanced resistance to soybean cyst nematode *Heterodera glycines* in transgenic soybean by silencing putative CLE receptors. *Plant Biotechnol. J.* **2015**, *13*, 801–810. [CrossRef]

98. Ali, M.A.; Plattner, S.; Radakovic, Z.; Wieczorek, K.; Elashry, A.; Grundler, F.M.; Ammelburg, M.; Siddique, S.; Bohlmann, H. An Arabidopsis ATPase gene involved in nematode-induced syncytium development and abiotic stress responses. *Plant J.* **2013**, *74*, 852–866. [CrossRef]

99. Ali, M.A.; Wieczorek, K.; Kreil, D.P.; Bohlmann, H. The beet cyst nematode *Heterodera schachtii* modulates the expression of WRKY transcription factors in syncytia to favour its development in Arabidopsis roots. *PLoS ONE* **2014**, *9*, e102360. [CrossRef]

100. Chen, C.L.; Liu, S.S.; Liu, Q.; Niu, J.H.; Liu, P.; Zhao, J.L.; Jian, H. An ANNEXIN-like protein from the cereal cyst nematode *Heterodera avenae* suppresses plant defense. *PLoS ONE* **2015**, *10*, e0122256. [CrossRef]

101. Luo, S.J.; Liu, S.M.; Kong, L.A.; Peng, H.; Huang, W.K.; Jian, H.; Peng, D.L. Two venom allergen-like proteins, HaVAP1 and HaVAP2, are involved in the parasitism of *Heterodera avenae. Mol. Plant Pathol.* **2018**, in press. [CrossRef]

102. Kumar, M.; Gantasala, N.P.; Roychowdhury, T.; Thakur, P.K.; Banakar, P.; Shukla, R.N.; Jones, M.G.K.; Rao, U. *De novo* transcriptome sequencing and analysis of the cereal cyst nematode, *Heterodera avenae. PLoS ONE* **2014**, *9*, e96311. [CrossRef]

103. Jones, M.G.K.; Fosu-Nyarko, J. Molecular biology of root lesion nematodes (*Pratylenchus* spp.) and their interaction with host plants. *Ann. Appl. Biol.* **2014**, *164*, 163–181. [CrossRef]

104. Lilley, C.J.; Devlin, F.; Urwin, P.E.; Atkinson, H.J. Parasitic nematodes, proteinases and transgenic plants. *Parasitol. Today* **1999**, *15*, 414–417. [CrossRef]

105. Vishnudasan, D.; Tripathi, M.N.; Rao, U.; Khurana, P. Assessment of nematode resistance in wheat transgenic plants expressing potato proteinase inhibitor (*PIN2*) gene. *Transgenic Res.* **2005**, *14*, 665–675. [CrossRef]

106. Winter, M.D.; McPherson, M.J.; Atkinson, H.J. Neuronal uptake of pesticides disrupts chemosensory cells of nematodes. *Parasitology* **2002**, *125*, 561–565.

107. Liu, B.; Hibbard, J.K.; Urwin, P.E.; Atkinson, H.J. The production of synthetic chemodisruptive peptides in planta disrupts the establishment of cyst nematodes. *Plant Biotechnol. J.* **2005**, *3*, 487–496. [CrossRef]

108. Costa, J.C.; Lilley, C.J.; Atkinson, H.J.; Urwin, P.E. Functional characterisation of a cyst nematode acetylcholinesterase gene using *Caenorhabditis elegans* as a heterologous system. *Int. J. Parasitol.* **2009**, *39*, 849–858. [CrossRef]

109. Wang, D.; Jones, L.M.; Urwin, P.E.; Atkinson, H.J. A synthetic peptide shows retro- and anterograde neuronal transport before disrupting the chemosensation of plant-pathogenic nematodes. *PLoS ONE* **2011**, *6*, e17475. [CrossRef]

110. Tripathi, L.; Babirye, A.; Roderick, H.; Tripathi, J.N.; Changa, C.; Urwin, P.E.; Tushemereirwe, W.K.; Coyne, D.; Atkinson, H.J. Field resistance of transgenic plantain to nematodes has potential for future African food security. *Sci. Rep.* **2015**, *5*, 8127. [CrossRef]

111. Green, J.; Wang, D.; Lilley, C.J.; Urwin, P.E.; Atkinson, H.J. Transgenic potatoes for potato cyst nematode control can replace pesticide use without impact on soil quality. *PLoS ONE* **2012**, *7*, e30973. [CrossRef]

112. Tripathi, L.; Tripathi, J.N.; Roderick, H.; Atkinson, H.J. Engineering nematode resistant plantains for sub-Saharan Africa. *Acta Hort. (ISHS)* **2013**, *974*, 99–107. [CrossRef]

113. Roderick, H.; Tripathi, L.; Babirye, A.; Wang, D.; Tripathi, J.; Urwin, P.E.; Atkinson, H.J. Generation of transgenic plantain (*Musa* spp.) with resistance to plant pathogenic nematodes. *Mol. Plant Pathol.* **2012**, *13*, 842–851. [CrossRef]

114. Chan, Y.-L.; He, Y.; Hsiao, T.-T.; Wang, C.-J.; Tian, Z.; Yeh, K.-W. Pyramiding taro cystatin and fungal chitinase genes driven by a synthetic promoter enhances resistance in tomato to root-knot nematode *Meloidogyne incognita. Plant Sci.* **2015**, *231*, 74–81. [CrossRef]

115. Tripathi, L.; Atkinson, H.; Roderick, H.; Kubiriba, J.; Tripathi, J.N. Genetically engineered bananas resistant to *Xanthomonas* wilt disease and nematodes. *Food Energy Secur.* **2017**, *6*, 37–47. [CrossRef]

116. Waltz, E. Nonbrowning GM apple cleared for market. *Nat. Biotechnol.* **2015**, *33*, 326–327. [CrossRef]

117. Kumar, V.; Jain, M. The CRISPR–Cas system for plant genome editing: Advances and opportunities. *J. Exp. Bot.* **2015**, *66*, 47–57. [CrossRef]

118. Belhaj, K.; Chaparro-Garcia, A.; Kamoun, S.; Nekrasov, V. Plant genome editing made easy: Targeted mutagenesis in model and crop plants using the CRISPR/Cas system. *Plant Methods* **2013**, *9*, 39. [CrossRef]

119. Ali, Z.; Abulfaraj, A.; Idris, A.; Ali, S.; Tashkandi, M.; Mahfouz, M.M. CRISPR/Cas9-mediated viral interference in plants. *Genome Biol.* **2015**, *16*, 238. [CrossRef]

120. Baltes, N.J.; Hummel, A.W.; Konecna, E.; Cegan, R.; Bruns, A.N.; Bisaro, D.M.; Voytas, D.F. Conferring resistance to geminiviruses with the CRISPR–Cas prokaryotic immune system. *Nat. Plants* **2015**, *1*, 15145. [CrossRef]

121. Lozano-Juste, J.; Cutler, S.R. Plant genome engineering in full bloom. *Trends Plant Sci.* **2014**, *19*, 284–287. [CrossRef] [PubMed]

122. Dickinson, D.J.; Goldstein, B. CRISPR-based methods for *Caenorhabditis elegans* genome engineering. *Genetics* **2016**, *202*, 885–901. [CrossRef] [PubMed]

123. Paix, A.; Folkmann, A.; Seydoux, G. Precision genome editing using CRISPR-Cas9 and linear repair templates in *C. elegans*. *Methods* **2017**, *121–122*, 86–93. [CrossRef] [PubMed]

124. Kang, J. *Application of CRISPR/Cas9-Mediated Genome Editing for Studying Soybean Resistance to Soybean Cyst Nematode*; University of Missouri-Columbia, University of Missouri-Columbia: Columbia, MO, USA, 2016.

125. Mayer, K.F.X.; Rogers, J.; Doleel, J.; Pozniak, C.; Eversole, K.; Feuillet, C.; Gill, B.; Friebe, B.; Lukaszewski, A.J.; Sourdille, P.; et al. A chromosome-based draft sequence of the hexaploid bread wheat (*Triticum aestivum*) genome. *Science* **2014**, *345*, 1251788.

126. Consortium, I.B.G.S. A physical, genetic and functional sequence assembly of the barley genome. *Nature* **2012**, *491*, 711–716.

127. International Wheat Genome Sequencing Consortium. Shifting the limits in wheat research and breeding using a fully annotated reference genome. *Science* **2018**, *361*, eaar7191. [CrossRef]

128. Coskun, D.; Deshmukh, R.; Sonah, H.; Menzies, J.G.; Reynolds, O.; Ma, J.F.; Kronzucker, H.J.; Belanger, R.R. The controversies of silicon's role in plant biology. *New Phytol.* **2019**, *221*, 67–85. [CrossRef]

129. Fosu-Nyarko, J.; Jones, M.G.K. Application of biotechnology for nematode control in crop plants. In *Advances in Botanical Research*; Carolina, E., Carmen, F., Eds.; Academic Press: Cambridge, MA, USA, 2015; Volume 73, pp. 339–376.

130. Naz, F.; Fosu-Nyarko, J.; Jones, M. Improving the effectiveness and delivery of gene silencing triggers to control plant nematode pests. In *Proc. 32nd Symp. Eur. Soc. Nematol.*; Murdoch University: Braga, Portugal, 2016; p. 69.

131. Huang, Q.; Li, L.; Zheng, M.; Chen, F.; Long, H.; Deng, G.; Pan, Z.; Liang, J.; Li, Q.; Yu, M.; et al. The Tryptophan decarboxylase 1 gene from *Aegilops variabilis* No.1 regulate the resistance against cereal cyst nematode by altering the downstream secondary metabolite contents rather than auxin synthesis. *Front. Plant Sci.* **2018**, *9*, 1297. [CrossRef]

© 2019 by the authors. Licensee MDPI, Basel, Switzerland. This article is an open access article distributed under the terms and conditions of the Creative Commons Attribution (CC BY) license (http://creativecommons.org/licenses/by/4.0/).

International Journal of
Molecular Sciences

MDPI

Review

Host-Induced Gene Silencing: A Powerful Strategy to Control Diseases of Wheat and Barley

Tuo Qi [†], Jia Guo [†], Huan Peng, Peng Liu, Zhensheng Kang and Jun Guo *

State Key Laboratory of Crop Stress Biology for Arid Areas, College of Plant Protection,
Northwest A&F University, Yangling 712100, Shaanxi, China; 2014060057@nwsuaf.edu.cn (T.Q.);
guojia1889@126.com (J.G.); 18821675862@163.com (H.P.); wood319@126.com (P.L.);
kangzs@nwsuaf.edu.cn (Z.K.)
* Correspondence: guojunwgq@nwsuaf.edu.cn; Tel.: +86-0298-7082439
† These authors contributed equally to this work.

Received: 14 December 2018; Accepted: 3 January 2019; Published: 8 January 2019

Abstract: Wheat and barley are the most highly produced and consumed grains in the world. Various pathogens—viruses, bacteria, fungi, insect pests, and nematode parasites—are major threats to yield and economic losses. Strategies for the management of disease control mainly depend on resistance or tolerance breeding, chemical control, and biological control. The discoveries of RNA silencing mechanisms provide a transgenic approach for disease management. Host-induced gene silencing (HIGS) employing RNA silencing mechanisms and, specifically, silencing the targets of invading pathogens, has been successfully applied in crop disease prevention. Here, we cover recent studies that indicate that HIGS is a valuable tool to protect wheat and barley from diseases in an environmentally friendly way.

Keywords: HIGS; transgene; wheat; barley

1. Introduction

Cereal grains, including major cereal grains (e.g., wheat and rice) and other minor grains (e.g., barley and oats) have provided over 56% of the caloric and 50% of the protein requirements in human diets for thousands of years, since their domestication. Wheat is one of the most widely grown small-grain cereal crops around the world [1]. About 721 million tons of wheat annually was produced globally from 2012 to 2016, and the record of 760 million tons of global wheat production was reached in 2016, according to the Food and Agricultural Organization (FAO) of the United Nations (http://www.fao.org/worldfoodsituation/csdb/en/, access on 17 October 2018). China, India, the United States of America, the Russian Federation, and France are the main producers in the world and provide the production of half of the world's wheat. Barley is also one of the most important minor grains. According to the FAO, 142 million tons of barley was produced annually worldwide from 2012 to 2016. The Russian federation, France, Germany, and Australia provide one third of the total production. Barley can be used in beer brewing and is also an important feed for the livestock industry. The yields of the global barley supply were related to beer consumption [2]. Although the development of modern agricultural science and technology greatly reduced the yield loss, an average of 10–15% of the global crop production (more than 300 million tons) is still threatened by plant diseases [3,4]. With the increasing world population, the demand for crop products, combined with food security and balanced nutrition, are rapidly increasing [5]. High-yielding and disease resistant varieties are required at unprecedented levels.

Crop plants are subject to diseases caused by parasitic insects, nematode parasites, pathogenic viruses, bacteria, oomycetes, and fungi. Crop yields and their associated economic losses are major global concerns in modern agriculture. Aphids infesting wheat and barley, including grain aphid

(*Sitobion avenae* F.), Russian wheat aphid (*Diuraphis noxia*), and greenbug (*Schizaphis graminum* Rondani), are major agricultural pests in crop plants, not only because of feeding injury, but also because of vectoring viruses (i.e., *Barley yellow dwarf virus* (BYDV) and *Triticum mosaic virus* (TriMV)). Coupled with the presence of parasitic nematodes (*Heterodera* spp.) on wheat and barley, hundreds of millions of dollars are lost every year. In addition, the fungal diseases, including the rusts, blotches, powdery mildew, and head blight/scab, of wheat and barley are currently prominent threats. New virulent races of stem rust fungus, such as strain Ug99 in Africa and the Middle East and V26 in China [6], have caused tremendous yield losses in wheat production. Wheat blast appeared suddenly in 2016 and destroyed wheat fields in Bangladesh [7]. *Fusarium graminearum* ravaged fields of cereal crops and produced mycotoxins that threatened food security. The continual pursuit for yield and quality is a big challenge for breeds, and the constant loss of suitable farm land, unpredictable natural calamity, and the epidemic of multitudinous disease severely hinder the production of wheat and barley. Widely cultivated high-performing varieties lead to disease pandemics and to the loss of genetic diversity. Thus, so many examples remind us that disease resistance breeding and selection are urgently needed.

An array of approaches has been applied to manage crop diseases, such as agrochemical applications, biological control, host resistant selection and breeding, and crop management strategies. The most effective strategy is the use of resistant cultivars combined with reasonable management methods. Because of the rapid emergence of virulent races and fungicide-resistant pathogen strains, traditional breeding is insufficient to combat the abundance of crop diseases. The development of biotechnological approaches provides a novel approach to obtain disease-resistant plants that not only display a high resistance to multiple pathogens, but that are environmentally safe and sound. Based on the knowledge of the molecular patterns involved in plant–microbe interactions, genetically modified plants via transgene-based host-induced gene silencing (HIGS) may be a new effective, environmentally-friendly approach to controlling the crop diseases caused by parasitic pests, nematodes, viruses, and fungi.

2. General Mechanism of HIGS

Plants naturally develop an immune system, based on the RNA silencing machinery, to defend against invading viruses [8–10]. This feature has been utilized to develop host-induced gene silencing technology (HIGS) to control other plant pathogens [11]. HIGS is a further development of virus induced gene silencing (VIGS) [9], which allows for the silencing of genes in plant pathogens by expressing an RNA interference (RNAi) construct against specific genes endogenous to the pathogen in the host plant. HIGS is an RNAi-based process where small RNAs made in the plant silence the genes of the pests or pathogens that attack the plant. The small RNAs are generally made by producing double stranded RNA (dsRNA) in transgenic plants, but for experimental purposes, the dsRNA can be introduced into the plant cells with agrobacterium or viruses that replicate through dsRNA. RNAi, discovered in the expression of a chimeric chalcone synthase gene in petunia, led to a reversible homologous gene suppression [12]. RNA silencing is a highly-conserved mechanism that operates in most eukaryotes, including plants, animals, and fungi. Andrew Z. Fire and Craig C. Mello received the Nobel Prize in Physiology or Medicine in 2006 for their contribution to RNA interference [13]. One of the major features of RNA silencing is the production of small RNAs of 21–30 nucleotides (nt) in length, which can regulate gene expression in a sequence-specific manner. Small RNAs generated from dsRNA guide transcriptional gene silencing (TGS) and post-transcriptional gene silencing (PTGS). PTGS involves the recognition and silencing of mRNA in the cytoplasm, whereas TGS involves RNA-mediated DNA methylation in the promoter region, which suppresses the specific gene expression. Typically, Dicer or Dicer-like (DCL) proteins recognize a double-stranded RNA and process it into smaller RNAs. These sequence-specific small RNAs are hired by Argonaute (AGO) protein and are processed into an RNA-induced multi-subunit silencing complex (RISC). The RISC regulates the target genes to achieve gene silencing in a sequence-specific manner.

With the deep insight of the RNAi mechanism, researchers found that RNAi can be triggered by the dsRNA or DNA from a virus in higher plants, the foreign dsRNA that can be silenced endogenously is a natural immunity to defense against virus [9], the specialty of RNAi that it has sequence specificity and is triggered by introducing exogenous dsRNA makes RNAi a perfect tool to analyze the genomic function in the parasitic system. VIGS is based on the RNAi-mediated natural immunity to defense against viruses, by importing the foreign gene in the viral vector. The initial long RNA precursors at the target sites, transcribed by RNA polymerase IV, and as a template for RNA-dependent RNA polymerase 2 (RDR2) to produce dsRNA, dsRNA is processed into 24 nt hetsiRNAs by DCL3. HetsiRNAs, which represent secondary siRNAs (hetsiRNAs), are the most abundant class of siRNAs in plants. HetsiRNAs establish and maintain silencing through the epigenetic pathway. Endogenous expressing siRNA in plants always need TGS or RNA-directed DNA methylation (RdDM), which is an epigenetic regulation essential for silencing. dsRNA is also processed by DCL2 and DCL4 to 21–22 nt siRNAs. The siRNAs induced cleavage of target mRNA can trigger the production of secondary siRNAs (secsiRNAs). Most secsiRNAs are related to PTGS and the spreading of RNA silencing locally or systemically. SecsiRNAs is important for the amplification of RNA silencing in plants, especially the VIGS and HIGS. HIGS is a transgenic modification technology that relies on the instruction of an inverted repeat sequence into the plant genome. The sequence homolog to the gene from pathogens integrated into the plant genome express the siRNA targeted gene and result in the gene silencing of pathogens. Various HIGS vectors have been constructed to drive the long dsRNA as well as hairpin sequence through an inverted repeated gene or an inverted promotor sequences in to the plant genome. Foreign dsRNA or hairpin sequence are introduced into the plant genome and are cleaved by endonuclease-type DICER enzymes into siRNAs, with the amplification of the hetsiRNAs and secsiRNAs, and these siRNAs are transported systemically and taken up by the fungal cells or other parasites (Figure 1) to silence the fungal gene, which is critical for the invasion [14]. In higher plants, RNAi not only regulates the endogenous gene expression, but the instruction of foreign dsRNA in plants has also recently been a conventional agriculture plant protection.

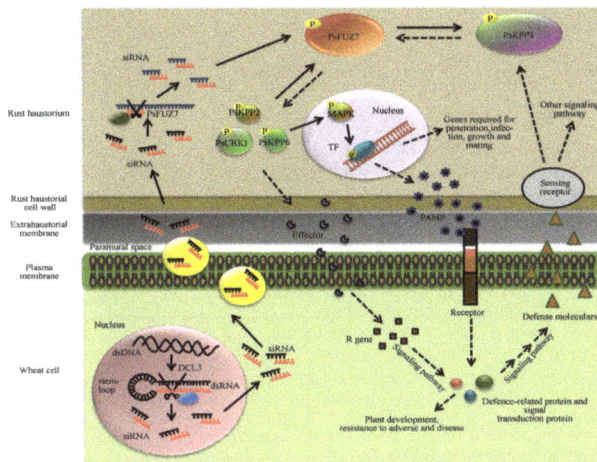

Figure 1. Schematic presentation of possible molecular dialogues between transgenic lines carrying RNAi constructs and colonizing *Pst*. Fungal dsRNA, produced inside transgenic wheat cells, is cleaved by the plant silencing machinery using endonuclease-type dicer enzymes into small silencing molecules (siRNAs). These siRNAs are trapped by a complex of proteins, and are transported to the paramural space. After passing the haustorial cell wall, the silencing molecules trigger the RNAi of their mRNA targets, and may act as primers leading to the activation of systemic silencing signals, thus inducing the immune system of transgenic wheat by mechanisms, including PAMP triggered immunity and Effector triggered immunity [14].

Common wheat (*Triticum aestivum*) is an AABBDD hexaploid (AA from *Triticum uratu*, BB from *Aegilops* species and DD from *Triticum tauschii*) with a complex genome containing approximately 17 billion (17 Gb) nucleotides [15]. Barley (*Hordeum vulgare* L.) is a diploid monocotyledon with seven chromosomes containing about 5.1 billion (5.1 Gb) nucleotides [16]. The huge and complex genomes of wheat and barley complicate the isolation and functional description of the individual genes. HIGS is an effective and convenient tool to explore the molecular genetics of disease resistance in crops, and a series of studies have been performed in wheat and barley using this technology. For wheat and barley, BSMV-induced gene silencing enables the analysis of large-scale functional genomics [17]. Moreover, transgene-induced RNAi could be a useful tool for genetically modified plants to achieve high-value crops [18]. Basically, trigger selection is the primary concern for RNAi. The trigger size and sequence identity between the trigger and its target genes must be considered before RNAi experiments can be designed [19–21]. Trigger sizes from 200–550-bp have been successfully used in wheat [22]. A perfect identity of the trigger and its targets is necessary for a continuous stretch of siRNA [23–25]. Reducing the off-target effects is another factor that must be considered. Bioinformatics analysis, such as Blast and SIFI software, are of considerable value. Overall, the RNAi modification of wheat and barley genes is an environmentally friendly and effective strategy to engineer novel phenotypes, and the perfect parameters suitable for this system will be amended with caution in the future.

3. HIGS Used for Controlling Diseases of Wheat and Barely

Wheat and barley are susceptible to various biotic stresses, such as grain aphid (*Sitobion avenae* F.), Russian wheat aphid (*Diuraphis noxia*), greenbug (*Schizaphis graminum* Rondani), *Barley yellow dwarf virus* (BYDV), *Triticum mosaic virus* (TriMV), parasitic nematodes (*Heterodera* spp., *Tylenchus tritici*), and bacterial pathogens (*Bacterium tritici*). Fungal pathogens are major threats to wheat and barley, causing huge economic losses. Take-all, caused by *Gaeumannomyces graminis* var. *tritici*, is a serious disease of barley and wheat. Powdery mildew, caused by *Blumeria graminis*, also devastates both hosts. Fusarium head blight, also known as scab, caused by *Fusarium* species and rust fungi (*Puccinia species*), causes disastrous yield and quality losses in cereals. Other pathogens, such as *Zymoseptoria tritici*, *Parastagonospora nodorum*, *Pyricularia graminis-tritici*, and *Pyrenophora tritici-repentis*, threaten crop production and food safety. The success of traditional disease management strategies (e.g., irrigation management, chemical control, and agronomic practices) are largely influenced by the environment, whereas RNAi-based HIGS technology provides a novel and innovative approach to control crop diseases. Here, we summarize the recent studies of HIGS in wheat and barley (Table 1), with consideration of the key diseases affecting crop production, and briefly address the current status of the RNAi system used in gaining an understanding of the molecular mechanism of each interaction.

Table 1. Summary of study test in wheat and barley by using host-induced gene silencing.

Species		Host	Target Gene	Remark	Major Phenotype	Ref.
Insects	*Sitobion avenae*	Wheat	*CbE E4*	Carboxylesterase gene	Impaired tolerance of phoxim insecticides	[26]
	Sitobion avenae	Barley	*shp*	Structural sheath protein	Reduced fecundity and inhibited feeding behavior	[27]
Nematodes	*Meloidogyne incognita*	Wheat	*HSP90, ICL,* and *Mi-cpl-1*	Heat-shock protein 90, isocitrate lyase, and Mi-cpl-1	Reduced reproduction	[28]
	Pratylenchus spp.	Wheat	*pat-10, unc-87*	Troponin C (Pat-10) and Calponin (unc-87)	Reduced reproduction	[29]
Viruses	*Wheat streak mosaic virus* (WSMV)	Wheat	pre-miR395	Artificial microRNA (amiRNA)	Stable resistance	[30]
	Wheat dwarf virus (WDV)	Barley	amiR1, amiR6, and amiR8	amiRNAs	Highly efficient resistance	[31]
	WSMV	Wheat	coat protein gene	Coat protein	Consistent resistance	[32]
Fungi	*Blumeria graminis*	Wheat, Barley	*Avra10*	Virulence effector	Reduced virulence	[33]
	Blumeria graminis	Wheat, Barley	*BEC1011, BEC1054, BEC1038, BEC1016, BEC1005, BEC1019, BEC1040,* and *BEC1018*	Effectors	Reduced virulence	[34]
	Puccinia striiformis f. sp. *tritici*	Wheat	*PsCPK1*	PKA catalytic subunit	Stable resistance	[35]
	Puccinia striiformis f. sp. *tritici*	Wheat	*PsFuz7*	MAP kinase kinase	Stable resistance	[14]
	Puccinia triticina	Wheat	*PtMAPK1* and *PtCYC1*	MAP kinase, cyclophilin	Enhanced resistance	[36]
	Fusarium species	Barley	*CYP51*	Cytochrome P450 lanosterol C14-demethylase	Enhanced resistance	[37]
	Fusarium graminearum	Wheat	*Chs3b*	Chitin synthase 3b	Stable resistance	[38]
	Fusarium culmorum	Wheat	*FcGls1*	β-1, 3-glucan synthase	Enhanced resistance	[39]

3.1. HIGS against Insect Pests

The expression of siRNA in transgenic plants provides protection to wheat, barley, and other small grains against individual insect species by targeting the genes in aphids and other parasitic pests that threaten crops. Transgenic siRNA-expressing plants like maize and corn have proven to be effective in insect control. For example, *Bacillus thuringiensis* (Bt) has been successfully used for the management of cotton insect pests [40]. Four targets from 66 unigenes were isolated for RNAi against grain aphids in wheat [41]. In addition, five potential RNAi targets have been identified from 5490 unigenes of grain aphids in wheat [42]. Several studies have confirmed that heat-shock protein 90 (HSP90) is an ideal target, because of its crucial role, ubiquity, and conservation [43,44]. The silencing of the chitin synthase gene prevented the insects from hatching [45], and from knocking down the segmentation gene, hairy, which prevented insect feeding [46]. The silencing of the essential genes (e.g., ecdysone receptor (EcR) and ultraspiracle protein (USP)) in grain aphids (*Sitobion avenae* F.) reduced the survival and fecundity of wheat, providing a persistent and transgenerational method for improving wheat resistance [47]. Silencing the matrix metalloproteinase MMP-2 is lethal, because of its importance in gut development [48]. Additionally, several potential RNAi targets were confirmed through feeding or injection in the grain aphid *Sitobion avenae*, such as genes encoding catalase, acetylcholinesterase1, cytochrome c oxidase subunit VIIc precursor, zinc finger protein, secreted salivary peptide DSR32, salivary protein DSR33, serine protease 1 DSR48, and olfactory co-receptor [49]. Not only can individual genes be targeted by RNAi, their promoters are perfect elements for controlling gene expression, and, therefore, can be used in the RNAi strategy. Replacing the promoter of the HIGS constructs to the *SUC2* expressed along the sieve tubes to the transport solutes can help control tissue-specific feeding insect pests [50]. The first genomic modification of plants against pests was produced in 1983 [51]. Transgenic wheat expressing dsRNA to knock down the CarE gene (*CbE E4*) of *Sitobion avenae* delayed larval growth and reduced resistance to pesticides [26]. The transgenic expression of siRNA of the structural sheath protein (SHP), which is an essential component of the leaf sheath in barley, strongly reduced the feeding and spread of aphids [27,52], and the effects were sustained for seven generations [52]. Coupled with the dsRNA in transgenic plants, which are also heritable, the control of crop diseases would be continuous. HIGS for controlling the insect pests of crop plants is becoming an environmentally friendly and convenient commercial product.

3.2. HIGS against Nematodes

The damage caused by nematodes hinges mainly on the migratory or sedentary phases of this species. The exo-parasitic species live in the soil and use long stylets to feed on epidermal cells; the endo-parasitic, mobile nematodes live inside the plant tissue and feed on non-lignified cells; other species are the sedentary nematodes, which are effectively managed through RNAi [53]. The yield losses of wheat in the presence of nematodes are caused mainly by cyst nematodes (*Heterodera* spp.), which also threaten the production of barley [54]. RNA interference in several nematode species (e.g., cyst nematodes, root knot nematodes, root lesion nematodes, and other ectoparasitic nematodes) has been explored through feeding, soaking, injection, or in planta delivery [28,29,55]. The knockdown of the expression of *pat-10* and *unc-87* of *Pratylenchus thornei*, which attacks the wheat roots, reduces reproduction by 77–81% [29]. Cysteine proteinase, C-type lectin, major sperm protein, chitin synthase, aminopeptidase, β-1,4-endoglucanase, secreted amphid protein, FMRF amide-like peptides (flp-14 and flp-18), pectate lyase, chorismate mutase, secreted peptide SYV46, dual oxidase, splicing factor, integrase, and secreted peptide 16D 10 were reviewed as potential RNAi targets by Lilley et al. (2007) [28]. The phenotypic effects of these RNA interference experiments were commonly a reduced number of established nematodes or an increase in the male population, which indicates that juveniles experience adverse conditions [28]. The first successful application of HIGS in 2006 was used to confer nematode resistance by protecting the host from infection [56]. Several factors in nematodes, such as Cpn-1, Y25, Prp-17, tyrosine phosphatase, calreticuline Mi-CRT, a parasitism gene 8D05, ribosome 3a, 4, synaptobrevin, and a spliceosomal SR protein have been shown to be good candidates for HIGS for improving resistance to nematodes [57]. These reports provide the foundation and confidence to control nematodes through RNAi strategies.

3.3. HIGS against Viruses

To date, no technologies are available to cure virus-infected plants, and the only way to limit the spread of viruses is via the application of costly chemical treatments eliminating the virus-transmitting organisms. The selection of virus-resistant cultivars has been the most effective solution, but natural resources of resistance are not sufficient. In the past few years, HIGS against viruses has proven to be an efficient technology. *Wheat streak mosaic virus* (WSMV) infection is inhibited using the full-length sequence of the viral replicase (Nib) in transgenic wheat. Interestingly, only one of six lines can detect the transgene mRNA, which did not provide resistance to the virus [58], suggesting that post-transcriptional gene silencing was involved in the transgenic wheat resistance. Furthermore, the full length of the WSMV coat protein gene transformed into wheat, conferred resistance to WSMV by two groups, respectively [59,60]. Li and co-authors (2014) found a strong resistance phenotype in some transgene lines for the first generation, but the T2 and T3 generations lost the phenotype. In turn, the use of hairpin-forming transgenes of the WSMV CP protein elicits efficient resistance. This is also the first report of stable WSMV-resistant wheat, which has stable resistance lines through the T5 generation [32]. The second report of the stable transgene (T6 generation) of RNAi-mediated resistance in wheat, and also the first report of the *Triticum mosaic virus* (TriMV), was accomplished by constructing a hairpin sequence of the TriMV CP gene [61]. Polycistronic artificial miRNA-mediated resistance was also demonstrated as an efficient tool. The artificial microRNAs (amiRNAs) can be designed by targeting the different conserved sequence elements of the viruses. Potential amiRNA sequences were selected to minimize the off-target effects. Fahim and colleagues (2016) concluded that polycistronic amiRNAs can be utilized to induce virus resistance [30]. In addition, amiRNA technology was also used for the introduction of highly efficient resistance in barley against *Wheat dwarf virus* (WDV), a DNA virus belonging to the *Geminiviridae* family, and the resistance is effective at low temperatures [31].

3.4. HIGS against Fungi

Fungal pathogens cause more than 70% of the yield loss of crops worldwide, and RNAi strategies have been widely used to characterize the gene function and to construct transgenic lines against the fungal pathogens of wheat and barley. Resistance genes are not always useful for some fungi (e.g., *Fusarium* species and *Puccinia* species), because of the associated toxins and rapid evolution of virulent strains. RNAi-based HIGS provide a novel, innovative approach to control the crop diseases caused by fungi. Here, we review the major diseases [62] that threaten the production of small grains.

The powdery mildew of barley caused by *Blumeria graminis* f. sp. *hordei* and wheat caused by *B. graminis* f. sp. *tritici* are serious diseases. Nowara et al. (2010) showed that dsRNA targeting the avirulence gene *Avra10*, which is recognized by the resistance gene *Mla10*, significantly reduced fungal development in the absence of *Mla10*, and the silencing of 1,3-b-glucanosyltransferases (*BgGTF1* and *BgGTF2*) reduced the early development of the pathogen [33]. Pliego et al. (2013) screened eight targets (e.g., β-1,3 glucosyltransferases, metallo-proteases, microbial secreted ribonucleases, and proteins of a known function) from fifty *Blumeria* effector candidates, that contribute to fungal development through the transient expressing inverted repeat sequence of these gene of effectors into barley leaves [34].

Rust disease, caused by *Puccinia* species, which are critical obligate biotrophic organisms [63], threaten the world's crop production and have caused enormous losses ranging from 4.3 to 5 billion US dollars annually, especially for wheat and barley [64]. Yin et al. (2010) [65] established a *Barley stripe mosaic virus* (BSMV)-induced gene silencing system to knockdown *Puccinia striiformis* f. sp. *tritici* (*Pst*) genes. BSMV-HIGS provide a way to analyze the function and to screen RNAi targets for the control of rust diseases. Panwar et al. [66] transiently silenced the genes encoding mitogen-activated protein kinase 1 (*PtMAPK1*), cyclophilin (*PtCYC1*), and calcineurin B (*PtCNB*) from *Puccinia triticina*

(*Pt*) through *A. tumefaciens*-mediated transformation, and the disease symptom significantly changed and the sporulation decreased macroscopically. The fungal biomass and emergence of uredinia also declared that silencing *PtMAPK1*, *PtCYC1*, and *PtCNB* reduced the virulence and restricted the development of rust fungi, including *Puccinia graminis* and *Puccinia striiformis*. Many factors in the *Puccinia* species have been successfully silenced by BSMV-HIGS, such as protein kinase (*PsSRPKL*) [67], small GTP-binding protein (*PsRan*) [68], distinct *Ras* genes (*PsRas1* and *PsRas2*) [69], MADX-box transcription factor (*PstMCM1-1*) [70], transcription factor *PstSTE12* [71], MAP kinase kinase kinase *PsKPP4* [72], secreted protein gene *Pst_8713* [73], effector (*PSTHa5a23*) [74], effector *PEC6* [75], and Zn-only superoxide dismutase (*PsSOD1*) [76]. Most of these genes (e.g., *PsSOD1*, *PsSRPKL*, and *PsRas2*) contribute to the virulence, and the knockdown them results in a substantial restriction in the growth and spread of the hyphae of *Pst*, the portion of these genes (e.g., *PsRan* and *PsSRPKL*) silenced through BSMV-HIGS leading to the morphological abnormalities, such as the hyphal expansion or haustorium shrink of rust fungi. Qi et al. (2018) produced siRNA targeting the gene of the catalytic subunit of protein kinase A (*PsCPK1*) in transgenic wheat and observed resistance against *Pst* through the fourth generation, and showed that the cAMP signaling pathway genes were down-regulated [35]. Additionally, Zhu et al. (2017) produced transgenic wheat lines expressing a RNAi construct against the MAP kinase kinase gene (*PsFUZ7*), and proved them to be significantly resistant to *Pst*. Infection by the virulent race was strongly restricted, and only rarely produced urediniospores or caused substantial necrosis on transgenic wheat leaves [14]. The transgenic wheat, which stably expressed the hairpin RNAi constructs of the homologous gene of MAP kinase (*PtMAPK1*) and the cyclophilin (*PtCYC1*) of leaf rust, restricted the fungal development and dramatically reduced the fungal biomass [36].

Wheat scab (also called Fusarium head blight) is a devastating disease mainly caused by *F. graminearum*. HIGS technology has also been used to control this disease. Koch et al. (2013) constructed dsRNA in barley expressing the siRNA targeting fungal sterol cytochrome P450 lanosterol C-14a-demethylase gene (*CYP51*), and found that it restricted fungal growth and alleviated disease symptoms. Furthermore, the RNAi transgenic barley and *Arabidopsis* containing three constructs (*CYP51B*, *CYP51A*, and *CYP51C*) that target the three paralogs of CYP51 are completely immune against *F. graminearum* [37]. The RNA interference of the chitin synthase (Chs) gene conferred stable resistance for five generations and reduced the deoxynivalenol (DON) production when three hairpin RNAs of *Chs* were transferred into two wheat cultivars (Yangmai15 and Sumai3) [38]. Chen et al. (2016) stably expressed the dsRNA of β-1,3-glucan synthase gene *FcGls1* in wheat against *F. culmorum* and found that it enhanced the Fusarium head blight (FHB) resistance in the leaf and spike [39]. Machado et al. (2018) covered the recent advances in the RNAi-based methods in the control of FHB, and highlighted that despite some of these disadvantages of the RNAi approach, HIGS has emerged as a promising new approach to control fungal plant diseases [77]. Wegulo et al. (2015) reviewed strategies, including planting resistant or tolerant cultivars, spraying chemicals, improving cultivation techniques, and harvesting strategies to manage FHB and reduce the toxin content, and the author declared that biological control agents may be the most effective application, and that new strategies and new sources of resistance introgression into new cultivars need to be developed for the management of FHB and DON more effectively in the future [78], and that HIGS technology would be a novel and effective way to control head blight—the use of HIGS on a commercial scale appears possible in the near future.

These studies provide a proof-of-concept that HIGS is an effective strategy to control the fungal diseases of small grains, although perfect targets and conditions still must be determined in order to produce resistance in wheat and barley, especially to wheat blast (WB) caused by Magnaporthe oryzae Triticum (MoT); take-all caused by *Gaeumannomyces graminis* var. *tritici*; and blotch diseases caused by Zymoseptoria tritici, Parastagonospora nodorum, and Pyrenophora tritici-repentis. Breeding for and sustaining multiple disease resistance against complex and mixed disease situations is a challenging task.

4. Current Challenges of HIGS

HIGS technology as a novel plant genomic modification tool that has been used to improve resistance against various diseases, breed high quality crops, develop seedless plant varieties, eliminate fungal toxins, improve stress tolerance, and develop male sterility varieties for breeding [18]. Significant challenges must be overcome before RNAi-mediated transgenic plants can be deployed in the field. Parasites and pathogens deliver small RNAs into their host to modify the host defense responses. Small RNAs also can be transferred into pathogens and pests to depress the invasion through feeding on plants and parasitism [79]. Researchers have proved that siRNAs and dsRNA can be transferred into fungal cells efficiently. External incubation with siRNA leads to the down regulation of a specific gene in *Aspergillus*, which proved the absorption of siRNA [80], and the silencing of the specific fungal gene was also interfered by the external treatment of siRNAs and dsRNAs in *Botrytis cinerea* [81]. However, it is still shrouded in mist whether and how siRNAs and/or dsRNAs are transported, and whether the intact RNAi machineries of the parasite are necessary or not. Recent studies report that exogenous plant miRNAs are sufficient to regulate animal target genes by feeding on the transgenic plants [82]. These findings alter the vistas for the safety of transgenic plants (not only the target gene, but also the selection makers), and should be interpreted with caution. The methods of selecting an appropriate target and the most suitable fragment of HIGS to delay or even entirely restrict disease still need to be defined. The efficiency of HIGS relies on enough supplements and on the successful transportation of siRNA between the two organisms. Accordingly, HIGS cannot be used against necrotrophic fungi, as they absorb nutrition and other metabolites from dead host tissue, which could not supply sufficient amounts of siRNAs. In addition, silencing an individual gene of the pathogens may not be sufficient to control disease, because of functional redundancy, and the incomplete silencing of mRNA levels does not guarantee the deactivation of the protein. This drawback can be overcome by using the transient silencing system for high-throughput screening. Off-target effects are also serious problems when we construct transgenic plants and try to avoid the off-target influence of unwanted symptoms or effects on agronomic traits. Bioinformatic tools (e.g., SIFI and Blastn) are very useful. In addition, HIGS is not always available in certain tissues (e.g., root and fruit) or systems (e.g., soil-borne fungal pathogens and insect groups that block siRNA uptake), and many crop species are not suitable for genomic transformation [83,84]. Although recent studies have attempted to express dsRNA in chloroplasts or other plastids so as to avoid the disadvantage of expressing dsRNA in the nucleus, the results show that it is obvious that the artificial constructs would be limited to being spread with the absence of chloroplasts when there is random mating in natural conditions [84]. The development of chloroplast transformation protocols for cereals (e.g., rice, wheat, maize, and barley) will be more commercially and environmentally attractive for plant protection against certain pests and pathogens.

A spray-induced gene silencing (SIGS) strategy is another innovative RNAi-based disease control strategy, which has been successfully used on both monocots and dicots, against fungi and pest infection. Koch et al. (2016) silenced the three fungal genes encoding cytochrome P450 lanosterol C-14α-demethylases in *F. graminearum* through the RNA spraying method on barley, and successfully restricted fungal development [37,85]. There may be some new theory in the siRNAs transportation. During HIGS against insects, systemic RNA interference deficient (SID) in *Caenorhabditis elegans* functions as a dsRNA transport protein and binds to the specific long dsRNA [86], while homologous genes or similar proteins responsible for the RNA transporters in fungi are not reported yet. While how RNAs are transported from the host tissue into the parasite, especially fungi, are unclear, speculation exists that siRNAs are transferred via specific transporters or some extracellular vesicles [85]. Li et al. (2015) mentioned that fungi produced vesicles to transfer siRNAs to be taken up into plant cells [87]. Recently, the necrotrophic pathogen, plant pathogenic bacteria, and parasitic nematodes delivered vesicles that contain siRNAs into the host tissue and modulated host immunity [88–90]. The vesicles are highly related to the RNA communication between the organism and pathogens. However, there is still lack evidence that plant-derived vesicles transmit RNAs to pathogenic fungi.

Every organism has its specific and stable genome for genetic information storage and for transfer to future generations. However, how does the genome of the organism remain stable, inheritable, and variable for millions of years? The discovery and research of RNA silencing provides a new perspective on this issue. RNAi-derived silencing can inhibit the invasion of foreign DNA (virus and transgenic ways). Maintaining a low activity of endogenous transposons and repeats is an important immune mechanism of gene expression during plant development, while the regulatory elements and regulation directing the signal pathway have not been found. How does the plant endogenous system identify itself and the "exotic" DNA molecules, and consequently the blocking and silencing? This still needs to be explored. Under the supervision of the endogenous RdDM silent epigenetic regulation, the over-expression function difficultly and often produces co-suppression. Conversely, RNAi construction for knocking down the function of endogenous genes can only be made at the site of initiation, and is difficult to pass to the surrounding tissues. The systems used to study the gene silencing pathways are often associated with transgene technology and need to be rethought and verified by future generations. With deeper insights into the mechanism of RNAi, the HIGS strategy of disease control opens novel avenues to improve crop yields.

5. Conclusions and Future Prospects

HIGS is a powerful and effective tool in gaining disease resistant transgenic plants and for the functional characterization of vital genes. In the near future, the major question for HIGS strategy will be answered. New targets and fragment selection methods, highly efficient transformation constructs, stable transgenic systems, and other new technologies will enhance the RNAi-derived strategy to develop durable disease resistant plants. RNAi-based technology provides a precise avenue to select high quality traits of varieties. The HIGS strategy has proven to be a novel approach in defending abiotic or biotic stress and quality improvement in an eco-friendly and sustainable manner. Recently, genome editing technologies have progressed rapidly and, in particular, the application of clustered regularly interspaced short palindrome repeats (CRISPR)/CRISPR-associated protein 9 (Cas9) editing has become a powerful tool for the enhancement of pathogen resistance in model plants and important crops [91]. In the near future, through the combined application of HIGS and CRISPR/Cas9, it will become much easier to achieve durable control of the diseases of wheat and barley.

Author Contributions: T.Q. and J.G. (Jun Guo) conceived and wrote the review; J.G. (Jun Guo), Z.K., and T.Q. contributed to the conception and design of the study; T.Q., P.L., H.P., and J.G. (Jia Guo) performed the data collection and interpretation, contributed with text and figure design, and drafted the paper; all of the authors discussed and approved the manuscript.

Funding: This study was supported by the National Key Research and Development Program of China (2018YFD0200402), the Natural Science Basic Research Plan in Shaanxi Province of China (2017JM3007), and the National Natural Science Foundation of China (31620103913).

Conflicts of Interest: The authors declare no conflict of interest.

References

1. Blanco, A.; Cenci, A.; Simeone, R.; Gadaleta, A.; Pignone, D.; Galasso, I. The cytogenetics and molecular characteristics of a translocated chromosome 1AS.1AL-1DL with a Glu-D1 locus in durum wheat. *Cell. Mol. Biol. Lett.* **2002**, *7*, 559–567. [PubMed]
2. Xie, W.; Xiong, W.; Pan, J.; Ali, T.; Cui, Q.; Guan, D.; Meng, J.; Mueller, N.D.; Lin, E.; Davis, S.J. Decreases in global beer supply due to extreme drought and heat. *Nat. Plants* **2018**, *4*, 964–973. [CrossRef] [PubMed]
3. Chakraborty, S.; Newton, A.C. Climate change, plant diseases and food security: An overview. *Plant Pathol.* **2011**, *60*, 2–14. [CrossRef]
4. Fisher, M.C.; Henk, D.A.; Briggs, C.J.; Brownstein, J.S.; Madoff, L.C.; McCraw, S.L.; Gurr, S.J. Emerging fungal threats to animal, plant and ecosystem health. *Nature* **2012**, *484*, 186–194. [CrossRef] [PubMed]

5. Godfray, H.C.J.; Crute, I.R.; Haddad, L.; Lawrence, D.; Muir, J.F.; Nisbett, N.; Pretty, J.; Robinson, S.; Toulmin, C.; Whiteley, R. The future of the global food system. *Philos. Trans. R. Soc. Lond.* **2010**, *365*, 2769–2777. [CrossRef]

6. Liu, T.G.; Peng, Y.L.; Chen, W.Q.; Zhang, Z.Y. First detection of virulence in *Puccinia striiformis* f. sp. *tritici* in China to resistance genes *Yr24* (=*Yr26*) present in wheat cultivar *Chuanmai* 42. *Plant Dis.* **2010**, *94*, 1163. [CrossRef]

7. Keller, B.; Wicker, T.; Krattinger, S.G. Advances in wheat and pathogen genomics: Implications for disease control. *Annu. Rev. Phytopathol.* **2018**, *56*, 67–87. [CrossRef]

8. Csorba, T.; Pantaleo, V.; Burgyan, J. RNA silencing: An antiviral mechanism. *Adv. Virus Res.* **2009**, *75*, 35–71.

9. Harvey, J.J.W.; Lewsey, M.G.; Patel, K.; Westwood, J.; Heimstädt, S.; Carr, J.P.; Baulcombe, D.C. An antiviral defense role of AGO2 in plants. *PLoS ONE* **2011**, *6*, e14639. [CrossRef]

10. Hu, Q.; Niu, Y.; Zhang, K.; Liu, Y.; Zhou, X. Virus-derived transgenes expressing hairpin RNA give immunity to *Tobacco mosaic virus* and *Cucumber mosaic virus*. *Virol. J.* **2011**, *8*, 41. [CrossRef]

11. Guozhong, H.; Rex, A.; Davis, E.L.; Baum, T.J.; Hussey, R.S. Engineering broad root-knot resistance in transgenic plants by RNAi silencing of a conserved and essential root-knot nematode parasitism gene. *Proc. Natl. Acad. Sci. USA* **2006**, *103*, 14302–14306.

12. Napoli, C.; Lemieux, C.; Jorgensen, R. Introduction of a chimeric chalcone synthase gene into petunia results in reversible co-suppression of homologous genes in trans. *Plant Cell* **1990**, *2*, 279–289. [CrossRef]

13. Jorgensen, R. Plants, RNAi, and the Nobel Prize. *Science* **2006**, *314*, 1242–1243. [CrossRef] [PubMed]

14. Zhu, X.; Qi, T.; Yang, Q.; He, F.; Tan, C.; Ma, W.; Voegele, R.T.; Kang, Z.; Guo, J. Host-induced gene silencing of the MAPKK gene *PsFUZ7* confers stable resistance to wheat stripe rust. *Plant Physiol.* **2017**, *175*, 1853–1863. [CrossRef] [PubMed]

15. Appels, R.; Eversole, K.; Feuillet, C.; Keller, B.; Rogers, J.; Stein, N.; Pozniak, C.J.; Stein, N.; Choulet, F.; Distelfeld, A.; et al. Shifting the limits in wheat research and breeding using a fully annotated reference genome. *Science* **2018**, *361*, 6403.

16. Wicker, T.; Schulman, A.H.; Tanskanen, J.; Spannagl, M.; Twardziok, S.; Mascher, M.; Springer, N.M.; Li, Q.; Waugh, R.; Li, C.; et al. The repetitive landscape of the 5100 Mbp barley genome. *Mob. DNA* **2017**, *8*, 22. [CrossRef] [PubMed]

17. Holzberg, S.; Brosio, P.; Gross, C.; Pogue, G.P. *Barley stripe mosaic virus*-induced gene silencing in a monocot plant. *Plant J.* **2002**, *30*, 315–327. [CrossRef]

18. Saurabh, S.; Vidyarthi, A.S.; Prasad, D. RNA interference: Concept to reality in crop improvement. *Planta* **2014**, *239*, 543–564. [CrossRef]

19. Llave, C. Virus-derived small interfering RNAs at the core of plant-virus interactions. *Trends Plant Sci.* **2010**, *15*, 701–707. [CrossRef]

20. Tang, G.; Reinhart, B.J.; Bartel, D.P.; Zamore, P.D. A biochemical framework for RNA silencing in plants. *Genes Dev.* **2003**, *17*, 49–63. [CrossRef]

21. Qi, Y.; Hannon, G.J. Uncovering RNAi mechanisms in plants: Biochemistry enters the foray. *FEBS Lett.* **2005**, *579*, 5899–5903. [CrossRef] [PubMed]

22. Fu, D.; Uauy, C.; Blechl, A.; Dubcovsky, J. RNA interference for wheat functional gene analysis. *Transgenic Res.* **2007**, *16*, 689–701. [CrossRef] [PubMed]

23. Miki, D. RNA silencing of single and multiple members in a gene family of rice. *Plant Physiol.* **2005**, *138*, 1903–1913. [CrossRef] [PubMed]

24. Yue, S.J.; Li, H.; Li, Y.W.; Zhu, Y.F.; Guo, J.K.; Liu, Y.J.; Chen, Y.; Jia, X. Generation of transgenic wheat lines with altered expression levels of 1Dx5 high-molecular weight glutenin subunit by RNA interference. *J. Cereal Sci.* **2008**, *47*, 153–161. [CrossRef]

25. McGinnis, K.; Murphy, N.; Carlson, A.R.; Akula, A.; Akula, C.; Basinger, H.; Carlson, M.; Hermanson, P.; Kovacevic, N.; McGill, M.A.; et al. Assessing the efficiency of RNA interference for maize functional genomics. *Plant Physiol.* **2007**, *143*, 1441–1451. [CrossRef] [PubMed]

26. Xu, L.; Duan, X.; Lv, Y.; Zhang, X.; Nie, Z.; Xie, C.; Ni, Z.; Liang, R. Silencing of an aphid carboxylesterase gene by use of plant-mediated RNAi impairs *Sitobion avenae* tolerance of Phoxim insecticides. *Transgenic Res.* **2014**, *23*, 389–396. [CrossRef] [PubMed]

27. Abdellatef, E.; Will, T.; Koch, A.; Imani, J.; Vilcinskas, A.; Kogel, K.H. Silencing the expression of the salivary sheath protein causes transgenerational feeding suppression in the aphid *Sitobion avenae*. *Plant Biotechnol. J.* **2015**, *13*, 849–857. [CrossRef] [PubMed]

28. Lilley, C.J.; Bakhetia, M.; Charlton, W.L.; Urwin, P.E. Recent progress in the development of RNA interference for plant parasitic nematodes. *Mol. Plant Pathol.* **2007**, *8*, 701–711. [CrossRef]

29. Tan, J.C.H.; Jones, M.G.K.; Fosu-Nyarko, J. Gene silencing in root lesion nematodes (*Pratylenchus* spp.) significantly reduces reproduction in a plant host. *Exp. Parasitol.* **2013**, *133*, 166–178. [CrossRef]

30. Fahim, M.; Millar, A.A.; Wood, C.C.; Larkin, P.J. Resistance to *Wheat streak mosaic virus* generated by expression of an artificial polycistronic microRNA in wheat. *Plant Biotechnol. J.* **2012**, *10*, 150–163. [CrossRef]

31. Kis, A.; Tholt, G.; Ivanics, M.; Várallyay, É.; Jenes, B.; Havelda, Z. Polycistronic artificial miRNA-mediated resistance to *Wheat dwarf virus* in barley is highly efficient at low temperature. *Mol. Plant Pathol.* **2016**, *17*, 427–437. [CrossRef] [PubMed]

32. Cruz, L.F.; Rupp, J.L.S.; Trick, H.N.; Fellers, J.P. Stable resistance to Wheat streak mosaic virus in wheat mediated by RNAi. *In Vitro Cell. Dev. Plant* **2014**, *50*, 665–672. [CrossRef]

33. Nowara, D.; Gay, A.; Lacomme, C.; Shaw, J.; Ridout, C.; Douchkov, D.; Hensel, G.; Kumlehn, J.; Schweizer, P. HIGS: Host-induced gene silencing in the obligate biotrophic fungal pathogen *Blumeria graminis*. *Plant Cell* **2010**, *22*, 3130–3141. [CrossRef] [PubMed]

34. Pliego, C.; Nowara, D.; Bonciani, G.; Gheorghe, D.M.; Xu, R. Host-induced gene silencing in barley powdery mildew reveals a class of ribonuclease-like effectors. *Mol. Plant Microbe Interact.* **2013**, *26*, 633–642. [CrossRef] [PubMed]

35. Qi, T.; Zhu, X.; Tan, C.; Liu, P.; Guo, J.; Kang, Z.; Guo, J. Host-induced gene silencing of an important pathogenicity factor PsCPK1 in *Puccinia striiformis* f. sp. *tritici* enhances resistance of wheat to stripe rust. *Plant Biotechnol. J.* **2017**, *16*, 797–807. [CrossRef] [PubMed]

36. Panwar, V.; Jordan, M.; McCallum, B.; Bakkeren, G. Host-induced silencing of essential genes in *Puccinia triticina* through transgenic expression of RNAi sequences reduces severity of leaf rust infection in wheat. *Plant Biotechnol. J.* **2018**, *16*, 1013–1023. [CrossRef] [PubMed]

37. Koch, A.; Kumar, N.; Weber, L.; Keller, H.; Imani, J.; Kogel, K.H. Host-induced gene silencing of cytochrome P450 lanosterol C14 -demethylase-encoding genes confers strong resistance to *Fusarium* species. *Proc. Natl. Acad. Sci. USA* **2013**, *110*, 19324–19329. [CrossRef]

38. Cheng, W.; Song, X.; Li, H.; Cao, L.; Sun, K.; Qiu, X.; Xu, Y.; Yang, P.; Huang, T.; Zhang, J.; et al. Host-induced gene silencing of an essential chitin synthase gene confers durable resistance to *Fusarium* head blight and seedling blight in wheat. *Plant Biotechnol. J.* **2015**, *13*, 1335–1345. [CrossRef]

39. Chen, W.; Kastner, C.; Nowara, D.; Oliveira-Garcia, E.; Rutten, T.; Zhao, Y.; Deising, H.B.; Kumlehn, J.; Schweizer, P. Host-induced silencing of *Fusarium culmorum* genes protects wheat from infection. *J. Exp. Bot.* **2016**, *67*, 4979–4991. [CrossRef]

40. Baum, J.A.; Bogaert, T.; Clinton, W.; Heck, G.R.; Feldmann, P.; Ilagan, O.; Johnson, S.; Plaetinck, G.; Munyikwa, T.; Pleau, M. Control of coleopteran insect pests through RNA interference. *Nat. Biotechnol.* **2007**, *25*, 1322–1326. [CrossRef]

41. Wang, D.; Liu, Q.; Li, X.; Sun, Y.; Wang, H.; Xia, L. Double-stranded RNA in the biological control of grain aphid (*Sitobion avenae* F.). *Funct. Integr. Genom.* **2015**, *15*, 211–223. [CrossRef] [PubMed]

42. Zhang, M.; Zhou, Y.; Wang, H.; Jones, H.D.; Gao, Q.; Wang, D.; Ma, Y.; Xia, L. Identifying potential RNAi targets in grain aphid (*Sitobion avenae* F.) based on transcriptome profiling of its alimentary canal after feeding on wheat plants. *BMC Genom.* **2013**, *14*, 560. [CrossRef]

43. Richter, K.; Buchner, J. Hsp90: Chaperoning signal transduction. *J. Cell. Physiol.* **2010**, *188*, 281–290. [CrossRef] [PubMed]

44. Will, T.; Vilcinskas, A. Aphid-Proof Plants: Biotechnology-based approaches for aphid control. *Adv. Biochem. Eng. Biotechnol.* **2013**, 179–203. [CrossRef]

45. Arakane, Y.; Specht, C.A.; Kramer, K.J.; Muthukrishnan, S.; Beeman, R.W. Chitin synthases are required for survival, fecundity and egg hatch in the red flour beetle, *Tribolium castaneum*. *Insect Biochem. Mol.* **2008**, *38*, 959–962. [CrossRef] [PubMed]

46. Aranda, M.; Marquessouza, H.; Bayer, T.; Tautz, D. The role of the segmentation gene hairy in *Tribolium*. *Dev. Genes Evol.* **2008**, *218*, 465–477. [CrossRef] [PubMed]

47. Yan, T.; Chen, H.; Sun, Y.; Yu, X.; Xia, L. RNA interference of the ecdysone receptor genes *EcR* and *USP* in grain aphid (*Sitobion avenae* F.) affects its survival and fecundity upon feeding on wheat plants. *Int. J. Mol. Sci.* **2016**, *17*, 2098. [CrossRef]

48. Eileen, K.; Henrike, S.; Andreas, V.; Boran, A. MMPs regulate both development and immunity in the tribolium model insect. *PLoS ONE* **2009**, *4*, e4751.

49. Yu, X.D.; Liu, Z.C.; Huang, S.L.; Chen, Z.Q.; Sun, Y.W.; Duan, P.F.; Ma, Y.Z.; Xia, L.Q. RNAi-mediated plant protection against aphids. *Pest Manag. Sci.* **2016**, *72*, 1090–1098. [CrossRef]

50. Christina, K.; Mohammad-Reza, H.; Alisdair, R.F.; Ute, R.; Tomasz, C.; Brigitte, H.; Wolf, B.F. The sucrose transporter StSUT1 localizes to sieve elements in potato tuber phloem and influences tuber physiology and development. *Plant Physiol.* **2003**, *131*, 102–113.

51. Fraley, R.T.; Rogers, S.G.; Horsch, R.B.; Sanders, P.R.; Flick, J.S.; Adams, S.P.; Bittner, M.L.; Brand, L.A.; Fink, C.L.; Fry, J.S. Expression of bacterial genes in plant cells. *Proc. Natl. Acad. Sci. USA* **1983**, *80*, 4803–4807. [CrossRef] [PubMed]

52. Carolan, J.C.; Fitzroy, C.I.J.; Ashton, P.D.; Douglas, A.E.; Wilkinson, T.L. The secreted salivary proteome of the pea aphid *Acyrthosiphon pisum* characterised by mass spectrometry. *Proteomics* **2010**, *9*, 2457–2467. [CrossRef] [PubMed]

53. Mashela, P.W.; Ndhlala, A.R.; Pofu, K.M.; Dube, Z.P. Phytochemicals of nematode-resistant transgenic plants. *Transgenes. Second. Metab.* **2017**, 553–568. [CrossRef]

54. Bernard, G.; Egnin, M.; Bonsi, C. The impact of plant-parasitic nematodes on sgriculture and methods of control. *Intechopen* **2017**, e68958. [CrossRef]

55. Joseph, S.; Gheysen, G.; Subramaniam, K. RNA interference in *Pratylenchus coffeae*: Knock down of Pc-pat-10 and Pc-unc-87 impedes migration. *Mol. Biochem. Parasitol.* **2012**, *186*, 51–59. [CrossRef] [PubMed]

56. Yadav, B.C.; Veluthambi, K.; Subramaniam, K.; Yadav, B.C.; Veluthambi, K.; Subramaniam, K. Host-generated double stranded RNA induces RNAi in plant-parasitic nematodes and protects the host from infection. *Mol. Biochem. Parasitol.* **2006**, *148*, 219–222. [CrossRef] [PubMed]

57. Ghag, S.B. Host induced gene silencing, an emerging science to engineer crop resistance against harmful plant pathogens. *Physiol. Mol. Plant Pathol.* **2017**, *100*. [CrossRef]

58. Sivamani, E.; Brey, C.W.; Dyer, W.E.; Talbert, L.E.; Qu, R. Resistance to wheat streak mosaic virus in transgenic wheat expressing the viral replicase (NIb) gene. *Mol. Breed.* **2000**, *6*, 469–477. [CrossRef]

59. Sivamani, E.; Brey, C.W.; Talbert, L.E.; Young, M.A.; Dyer, W.E.; Kaniewski, W.K.; Qu, R. Resistance to wheat streak mosaic virus in transgenic wheat engineered with the viral coat protein gene. *Transgenic Res.* **2002**, *11*, 31–41. [CrossRef]

60. Li, Z.; Liu, Y.; Berger, P.H. Transgenic silencing in wheat transformed with the WSMV-CP gene. *Biotechnology* **2005**, *4*, 62–68.

61. Shoup Rupp, J.L. RNAi-mediated, stable resistance to *Triticum mosaic virus* in wheat. *Crop. Sci.* **2016**, *56*, 1602–1610. [CrossRef]

62. Brown, J.K.M. Durable resistance of crops to disease: A darwinian perspective. *Annu. Rev. Phytopathol.* **2015**, *53*, 513–539. [CrossRef] [PubMed]

63. Larkin, P.J.; Banks, P.M.; Lagudah, E.S.; Appels, R.; Xiao, C.; Zhiyong, X.; Ohm, H.W.; McIntosh, R.A. *Disomic Thinopyrum* intermedium addition lines in wheat with *Barley yellow dwarf virus* resistance and with rust resistances. *Genome* **1995**, *38*, 385–394. [CrossRef] [PubMed]

64. Roelfs, A.P.; McCallum, B.; McVey, D.V.; Groth, J.V. Comparison of virulence and isozyme phenotypes of Pgt-QCCJ and great plains races of *Puccinia graminis* f. sp. *tritici. Phytopathology* **1997**, *87*, 910–914. [CrossRef] [PubMed]

65. Yin, C.; Jurgenson, J.E.; Hulbert, S.H. Development of a host-induced RNAi system in the wheat stripe rust fungus *Puccinia striiformis* f. sp. *tritici. Mol. Plant Microbe Interact.* **2011**, *24*, 554–561. [CrossRef] [PubMed]

66. Panwar, V.; McCallum, B.; Bakkeren, G. Endogenous silencing of *Puccinia triticina* pathogenicity genes throughin planta-expressed sequences leads to the suppression of rust diseases on wheat. *Plant J.* **2013**, *73*, 521–532. [CrossRef] [PubMed]

67. Cheng, Y.; Wang, X.; Yao, J.; Voegele, R.T.; Zhang, Y.; Wang, W.; Huang, L.; Kang, Z. Characterization of protein kinase PsSRPKL, a novel pathogenicity factor in the wheat stripe rust fungus. *Environ. Microbiol.* **2015**, *17*, 2601–2617. [CrossRef]

68. Cheng, Y.; Yao, J.; Zhang, Y.; Li, S.; Kang, Z. Characterization of a Ran gene from *Puccinia striiformis* f. sp. *tritici* involved in fungal growth and anti-cell death. *Sci. Rep.* **2016**, *6*, 35248. [CrossRef]

69. Cheng, Y.; Wang, W.; Yao, J.; Huang, L.; Voegele, R.T.; Wang, X.; Kang, Z. Two distinct Ras genes from *Puccinia striiformis* exhibit differential roles in rust pathogenicity and cell death. *Environ. Microbiol.* **2016**, *18*, 3910–3922. [CrossRef]

70. Zhu, X.; Jiao, M.; Guo, J.; Liu, P.; Tan, C.; Yang, Q.; Zhang, Y.; Thomas Voegele, R.; Kang, Z.; Guo, J. A novel MADS-box transcription factor PstMCM1-1 is responsible for full virulence of *Puccinia striiformis* f. sp. *tritici*. *Environ. Microbiol.* **2018**, *20*, 1452–1463. [CrossRef]

71. Zhu, X.; Liu, W.; Chu, X.; Sun, Q.; Tan, C.; Yang, Q.; Jiao, M.; Guo, J.; Kang, Z. The transcription factor PstSTE12 is required for virulence of *Puccinia striiformis* f. sp. *tritici*. *Mol. Plant Pathol.* **2017**, *19*, 961–974. [CrossRef] [PubMed]

72. Zhu, X.; Guo, J.; He, F.; Zhang, Y.; Tan, C.; Yang, Q.; Huang, C.; Kang, Z.; Guo, J. Silencing *PsKPP4*, a MAP kinase kinase kinase gene, reduces pathogenicity of the stripe rust fungus. *Mol. Plant Pathol.* **2018**, *19*, 2590–2602. [CrossRef] [PubMed]

73. Zhao, M.; Wang, J.; Ji, S.; Chen, Z.; Xu, J.; Tang, C.; Chen, S.; Kang, Z.; Wang, X. Candidate effector Pst_8713 impairs the plant immunity and contributes to virulence of *Puccinia striiformis* f. sp. *tritici*. *Front. Plant Sci.* **2018**, *9*, 1294. [CrossRef] [PubMed]

74. Cheng, Y.; Wu, K.; Yao, J.; Li, S.; Wang, X.; Huang, L.; Kang, Z. PSTha5a23, a candidate effector from the obligate biotrophic pathogen *Puccinia striiformis* f. sp. *tritici*, is involved in plant defense suppression and rust pathogenicity. *Environ. Microbiol.* **2017**, *19*, 1717–1729. [CrossRef] [PubMed]

75. Liu, C.; Pedersen, C.; Schultz Larsen, T.; Aguilar, G.B.; Amp, M.O.; Ana, K. The stripe rust fungal effector PEC6 suppresses pattern-triggered immunity in a host species-independent manner and interacts with adenosine kinases. *New Phytol.* **2016**, *7*, 14034. [CrossRef]

76. Liu, J.; Guan, T.; Zheng, P.; Chen, L.; Yang, Y.; Huai, B.; Li, D.; Chang, Q.; Huang, L.; Kang, Z. An extracellular Zn-only superoxide dismutase from *Puccinia striiformis* confers enhanced resistance to host-derived oxidative stress. *Environ. Microbiol.* **2016**, *18*, 4118–4135. [CrossRef]

77. Machado, A.K.; Brown, N.A.; Urban, M.; Kanyuka, K.; Hammond-Kosack, K.E. RNAi as an emerging approach to control *Fusarium* head blight disease and mycotoxin contamination in cereals. *Pest Manag. Sci.* **2018**, *74*, 790–799. [CrossRef]

78. Wegulo, S.N.; Baenziger, P.S.; Hernandez Nopsa, J.; Bockus, W.W.; Hallen-Adams, H. Management of *Fusarium* head blight of wheat and barley. *Crop. Prot.* **2015**, *73*, 100–107. [CrossRef]

79. Cai, Q.; Qiao, L.; Wang, M.; He, B.; Lin, F.M.; Palmquist, J.; Huang, H.D.; Jin, H. Plants send small RNAs in extracellular vesicles to fungal pathogen to silence virulence genes. *Science* **2018**, *360*, 1126–1129. [CrossRef]

80. Khatri, M.; Rajam, M.V. Targeting polyamines of *Aspergillus nidulans* by siRNA specific to fungal ornithine decarboxylase gene. *Med. Mycol.* **2007**, *45*, 211–220. [CrossRef]

81. Wang, M.; Weiberg, A.; Lin, F.M.; Thomma, B.P.; Huang, H.D.; Jin, H. Bidirectional cross-kingdom RNAi and fungal uptake of external RNAs confer plant protection. *Nat. Plants* **2016**, *2*, 16151. [CrossRef] [PubMed]

82. Lin, Z.; Dongxia, H.; Xi, C.; Donghai, L.; Lingyun, Z.; Yujing, Z.; Jing, L.; Zhen, B.; Xiangying, L.; Xing, C. Exogenous plant MIR168a specifically targets mammalian LDLRAP1: Evidence of cross-kingdom regulation by microRNA. *Cell Res.* **2012**, *22*, 107–126.

83. Yin, C.; Hulbert, S. Host Induced Gene Silencing (HIGS), a promising strategy for developing disease resistant crops. *In Omics Int.* **2015**, *4*, 1–2. [CrossRef]

84. Zhang, J.; Khan, S.A.; Heckel, D.G.; Bock, R. Next-generation insect-resistant plants: RNAi-mediated crop protection. *Trends Biotechnol.* **2017**, *35*, 871–882. [CrossRef] [PubMed]

85. Koch, A.; Biedenkopf, D.; Furch, A.; Weber, L.; Rossbach, O.; Abdellatef, E.; Linicus, L.; Johannsmeier, J.; Jelonek, L.; Goesmann, A. An RNAi-based control of *Fusarium graminearum* infections through spraying of long dsRNAs involves a plant passage and is controlled by the fungal silencing machinery. *PLoS Pathog.* **2016**, *12*, e1005901. [CrossRef]

86. Koch, A.; Kogel, K.H. New wind in the sails: Improving the agronomic value of crop plants through RNAi-mediated gene silencing. *Plant Biotechnol. J.* **2015**, *12*, 821–831. [CrossRef]

87. Li, W.; Koutmou, K.S.; Leahy, D.J.; Li, M. Systemic RNA Interference Deficiency-1 (SID-1) extracellular domain selectively binds long double-stranded RNA and is required for RNA transport by SID-1. *J. Biol. Chem.* **2015**, *290*, 18904–18913. [CrossRef]

88. Arne, W.; Ming, W.; Feng-Mao, L.; Hongwei, Z.; Zhihong, Z.; Isgouhi, K.; Hsien-Da, H.; Hailing, J. Fungal small RNAs suppress plant immunity by hijacking host RNA interference pathways. *Science* **2013**, *342*, 118–123.

89. Quintana, J.F.; Babayan, S.A.; Buck, A.H. Small RNAs and extracellular vesicles in filarial nematodes: From nematode development to diagnostics. *Parasite Immunol.* **2016**, *39*. [CrossRef]

90. Katsir, L.; Bahar, O. Bacterial outer membrane vesicles at the plant-pathogen interface. *PLoS Pathog.* **2017**, *13*, e1006306. [CrossRef]

91. Borrelli, V.M.G.; Brambilla, V.; Rogowsky, P.; Marocco, A.; Lanubile, A. The enhancement of plant disease resistance using CRISPR/Cas9 technology. *Front. Plant Sci.* **2018**, *9*, 1245. [CrossRef] [PubMed]

© 2019 by the authors. Licensee MDPI, Basel, Switzerland. This article is an open access article distributed under the terms and conditions of the Creative Commons Attribution (CC BY) license (http://creativecommons.org/licenses/by/4.0/).

International Journal of
Molecular Sciences

MDPI

Article

Overexpression of Arabidopsis *OPR3* in Hexaploid Wheat (*Triticum aestivum* L.) Alters Plant Development and Freezing Tolerance

Alexey V. Pigolev [1,†], Dmitry N. Miroshnichenko [1,2,†], Alexander S. Pushin [1,2], Vasily V. Terentyev [1], Alexander M. Boutanayev [1], Sergey V. Dolgov [2] and Tatyana V. Savchenko [1,*]

1   Institute of Basic Biological Problems RAS, Pushchino 142290, Russia; alexey-pigolev@rambler.ru (A.V.P.); miroshnichenko@bibch.ru (D.N.M.); aspushin@gmail.com (A.S.P.); v.v.terentyev@gmail.com (V.V.T.); boutanaev@mail.ru (A.M.B.)
2   Branch of Shemyakin and Ovchinnikov Institute of Bioorganic Chemistry RAS, Pushchino 142290, Russia; dolgov@bibch.ru
*   Correspondence: savchenko_t@rambler.ru; Tel.: +7-496-773-3601
†   These authors contributed equally to this work.

Received: 22 October 2018; Accepted: 8 December 2018; Published: 11 December 2018

Abstract: Jasmonates are plant hormones that are involved in the regulation of different aspects of plant life, wherein their functions and molecular mechanisms of action in wheat are still poorly studied. With the aim of gaining more insights into the role of jasmonic acid (JA) in wheat growth, development, and responses to environmental stresses, we have generated transgenic bread wheat plants overexpressing *Arabidopsis 12-OXOPHYTODIENOATE REDUCTASE 3* (*AtOPR3*), one of the key genes of the JA biosynthesis pathway. Analysis of transgenic plants showed that *AtOPR3* overexpression affects wheat development, including germination, growth, flowering time, senescence, and alters tolerance to environmental stresses. Transgenic wheat plants with high *AtOPR3* expression levels have increased basal levels of JA, and up-regulated expression of *ALLENE OXIDE SYNTHASE*, a jasmonate biosynthesis pathway gene that is known to be regulated by a positive feedback loop that maintains and boosts JA levels. Transgenic wheat plants with high *AtOPR3* expression levels are characterized by delayed germination, slower growth, late flowering and senescence, and improved tolerance to short-term freezing. The work demonstrates that genetic modification of the jasmonate pathway is a suitable tool for the modulation of developmental traits and stress responses in wheat.

Keywords: transgenic wheat; 12-oxophytodienoate reductase; jasmonates; freezing tolerance

---

1. Introduction

Plant growth, development and responses to environmental signals are coordinated by complex multicomponent signaling networks. Jasmonates, derived from fatty acids, are important components of this regulatory system. Jasmonates biosynthesis and signal transduction pathways have been thoroughly studied in model plant Arabidopsis [1–4]. The biosynthesis is initiated in chloroplasts, where lipoxygenases oxidize the 13th carbon atom of the α-linolenic acid, forming 13(S)-hydroperoxy-9,11,15-octadecatrienoic acid (13-HPOT). The 13-HPOT is further converted into (9S,13S)-12-oxo-phytodienoic acid (12-OPDA) by the consequent action of allene oxide synthase and allene oxide cyclase, and then 12-OPDA is transported to peroxisomes. Inside the peroxisomes, the double bond of the cyclopentenone ring in 12-OPDA is reduced by 12-OPDA reductase [5], and three cycles of β-oxidation of the carboxylic acid side chain occur, resulting in the formation of JA. JA itself can be further modified in the cytoplasm through the formation of a conjugate with the amino acid isoleucine, the signaling ligand, which is responsible for the regulation of majority of JA-dependent processes. Several metabolites of the pathway, 12-OPDA, JA, jasmonate–isoleucine conjugate, and methyl jasmonate display biological activity with only a partial overlap in functions. To date, most of the information on jasmonates biosynthesis and signaling comes from studies of dicotyledonous species, mainly Arabidopsis [1–5]. In monocots, this pathway is less well studied, and the information is mainly limited to maize and rice [6]. Biosynthesis, functions, and molecular mechanisms of action of jasmonates in wheat, one of the dominant cereal crops worldwide, remain mostly uncharacterized.

Unfavorable environmental conditions affect wheat crop yield globally [7]; therefore, the improvement of wheat tolerance to environmental stresses is of particular importance. The modification of individual steps of oxylipin biosynthesis pathway results in the alteration of plant tolerance to environmental challenges, as shown for Arabidopsis [8–12] and other species, including a few monocots [6,13,14]. For wheat, a protective effect of exogenously applied jasmonates under conditions of biotic and abiotic stresses was demonstrated in numerous studies [15–17], suggesting that the modulation of the activity of jasmonate-based signaling systems is a promising approach for the improvement of wheat tolerance to environmental stresses. The effect of the genetic modification of the jasmonate pathway on wheat development and stress response has not been sufficiently investigated, probably due to the fact that hexaploid wheat remains recalcitrant in tissue culture [18]. With the purpose of altering the level of jasmonic acid and studying the effect of such alterations on plant growth, development, and stress responses, we have generated transgenic hexaploid (*Triticum aestivum* L.) wheat plants constitutively overexpressing the *12-OXOPHYTODIENOATE REDUCTASE 3* gene from *Arabidopsis thaliana* (*AtOPR3*). 12-Oxophytodienoate reductases are classified into OPRI and OPRII subgroups, and only OPRII subgroup genes take part in the biosynthesis of jasmonic acid. Out of three oxophytodienoate reductases in Arabidopsis, only *OPR3* belongs to the OPRII subgroup, and it is involved in the jasmonate biosynthesis pathway [19]. *AtOPR3* is a well-characterized gene, coding for the enzyme controlling the levels of major metabolites of the jasmonate pathway, the upstream substrate 12-OPDA, and downstream products of the biosynthetic pathway, JA, and jasmonic acid derivatives [5,12,20,21]. The role of *AtOPR3* in plant growth and stress responses has been demonstrated [5,22,23], and the crystal structure of the OPR3 protein has been determined [24].

2. Results

2.1. Generation of Wheat Plants Overexpressing AtOPR3

The hexaploid spring bread wheat variety 'Saratovskaya-60' (Sar-60) was previously thoroughly characterized and used as a parent line for the generation of a number of commercial wheat varieties grown in Russia [25]. In the present study, Sar-60 was genetically transformed with the pBAR-GFP.UbiOPR3 vector, enabling the expression of the *Arabidopsis thaliana OPR3* gene under the control of a strong constitutive promoter (Figure 1A). A number of transgenic plants of various generations (T0–T3) were produced to study the effect of *AtOPR3* overexpression in wheat. A total of 23 putative transgenic plantlets produced from 20 independent explants, designated as Tr-1 to Tr-20, were established in the greenhouse. In a few cases, two plants were regenerated from one explant (1a and 1b, 3a and 3b, 6a and 6b). Sixteen primary independent lines were identified as being transgenic by PCR, with a pair of primers targeting a 600 bp fragment of *GFP* (Figure 1B, upper panel), and a 295 bp fragment of *AtOPR3* (Figure 1B, lower panel). *AtOPR3* was detected in all transgenic lines, except line 10. End-point RT-PCR was performed on leaf samples collected from plants at the heading stage, to confirm the expression of the introduced genes in *GFP*-positive primary transgenic lines. The *TaWIN1* gene [26] was used as a control for the expression analysis (Figure 1C, upper panel). The *GFP* transcript was detected in all studied plants except for non-transformed Sar-60 control (Figure 1C, middle panel). The amplification bands corresponding to *AtOPR3* were specifically detected in the nine primary transgenic lines, but not in non-transformed Sar-60 plants and several green fluorescent protein (GFP)-positive transgenic lines (Figure 1C, lower panel).

T0 plants from eight *AtOPR3*-positive events produced T1 seeds, and segregation analysis was conducted on their progeny via GFP fluorescence, as described in [27]. The embryos produced by self-pollination of individual T1 plants were used for the detection of homozygous populations. At first, we analyzed the pollen produced by T1 plants, collected immediately after the anthers appeared (Figure S1A). If the pollen displayed a homozygous pattern, T2 zygotic embryos were additionally analyzed for GFP fluorescence (Figure S1B). Using this method, we were able to identify homozygous offspring, derived from six self-pollinated independent primary (T0) lines. The successful expression of transgenes in homozygous plants of the selected T2 population was confirmed by end-point RT-PCR (Figure S1C). Homozygous lines with confirmed expression of *AtOPR3* in the T2 generation, which produced sufficient amount of seeds, including Tr-3 (the progeny from 3b line), Tr-15, Tr-18, Tr-19, and Tr-20, were used for further analysis.

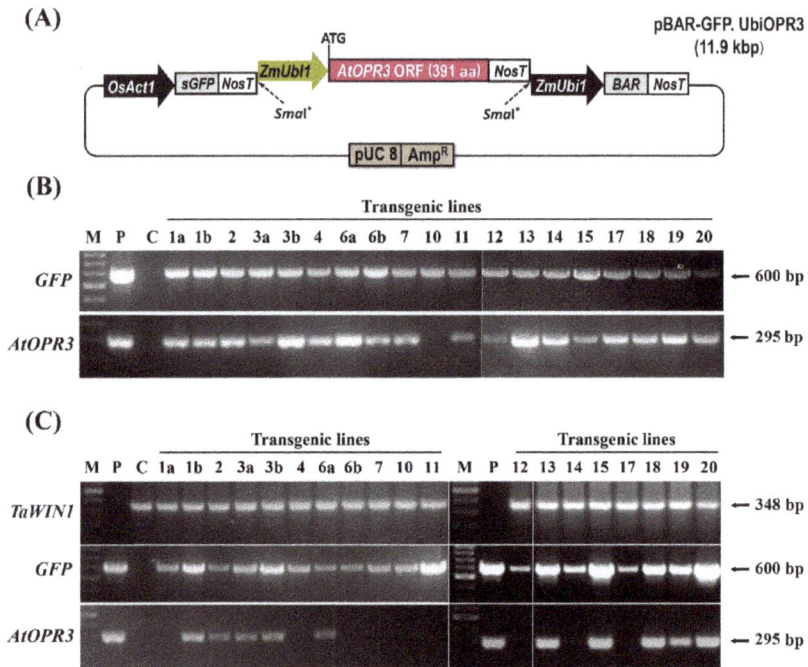

Figure 1. Generation and molecular analysis of transgenic plants. (**A**) Schematic representation of the pBAR-GFP.UbiOPR3 expression cassette used for Sar-60 transformation. *OsAct1*, rice Actin 1 promoter; *ZmUbi1*, maize ubiquitin 1 promoter; *NosT*, nopaline synthase terminator; *GFP*, Green Fluorescent Protein gene; *BAR*, BASTA resistance gene (phosphinothricin acetyl transferase); *AmpR*, ampicillin resistance gene. (**B**) PCR analysis performed on the genomic DNA of T0 plants for the insertion of *GFP* gene (upper panel) and *AtOPR3* (bottom panel). (**C**) End-point RT-PCR analysis on the total RNA of T0 plants (at heading stage) for the expression of the reference gene *TaWIN1* (top panel), *GFP* gene (middle panel) and *AtOPR3* (bottom panel). Lane M, DNA ladder as a molecular weight marker; Lane P, plasmid DNA pBAR-GFP.UbiOPR3; Lane C, non-transgenic wheat plant Sar-60; Lanes Tr-1–Tr-20 represent putative transgenic wheat plants.

2.2. Transgenic Wheat Plants Overexpressing the Arabidopsis OPR3 Gene Display an Altered Growth and Development Phenotype

To find out if there is an effect of *AtOPR3* gene overexpression on wheat plants growth and development, we carefully examined the germination, growth, flowering time, and senescence of generated transgenic lines. The observation allowed us to identify two distinct phenotypes among stable transgenic lines. The lines Tr-3 and Tr-18 (and to a lesser extent Tr-19) were characterized by delayed seed germination, slower growth, delayed flowering, and senescence, while the plants of Tr-15 and Tr-20, on the contrary, are characterized by earlier germination, faster growth, earlier flowering, and senescence (Figures 3–5, and Figures S2–S4). Quantitative analysis of *AtOPR3* gene expression in T3 homozygous plants revealed a significant difference in the expression levels between transgenic lines (Figure 2A). *AtOPR3* expression level was highest in the Tr-3 line; in Tr-18, this level was lower by almost three times in comparison to Tr-3, and in Tr-20, it was lower by 40 times. The JA level was also highest in Tr-3 and lowest in Tr-20 and Tr-15 (Figure 2B). Interestingly, JA levels in Tr-20 and Tr-15 were decreased, even in comparison to Sar-60. The JA level in Sar-60 leaves was about 3 ng/g fresh weight; in Tr-3, this value was increased by 70%, in line Tr-18 the increase was about 21%, in lines Tr-20 and Tr-15 JA level was decreased by 23% (Table S1). We also analyzed the expression levels of three genes involved in jasmonates biosynthesis and signaling, *ALLENE OXIDE SYNTHASE* (*TaAOS*) (GenBank: KJ001800.1), *12-OPDA REDUCTASE 2* (*TaOPR2*) (TraesCS7B02G311600), and *CORONATINE INSENSITIVE 1-LIKE* (*TaCOI-1*) (TraesCS1A02G279100) (Figure 2C). The analysis revealed a significant increase in the level of *TaAOS* gene expression in Tr-3 (16-fold increase) and Tr-18 (4-fold increase) in comparison to Sar-60. Expression level of *TaOPR2* was slightly increased in Tr-3, while *TaCOI-1* levels remained unchanged in all analyzed lines.

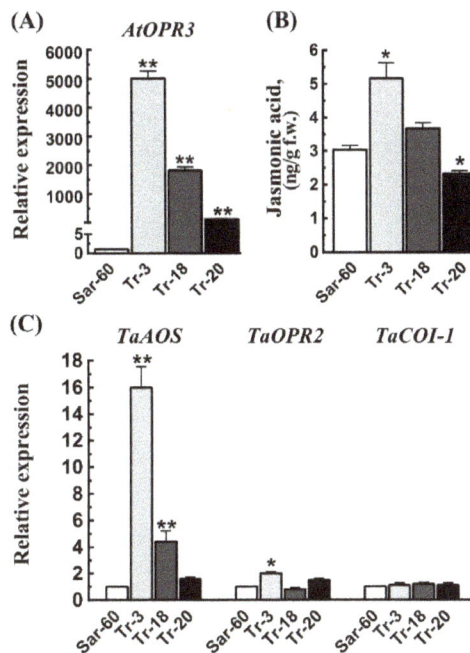

Figure 2. Gene expression analysis and JA levels in transgenic plants. The relative expression of *AtOPR3* (**A**), and endogenous wheat jasmonate biosynthesis and signaling pathways genes *TaAOS*, *TaOPR2*, and *TaCOI-1* (**C**). Total RNA was extracted from leaves (the second leaf of T3 plants at four leaves stage) and subjected to real-time quantitative RT-PCR analysis. The transcript levels of each gene were normalized to *TaWIN1* and *TaUbi*, as measured in the same samples. Data are means of four biological replicates represented by 6–8 leaves ± SE. (**B**) JA levels in second leaves of T3 plants collected at the four-leaf stage. Stars above the graphs indicate statistically significant differences (* $p \leq 0.05$, ** $p \leq 0.001$).

We examined germination and early growth of immature embryos of Tr-3 and Tr-20, the lines with pronounced phenotype on the synthetic medium. The analysis showed that Tr-3 embryos germinated one day later than non-transformed Sar-60, while Tr-20 germinated one day earlier. The representative Petri plates with seedlings (four days after culture initiation) are shown in Figure 3A. The significant difference between the seedlings was further confirmed by measurements of the length of coleoptiles, radicles, and lateral seminal roots (Figure 3B). To confirm that observed differences between the transgenic lines were indeed associated with altered JA level, we studied the effect of exogenous JA treatment on the germination of non-transformed Sar-60 seeds. The analysis showed that JA treatment indeed led to delays in seed germination, and the slower growth of coleoptiles and roots (Figure 3C).

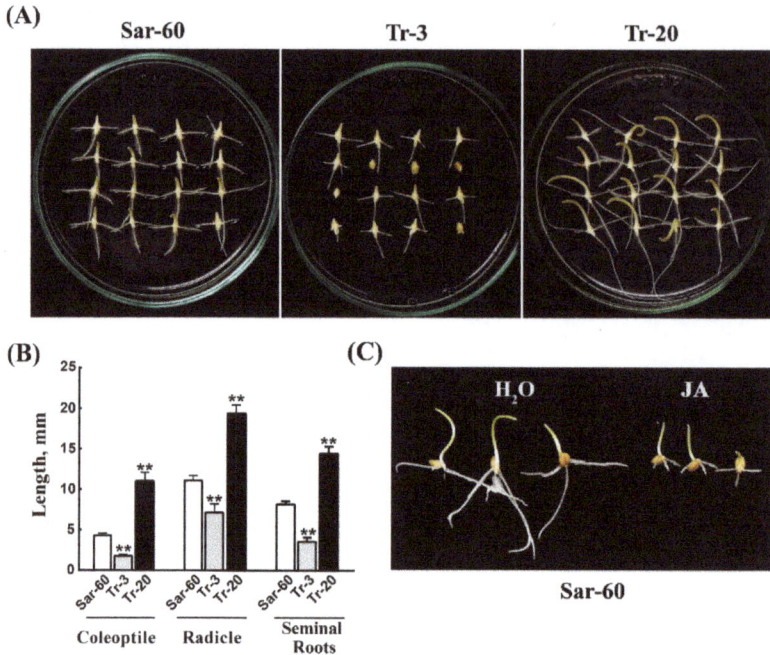

Figure 3. Jasmonate suppresses germination and early growth of wheat. (**A**) The representative Petri plates with seedlings grown from immature embryos of Sar-60, Tr-3, and Tr-20 T3 seeds, four days after culture initiation. (**B**) Length of coleoptiles, radicle and lateral seminal roots of four-day-old seedlings of Sar-60 (white bars), Tr-3 (grey bars), and Tr-20 (black bars). The means of 16 measurements ±SE are presented; stars above the graphs indicate statistically significant difference between genotypes ($p \leq 0.001$). (**C**) Effects of exogenously applied MeJA on the germination and growth of Sar-60 seeds. Seeds were placed on filter paper soaked in distilled water (left side) or a 0.1 mM solution of MeJA (right side); the picture was taken after four days of seed imbibition.

The difference in the growth rate between the transgenic plants and the non-transgenic Sar-60 was obvious in soil-grown plants (Figure 4). As seen in Figure 4A,B, the growth of the above-ground parts of Tr-3 and Tr-18 plants was retarded, while Tr-20 plants grew faster. The difference in roots growth was not observed at this growth stage under the given experimental conditions. The difference in plants height was also noticeable during later developmental stages (Figure S2). Further, we found that the flowering time was also affected in the transgenic lines (Figure 4C,D). Anthesis occurred later in Tr-3 plants in comparison to Sar-60, as observed in first, second, and third spikes. As expected, Tr-20 plants flowered earlier.

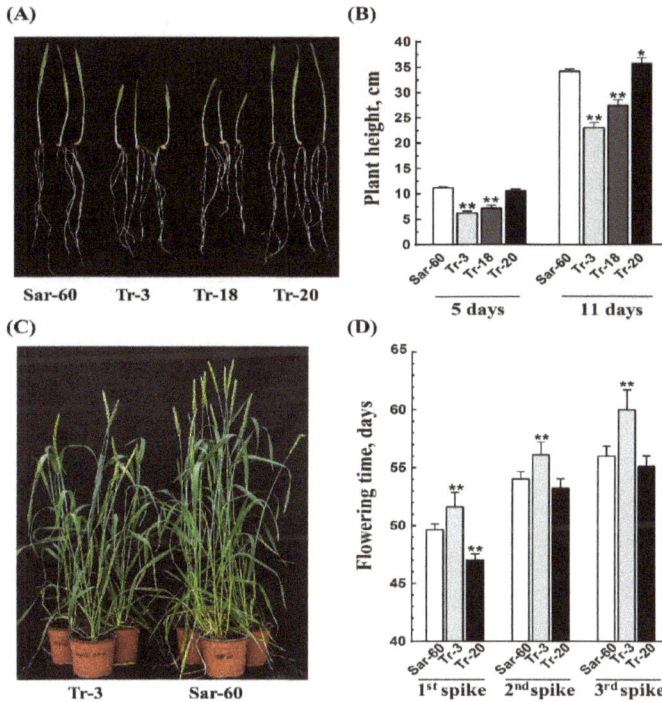

Figure 4. Transgenic wheat plants overexpressing *AtOPR3* gene have altered growth and flowering time phenotype. (**A**) The picture of soil-grown five-day-old homozygous T2 plants, Sar-60, Tr-3, Tr-18, and Tr-20. (**B**) The height of plants in 5 and 11 days after sowing. Means of 25 measurements ± SE are presented; stars above graphs indicate statistically significant differences (* $p \leq 0.05$, ** $p \leq 0.001$). (**C**) 55-day-old plants, Tr-3 and Sar-60. (**D**) Time (in days) of appearance of first, second, and third spikes on Sar-60 (white bars), Tr-3 (grey bars), and Tr-20 (black bars) plants. Means of 45 plants analyzed in three independent experiments ±SE are presented. Stars above the graphs indicate statistically significant differences ($p \leq 0.001$).

Leaf senescence symptoms (leaf yellowing) also appear earlier in Tr-20, and later in Tr-3 plants, in comparison to Sar-60. For the quantitative estimation of senescence in the studied plant lines, we have performed the measurements of variable chlorophyll fluorescence parameters, reflecting functional activity of Photosystem II (PS II). The analysis of the functional activity of PS II is a simple and reliable method for the evaluation of leaf senescence when performed with the necessary precautions. It is important to remember that photosynthetic parameters show heterogeneous spatial patterns within the leaf blade, and that individual leaf senescence starts at the tip of the leaf, and then the senescent area enlarges progressively toward the leaf sheath [28]. For the measurements of light-induced chlorophyll fluorescence in leaves of studied plants the area located 10 cm away from the tip of leaf blade was chosen. In preliminary experiments, it has been established that the measured parameter does not differ in young leaves of different genotypes. Results of the analysis confirmed that the senescence time was altered in the studied transgenic lines (Figure 5). Analysis of PS II operating efficiency in the fourth and flag leaves of plants (at maturation stage) in combination with visual assessment (yellowness of leaves) confirmed delayed senescence in Tr-3 and accelerated senescence in Tr-20. Interestingly, detached same-aged leaves of Tr-3 plants also displayed later senescence in comparison to Sar-60 (Figure 5B). Similar trends were observed in independent transgenic lines with altered JA levels: growth and development lines with increased JA levels were slowed down, while in lines with decreased JA level, they were accelerated (Figures S3 and S4).

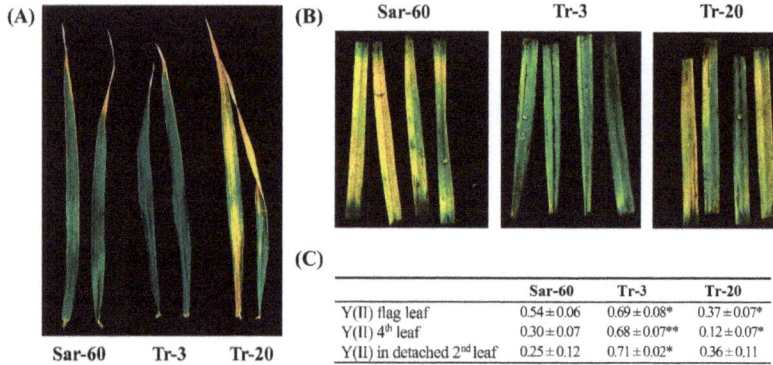

Figure 5. The timing of senescence is altered in wheat plants overexpressing *AtOPR3*. (**A**) Representative flag leaves of two-month-old homozygous T2 plants, Sar-60, Tr-3, and Tr-20. (**B**) The senescence of detached leaves. Same-aged detached leaves of Sar-60, Tr-3, and Tr-20 were kept floating on distilled water in Petri dishes under photoperiodic illumination 16/8 h (light/dark). The picture was taken on the fifth day of the experiment. (**C**) Analysis of PS II operating efficiency in flag (upper line) and fourth (middle line) leaves of two-month-old Sar-60, Tr-3, and Tr-20 plants growing in the soil, and detached leaves kept for five days under photoperiodic illumination (bottom line). Means of 13–16 measurements ± SE are presented. Stars indicate statistically significant differences from Sar-60 (* $p \leq 0.05$, ** $p \leq 0.001$).

2.3. AtOPR3 Overexpression Alters Wheat Tolerance to Freezing Stress

To estimate the stress tolerance of transgenic plants with altered JA level, we have performed short-term freezing treatment of plants of different transgenic lines as described in the Materials and Methods section. The freezing-induced damages were estimated by two methods—analysis of PS II activity by measuring light-induced chlorophyll fluorescence parameters and estimation of electrolytes leakage (EL), reflecting the degree of cell membranes damages in leaves. Both analyses showed that stress treatment causes the least amount damage to the leaves of Tr-3 and Tr-18 lines, while leaves of Tr-20 and Tr-15 plants displayed higher sensitivities to given stresses, even in comparison to Sar-60 (Figure 6A,B, and Figure S5). At the same time, no significant differences between the studied lines were observed when plants have been grown under low positive temperature conditions for prolonged periods of time (up to 30 days) (Figure S6).

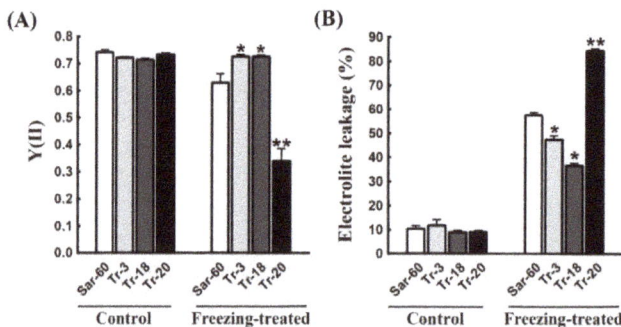

Figure 6. Freezing tolerance test of transgenic wheat plants overexpressing *AtOPR3*. Freezing treatments were performed on homozygous T2 plants, as described in the Materials and Methods section, and then the degree of damage was assessed by measurements of PS II operating efficiency (**A**) and electrolytes leakage (**B**) in control and treated leaves of Sar-60, Tr-3, Tr-18, and Tr-20. Means of 20 measurements ± SE are presented. Stars indicate statistically significant differences from Sar-60 (* $p \leq 0.05$, ** $p \leq 0.001$).

3. Discussion

In this work, we have implemented the transformation of hexaploid bread wheat with the Arabidopsis *OPR3* gene coding for the enzyme catalyzing the reduction of the cyclopentenone ring of 12-OPDA. The expression level of *AtOPR3* varied significantly between different transgenic lines, wherein plants with high *AtOPR3* expression levels have elevated levels of JA. Besides, the expression of *TaAOS*, a jasmonate biosynthesis pathway gene known to be regulated by a positive feedback loop that maintains and boosts JA levels, is also significantly up-regulated in these plants. Increase in *ALLENE OXIDE SYNTHASE* gene expression level in response to the up-regulation of the jasmonate pathway was demonstrated previously [29]. There are three homeologs of *12-OPDA REDUCTASE*, coding for OPRII subgroup proteins, which are potentially involved in the biosynthesis of endogenous jasmonic acid in the wheat genome (one homeolog in each of the A, B, and D subgenomes), TraesCS7A02G412400, TraesCS7B02G311600, and TraesCS7D02G405500. These genes share high similarities at the amino acid level with each other and *AtOPR3* (Figure S7). The expression level of wheat *12-OPDA REDUCTASE* (*TaOPR2*) encoded in the B subgenome, previously shown to be induced by methyl jasmonate [30], was slightly increased in Tr-3, and remained unaltered in other transgenic lines (Figure 2C). We could not detect the expression of *TaOPR2* homeolog genes from the A and D genomes, either because of the low expression level of these genes in leaf tissue, or because of the sequence difference between Saratovskaya-60 and the sequenced wheat varieties. The expression level of the jasmonate receptor COI-1 (*TaCOI-1*) [31] was not altered in transgenic lines (Figure 2C). The gene expression analysis demonstrates that *AtOPR3* gene overexpression selectively affects some genes of the endogenous jasmonate system, while the expression of other genes remains unaltered. The growth and development phenotypes of transgenic wheat plants with high *AtOPR3* expression levels were altered noticeably. These plants are characterized by delayed germination, slower growth, and delayed flowering time and senescence. The obtained results are in accordance with several previously published studies demonstrating the role of jasmonates in the regulation of plant growth and development, including seed germination, root growth, blossoming, fruit ripening, and senescence [13,32,33]. It is known, that JA-mediated suppression of plant growth is implemented through the regulation of cell cycle progression [34]. In Arabidopsis, jasmonate caused a touch-induced rosette diameter reduction and delay, in flowering [35]. The decreased size was also observed in rice plants with increased JA levels caused by the inactivation of the parallel competing hydroperoxide lyase branch of the oxylipin biosynthesis pathway [36]. Transgenic rice plants accumulating high levels of methyl jasmonate (MeJA) as a result of overexpression of the Arabidopsis jasmonic acid carboxyl methyltransferase gene were slightly smaller than the non-transformed controls. Flowering time was not affected in these plants, but grain yield was reduced [37]. Rice mutants *cpm2* (*coleoptile photomorphogenesis2*)/*hebiba* (defective in allene oxide cyclase) and *eg1* (*extra glume1*) (defective in a lipase involved in JA biosynthesis) have longer leaves [38,39]. A rice mutant with low JA levels resulting from a mutation in the *OsAOS1* gene had an accelerated juvenile–adult phase change, and flowered approximately five days earlier than did the wild type [40]. Leaf size was also affected in the rice *aos1* mutant. Increased endogenous JA levels in transgenic soybeans led to alterations in organogenesis: the leaves of the transgenic plants were slightly elongated in length, but they dramatically narrowed in width compared with the non-transformed plants. In addition, the elongation of the primary root was inhibited in the transgenic soybean plantlets, whereas the development of lateral root was stimulated [41]. In *Pharbitis nil*, high concentrations of MeJA inhibited root and shoot growth, while low concentrations of this metabolite, in contrast, enhanced plant growth [42]. MeJA treatment also inhibited growth and flowering in *Chenopodium rubrum* plants [43]. Transgenic potato plants overexpressing the gene coding for jasmonic acid carboxyl methyltransferase, in contrast, showed a significant increase in size and tuber yield [44]. A role for MeJA in the modulation of vernalization and flowering time in einkorn wheat was demonstrated [45]. Bread wheat plants overexpressing the allene oxide cyclase *TaAOC1* developed shorter roots and exhibited enhanced tolerance to salinity [13].

We estimated the senescence in growing plants and detached leaves by measuring the activity of PS II. A decline in PS II activity is one of the earliest events during leaf senescence, and visible senescence observed on individual leaves correlates with a decline in PSII activity [46]. Results demonstrate that increased JA levels in wheat plants leads to delay in senescence (brightly manifested in Tr-3), while the decreased level of the hormone leads to earlier senescence (Figure 5). Jasmonate roles in leaf senescence have been studied extensively [47]. Similarly to our observation, JA involvement in the regulation of barley leaf senescence was previously described [48].

As expected, the alteration of JA level also affected wheat tolerance to abiotic stress. It was established in our experiments that Tr-3 and Tr-18 have improved tolerances to short-term freezing treatment (Figure 6), although these plants' growth at low positive temperatures is not altered (Figure S6). Freezing temperatures are one of the most important abiotic factors controlling the geographical distribution of plants and restricting the use of agricultural lands. Short spring freezing episodes may cause injuries to young wheat plants, affecting the yield of this economically important crop. The role of jasmonates in the regulation of plant responses to abiotic stresses has been established [18,49]. An increase in JA level in response to stress factors was demonstrated for many plant species, including those phylogenetically closely related to wheat barley [50]. Exogenous application of jasmonates also improves plant tolerance to abiotic stresses [51,52]. Numerous studies have shown that jasmonates protect plants from damage caused by low temperatures when it accumulates in tissue or it is applied exogenously [53–56]. Molecular mechanisms underlying the protective effect of jasmonates have also been characterized. It is known that the transcription pathway *ICE-CBF/DREB1* (Inducer of CBF Expression—C-Repeat Binding Factor/DRE Binding Factor 1), playing a key role in plant adaptation to low temperatures, is regulated by jasmonates [57,58].

Interestingly, in two independent transgenic lines, Tr-15 and Tr-20, constitutive expression of *AtOPR3* leads to a decrease in basal JA level in comparison to Sar-60. Gene expression analysis showed that the *AtOPR3* expression level is significantly lower in these plants (about a 40-fold difference between Tr-20 and Tr-3) (Figure 2A). In all assays, these plants displayed phenotypes that were opposite to Tr-3 and Tr-18. Tr-20 showed an earlier germination phenotype, faster growth and development, and higher susceptibility to short-term freezing (Figures 3–6). Similar traits, such as faster growth, early senescence, and sensitivity to short-term freezing were also observed in Tr-15 (Figures S3–S5). Currently, we are unable to explain the mechanisms leading to a decrease in JA level upon overexpression of the *AtOPR3* gene, but similar results have been previously obtained in several studies describing transgenic plants overexpressing jasmonates biosynthesis pathway genes [12,59]. Transgenic tobacco plants overexpressing few JA pathway genes from Arabidopsis, including *AtOPR3*, had a decreased level of jasmonate–isoleucine conjugate, the most active metabolite of the pathway [59]. Interestingly, the expression level of jasmonate-regulated genes involved in nicotine biosynthesis was increased in these plants, despite the lack of an increase in the JA–isoleucine conjugate level known to regulate the expression of the mentioned genes. Our previous attempts to manipulate the jasmonate branch via overexpression of *AtAOS* in Arabidopsis led to the silencing of the pathway and the generation of sterile plants [12], a clear physiological response in plants depleted of jasmonates [5,60]. In another research, an attempt to overexpress *AtAOS* in tobacco resulted in the reduction of basal JA level by nearly 10-fold [61]. For transgenic wheat plants overexpressing allene oxide cyclase *TaAOC1*, another gene of the jasmonate biosynthesis pathway, authors reported the increase in wounding-induced levels of JA, while the basal levels of the hormone remained unaltered [13]. The dimerization of OPR proteins leading to the inhibition of enzymatic activity could be another possible explanation for the decrease in JA level upon *AtOPR3* overexpression in wheat, as shown previously for OPR3 from tomato [62]. The jasmonate level in plant tissue is controlled by the intricate regulatory system with positive and negative feedback loops [63–65]. Reduction of cyclopentenone catalyzed by AtOPR3 is also a highly regulated process [19]. In wheat, the regulation of the pathway activity may be even more complicated, due to polyploidy [66]. Apparently, jasmonate pathway-related genes are represented by large gene families in hexaploid wheat. According to WheatExp, a homologue-specific database of gene expression profiles [67], there are 12 expressed allene oxide synthase genes; based on the available published data, there are 18 COI genes [68] and 14 JASMONATE-ZIM DOMAIN transcription repressor genes in the wheat genome [69]. Further expression profiling study of the generated transgenic lines may shed light onto the jasmonate-dependent regulation of homolog gene expression in wheat.

Altogether, the results obtained in this study prove that the modulation of jasmonate pathway activity can be used as a tool for the modification of developmental characteristics and tolerance to freezing in wheat. Constitutive *AtOPR3* expression causes a pleiotropic phenotype in transgenic wheat, including the manifestation of beneficial traits and characteristics, which could be disadvantageous to wheat growers. On the one hand, transgenic plants with high *AtOPR3* expression levels are characterized by freezing tolerance and reduced height, which is usually associated with increased resistance against lodging, and enables plants to support high grain yields [70]. On the one hand, these plants display late germination and slower developmental phenotypes, which under field conditions could lead to plant exposure to adverse temperatures and water deficiency during late development [71]. The use of development stage-specific and inducible promoters [72,73] will allow the desired combination of characteristics to be achieved through the fine-tuning of the activity of the jasmonate-based system.

4. Materials and Methods

4.1. Plants and Growth Conditions

Non-transgenic and transgenic plants of spring bread wheat 'Saratovskaya-60' (Sar-60) (*Triticum aestivum* L. 2n = 6x = 42, ABD genome) were grown at greenhouse conditions (25 ± 2 °C during the day and 20 ± 2 °C at night). Plants were grown in potted soil under a 16-h photoperiod with additional lighting during the winter period for providing light intensity of up to 150 μmol/m^{2}s. For the germination test and segregation analysis, mature seeds or embryos of immature seeds were plated under sterile conditions on Murashige and Skoog medium, or on wet filter paper in a Petri dish.

4.2. Generation of Transgenic Wheat Plants

The full-length coding sequence for the *AtOPR3* gene (AT2G06050, UniProt ID Q9FUP0) (including the sequence for peroxisome targeting signal) was obtained by reverse transcription polymerase chain reaction (RT-PCR) using total RNA isolated from *Arabidopsis thaliana* leaves as a template. *Sma*I and *Sac*I restriction sites were added to gene-specific primer sequences for cloning (Table S2). The *AtOPR3* gene sequence was introduced into pUC19-based vectors between the *Ubi1* promoter from maize and the *Tnos* terminator from *Agrobacterium tumefaciens*. the fragment with the sequence of the target gene with regulatory sequences was then cut out using *Pvu*II, and the PUbi-*AtOPR3*-Tnos fragment was transferred into the *Sma*I site of the psGFP-BAR vector [74], containing two selective marker genes, *GFP* and *BAR*, coding respectively for green fluorescent protein, and phosphinothricin acetyl transferase, conferring resistance to herbicide phosphinotricin. It was established in our previous work that wheat transformation with the psGFP-BAR vector does not result in any phenotypical alterations [27]. For all constructs, sequences were checked after cloning to ensure fidelity and integrity. The generated construct pBAR-GFP.UbiOPR3 was used for biolistic delivery by a particle inflow gun (PIG) into cells of wheat morphogenic callus induced from immature embryos. The vector DNA was bombarded into 377 immature embryo-derived calli of Sar-60 to generate primary transgenic wheat plants (T0). Primary transformants (T0) were selected on medium supplemented with herbicide glufosinate ammonium, and by monitoring the fluorescence of GFP, as described [75]. The herbicide-resistant plantlets with a height of at least 10 cm were transferred to the soil, as described [76], and grown in an environmentally controlled greenhouse, as described in Section 4.1.

Leaf samples collected from plants at heading stage were analyzed for the presence of the insert by PCR. The expression of the transferred genes was confirmed by end-point RT-PCR using RNA isolated from the studied plants as a template. RNA isolation and reverse transcription was carried out as described below (Section 4.4). The primers used for the target sequences amplification are shown in Table S2.

T1 and T2 progenies of *AtOPR3* overexpressing plants produced by self-pollination of primary T0 transgenic plants were assessed for transgene inheritance. To speed up the segregation analysis and the selection of homozygous transgenic progeny, we analyzed T0→T1→T2 plants, as was described earlier [27,75]. Zygotic T1 embryos were isolated from immature T0 spikes, placed on germination medium for 7–9 days, and then GFP fluorescence was monitored in the seedlings using a ZEISS SteREO Discovery.V12 microscope equipped with a PentaFluar S 120 vertical illuminator. Two commercially available filter sets 38 GFP BP (EX BP 470/40, BS FT 495, EM BP 525/50) and 57 GFP BP (EX BP 470/40, BS FT 495, EM LP 550) (Carl Zeiss MicroImaging GmbH, Jena, Germany) were used to examine GFP expression. Seedlings with a high level of GFP fluorescence were transferred to soil and placed in greenhouse conditions where mature plants were grown. To identify homozygous plants, the pollen from individual mature T1 plants and immature embryos were analyzed for GFP fluorescence (Figure S1A,B), as described previously [27]. Individual homozygous T1 plants were grown in a greenhouse to produce T2 homozygous wheat seeds. T2 homozygous transgenic plants and the T3 progeny obtained from T2 transgenic plants were used for further analysis.

4.3. DNA Extraction and PCR Analysis

Total genomic DNA was extracted from leaf tissues at heading stage, in accordance with the method described [77]. PCR analysis was carried in a thermal cycler, MJ Mini (BIO-RAD, Hercules, CA, USA). Each PCR reaction mixture of 25 μL consisted of 2.5 μL of 10 × buffer for Encyclo DNA polymerase (Evrogen, Moscow, Russia), 0.5 μL of 10 mM dNTPs, 1.25 μL of reverse and forward primers each at 10 μM, 0.5 μL of Encyclo DNA polymerase (Evrogen, Russia), 1 μL DMSO (≥99.5%), 90 ng (1.5 μL) of DNA template, and 16.5 μL of deionized water. Primers are listed in Table S2.

4.4. RNA Extraction and Gene Expression Analysis

Real-time quantitative reverse transcription PCR was used for the analysis of the expression of selected genes: *ALLENE OXIDE SYNTHASE* (*TaAOS*) (GenBank: KJ001800.1), *12-OPDA REDUCTASE 2* (*TaOPR2*) (TraesCS7B02G311600), and *CORONATINE INSENSITIVE 1-LIKE* (*TaCOI-1*) (TraesCS1A02G279100). Gene sequences were obtained from the GenBank and Ensemble Plants databases [78]. Total RNA from the second leaf of T3 plants at the four-leaf stage was isolated by Extract RNA reagent (Evrogen, Moscow, Russia), and further purified using the Clean RNA Standard kit (Evrogen, Moscow, Russia) with on-column DNase I treatment (Thermo Fisher Scientific, Welhem, MA, USA) to eliminate DNA contamination. RNA was reverse-transcribed using the MMLV Reverse Transcriptase kit (Eurogen, Russia). Real-time quantitative PCR was conducted in 20 μL reactions containing complementary DNA (cDNA) synthesized from 10 ng of total RNA, SYBR Green I qPCR Master Mix, (Syntol, Moscow, Russia), and 200 nM for each primer. Amplification was performed using Roche LightCycler®96 System according to the manufacturer's instructions. *TaWIN1* and *TaUbi* genes were used as reference genes for the internal controls, as previously described for transcript normalization [26]. Primers are listed in Table S2.

4.5. Jasmonic Acid Extraction and Quantification

The second leaves of homozygous T3 plants at the four-leaf stage were used for the analysis of JA level. Jasmonic acid extraction was carried out similarly to the procedure previously described [8]. Dihydrojasmonic acid (Merck KGaA, Darmstadt, Germany) was used as an internal standard. The formed methyl ester derivatives of jasmonic acid (JA-ME) and dehydrojasmonic acid were analyzed on a gas chromatograph Bruker 436-GC coupled with a mass spectrometer detector Bruker Scion SQ operating in electron ionization mode. One microliter volume samples were injected by autosampler Bruker CP-8400 in splitless mode at 250 °C, and separated on a column DB-5MS UI (length 30 m, inner diameter 0.25 mm, film thickness 0.25 μm) with helium being used as a carrier gas (constant flow

0.7 mL/min). Oven temperature programming was as follows: hold at 40 °C for 1 min after injection, ramp at 15 °C/min to 150 °C, then increase at 10 °C/min to 250 °C, hold for 10 min. Mass spectral analysis was done in selective ion monitoring mode (SIM). The fragment ions monitored were as follows: JA-ME 224, 151, 83; dihydro-JA-ME 96, 83. The Bruker Workstation 8.2.1. program was used for the data analysis.

4.6. Electrolyte Leakage

The second leaves of homozygous T2 plants at the four-leaf stage were used for the analysis. A total of 2 g of leaves cut into 4 cm pieces were rinsed with deionized water, placed into 50 mL of deionized water, and left on an orbital shaker at room temperature. The conductivity of the solution was measured after 20 h of gentle shaking using the conductivity meter OK-102/1 (Radelkis, Budapest, Hungary). Total conductivity was determined on the same samples after they were autoclaved for 30 min at 121 °C. The samples were cooled down to room temperature prior to measurements. Results were expressed as a percentage of total conductivity.

4.7. Chlorophyll Fluorescence Measurements

Light-induced chlorophyll fluorescence measurements were carried out on leaves of growing plants, or on detached leaves, using a FluorPen FP 100 fluorimeter (Photon Systems Instruments, Drasov, Czech Republic). Photosystem II operating efficiency was measured in light-adapted samples according to the FluorPen FP 100 fluorimeter manual, and calculated according to equations: $Y(II) = (F'm - Fs)/F'm$, where Fs—stationary level and F'm—light-induced maximum level of chlorophyll fluorescence in light-adapted leaves [79]. Taking into account the heterogeneity in photosynthetic parameters within the leaf blade, all measurements were carried out in the area 10 cm away from the tip of the leaf blade. Light-induced chlorophyll fluorescence measurements were performed on the flag and fourth leaves of homozygous T2 plants at the maturation stage to estimate senescence. For the detached leaf senescence analysis, the second leaves of homozygous T2 plants at the four-leaf stage were used.

4.8. Exogenous Methyl Jasmonate Treatment

For exogenous methyl jasmonate treatment, 100 μM solution of MeJA was applied to filter paper lining Petri dishes with germinating seeds of Sar-60.

4.9. Freezing-Tolerance Test

Three-week-old homozygous T2 plants at the four-leaf stage growing in pots were placed into a growth chamber at 4 °C for 24 h for cold acclimation. Then, leaves of the low temperature-adapted plants were cut, wrapped into plastic bags, and placed into a chilled vacuum-insulated vessel. Leaves in the vessel were subjected to decreasing temperatures for 30, 60, or 90 min (the temperature was controlled by a wireless thermometer with a remote sensor; the decrease rate was 1 °C/10 min), then leaves in the vacuum-insulated vessel were placed back at 4 °C for another 24 h prior to assessment of leaf damage caused by the freezing treatment by the means of light-induced chlorophyll fluorescence and electrolyte leakage measurements.

4.10. Statistical Analysis

To determine statistically significant differences between genotypes, a one-way analysis of variance (ANOVA) followed by post hoc comparisons by Tukey's test and the Student's t-test for independent means were performed.

Supplementary Materials: Supplementary materials can be found at http://www.mdpi.com/1422-0067/19/12/3989/s1.

Author Contributions: T.V.S., S.V.D., A.V.P., and D.N.M. conceived the research, A.V.P., D.N.M., A.S.P., V.V.T., A.M.B., and T.V.S. carried out the experiments, T.V.S. wrote the manuscript with support from A.V.P. and D.N.M.

Funding: This work was supported by the Russian Science Foundation, grant No 16-14-10155 for T.S., the results presented in Figure 2 were obtained with support from the Ministry of Education and Science of the Russian Federation (theme AAAA-A17-117030110136-8).

Acknowledgments: We thank Alexey Nezvetckiy (ANO "Test-Pushchino") for the analysis of jasmonates.

Conflicts of Interest: The authors declare no conflict of interest.

Abbreviations

13-HPOT	13(S)-hydroperoxy-9,11,15-octadecatrienoic acid
12-OPDA	(9S,13S)-12-oxo-phytodienoic acid
JA	Jasmonic acid
MeJA	Methyl jasmonate

References

1. Wasternack, C.; Song, S. Jasmonates: Biosynthesis, metabolism, and signaling by proteins activating and repressing transcription. *J. Exp. Bot.* **2017**, *68*, 1303–1321. [CrossRef]
2. Wasternack, C.; Strnad, M. Jasmonates: News on occurrence, biosynthesis, metabolism and action of an ancient group of signaling compounds. *Int. J. Mol. Sci.* **2018**, *19*, 2539. [CrossRef]
3. Schaller, A.; Stintzi, A. Enzymes in jasmonate biosynthesis—Structure, function, regulation. *Phytochemistry* **2009**, *70*, 1532–1538. [CrossRef]
4. Feussner, I.; Wasternack, C. The lipoxygenase pathway. *Annu. Rev. Plant Biol.* **2002**, *53*, 275–297. [CrossRef] [PubMed]
5. Stintzi, A.; Browse, J. The Arabidopsis male-sterile mutant, opr3, lacks the 12-oxophytodienoic acid reductase required for jasmonate synthesis. *Proc. Natl. Acad. Sci. USA* **2000**, *97*, 10625–10630. [CrossRef] [PubMed]
6. Lyons, R.; Manners, J.M.; Kazan, K. Jasmonate biosynthesis and signaling in monocots: A comparative overview. *Plant Cell Rep.* **2013**, *32*, 815–827. [CrossRef]
7. Afzal, F.; Chaudhari, S.K.; Gul, A.; Farooq, A.; Ali, H.; Nisar, S.; Sarfraz, B.; Shehzadi, K.J.; Mujeeb-Kazi, A. Bread wheat (*Triticum aestivum* L.) under biotic and abiotic stresses: An overview. In *Crop Production and Global Environmental Issues*; Hakeem, K., Ed.; Springer International Publishing: Cham, Switzerland, 2015; pp. 293–317.
8. Chehab, E.W.; Kaspi, R.; Savchenko, T.; Rowe, H.; Negre-Zakharov, F.; Kliebenstein, D.; Dehesh, K. Distinct roles of jasmonates and aldehydes in plant-defense responses. *PLoS ONE* **2008**, *3*, e1904. [CrossRef]
9. Savchenko, T.; Dehesh, K. Insect herbivores selectively mute GLV production in plants. *Plant Signal. Behav.* **2013**, *8*, e24136. [CrossRef]
10. Savchenko, T.; Yanykin, D.; Khorobrykh, A.; Terentyev, V.; Klimov, V.; Dehesh, K. The hydroperoxide lyase branch of the oxylipin pathway protects against photoinhibition of photosynthesis. *Planta* **2017**, *245*, 1179–1192. [CrossRef] [PubMed]
11. Xiao, Y.; Savchenko, T.; Baidoo, E.E.; Chehab, W.E.; Hayden, D.M.; Tolstikov, V.; Corwin, J.A.; Kliebenstein, D.J.; Keasling, J.D.; Dehesh, K. Retrograde signaling by the plastidial metabolite MEcPP regulates expression of nuclear stress-response genes. *Cell* **2012**, *149*, 1525–1535. [CrossRef] [PubMed]
12. Savchenko, T.; Kolla, V.A.; Wang, C.Q.; Nasafi, Z.; Hicks, D.R.; Phadungchob, B.; Chehab, W.E.; Brandizzi, F.; Froehlich, J.; Dehesh, K. Functional convergence of oxylipin and abscisic acid pathways controls stomatal closure in response to drought. *Plant Physiol.* **2014**, *164*, 1151–1160. [CrossRef]
13. Zhao, Y.; Dong, W.; Zhang, N.; Ai, X.; Wang, M.; Huang, Z.; Xiao, L.; Xia, G. A wheat allene oxide cyclase gene enhances salinity tolerance via jasmonate signaling. *Plant Physiol.* **2014**, *164*, 1068–1076. [CrossRef] [PubMed]
14. Sharma, V.K.; Monostori, T.; Gobel, C.; Hansch, R.; Bittner, F.; Wasternack, C.; Feussner, I.; Mendel, R.R.; Hause, B.; Schulze, J. Transgenic barley plants overexpressing a 13-lipoxygenase to modify oxylipin signature. *Phytochemistry* **2006**, *67*, 264–276. [CrossRef]

15. Qiu, Z.; Guo, J.; Zhu, A.; Zhang, L.; Zhang, M. Exogenous jasmonic acid can enhance tolerance of wheat seedlings to salt stress. *Ecotoxicol. Environ. Saf.* **2014**, *104*, 202–208. [CrossRef] [PubMed]

16. Liu, H.; Carvalhais, L.C.; Schenk, P.M.; Dennis, P.G. Effects of jasmonic acid signalling on the wheat microbiome differ between body sites. *Sci. Rep.* **2017**, *7*, 41766. [CrossRef]

17. Ahmad, P.; Rasool, S.; Gul, A.; Sheikh, S.A.; Akram, N.A.; Ashraf, M.; Kazi, A.M.; Gucel, S. Jasmonates: Multifunctional roles in stress tolerance. *Front. Plant Sci.* **2016**, *7*, 813. [CrossRef]

18. Alikina, O.; Chernobrovkina, M.; Dolgov, S.; Miroshnichenko, D. Tissue culture efficiency of wheat species with different genomic formulas. *Crop Breed. Appl. Biotechnol.* **2016**, *16*, 307–314. [CrossRef]

19. Strassner, J.; Schaller, F.; Frick, U.B.; Howe, G.A.; Weiler, E.W.; Amrhein, N.; Macheroux, P.; Schaller, A. Characterization and cDNA-microarray expression analysis of 12-oxophytodienoate reductases reveals differential roles for octadecanoid biosynthesis in the local versus the systemic wound response. *Plant J.* **2002**, *32*, 585–601. [CrossRef]

20. Chehab, E.W.; Kim, S.; Savchenko, T.; Kliebenstein, D.; Dehesh, K.; Braam, J. Intronic T-DNA insertion renders Arabidopsis opr3 a conditional jasmonic acid-producing mutant. *Plant Physiol.* **2011**, *156*, 770–778. [CrossRef]

21. Wasternack, C.; Hause, B. A bypass in jasmonate biosynthesis—The OPR3-independent formation. *Trends Plant Sci.* **2018**, *23*, 276–279. [CrossRef]

22. Stintzi, A.; Weber, H.; Reymond, P.; Browse, J.; Farmer, E.E. Plant defense in the absence of jasmonic acid: The role of cyclopentenones. *Proc. Natl. Acad. Sci. USA* **2001**, *98*, 12837–12842. [CrossRef] [PubMed]

23. Zheng, H.; Pan, X.; Deng, Y.; Wu, H.; Liu, P.; Li, X. AtOPR3 specifically inhibits primary root growth in Arabidopsis under phosphate deficiency. *Sci. Rep.* **2016**, *6*, 24778. [CrossRef] [PubMed]

24. Han, B.W.; Malone, T.E.; Kim, D.J.; Bingman, C.A.; Kim, H.-J.; Fox, B.G.; Phillips, G.N. Crystal structure of Arabidopsis thaliana 12-oxophytodienoate reductase isoform 3 in complex with 8-iso prostaglandin A(1). *Proteins* **2011**, *79*, 3236–3241. [CrossRef] [PubMed]

25. Gagarinsky, E.L.; Stepanov, S.A.; Signaevsky, V.D. Microevolution of elements productivity of shoot spring soft wheat of the Saratov selection. *Izv. Saratov Univ. New Ser. Ser. Earth Sci.* **2015**, *13*, 171–181. (In Russian)

26. Tenea, G.N.; Peres Bota, A.; Cordeiro Raposo, F.; Maquet, A. Reference genes for gene expression studies in wheat flag leaves grown under different farming conditions. *BMC Res. Notes* **2011**, *4*, 373. [CrossRef] [PubMed]

27. Miroshnichenko, D.N.; Poroshin, G.N.; Dolgov, S.V. Genetic transformation of wheat using mature seed tissues. *Appl. Biochem. Microbiol.* **2011**, *47*, 767–775. [CrossRef]

28. Feller, U.; Keist, M. Senescence and nitrogen metabolism in annual plants. In *Fundamental, Ecological and Agricultural Aspects of Nitrogen Metabolism in Higher Plants. Developments in Plant and Soil Sciences*; Lambers, H., Neeteson, J.J., Stulen, I., Eds.; Springer: Dordrecht, The Netherlands, 1986; Volume 19, pp. 219–234.

29. Liu, H.; Carvalhais, L.C.; Kazan, K.; Schenk, P.M. Development of marker genes for jasmonic acid signaling in shoots and roots of wheat. *Plant Signal. Behav.* **2016**, *11*, e1176654. [CrossRef]

30. Wang, Y.; Yuan, G.; Yuan, S.; Duan, W.; Wang, P.; Bai, J.; Zhang, F.; Gao, S.; Zhang, L.; Zhao, C. TaOPR2 encodes a 12-oxo-phytodienoic acid reductase involved in the biosynthesis of jasmonic acid in wheat (*Triticum aestivum* L.). *Biochem. Biophys. Res. Commun.* **2016**, *470*, 233–238. [CrossRef]

31. Liu, X.; Wang, J.; Fan, B.; Shang, W.; Sun, Y.; Dang, C.; Xie, C.; Wang, Z.; Peng, Y. A *COI1* gene in wheat contributes to the early defence response against wheat powdery mildew. *J. Phytopathol.* **2018**, *166*, 116–122. [CrossRef]

32. Wasternack, C. Jasmonates: An update on biosynthesis, signal transduction and action in plant stress response, growth and development. *Ann. Bot.* **2007**, *100*, 681–697. [CrossRef]

33. Creelman, R.A.; Mullet, J.E. Biosynthesis and action of jasmonates in plants. *Annu. Rev. Plant Physiol. Plant Mol. Biol.* **1997**, *48*, 355–381. [CrossRef] [PubMed]

34. Noir, S.; Bomer, M.; Takahashi, N.; Ishida, T.; Tsui, T.L.; Balbi, V.; Shanahan, H.; Sugimoto, K.; Devoto, A. Jasmonate controls leaf growth by repressing cell proliferation and the onset of endoreduplication while maintaining a potential stand-by mode. *Plant Physiol.* **2013**, *161*, 1930–1951. [CrossRef] [PubMed]

35. Chehab, E.W.; Yao, C.; Henderson, Z.; Kim, S.; Braam, J. Arabidopsis touch-induced morphogenesis is jasmonate mediated and protects against pests. *Curr. Biol.* **2012**, *22*, 701–706. [CrossRef] [PubMed]

36. Liu, X.; Li, F.; Tang, J.; Wang, W.; Zhang, F.; Wang, G.; Chu, J.; Yan, C.; Wang, T.; Chu, C.; et al. Activation of the jasmonic acid pathway by depletion of the hydroperoxide lyase OsHPL3 reveals crosstalk between the HPL and AOS branches of the oxylipin pathway in rice. *PLoS ONE* **2012**, *7*, e50089. [CrossRef] [PubMed]

37. Kim, E.H.; Kim, Y.S.; Park, S.H.; Koo, Y.J.; Choi, Y.D.; Chung, Y.Y.; Lee, I.J.; Kim, J.K. Methyl jasmonate reduces grain yield by mediating stress signals to alter spikelet development in rice. *Plant Physiol.* **2009**, *149*, 1751–1760. [CrossRef] [PubMed]

38. Riemann, M.; Muller, A.; Korte, A.; Furuya, M.; Weiler, E.W.; Nick, P. Impaired induction of the jasmonate pathway in the rice mutant hebiba. *Plant Physiol.* **2003**, *133*, 1820–1830. [CrossRef] [PubMed]

39. Cai, Q.; Yuan, Z.; Chen, M.; Yin, C.; Luo, Z.; Zhao, X.; Liang, W.; Hu, J.; Zhang, D. Jasmonic acid regulates spikelet development in rice. *Nat. Commun.* **2014**, *5*, 3476. [CrossRef]

40. Hibara, K.; Isono, M.; Mimura, M.; Sentoku, N.; Kojima, M.; Sakakibara, H.; Kitomi, Y.; Yoshikawa, T.; Itoh, J.; Nagato, Y. Jasmonate regulates juvenile-to-adult phase transition in rice. *Development* **2016**, *143*, 3407–3416. [CrossRef] [PubMed]

41. Xue, R.; Zhang, B. Increased endogenous methyl jasmonate altered leaf and root development in transgenic soybean plants. *J. Genet. Genom.* **2007**, *34*, 339–346. [CrossRef]

42. Maciejewska, B.; Kopcewicz, J. Inhibitory effect of methyl jasmonate on flowering and elongation growth in Pharbitis nil. *J. Plant. Growth. Regul.* **2002**, *21*, 216–223. [CrossRef]

43. Albrechtova, J.T.P.; Ullmann, J. Methyl jasmonate inhibits growth and flowering in *Chenopodium-Rubrum*. *Biol. Plant.* **1994**, *36*, 317–319. [CrossRef]

44. Sohn, H.B.; Lee, H.Y.; Seo, J.S.; Jung, C.; Jeon, J.H.; Kim, J.H.; Lee, Y.W.; Lee, J.S.; Cheong, J.J.; Do Choi, Y. Overexpression of jasmonic acid carboxyl methyltransferase increases tuber yield and size in transgenic potato. *Plant Biotechnol. Rep.* **2011**, *5*, 27–34. [CrossRef]

45. Diallo, A.O.; Agharbaoui, Z.; Badawi, M.A.; Ali-Benali, M.A.; Moheb, A.; Houde, M.; Sarhan, F. Transcriptome analysis of an mvp mutant reveals important changes in global gene expression and a role for methyl jasmonate in vernalization and flowering in wheat. *J. Exp. Bot.* **2014**, *65*, 2271–2286. [CrossRef] [PubMed]

46. Wingler, A.; Mares, M.; Pourtau, N. Spatial patterns and metabolic regulation of photosynthetic parameters during leaf senescence. *New Phytol.* **2004**, *161*, 781–789. [CrossRef]

47. Hu, Y.; Jiang, Y.; Han, X.; Wang, H.; Pan, J.; Yu, D. Jasmonate regulates leaf senescence and tolerance to cold stress: Crosstalk with other phytohormones. *J. Exp. Bot.* **2017**, *68*, 1361–1369. [CrossRef] [PubMed]

48. Weidhase, R.A.; Kramell, H.M.; Lehmann, J.; Liebisch, H.W.; Lerbs, W.; Parthier, B. Methyl jasmonate-induced changes in the polypeptide pattern of senescing barley leaf segments. *Plant Sci.* **1987**, *51*, 177–186. [CrossRef]

49. Kazan, K. Diverse roles of jasmonates and ethylene in abiotic stress tolerance. *Trends Plant Sci.* **2015**, *20*, 219–229. [CrossRef]

50. Walia, H.; Wilson, C.; Condamine, P.; Liu, X.; Ismail, A.M.; Close, T.J. Large-scale expression profiling and physiological characterization of jasmonic acid-mediated adaptation of barley to salinity stress. *Plant Cell Environ.* **2007**, *30*, 410–421. [CrossRef]

51. Sharma, M.; Laxmi, A. Jasmonates: Emerging players in controlling temperature stress tolerance. *Front. Plant Sci.* **2016**, *6*, 1129. [CrossRef]

52. Wang, S.Y. Methyl jasmonate reduces water stress in Strawberry. *J. Plant Growth Regul.* **1999**, *18*, 127–134. [CrossRef]

53. Yoshikawa, H. Effect of low-temperature stress on abscisic acid, jasmonates, and polyamines in apples. *Plant Growth Regul.* **2007**, *52*, 199–206. [CrossRef]

54. Ding, C.K.; Wang, C.Y.; Gross, K.C.; Smith, D.L. Jasmonate and salicylate induce the expression of pathogenesis-related-protein genes and increase resistance to chilling injury in tomato fruit. *Planta* **2002**, *214*, 895–901. [CrossRef] [PubMed]

55. Kondo, S.; Jitratham, A.; Kittikorn, M.; Kanlayanarat, S. Relationships between jasmonates and chilling injury in mangosteens are affected by spermine. *HortScience* **2004**, *39*, 1346–1348.

56. Pedranzani, H.; Sierra-de-Grado, R.; Vigliocco, A.; Miersch, O.; Abdala, G. Cold and water stresses produce changes in endogenous jasmonates in two populations of *Pinus pinaster* Ait. *Plant Growth Regul.* **2007**, *52*, 111–116. [CrossRef]

57. Hannah, M.A.; Wiese, D.; Freund, S.; Fiehn, O.; Heyer, A.G.; Hincha, D.K. Natural genetic variation of freezing tolerance in Arabidopsis. *Plant Physiol.* **2006**, *142*, 98–112. [CrossRef] [PubMed]

58.   Hu, Y.; Jiang, L.; Wang, F.; Yu, D. jasmonate regulates the inducer of cbf expression-C-repeat binding factor/DRE binding factor1 cascade and freezing tolerance in Arabidopsis. *Plant Cell* **2013**, *25*, 2907–2924. [CrossRef]

59.   Chen, H.; Wang, B.; Geng, S.; Arellano, C.; Chen, S.; Qu, R. Effects of overexpression of jasmonic acid biosynthesis genes on nicotine accumulation in tobacco. *Plant Direct* **2018**, *2*, e00036. [CrossRef]

60.   Park, J.H.; Halitschke, R.; Kim, H.B.; Baldwin, I.T.; Feldmann, K.A.; Feyereisen, R. A knock-out mutation in allene oxide synthase results in male sterility and defective wound signal transduction in Arabidopsis due to a block in jasmonic acid biosynthesis. *Plant J.* **2002**, *31*, 1–12. [CrossRef]

61.   Laudert, D.; Schaller, F.; Weiler, E.W. Transgenic Nicotiana tabacum and Arabidopsis thaliana plants overexpressing allene oxide synthase. *Planta* **2000**, *211*, 163–165. [CrossRef]

62.   Breithaupt, C.; Kurzbauer, R.; Lilie, H.; Schaller, A.; Strassner, J.; Huber, R.; Macheroux, P.; Clausen, T. Crystal structure of 12-oxophytodienoate reductase 3 from tomato: Self-inhibition by dimerization. *Proc. Natl. Acad. Sci. USA* **2006**, *103*, 14337–14342. [CrossRef]

63.   Hickman, R.; Van Verk, M.C.; Van Dijken, A.J.H.; Mendes, M.P.; Vroegop-Vos, I.A.; Caarls, L.; Steenbergen, M.; Van der Nagel, I.; Wesselink, G.J.; Jironkin, A.; et al. Architecture and dynamics of the jasmonic acid gene regulatory network. *Plant Cell* **2017**, *29*, 2086–2105. [CrossRef] [PubMed]

64.   Woldemariam, M.G.; Onkokesung, N.; Baldwin, I.T.; Galis, I. Jasmonoyl-L-isoleucine hydrolase 1 (JIH1) regulates jasmonoyl-L-isoleucine levels and attenuates plant defenses against herbivores. *Plant J.* **2012**, *72*, 758–767. [CrossRef] [PubMed]

65.   Abdelrahman, M.; Suzumura, N.; Mitoma, M.; Matsuo, S.; Ikeuchi, T.; Mori, M.; Murakami, K.; Ozaki, Y.; Matsumoto, M.; Uragami, A.; et al. Comparative de novo transcriptome profiles in *Asparagus officinalis* and *A. kiusianus* during the early stage of *Phomopsis asparagi* infection. *Sci. Rep.* **2017**, *7*, 2608. [CrossRef] [PubMed]

66.   Powell, J.J.; Fitzgerald, T.L.; Stiller, J.; Berkman, P.J.; Gardiner, D.M.; Manners, J.M.; Henry, R.J.; Kazan, K. The defence-associated transcriptome of hexaploid wheat displays homoeolog expression and induction bias. *Plant Biotechnol. J.* **2017**, *15*, 533–543. [CrossRef] [PubMed]

67.   Pearce, S.; Vazquez-Gross, H.; Herin, S.Y.; Hane, D.; Wang, Y.; Gu, Y.Q.; Dubcovsky, J. WheatExp: An RNA-seq expression database for polyploid wheat. *BMC Plant Biol.* **2015**, *15*, 299. [CrossRef]

68.   Bai, J.F.; Wang, Y.K.; Wang, P.; Yuan, S.H.; Gao, J.G.; Duan, W.J.; Wang, N.; Zhang, F.T.; Zhang, W.J.; Qin, M.Y.; et al. Genome-wide identification and analysis of the COI gene family in wheat (*Triticum aestivum* L.). *BMC Genom.* **2018**, *19*, 754. [CrossRef] [PubMed]

69.   Wang, Y.; Qiao, L.; Bai, J.; Wang, P.; Duan, W.; Yuan, S.; Yuan, G.; Zhang, F.; Zhang, L.; Zhao, C. Genome-wide characterization of JASMONATE-ZIM DOMAIN transcription repressors in wheat (*Triticum aestivum* L.). *BMC Genom.* **2017**, *18*, 152. [CrossRef] [PubMed]

70.   Piñera-Chavez, F.J.; Berry, P.M.; Foulkes, M.J.; Molero, G.; Reynolds, M.P. Avoiding lodging in irrigated spring wheat. II. Genetic variation of stem and root structural properties. *Field Crop. Res.* **2016**, *196*, 64–74. [CrossRef]

71.   Harris, D.; Joshi, A.; Khan, P.A.; Gothkar, P.; Sodhi, P.S. On-farm seed priming in semi-arid agriculture: Development and evaluation in maize, rice and chickpea in India using participatory methods. *Exp. Agric.* **1999**, *35*, 15–29. [CrossRef]

72.   Tang, W.; Luo, X.; Samuels, V. Regulated gene expression with promoters responding to inducers. *Plant Sci.* **2004**, *166*, 827–834. [CrossRef]

73.   Muthusamy, S.K.; Dalal, M.; Chinnusamy, V.; Bansal, K.C. Differential regulation of genes coding for organelle and cytosolic ClpATPases under biotic and abiotic stresses in wheat. *Front. Plant Sci.* **2016**, *7*, 9. [CrossRef] [PubMed]

74.   Richards, H.A.; Rudas, V.A.; Sun, H.; McDaniel, J.K.; Tomaszewski, Z.B.; Conger, V. Construction of a GFP-BAR plasmid and its use for switchgrass transformation. *Plant Cell Rep.* **2001**, *20*, 48–54. [CrossRef]

75.   Miroshnichenko, D.; Ashin, D.; Pushin, A.; Dolgov, S. Genetic transformation of einkorn (*Triticum monococcum* L. ssp. *monococcum* L.), a diploid cultivated wheat species. *BMC Biotechnol.* **2018**, *18*, 68. [CrossRef] [PubMed]

76.   Miroshnichenko, D.; Chernobrovkina, M.; Dolgov, S. Somatic embryogenesis and plant regeneration from immature embryos of *Triticum timopheevii* Zhuk. and *Triticum kiharae* Dorof. et Migusch, wheat species with G. genome. *Plant Cell Tiss. Org. Cult.* **2016**, *125*, 495–508. [CrossRef]

77.   Rogers, S.O.; Bendich, A.J. Extraction of DNA from milligram amounts of fresh herbarium and mummified plant tissues. *Plant Mol. Biol.* **1985**, *5*, 69–76. [CrossRef] [PubMed]

78. Bolser, D.M.; Staines, D.M.; Perry, E.; Kersey, P.J. Ensembl Plants: Integrating tools for visualizing, mining, and analyzing plant genomic data. *Methods Mol. Biol.* **2017**, *1533*, 1–31. [CrossRef]

79. Genty, B.; Harbinson, J.; Cailly, A.; Rizza, F. *Fate of Excitation at PS II in Leaves: The Non-Photochemical Side. Proceedings of the Third BBSRC Robert Hill Symposium on Photosynthesis, 31 March–3 April 1996*; University of Sheffield, Department of Molecular Biology and Biotechnology, Western Bank: Sheffield, UK, 1996; Abstract no. P28.

© 2018 by the authors. Licensee MDPI, Basel, Switzerland. This article is an open access article distributed under the terms and conditions of the Creative Commons Attribution (CC BY) license (http://creativecommons.org/licenses/by/4.0/).

International Journal of
Molecular Sciences

MDPI

Article

Overexpression of a Wheat Aquaporin Gene, *Td*PIP2;1, Enhances Salt and Drought Tolerance in Transgenic Durum Wheat cv. Maali

Malika Ayadi, Faiçal Brini and Khaled Masmoudi *,†

Biotechnology and Plant Improvement Laboratory, Centre of Biotechnology of Sfax (CBS), University of Sfax, B.P 1177, 3018 Sfax, Tunisia; malika.ayadi@cbs.rnrt.tn (M.A.); faical.brini@cbs.rnrt.tn (F.B.)
* Correspondence: khaledmasmoudi@uaeu.ac.ae
† Present address: Arid Land Department, College of Food and Agriculture, United Arab Emirates University, P.O. Box 15551, Al Ain, UAE.

Received: 16 April 2019; Accepted: 9 May 2019; Published: 14 May 2019

Abstract: In this study, we generated transgenic durum wheat cv. Maali overexpressing the wheat plasma membrane aquaporin *Td*PIP2;1 gene under the control of Pr*Td*PIP2;1 promoter or under the constitutive PrCaMV35S promoter. Histochemical analysis of the fusion Pr*Td*PIP2;1::*Td*PIP2;1::GusA in wheat plants showed that the β-glucuronidase (GUS) activity was detected in the leaves, stems and roots of stably transformed wheat T3 plants. Our results showed that transgenic wheat lines overexpressing the *Td*PIP2;1 gene exhibited improved germination rates and biomass production and retained low Na^+ and high K^+ concentrations in their shoots under high salt and osmotic stress conditions. In a long-term study under greenhouse conditions on salt or drought stress, transgenic *Td*PIP2;1 lines produced filled grains, whereas wild-type (WT) plants either died at the vegetative stage under salt stress or showed drastically reduced grain filling under drought stress. Performing real time RT-PCR experiments on wheat plants transformed with the fusion Pr*Td*PIP2;1::GusA, we showed an increase in the accumulation of GusA transcripts in the roots of plants challenged with salt and drought stress. Study of the antioxidant defence system in transgenic wheat *Td*PIP2;1 lines showed that these lines induced the antioxidative enzymes Catalase (CAT) and Superoxide dismutase (SOD) activities more efficiently than the WT plants, which is associated with lower malondialdehyde and hydrogen peroxide contents. Taken together, these results indicate the high potential of the *Td*PIP2;1 gene for reducing water evaporation from leaves (water loss) in response to water deficit through the lowering of transpiration per unit leaf area (stomatal conductance) and engineering effective drought and salt tolerance in transgenic *Td*PIP2;1 lines.

Keywords: abiotic stress; antioxidant enzymes; aquaporin; *Td*PIP2;1; histochemical analysis; transgenic wheat; transpiration

---

1. Introduction

Wheat constitutes the most widely grown and consumed cereal in the world. There is a growing imbalance between its supply and demand. The loss of fertile lands for cereal production due to climate change and water scarcity is considered major obstacles against further increases in yield. Drought stress induces different types of responses in plants [1]. It promotes oxidative damage in chloroplasts [2,3], decreases photosynthesis and metabolic reactions [4–7], induces osmotically active compound (i.e., soluble sugars and glycine betaine) and signal molecules [8–10] and changes in cellular lipid composition [11]. To tolerate drought stress, plants have generated different strategies such as the formation of larger and deeper root systems [12], adjustment of stomatal closure by controlling turgor pressure changes in guard cells to reduce water loss [13], accumulation of organic metabolites of low molecular weight known as compatible solutes and protective proteins [14] and enhancement of antioxidative systems [15].

Water transport across the plasma membrane is very important for the plant cell. Many reports on the relationship between plant aquaporins (AQPs) and plant water have been published [16]. Earlier research has shown that AQPs play a major role in the transportation through the membrane of different physiologically important molecules such as water, glycerol, CO_2 and H_2O_2, as well as assisting with various physiological processes such as drought, salt and chilling responses [17–20]. This suggests that AQPs play an important role in transporting a large quantity of water with the minimum of energy expenditure and seem to regulate the trans-cellular transport of water [17].

AQPs are proteins belonging to the conserved family of major intrinsic proteins (MIPs) [16,21]. There are five subfamilies of plant MIPs and among them the Plasma Membrane Intrinsic Proteins, (PIPs) are the most represented. PIP proteins can be divided into two groups, PIP1 and PIP2. These proteins, when expressed in *Xenopus laevis* oocytes, exhibit different water channel activities and can interact physically, leading to an increase of the osmotic water permeability coefficient (*Pf*) of the oocyte membrane [22–24]. Previous reports on AQPs have demonstrated their contribution to the water permeability of root cortex cells and their involvement in osmotic water transport in the entire root system [25]. AQPs are key players in the transport of sap-assimilated elements into phloem, stomatal closure, control of cellular homeostasis and leaf movement. [21,24,26,27]. During transpiration, AQPs control the transport of water from the roots to the leaves [21,24,26,27]. Transgenic plants overexpressing AQP proteins provide promising strategies to explore the hydraulic conductance in roots and leaves, plant transpiration, stomatal aperture and gas exchange under water deficit.

Plant AQPs play an important role in water relations. Their activities regulation and gene expression, which depend on complex processes at the transcriptional, post-transcriptional, translational and post-translational levels, are considered part of the adaptation mechanisms to environmental constraints. Over the last decade, the physiological contributions of AQPs have been investigated in planta by reverse genetics. Assuming that overexpression or down-regulation of a gene can help identify its function, this kind of approach is of key importance in the field. Such studies have already provided results demonstrating the central role of AQPs in plant physiology [28]. Under abiotic stresses or other physiological or developmental changes, differential expression of AQP transcripts or proteins was observed and reported. In rice, under osmotic stress the *Os*PIP2 transcripts in roots increased substantially [29]. In contrast, under salinity stress, expression of *Os*PIP1;1 in rice increased in leaves but was reduced in roots [30]. In barley and under drought stress, the expression of *Hv*PIP2;1 was reduced in roots and enhanced in shoots [31]. Numerous studies have confirmed that modulating AQP expression in transgenic plants can increase resistance to stresses. Indeed, expression of AQP PIP1 of *Vicia faba* (*Vf*PIP1) in transgenic *Arabidopsis thaliana* enhanced drought resistance by the reduction of transpiration rates through stomatal closure [32]. Recently, it was shown that overexpression of a barley aquaporin gene, *Hv*PIP2;5, confers salt and osmotic stress tolerance in yeast and plants [33]. Moreover, transgenic banana plants overexpressing a native plasma membrane aquaporin *Musa*PIP1;2 present high levels of tolerance to different abiotic stresses [34]. In addition, it was reported that overexpression of *Mf*PIP2-7 from *Medicago falcata* promotes cold tolerance and growth under NO^{-3} deficiency in transgenic tobacco plants [35]. Conversely, the heterologous expression of *Ta*TIP2;2 in *Arabidopsis thaliana*, compromised its drought and salinity tolerance, suggesting that *Ta*TIP2;2 may be a negative regulator to abiotic stress. This was correlated with all down-regulated stress tolerance related genes acting in an ABA-independent manner, such as SOS1, SOS2, SOS3, CBF3 and DREB2A [36]. Moreover, it was reported that plasma membrane aquaporin overexpressed in transgenic tobacco increases plant vigour under favourable growth conditions but not under drought or salt stress [37]. The regulation of AQP activity and gene expression by various developmental and environmental factors, such as salinity and drought stress, relies on complex processes and signalling pathways [38].

In a previous work, we have isolated a durum wheat PIP2 gene, named *Td*PIP2;1 and have characterized its expression in *Xenopus* oocytes [39]. The generated transgenic tobacco plants overexpressing the *TdPIP2;1* gene showed a phenotype of tolerance towards drought and salt stress [39]. To understand this tolerance mechanism at the transcriptional level, we isolated and characterized the promoter region of the *TdPIP2;1* gene. When challenged with drought stress, the transgenic rice plants overexpressing Pr*TdPIP2;1* in fusion with the *TdPIP2;1* gene, showed enhanced drought stress tolerance, while WT plants were more sensitive and exhibited symptoms of wilting and chlorosis. These results suggest that expression of the *TdPIP2;1* gene regulated by its own promoter achieves enhanced drought tolerance in transgenic rice plants [40].

In this study, we ectopically overexpressed the *TdPIP2;1* gene under its own promoter and demonstrated that the promoter region contained all the regulatory elements required to mediate expression in transgenic durum wheat plants cv. Maali, resulting in improved tolerance to salt and drought stresses. We confirmed the role of the *TdPIP2;1* gene in developing tolerant-crops for specific abiotic stress without any penalty phenotype.

2. Results

2.1. Production of Transgenic Wheat TdPIP2;1 Plants

Transgenic wheat plants expressing the *TdPIP2;1* gene under a strong constitutive promoter (SP construct) or the native core promoter Pr*TdPIP2;1* (PR construct), were generated. Eight transgenic lines for each different construct were checked and approved by PCR for their transgenic status. For each construct, three transgenic T3 homozygous lines were selected for the further evaluation of GusA transcript accumulation and the analysis of phenotypic and physiological parameters for drought and salinity stress responses in durum wheat.

2.2. Analysis of GUS Activity in Relation to Stress Treatment

The histochemical staining of GUS activity for whole plants at an early developmental stage (7 day-old seedlings), grown under control conditions, uncovered a coloured product in the leaves and in roots (Figure 1a). To validate the status of stress-inducible expression of the Pr*TdPIP2;1*, GusA transcript accumulation was examined by quantitative real-time qPCR in WT, SP3, PR3 and PR6 lines grown under control conditions or subjected to salt stress treatment (150 mM NaCl) and osmotic stress (20% PEG). The changes in GusA gene expression levels, across multiple samples and stress, were carried out by comparative quantification based on the housekeeping Actin gene as a reference gene. After 48 h of stress treatment, the level of GusA transcripts in the roots was higher in transgenic lines expressing the GusA gene under the native promoter of the *TdPIP2;1* gene (PR6 and PR3), compared to the transgenic SP3 line expressing the GUS gene under the 35S constitutive promoter (Figure 1c). In leaves, the expression level of GusA transcripts was higher in the transgenic PR6 line compared to the two other transgenic PR3 and SP3 lines (Figure 1b). For each of the two responses (relative root expression and relative leaf expression), a Two-way analysis of variance with factors stress type and genotype was carried out. The results show that the genotype effect was highly significant (p-value < 0.001). There was a significant interaction between genotype and stress factors. The Tukey multiple comparisons procedure shows that there was a significant difference between the three genotypes, with the highest expression in roots for PR6, followed by PR3 and then SP3. For salt stress, there was no significant difference in gene expression in roots between the PR6 and the PR3 lines but both of them differed from the SP3 line. Similar results were obtained for drought stress. However, there was a significant difference for PR6 between salt and drought stress (Figure 1c). All these results suggest that the Pr*TdPIP2;1* promoter is an abiotic and a tissue stress inducible promoter.

a)

b)   c)

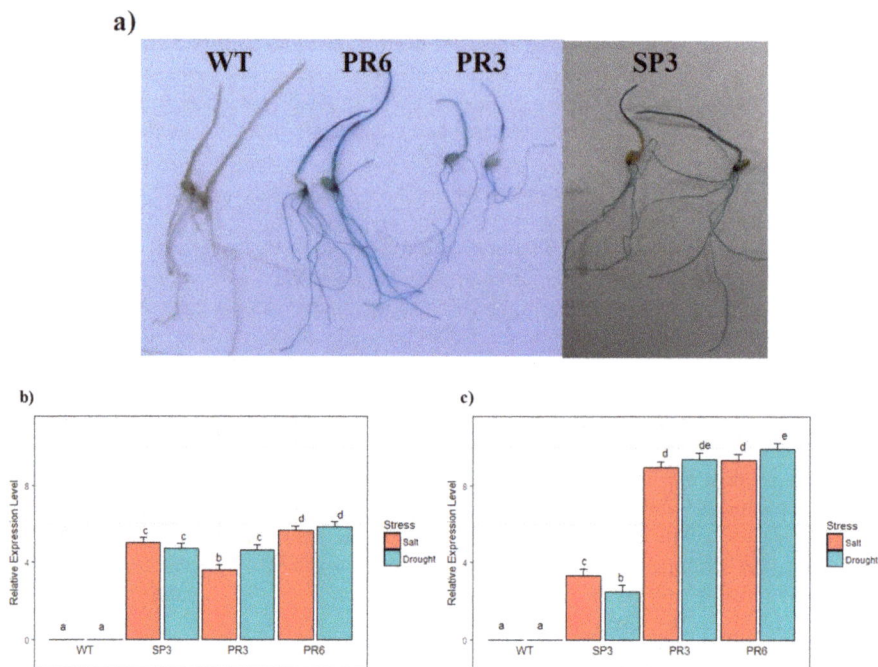

Figure 1. Histochemical analysis and GusA expression level of wheat transgenic lines overexpressing the *Td*PIP2;1 gene. (**a**) Histochemical GUS staining of 7 day-old wildtype (WT) and transgenic SP3, PR3 and PR6 wheat seedlings grown on MS medium. (**b,c**) qRT-PCR analysis of GusA gene expression in leaves and roots, respectively, of 15 day-old plantlets of WT, SP3, PR3 and PR6 lines subjected to salt (150 mM NaCl) and drought (20% PEG 6000) treatments over 48 h. Expression level of the GusA gene in control conditions was used as a reference. Bars show standard deviations of the replicates. Values represent the means of 5 replicates ($n = 5 \pm$ SD). Treatments with different letters have significant differences.

2.3. Evaluation of the Wheat TdPIP2;1 Gene for in Vitro Stress Tolerance

2.3.1. Effect of Salt Stress on Seed Germination

In a first trial, the homozygous T3 seedlings were tested in vitro for their tolerance to salt and osmotic (mannitol) stresses. Wheat seeds from WT and three homozygous transgenic lines, SP3, PR6 and PR3, were germinated in petri-dishes with two sheets of pre-wetted filter paper supplemented with 0 or 150 mM NaCl. Under a control condition, the germination rate increased from 4 to 7 days of culture for all tested lines (Figure 2), indicating that aquaporin gene expression has no adverse effect on seed germination. However, germination of WT and SP3 seeds was significantly reduced under stress conditions when compared to transgenic PR3 and PR6 seeds. In fact, when challenged with 150 mM NaCl, the WT and SP3 seeds did not germinate, whereas after 4 days, seeds of the PR3 and PR6 lines had a germination rate of 13.3 and 20%, respectively (Figure 2). Moreover, after 7 days of salt stress treatment, seeds of the PR3 and PR6 lines showed a germination rate of 56% and 63%, respectively (Figure 2). The Two-way analysis of variance shows that the effect of stress and genotype were highly significant. PR6 lines showed the highest germination rate, followed by PR3, then WT and SP3. From the germination experiments, we noticed that the transgenic seedlings had shorter leaves and roots when grown for 7 days under salt stress.

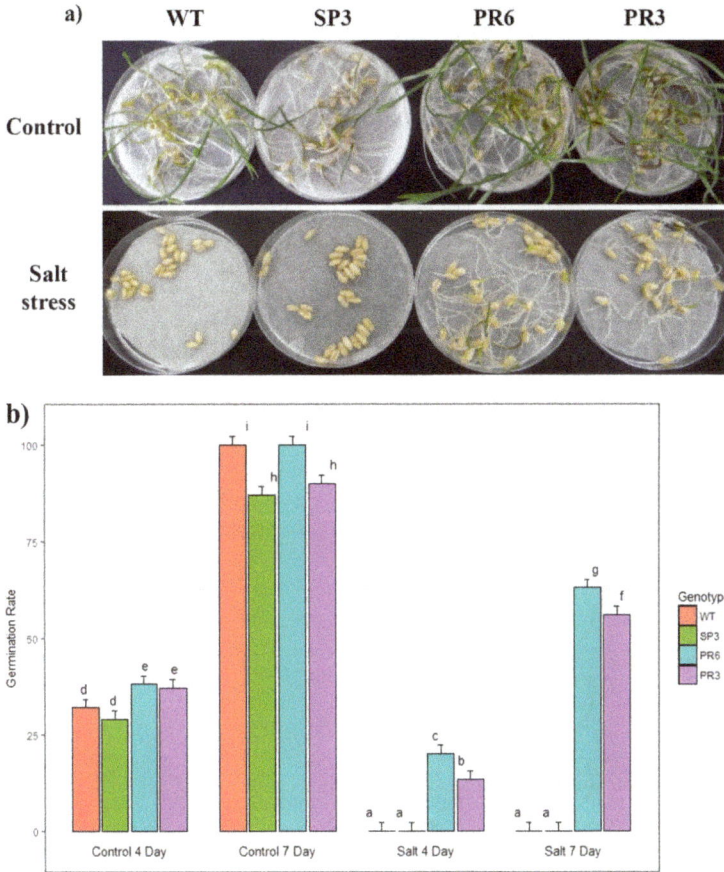

Figure 2. Effect of salt (NaCl 150 mM) on seed germination of transgenic wheat plants overexpressing *TdPIP2;1* gene. (**a**) Photographs taken one week after seed germination. (**b**) Percentage of seed germination in the presence of 0 and 150 mM NaCl at 4 and 7 days of WT and T3 homozygous transgenic plants (SP3, PR3 and PR6). Values are means of 3 replicates ± SD ($n = 3$) (20 seeds per repetition). Treatments with different letters have significant differences.

2.3.2. Effect of Salt and Osmotic Stresses on Root and Leaf Length

In order to further explore the consequences of salt and osmotic stress on the growth rate, 3 day-old WT and transgenic (SP3 and PR6) seedlings were transferred to MS, MSs (NaCl 150 mM) or MSm (mannitol 300 mM). After an additional 10 days, WT seedlings recorded a shorter leaf and root length compared to transgenic lines in control condition (Figure 3a,d,e). Under the same condition, WT plants showed severe root length reduction of 71.3% and 58% when compared to transgenic lines PR6 and SP3, respectively, while leaf length reduction in WT plants was only 34% and 38.5% when compared to transgenic lines SP3 and PR6, respectively (Figure 3d,e). Interestingly, inhibition of root elongation under salinity stress was 40% in WT, 23% and 34% in transgenic lines SP3 and PR6, respectively. Under NaCl treatment, leaf length showed non-significant difference among the three genotypes but when compared to the control MS media, the three genotypes showed significant leaf length reduction (Figure 3d). When grown in mannitol-supplemented culture media, WT and PR6 transgenic line did not show significant leaf length reduction when compared to control MS culture media. In contrast, the SP3 transgenic line showed significant leaf length reduction in comparison with the control. Under salinity treatment, root length differed significantly among the three genotypes and when compared to the control condition, they showed significant root length reduction (Figure 3e). Seedlings conducted with mannitol-supplemented culture showed a significant increase in root length for WT and SP3 lines, while the PR6 line showed a significant decrease when compared to control condition (Figure 3e). The reduction in root length in WT was 42% when compared to transgenic lines SP3 and PR6 (Figure 3a–c). It is noteworthy to notice that reduction in root length observed in transgenic plants challenged with NaCl or mannitol stress was compensated by an augmentation in the root number when compared to control conditions (Figure 3a–c). When challenged with NaCl or mannitol stress, transgenic seedlings produced two leaves, whereas WT seedlings showed inhibition and produced only one short leaf. Leaf length was inhibited by 30% under salinity stress in WT plants when compared to control condition, while no inhibition was observed in leaf length under mannitol stress (Figure 3d). Similarly, in transgenic line PR6, leaf length inhibition under salinity stress was about 45% but no inhibition was observed under mannitol stress (Figure 3b–e). The Two-way analysis of variance shows that the effect of stress and genotype were highly significant. PR6 lines showed the highest root and leaf length under control, salt and mannitol stress conditions, followed by SP3 and then WT. The difference in leaf or root length reduction or inhibition observed between WT and transgenic wheat plants overexpressing the *TdPIP2;1* gene under its own promoter, shows the ability of the transgenic plants to tolerate in vitro salinity and osmotic stresses.

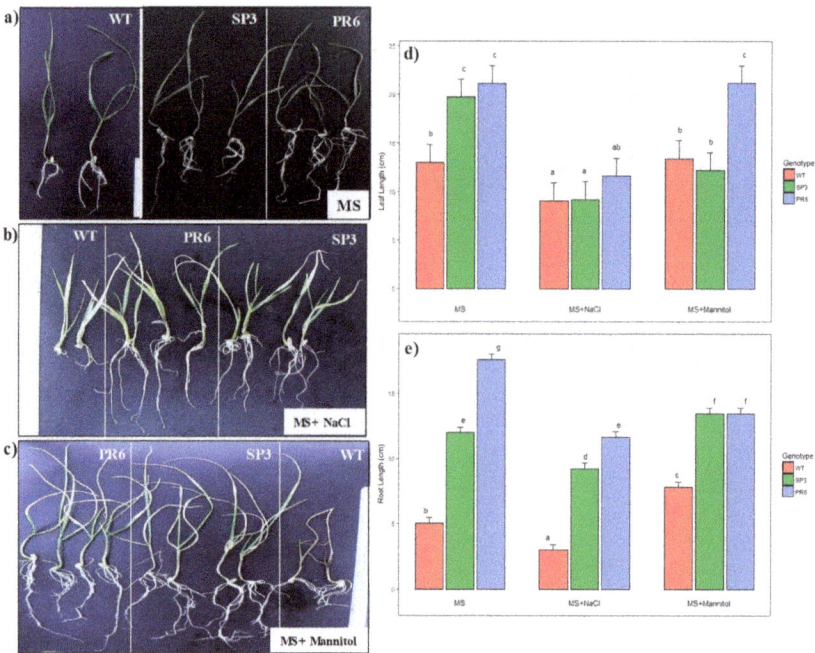

Figure 3. Effect of stress on growth rate of wild-type and *TdPIP12;1* transgenic lines (SP3, PR3 and PR6). (**a**) MS medium. (**b**): MS medium supplemented with 150 mM NaCl. (**c**): MS medium supplemented with 300 mM Mannitol. (**d,e**) Analysis of root and leaf length of WT, SP3 and PR6 transgenic plants cultured on MS, MS + 150 mM NaCl and MS + 300 mM Mannitol, respectively. The photographs were taken 15 days after stress application. The results are expressed as the means SE of measurements from three different experiments. Values are means SD (*n* = 5). Treatments with different letters have significant differences.

2.4. Evaluation of the Wheat TdPIP2;1 Gene for Stress Tolerance under Greenhouse Conditions

In vitro growth of the transgenic wheat plants overexpressing the *Td*PIP2;1 gene showed a promising level of tolerance under salt and osmotic stresses. It was worth studying the effect of this gene on biomass and grain production. Accordingly, we carried out salt and drought stress experiments with WT and the two T3 generation of transgenic lines (PR3, PR6 and SP3) grown in soil in a greenhouse.

2.4.1. Fraction of Transpirable Soil Water (FTSW)

As the soil dried progressively for about 20 days, the transpiration rate (TR) was recorded for the generated transgenic wheat plants overexpressing the Pr*Td*PIP2;1 promoter in fusion with the *Td*PIP2;1 gene. For each construct of the generated transgenic T3 homozygous lines (SP3, PR3, PR6), plants were evaluated in vitro and under greenhouse conditions for their growth (Figure 4). The Normalized transpiration rate (NTR) (transpiration of stressed plants /average transpiration of control plants = Gs/Gsmax) was calculated to reflect the daily transpiration rate and FTSW was calculated to reflect the soil-water content. The relationship between NTR and FTSW correlated perfectly to the plateau regression function (Figure 5). For the NTR decline, the FTSW threshold values were close for both transgenic wheat lines, SP3 and PR6. In contrast, the FTSW thresholds for the NTR decline were higher for the transgenic PR3 lines compared to PR6 lines. The threshold for the decline of transpiration (NTR) occurred when FTSW values of about 0.363, 0.909 and 0.983 were obtained for WT, SP3 and PR3, respectively. The FTSW threshold wherein transpiration rates began to decline was calculated for each line tested using a plateau regression procedure. The determined FTSW threshold values ranged from 0.363 to 0.909 for WT and PR3 (Table 1). There was no significant trend of change of the FTSW threshold for transpiration rate decrease between SP3 and PR6 lines. However, a significant linear ($r^2 = 0.656$) decline of FTSW threshold was observed with PR3. In this line, a preventive strategy to skip excessive water loss by stomata closure at high FTSW was observed. Accordingly, the PR3 line was shown to be the most tolerant to drought stress, whereas the WT line was the most sensitive to drought stress.

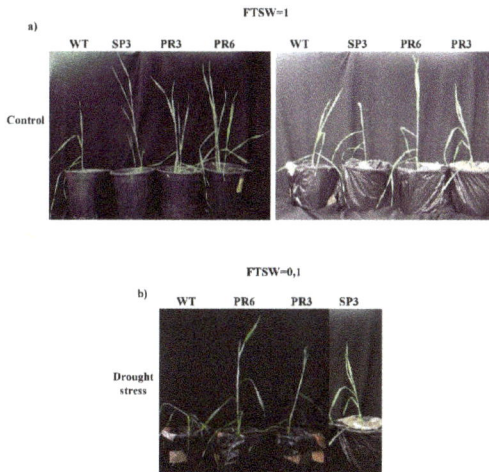

Figure 4. Phenotype and growth rate of WT and Pr*Td*PIP2;1 transgenic wheat lines PR6 and SP3 grown under (**a**) control conditions or subjected (**b**) to an average of 15 days of water withholding (FTSW = 0.1). Transgenic and WT wheat plants were gown in soil during 20 days in controlled greenhouse conditions. The photographs were taken 15 days after water stress application.

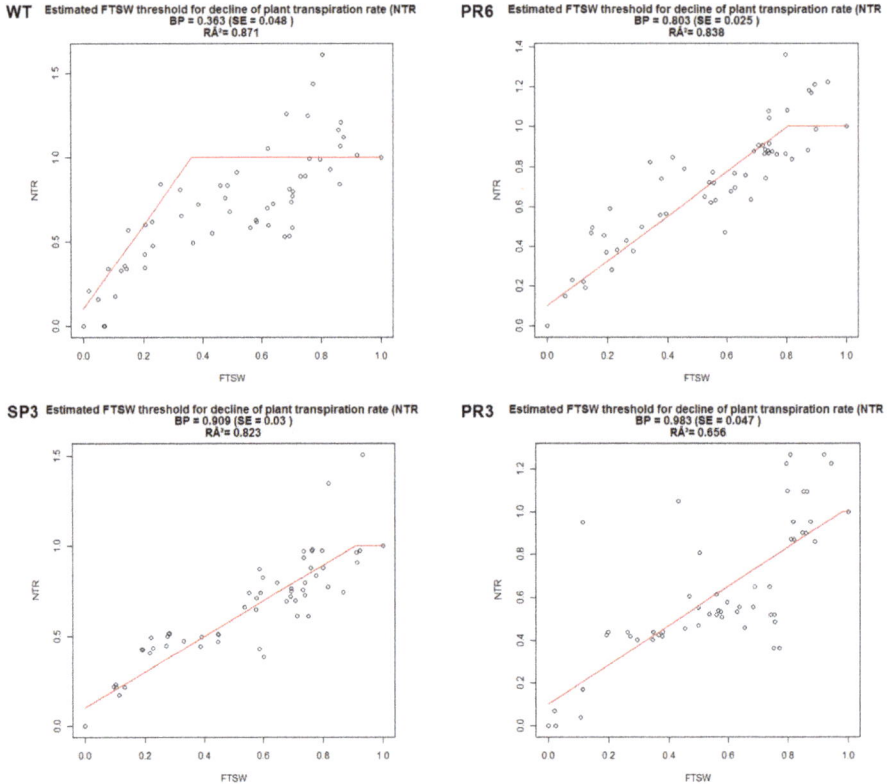

Figure 5. Variability for normalized transpiration rate (NTR) control under drought stress among the WT and transgenic wheat lines tested. Classification of wheat lines are from least sensitive WT, PR6 and SP3 to the most sensitive (stomata closure for a high FTSW) (PR3).

Table 1. FTSW threshold values.

	BP	(R^2)
WT	0.363	0.871
SP3	0.909	0.823
PR3	0.983	0.656
PR6	0.803	0.838

The same results were observed when these lines were subjected to salt stress.. When challenged with salt or drought stress, these lines exhibited a high level of tolerance. In fact, in response to water deficit, minimization of water loss constitutes a major aspect of drought tolerance and this can be reached by lowering the transpiration per unit leaf area (stomatal conductance). In this context, it is worthwhile elucidating the contribution of this protein in the tolerance mechanism to salt and drought stress by overexpressing the isolated *TdPIP2;1* gene under its own promoter in wheat plants.

2.4.2. Effect on Biomass and Grain Production

The increase in yield stability under harsh environmental and growth conditions with improved varieties adapted to salt and drought stresses would constitute a major breakthrough for farmers. To achieve this goal, we tested the generated transgenic wheat lines under greenhouse conditions for their capacity to produce normally filled grains under constant drought or salt stress. To this end, plants were grown under either optimal fresh water supply (100% Field capacity (FC)), drought stress or continuous salt stress using NaCl (150 mM) for irrigation until the end of the plant cycle.

The transgenic T3 generation plants of SP3, PR3 and PR6 clearly performed well and exhibited improvement of the growth parameters when compared to the WT plants under control conditions, confirming that aquaporin overexpression enhances growth and did not show any yield penalty in the transgenic plants (Figure 6).

Figure 6. Effect of salt and drought stresses on growth and size of grains of wild-type and transgenic *Td*PIP2;1 wheat lines (SP3 and PR6). (**a**) Photographs show WT and transgenic plants grown under either normal conditions (100% FC), a continuous presence of 150 mM NaCl or a continuous drought stress (40% FC) until the end of the plant cycle. (**b**) Grain aspects of WT and transgenic (SP3, PR3 and PR6) grown under control, drought and salinity conditions. (**c**) Mean weight of seeds produced per plant under control and stress conditions. Treatments with different letters have significant differences.

When challenged with drought stress, the transgenic wheat plants were able to continue to grow, reached maturity, flowered and set grains; whereas the WT plants' growth was strongly inhibited and the produced grains were poorly filled. Under salt stress, the WT plants were highly affected and showed chlorosis and a stunted phenotype, were unable to produce viable grains and ultimately died (Figure 6a,b). Under drought stress and when compared to the control condition, the decrease in the weight of 30 grains showed a non-significant difference for the WT, SP3 and PR3 lines, while the PR6 line showed a significant difference when compared to the control condition (Figure 6c). The recorded reduction in the weight of 30 grains was about 13% and 20% in transgenic lines SP3 and PR6, respectively in comparison to the obtained weight under control condition. In contrast, the PR3 line showed no reduction in the weight of 30 grains compared to the control condition. Under salinity stress, the weight of 30 grains was drastically affected in all genotypes and showed a significant difference when compared to the control condition. The reduction in the weight of 30 grains was above 70% in all genotypes compared to the control condition (Figure 6a–c).

2.5. Na^+ and K^+ Accumulation in Transgenic Wheat TdPIP2;1 Plants under Salt Stress Treatment

To understand the basic mechanism of salt and/or drought-tolerance, the endogenous Na^+ and K^+ contents in leaves of transgenic wheat lines overexpressing the *TdPIP2;1* gene were monitored. Under the control condition, there was no significant difference observed in Na^+ or K^+ accumulation or partitioning in leaves of WT and transgenic plants, except for PR6 plants where the accumulation was slightly increased for Na^+ (Figure 7a,b). Nevertheless, under drought or salt stress conditions, the increase of Na^+ accumulation in leaves showed a significant difference among all genotypes when compared to control condition. Hence, the increase of Na^+ concentration in leaves of the transgenic plants was approximately threefold that of WT (Figure 7a). Furthermore, K^+ accumulation was higher in control condition for all genotypes compared to stress conditions and the decrease in K^+ accumulation observed in all genotypes under drought or salt stress was highly significant compared to the control condition (Figure 7b). On the other hand, under drought or salt stress conditions, K^+ accumulation was greater in transgenic lines than in WT (range between 1.8 to 4.13 fold) (Figure 7b). The two-way analysis of variance shows that the effect of stress and genotype were highly significant. Overall, these results suggest that transgenic plants retained selectively more K^+ than Na^+ in their leaves and the Na^+/K^+ ratio, which is an important stress tolerance trait, decreased from 3.9 in WT plants to 2.85 in PR3 and PR6 transgenic lines and to 2.5 in transgenic SP3 lines, thus preventing the young photosynthetically active organs from Na^+ accumulation and toxicity.

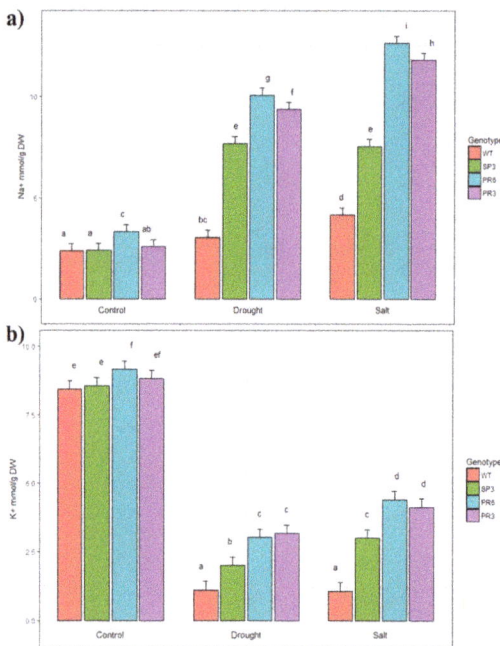

Figure 7. (**a**) Na^+ and (**b**) K^+ accumulation in leaves of the wild-type and transgenic lines overexpressing the *TdPIP2;1* gene and grown under control or continuous presence of 150 mM NaCl or drought stress (40% FC). Values are the mean ± standard deviation (SD) (*n* = 3). Treatments with different letters have significant differences.

2.6. Oxidative Stress Evaluation of Transgenic Wheat TdPIP2;1 Plants

In the present study, Malondialdehyde (MDA) content in the leaves of WT and transgenic wheat lines correlated with drought stress induced growth inhibition. Moreover, when compared to the transgenic lines, the MDA content increased significantly in WT plants under drought stress (Figure 8a). In contrast, MDA content decreased significantly in the two transgenic lines PR6 and PR3 compared to WT plants under drought stress conditions (Figure 8a). The decrease in MDA content under drought stress was significantly different for PR3 and PR6 lines compared to control condition, while for the SP3 line the difference was not significant. Hence, transgenic wheat *TdPIP2;1* lines with lower MDA content were shown to be more tolerant to drought.

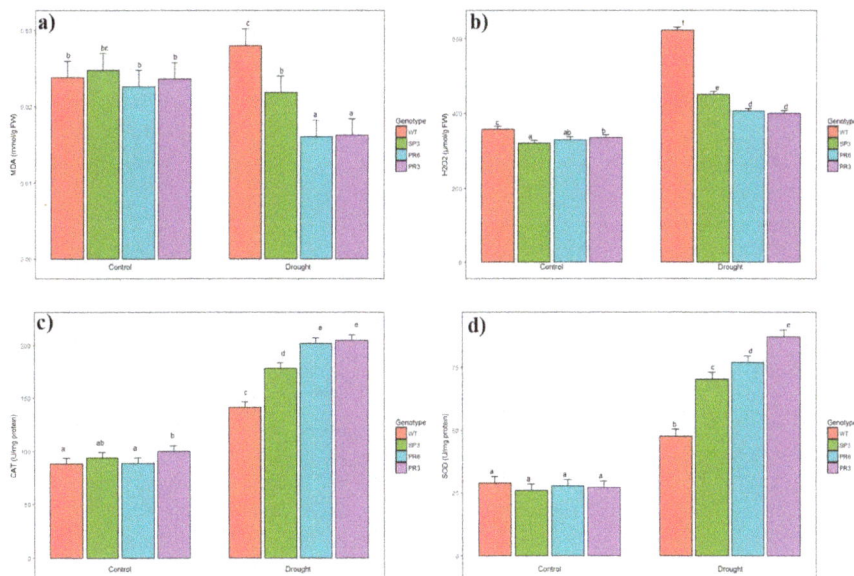

Figure 8. Analysis of enzymatic activities of (**a**) malondialdehyde (MDA), (**b**) hydrogen peroxide, (**c**) catalase (CAT) and (**d**) superoxide dismutase (SOD) in wild-type plants and transgenic lines (SP3, PR3 and PR6) subjected or not to drought stress. Values are means of 5 replicates ± SD (*n* = 5). Treatments with different letters have significant differences.

The oxidative damage was checked by measuring the accumulation of H_2O_2 in leaves of the control and stressed transgenic wheat lines. As shown in Figure 8b, drought stress treatment induced significant accumulation of H_2O_2 content in leaves of the WT plants compared to transgenic lines (1.53 fold). Under drought stress, the increase in H_2O_2 content in leaves showed significant difference for all genotypes when compared to control condition (Figure 8b).

The activities of SOD and CAT, two essential antioxidant enzymes, were analysed in the leaves of WT and transgenic wheat *TdPIP2;1* lines. When challenged with drought stress, the increase in CAT and SOD activities showed a significant difference for all genotypes compared to the control condition (Figure 8c,d). CAT and SOD activities increased significantly in transgenic lines, reaching about 2.2 and 3.2 fold, respectively relative to the non-treated plants (Figure 8c,d). However, this increase in CAT and SOD activities is lower in WT when comparing control and stress conditions (Figure 8c,d). The Two-way analysis of variance shows that the effect of stress and genotype were highly significant.

3. Discussion

In a large number of plant species, constitutive promoters were used to study the expression of transgenes. However, in some specific cases, constitutive expression could be damaging to the recipient plant, generating sterility, development retardation, atypical morphology, yield penalty, modified grain composition or transgene silencing [41]. To overcome the negative effects of the constitutive promoter, the use of a tissue-specific promoter is expected to be the solution for driving candidate gene expression restricted to tissues of interest and at given developmental stages. It is expected that utilization of stress inducible promoters will have a great impact on improving plant tolerance to either abiotic or biotic stress. Aquaporins (AQP) are ubiquitous membrane proteins with members exhibiting both tissue specific and inducible regulation. It has been shown that the *NtAQP1* gene is specifically expressed in the mesophyll tissue of *Arabidopsis thaliana* and plays an important role in increasing both net photosynthesis and mesophyll CO_2 conductance. Moreover, the *NtAQP1* gene targeted at the cells of the vascular envelope significantly improved the plants' stress response [42].

In this report, we demonstrated that the overexpression of the *TdPIP2;1* gene in transgenic durum wheat homozygous lines significantly enhanced tolerance under constant salt and drought stresses (Figure 4). These results affirm our previous findings in rice [40]. In fact, transgenic rice plants transformed with the Pr*TdPIP2;1* fused to the *TdPIP2;1* gene and treated with either salt or drought stress, exhibited growth and vigour enhancement when compared to wild-type plants and therefore showed higher tolerance to abiotic stresses [40]. Our results were similar to those observed in soybean, where the overexpression of *GmPIP2;9* under a native promoter increased tolerance to drought stress in both solution and soil plots. This was correlated with expanded net CO_2 assimilation of photosynthesis, stomata conductance and transpiration rate [43].

Characterization of transgenic (SP3) wheat plants transformed with the 35S promoter fused to the *TdPIP2;1* gene and cultivated in vitro under salt and drought stress showed enhanced growth and vigour compared to wild-type plants (Figure 3b,c). In our preceding work on the *TdPIP2;1* gene, similar results were reported where the expression of the *TdPIP2;1* gene was positively correlated with stress response pathways in transgenic Tobacco plants [39] and rice plants [40].

Our analysis using transcriptional (Pr*TdPIP2;1*::GUS) and translational (Pr*TdPIP2;1*:: *TdPIP2;1*::GUS) fusions demonstrated that the *TdPIP2;1* gene is expressed in the homologous transgenic wheat system. Using a histochemical staining technique, we were able to show that the Pr*TdPIP2;1* promoter can drive GusA reporter gene expression with inducible and tissue-specific patterns (Figure 1a). The differential expression of AQP transcripts or proteins in response to abiotic stress or other physiological or developmental changes was reported in a number of studies (reviewed in Reference [44]). However, the transcriptional regulation of these genes is still elusive. The analysis of the promoter sequences of a variety of PIP and TIP genes from rice, maize and *A. thaliana* has shown the presence of putative regulatory elements such as DREs (drought responsive elements), LTREs (low temperature responsive elements) and ABREs (ABA responsive elements) [44,45]. In wheat, the *TdPIP2;1* gene sequence revealed the presence of three highly conserved motifs in positions (261(+) CNGTTR; 435(+) CNGTTR; and 526(+) CNGTTR or CANNTG); identified at MYBCORE transcription factor binding sites responsive to water stress and induced by dehydration [36]. In addition, the sequence analysis of pro*TdPIP2;1* revealed the presence of WRKY transcription factor required for positive and negative regulation of abscisic acid signalling [40]. This indicates the regulation of these genes at a transcriptional level in response to abiotic stresses [44,45].

To confirm the results obtained in vitro for the transgenic wheat plants transformed with the *TdPIP2;1* gene driven by its own promoter (PR3 and PR6) or the constitutive 35S promoter (SP3) under salt and osmotic (mannitol) stress, we designed an experiment under greenhouse conditions and tested these transgenic plants for abiotic stress tolerance. Our results showed that transpiration rates were affected in control wheat plants under drought stress by closure of stomata and changes in leaf morphology. If drought stress was initiated at the vegetative phase, the first perceived response was a relatively quick decrease in leaf expansion, leaf rolling and stomata closure (Figure 4). Interestingly, the PR3 line, which is the most sensitive to stomata closure for a high FTSW, was the most tolerant line to drought stress compared to SP3 and PR6 (Figure 5). In this case, there was a significant linear decline of the FTSW threshold ($r^2 = 0.656$).

Under abiotic stress, oxidative damage can be induced in many plants such as pea, rice and tomato [46,47]. ROS degrades polyunsaturated lipids forming MDA which is considered a biomarker of the level of lipid peroxidation. Accordingly, our results indicated that drought stress induced oxidative stress in transgenic *Td*PIP2;1 wheat lines manifested by low MDA content and high H_2O_2 contents (Figure 8a,b). Nevertheless, under drought stress, the lipid peroxidation level and H_2O_2 content were higher in WT than in transgenic lines SP3, PR3 and PR6. These data suggest that the transgenic wheat lines were preserved against oxidative damage under drought stress.

For the antioxidant enzymes activities, drought stress induced up-regulation of the activities in WT and transgenic wheat lines. When comparing the enzymatic activity of these genotypes, higher activities were obtained in the transgenic lines than in WT (Figure 8c,d). This suggested that transgenic lines present a higher capacity for scavenging ROS than WT under stress conditions. Consequently, the reduction in the content of H_2O_2 in transgenic lines is the result of the SOD reaction, which is followed by an increased enzymatic activity to catalyse the dismutation of the superoxide radical. The correlation between abiotic stress tolerance and an increase in SOD activity has been reported in many previous works [48–52]. Overall, it seems that the salt tolerance phenotype of transgenic wheat lines could be the result of either a low production of ROS or a better capacity to prevent ROS than the WT plants.

4. Materials and methods

4.1. Plant Material

Wheat seeds *(Triticum turgidum* L. subsp. *durum)*, cv. Maali (New seed lot) were supplied by INRAT, Laboratoire de Physiologie Végétale (Tunis, Tunisia) and were used for genetic transformation.

4.2. Construction of the Binary Vector and Wheat Transformation Procedure

To perform gene expression and histochemical analysis, we used translational fusion with the reporter GUS gene, Pr*Td*PIP2;1::*Td*PIP2;1::GUS (named PR) and CaMV35S::*Td*PIP2;1::GUS (named SP) [40]. The resulting constructs were then transferred into *Agrobacterium tumefaciens* hyper-virulent strain EHA105 [53] using the freeze–thaw method [54] and finally used for durum wheat transformation experiments. The transformation method employed in this study was slightly modified from the applied protocol to bread wheat [55]. *A. tumefaciens* strains, harbouring the resulting constructs (PR and SP), were cultured on MG/L agar medium [56] containing Hygromycin (50 µg/mL) and Rifampicin (10 µg/mL) for two days at 28 °C. Bacteria were harvested when they reached an $OD_{600} = 1$ and were suspended in an R_2 co-cultivation liquid medium (R_2CL, pH 5.2) containing R_2 Basic medium [57], glucose 10 g/L and 100 µM acetosyringone. An aliquot of 5 mL from this culture was diluted five times with the same R_2CL medium containing the appropriate antibiotics and kept growing at 28 °C with shaking until OD_{600} reached 0.4–0.6. Durum wheat seeds cv. Maali were soaked first in 70% (*v/v*) ethanol for 3 min, then in sodium hypochlorite 1% (*v/v*) with gentle shaking for 20 min for sterilization. After that, seeds were washed five times with sterile distilled water. Seeds were additionally soaked at 4 °C in 0.1% (*v/v*) Pelt 44 (thiophanate-methyl at 450 g/L) fungicide (Bayer CropScience) and then germinated on wet filter paper with R2 basic medium at 25 °C for 2 days in the dark. A needle (Terumo Syringe 1 mL Insulin (U100) with 29 g × 0.5 needle ×100: BS-N1H2913) that had been previously filled with *A. tumefaciens* inoculum solution, was used to pierce twice, to a depth of approximately 1 mm, the embryonic apical meristem from the 2-d-old seedlings. The inoculated seeds were then co-cultivated on wet filter paper at 25 °C for an additional two days in the dark to avoid overgrowth of the Agrobacterium on the seeds. To eliminate the inoculum of Agrobacterium, seeds were dipped overnight at 4 °C in an aqueous solution containing Cefotaxime (1 g/L). Eventually, seeds were immersed in sodium hypochlorite 0.1% (*v/v*) for 15 min and then rinsed with sterile distilled water for five times. Ultimately, the inoculated wheat seeds were sown in pots filled with a mixture of sand (50%) and peat (50%), allowed to grow to maturity and to produce seeds (T1) under greenhouse conditions. The regenerated hygromycin-resistant wheat plants (T1) were numbered and named as indicated above (SP and PR). For the control, we used the WT wheat plants.

4.3. Histochemical GUS Staining

To help understand the presence of staining in tissues for GUS activity, a histochemical assay was carried out by incubating wheat seedlings under vacuum infiltration in 50 mM Na_2HPO_4 buffer (pH 7.0), 0.5 mM $K_3(Fe[CN]_6)$, 0.5 mM $K_4(Fe[CN]_6)$, 0.1% Triton X-100 and 1 mg/L X-Gluc (5-bromo-4-chloro-3-indolyl β-D-glucuronide cyclohexyl ammonium salt) for several minutes and then keeping them overnight at 37 °C. The chlorophyll and pigments were eliminated by soaking the stained seedlings tissues for several hours in 70% ethanol. Finally, the destained tissues were photographed. Transgenic plants overexpressing the *TdPIP2;1* gene driven by the CaMV35S promoter and WT wheat plants were considered the positive and negative controls, respectively.

4.4. GusA Expression Analysis by the Quantitative Real-Time qPCR

Transcription levels of the *GusA* gene under the control of the Pr*TdPIP2;1* or CaMV35S promoters were assessed by real-time qPCR in T3 homozygous transgenic wheat plants (leaves and roots of two-week-old plants) treated with 150 mM NaCl or 20% PEG 6000 for 48 h. The Trizol method (Invitrogen) was employed to extract total RNA from 100 mg plant tissues in accordance with the manufacturer's recommendations [39]. To check the concentration and the quality of the extracted RNA, agarose gel (1% RNase-free) and spectrophotometer measurements (260/280 nm) were performed. To synthesize first strand cDNA, 3 μg of total RNA was treated with DNase I (Promega, USA) and then reverse transcribed using oligo (dT)18 and SuperScript II reverse transcriptase (Invitrogen, Canada), per the supplier's recommendations. Primer 3 software was used to identify the primer pairs that amplify fragments of GusA and the housekeeping actin genes. The primers used for real-time qPCR were: qGusAF, qGusAR, q-Act-F and q-Act-R, (Table 2). PCR reactions were performed in a 10 μL final volume containing 4 μL cDNA (obtained from 40 ng of DNase-treated RNA), 0.5 μL of each primer (at 10 μM), 5 μL 2× SYBR Green I master mix and 1 μL of RNase-free water (Sigma, Canada). The reaction consisted of an initial denaturation step at 94 °C for 10 min followed by 45 cycles composed of 10 s at 94 °C, 10 s at 60 °C and 15 s at 72 °C, then a melting curve (5 s at 95 °C, 1 min at 65 °C and 5 min with temperature increasing from 65 °C to 97 °C). For each experimental condition three biological repetitions were performed, with three technical repetitions for each sample. The threshold cycle (CT) values of the triplicate PCRs were averaged and used for transcripts quantification at the end of the reaction. The relative expression ratio of the GusA gene was calculated by using the comparative CT method with the actin gene as an internal expression standard [58]. The relative expression level was calculated from triplicate measurements based on the 2-ΔΔCT, where ΔΔCT = (CT, Target gene—CT, Actin) stressed—(CT, Target gene—CT, Actin) control. Relative expression ratios from three independent experiments (three biological repetitions) were reported.

Table 2. List of primers used in real-time PCR.

Primers	Nucleotides Sequences (5'-3')
Act F:	TGCATAGAGGGAAAGCACG
Act R:	AACCCAAAAGCCAACAACAGAGA
q-GusF:	CACGCCGTATGTTATTGCCG
q-GusR:	TCTTGCCGTTTTCGTGGGTA

4.5. Evaluation of Transgenic Wheat Plants for Abiotic Stress Tolerance

4.5.1. In Vitro Assays

Seeds of wild type and transgenic homozygous T3 generation of durum wheat cv. Maali were employed for abiotic stress assays. For three experimental repetitions, thirty seeds from WT and transgenic plants were surface-sterilized by soaking in 70% ethanol for 1 min, washed with sterile distilled water and treated with 40% solution of sodium hypochlorite for 30 min. Finally, seeds were rinsed five times with sterile distilled water. Seeds were then incubated in sterile distilled water and maintained in a growth chamber (16 h of light per day, 500 μE m^{-2} S^{-1}, 28 °C/25 °C day/night) for germination and growth rate parameters. After 2 days, seeds from T3 transgenic lines (PR and SP) and WT plant were sown in Petri dishes on two layers of filter paper saturated with distilled water, supplemented with 0 or 150 mM NaCl. Plates were incubated at 25 °C in a growth chamber under a 16 h light/8 h dark photoperiod. Germination rates were recorded after one week of incubation. The seed germination experiment was repeated three times. Young seedlings were transplanted to a new medium supplemented with 150 mM NaCl or 300 mM Mannitol to estimate the growth rate under osmotic or salt stress conditions. These dishes were removed and placed vertically in a growth chamber. After 15 days of culture, root/shoot lengths were determined for WT and transgenic seedlings using the UTHSCSA Image Tool, a free image processing and analysis program used to acquire, display, edit and analyse images (available online: http://www.ddsdx.uthscsa.edu/dig/iTdesc.html).

4.5.2. Greenhouse Assays

To assess the effect of salt and drought treatments on biomass production, seeds from WT and transgenic lines (SP and PR) were germinated under greenhouse conditions. To initiate salt treatments, pots were irrigated with water supplemented with 150 mM NaCl. Soil moisture regimes were monitored daily by weighing the pots. For both salt and drought treatments, seeds were allowed to germinate and to grow for additional 20 days before harvesting the plants to measure their fresh weight, dry weight and plant transpiration rate. Seeds from T3 transgenic lines (PR and SP) and WT plants were sown in pots filled with peat moss and grown for 2 weeks before challenging them with stress treatments. Under control conditions, pots were irrigated weekly with fresh water to maintain plants at 100% FC. For salt stress treatment, the same irrigation program was used for watering the plants with addition of 150 mM NaCl. This NaCl concentration was maintained until the end of the plant cycle. Young leaves (top) and old leaves (bottom), were collected from salt-stressed and control plants and then dried at 80 °C for 24 h. Finally, the dried material was incubated in 0.5% HNO$_3$ for a week. The Na$^+$, Ca^{2+} and K$^+$ contents were analysed in the filtrate using atomic absorption spectrophotometry. For drought stress, plants were maintained at 40% FC until the end of plant cycle. This was achieved by refiling the exact amount of water lost by evapotranspiration.

4.5.3. Water Treatments

To monitor the soil water status, we used the fraction of transpirable soil water (FTSW). According to Sinclair et al. [59], when water stress is expressed as FTSW, plants respond to the progressive drying of soil in a similar manner.

Calculation of each pot's FTSW value was carried out according to Sinclair and Ludlow's method [60]. First, pots were fully watered daily and drained overnight until a day prior to initiating measurements. After one day of drainage, all pots were weighted to assess their initial water holding capacity. The total transpirable soil water (TTSW) was obtained by calculating the difference between initial pot capacities (Wi) and final pot weight after soil desiccation (Wf). The ratio of actual transpirable soil water (ATSW) to TTSW allowed the estimation of FTSW, where ATSW was the mass difference between daily (Wt) and final pot weight. A single drought cycle was implemented in half of the pots, when the plants had developed 8 leaves (five pots for each genotype).

$$FTSW = ATSW/TTSW = (Wt - Wf)/(Wi - Wf) \tag{1}$$

TTSW: total transpirable soil water; ATSW: actual transpirable soil water.

To restore the daily water loss, the remaining five control pots were fully watered in late afternoon. In contrast, for the stressed pots, water deficit was maintained by withholding irrigation and covering the pots with a plastic bag in order to prevent soil evaporation. Original FTSW values were assigned a value of 1. When a transpiration rate of less than 10% was reached for each stressed pot in comparison to the control, the experiment was terminated [60].

4.5.4. Plant Transpiration Rate

Plant transpiration (TRj) per unit leaf area (mmol·m^{-2}·s^{-1}) was determined every day for each pot as the mass difference between weights every 24 h (including the watering for unstressed pots) and were divided by plant total leaf area on the previous day. At the same time, stomatal conductance (mmol·m^{-2}·s^{-1}) was measured each morning between 9:00 a.m. and 12:00 p.m. on the last panicle leaf per plant. These measures were acquired with a porometer (type AP4-UM-3) on the control and treated plants. Conductance measurements were performed on leaves and on its abaxial face, where stomatal density is greater.

$$\text{Normalized transpiration rate (NTR)} = \text{transpiration of stressed plants/average transpiration of control plants} = (Gs/Gsmax) \tag{2}$$

Porometry was also used to determine the last day of measurements (when gs(stressed)/gs(control) was less than 0.1), allowing us to estimate the value of the TTSW of each pot. The mean TTSW was remarkably stable.

4.6. Whole Plant Response Modelling to Increasing Water Deficit

To monitor the daily transpiration rate fluctuation due to evaporative demand changes, the drought treatment values were divided by the corresponding mean values on the well-watered (control) treatment in order to calculate the relative TR values daily. When the soil was still moist, daily TR ratio values for FTSW > 0.6 were adjusted to a mean value of 1, in order to minimize the effect of variation in initial plant size variation. It was essential to normalize the calculation of plant transpiration (NTR) for the accuracy of a further linear model. A two-slope linear relation with one parameter (FTSWt) was used to meet the plant responses to water deficit, indicating the FTSW threshold below which conductance starts decreasing. By applying a two-slope linear model to the experimental data, the parameter FTSWt was thus estimated (Equation (3)):

$$
\begin{aligned}
&\text{if } FTSW \geq FTSWt \\
&\quad y = 1; \\
&\text{else } y = 1/FTSWt \times FTSW; \\
&\quad y \text{ being NTR}
\end{aligned} \tag{3}
$$

4.7. Lipid Peroxidation

By using the method of Ben Amor et al. [49] to calculate the amount of MDA in the leaves, the extent of lipid peroxidation was estimated. 1.0 mL of 0.5% (*w/v*) thiobarbituric acid (TBA) in 20% (*w/v*) TCA solution was added to an aliquot of 0.5 mL from the homogenate of fresh shoots in 0.1% (*w/v*) trichloroacetic acid (TCA) solution. Subsequently, the mixture was incubated at 90 °C for 30 min and then cooled on ice. By measuring absorbance at 532 and 600 nm in reference to an MDA standard curve, the equivalent MDA was calculated.

4.8. Quantitative H_2O_2 Measurement

The concentration of H_2O_2 was estimated according to the method described by Velikova et al. [61]. Fresh shoots tissue (0.5 g) was homogenized with 5 mL of 0.1% (*w/v*) TCA. This homogenate was then centrifuged at 12,000× *g* for 15 min. 10 mM phosphate buffer (pH 7.0) and 1 M potassium iodide were added to 0.5 mL of the supernatant. The mixture was then vortexed and its absorbance at 390 nm was recorded and H_2O_2 content was calculated using a standard curve with concentration ranging from 0.05 to 0.1 mM.

4.9. Enzyme Assays

Aliquots of frozen fresh shoot material (0.5 g) were ground to a fine powder with liquid nitrogen and homogenized in a cold solution containing 100 mM Tris–HCl buffer (pH 8.0), 10 mM EDTA (Ethylenediaminetetraacetic acid), 50 mM KCl, 20 mM $MgCl_2$, 0.5 mM PMSF (Phenyl methyl sulfonyl fluoride) and 2% (*w/v*) PVP (Polyvinylpyrolidone). The homogenate was centrifuged at 14,000× *g* for 30 min at 4 °C and the supernatant was used for determination of the antioxidative enzyme activities.

Total catalase (CAT) activity was measured according to the method of Aebi [62], by monitoring the decline in absorbance at 240 nm as H_2O_2 was consumed. An aliquot of crude enzyme extract was added to the reaction mixtures containing 50 mM phosphate buffer (pH 7), 30 mM H_2O_2. One unit of CAT was defined as 1 µmol mL^{-1} H_2O_2 decomposed per minute.

Total superoxide dismutase (SOD) activity was determined by measuring the percentage of inhibition of the pyrogallol autoxidation [63]. A crude enzyme extract aliquot was added to the reaction mixture containing 10 mM pyrogallol in Tris-cacodylic acidediethylene triamine penta acetic acid buffer (pH 7.4–8). One unit of SOD was defined as the enzyme quantity required to inhibit 50% of the pyrogallol autoxidation.

4.10. Statistical Analysis

All data analysis was carried out using the Two-way Analysis of Variance (ANOVA) with SPSS package. Means were compared using Tukey's (HSD) multiple comparison procedure.

5. Conclusions

In our previously published works, we reported that the expression of the *Td*PIP2;1 gene was strongly associated with abiotic stress response pathways in Tobacco [39] and rice [40]. In the present study, we found similar results, confirming that the expression of the *Td*PIP2;1 gene was closely related to abiotic stress response pathways in durum wheat, a staple crop, without causing any undesirable growth phenotypes or yield penalty. In addition, we found that mannitol, a major photosynthetic product in many higher plants, enhanced the activity of the *Td*PIP2;1 promoter, suggesting that Pr*Td*PIP2;1 may be involved in dehydration response. Our results provide insights for the *Td*PIP2;1 gene regulated by its own promoter which generates enhanced drought tolerance in transgenic wheat. In general, regulation of the activity and gene expression of aquaporins are considered part of the adaptation mechanisms to stress conditions through complex processes. Further studies are required to gain a deeper understanding of the function of this gene, particularly through transcriptome profiling and identification of *Td*PIP2;1 interacting proteins in plant PIPs. Recent advances in genome editing with CRISPR/Cas9 was efficiently used in combination with haploid induction (HI) to induce edits in nascent seeds of diverse monocot species [64]. HI enables widespread application of genome editing technology for crop improvement and offers a new path to understand the stress resistance mechanisms, when generating knockout of down-regulated AQP genes in wheat.

Author Contributions: K.M. conceived and designed the experiments. M.A., K.M. and F.B. performed the experiments, analysed the data and wrote the paper. All authors read and approved the manuscript.

Funding: This study was supported by a grant from the Ministry of Higher Education and Scientific Research of Tunisia and by a research grant from research office, United Arab Emirates University (Grant N°31F096 to KM).

Acknowledgments: We are thankful to the Plant Physiology Laboratory, INRAT, Tunisia for the provision of durum wheat seed lots of cv. Maali. We are thankful for the help with the editing of this manuscript by Elizabeth Whitehouse from the writing centre at UAEU. We are also thankful to Kilani Ghoudi from the Statistics Department at UAEU for his help with the analysis.

Conflicts of Interest: The authors declare that the research was conducted in the absence of any potential conflict of interest.

References

1. Chaves, M.M.; Maroco, J.P.; Pereira, J.S. Understanding plant response to drought-from genes to the whole plant. *Funct. Plant Biol.* **2003**, *30*, 239–264. [CrossRef]
2. Munne-Bosch, S.; Jubany-Mari, T.; Alegre, L. Drought-induced senescence is characterized by a loss of antioxidant defences in chloroplasts. *Plant Cell Environ.* **2001**, *24*, 1319–1327. [CrossRef]
3. Jubany-Mari, T.; Munne-Bosch, S.; Alegre, L. Redox regulation of water stress responses in field-grown plants. Role of hydrogen peroxide and ascorbate. *Plant Physiol. Biochem.* **2010**, *48*, 351–358. [CrossRef]
4. Kyparissis, A.; Drilias, P.; Manetas, Y. Seasonal fluctuations in photoprotective (xanthophyll cycle) and photoselective (chlorophylls) capacity in eight Mediterranean plant species belonging to two different growth forms. *Funct. Plant Biol.* **2000**, *27*, 265–272. [CrossRef]
5. Kyparissis, A.; Petropoulou, Y.; Manetas, Y. Summer survival of leaves in a soft-leaved shrub (*Phlomis fruticosa* L., Labiatae) under Mediterranean field conditions: Avoidance of photoinhibitory damage through decreased chlorophyll contents. *J. Exp. Bot.* **2000**, *46*, 1825–1831. [CrossRef]
6. Havaux, M.; Tardy, F. Loss of chlorophyll with limited reduction of photosynthesis as an adaptive response of Syrian barley landraces to high-light and heat stress. *Aust. J. Plant Physiol.* **1999**, *26*, 569–578. [CrossRef]
7. Beck, P.A.; Hutchison, S.; Gunter, A.S.; Losi, C.T.; Stewart, B.C.; Capps, K.P.; Phillips, J.M. Chemical composition and in situ dry matter and fiber disappearance of sorghum x Sudangrass hybrids. *J. Anim. Sci.* **2007**, *85*, 545–555. [CrossRef] [PubMed]
8. Pelleschi, S.; Rocher, J.P.; Prioul, J.L. Effect of water restriction on carbohydrate metabolism and photosynthesis in mature maize leaves. *Plant Cell Environ.* **1997**, *20*, 493–503. [CrossRef]
9. Pinheiro, C.; Chaves, M.M.; Ricardo, C.P. Alterations in carbon and nitrogen metabolism induced by water deficit in the stems and leaves of *Lupinus albus* L. *J. Exp. Bot.* **2001**, *52*, 1063–1070. [CrossRef]
10. Yang, J.; Zhang, J.; Wang, Z.; Zhu, Q.; Wang, W. Remobilization of carbon reserves in response to water-deficit during grain filling of rice. *Field Crops Res.* **2001**, *71*, 47–55. [CrossRef]
11. Toumi, I.; Gargouri, M.; Nouairi, I.; Moschou, P.; Salem-Fnayou, A.; Mliki, A.; Zarrouk, M.; Ghorbel, A. Water stress induced changes in the leaf lipid composition of four grapevine genotypes with different drought tolerance. *Biol. Plant.* **2008**, *52*, 161–164. [CrossRef]
12. Passioura, J. Increasing crop productivity when water is scarce- from breeding to field management. In Proceedings of the 4th International Crop Science Congress: New Directions for a Diverse Planet, Brisbane, Australia, 26 September–1 October 2004.
13. Cornic, G. Drought stress inhibits photosynthesis by decreasing stomatal aperture–not by affecting ATP synthesis. *Trends Plant Sci.* **2000**, *5*, 187–188. [CrossRef]
14. Chen, T.H.; Murata, N. Enhancement of tolerance of abiotic stress by metabolic engineering of betaines and other compatible solutes. *Curr. Opin. Plant Biol.* **2002**, *5*, 250–257. [CrossRef]
15. Zhang, J.; Kirkham, B. Antioxidant responses to drought in sunflower and sorghum seedlings. *New Phytol.* **1996**, *132*, 361–373. [CrossRef]
16. Chaumont, F.; Tyerman, S.D. Aquaporins: Highly regulated channels controlling plant water relations. *Plant Physiol.* **2014**, *164*, 1600–1618. [CrossRef]
17. Pandey, B.; Sharma, P.; Pandey, D.M.; Sharma, I.; Chatrath, R. Identification of New Aquaporin Genes and Single Nucleotide Polymorphism in Bread Wheat. *Evol. Bioinform.* **2013**, *9*, 437–452. [CrossRef]

18. Zhuang, L.; Liu, M.; Yuan, X.; Yang, Z.; Huang, B. Physiological Effects of Aquaporin in Regulating Drought Tolerance through Overexpressing of *Festuca arundinacea* Aquaporin Gene *FaPIP2;1*. *J. Am. Soc. Hort. Sci.* **2015**, *140*, 404–412. [CrossRef]

19. Kaldenhoff, R.; Kei, L.; Uehlein, N. Aquaporin and membrane diffusion of CO_2 in living organisms. *Biochim. Biophys. Acta-Gen. Subj.* **2013**, *1840*, 1592–1595. [CrossRef] [PubMed]

20. Bienert, G.P.; Schjoerring, J.K.; Jahn, T.P. Membrane transport of hydrogen peroxide. *Biochim. Biophys. Acta-Biomembr.* **2006**, *1758*, 994–1003. [CrossRef]

21. Maurel, C.; Verdoucq, L.; Luu, D.T.; Santoni, V. Plant aquaporins: Membrane channels with multiple integrated functions. *Ann. Rev. Plant Biol.* **2008**, *59*, 595–624. [CrossRef] [PubMed]

22. Chaumont, F.; Barrieu, F.; Jung, R.; Chrispeels, M.J. Plasma membrane intrinsic proteins from maize cluster in two sequence subgroups with differential aquaporin activity. *Plant Physiol.* **2000**, *122*, 1025–1034. [CrossRef] [PubMed]

23. Fetter, K.; van Wilder, V.; Moshelion, M.; Chaumont, F. Interactions between plasma membrane Aquaporins modulate their water channel activity. *Plant Cell* **2004**, *16*, 215–228. [CrossRef] [PubMed]

24. Chaumont, F.; Moshelion, M.; Daniels, M.J. Regulation of plant aquaporin activity. *Biol. Cell* **2005**, *97*, 749–764. [CrossRef]

25. Javot, H.; Maurel, C. The role of aquaporins in root water uptake. *Ann. Bot.* **2002**, *90*, 301–313. [CrossRef] [PubMed]

26. Maurel, C.; Javot, H.; Lauvergeat, V.; Gerbeau, P.; Tournaire, C.; Santoni, V. Molecular physiology of aquaporins in plants. *Int. Rev. Cytol.* **2002**, *215*, 105–148.

27. Heinen, R.B.; Ye, Q.; Chaumont, F. Role of aquaporins in leaf physiology. *J. Exp. Bot.* **2009**, *60*, 2971–2985. [CrossRef] [PubMed]

28. Hachez, C.; Zelazny, E.; Chaumont, F. Modulating the expression of aquaporin genes in planta: A key to understand their physiological functions? *Biochim. Biophys. Acta (BBA)-Biomembr.* **2006**, *1758*, 1142–1156. [CrossRef]

29. Guo, L.; Wang, Z.Y.; Lin, H.; Cui, W.E.; Chen, J.; Liu, M.H.; Chen, Z.L.; Qu, L.J.; Gu, H.Y. Expression and functional analysis of the rice plasma-membrane intrinsic protein gene family. *Cell Res.* **2006**, *16*, 277–286. [CrossRef]

30. Liu, C.; Fukumoto, T.; Matsumoto, T.; Gena, P.; Frascaria, D.; Kaneko, T.; Katsuhara, M.; Kitagawa, Y. Aquaporin *Os*PIP1;1 promotes rice salt resistance and seed germination. *Plant Physiol. Biochem.* **2013**, *63*, 151–158. [CrossRef]

31. Katsuhara, M.; Koshio, K.; Shibasaka, M.; Hayashi, Y.; Hayakawa, T.; Kasamo, K. Overexpression of a barley aquaporin increased the shoot/root ratio and raised salt sensitivity in transgenic rice plants. *Plant Cell Physiol.* **2003**, *44*, 1378–1383. [CrossRef]

32. Cui, X.H.; Hao, F.S.; Chen, H.; Chen, J.; Wang, X.C. Expression of the *Vicia faba VfPIP1* gene in Arabidopsis thaliana plants improves drought resistance. *J. Plant Res.* **2003**, *121*, 207–214. [CrossRef] [PubMed]

33. Alavilli, H.; Awasthi, J.P.; Rout, G.R.; Sahoo, L.; Lee, B.H.; Panda, S.K. Overexpression of a Barley Aquaporin gene, HvPIP2;5 confers salt and osmotic stress tolerance in yeast and plants. *Front. Plant Sci.* **2016**, *7*, 1566. [CrossRef]

34. Sreedharm, S.; Shekhawat, U.K.; Gamapathi, T.R. Transgenic Banana plants overexpressing a native plasma membrane aquaporin MusaPIP1;2 display high tolerance levels to different abiotic stresses. *Plant Biotechnol. J.* **2013**, *8*, 942–952. [CrossRef] [PubMed]

35. Zhuo, C.; Wang, T.; Guo, Z.; Lu, S. Overexpression of MfPIP2-7 from *Medicago falcata* promotes cold tolerance and growth under NO^-_3 deficiency in transgenic tobacco plants. *BMC Plant Biol.* **2016**, *16*, 138. [CrossRef] [PubMed]

36. Xu, C.; Wang, M.; Zhou, L.; Quan, T.; Xia, G. Heterologous expression of the wheat Aquaporin gene *TaPIP2;2* compromises the abiotic stress tolerance of Arabidopsis thaliana. *PLoS ONE* **2013**, *8*, e79618.

37. Aharon, R.; Shahak, Y.; Winniger, S.; Bendov, R.; Kapulink, Y.; Galili, G. Overexpression of a plasma membrane aquaporin in transgenic tobacco improves plant vigour under favourable growth conditions but not under drought or salt stress. *Plant Cell* **2003**, *15*, 439–447. [CrossRef]

38. Kapilan, R.; Vaziri, M.; Zwiazek, J.J. Regulation of aquaporins in plants under stress. *Biol. Res.* **2018**, *51*, 4. [CrossRef] [PubMed]

39. Ayadi, M.; Cavez, D.; Miled, N.; Chaumont, F.; Masmoudi, K. Identification and characterization of two plasma membrane aquaporins in durum wheat (*Triticum turgidum* L. *subsp. durum*) and their role in abiotic stress tolerance. *Plant Physiol. Biochem.* **2011**, *49*, 1029–1039. [CrossRef] [PubMed]

40. Ayadi, M.; Mieulet, D.; Fabre, D.; Verdeil, J.L.; Vernet, A.; Guiderdoni, E.; Masmoudi, K. Functional analysis of the durum wheat gene *TdPIP2; 1* and its promoter region in response to abiotic stress in rice. *Plant Physiol. Biochem.* **2014**, *79*, 98–108. [CrossRef]

41. Gago, J.; Grima-Pettenati, J.; Gallego, P.P. Vascular-specific expression of GUS and GFP reporter genes in transgenic grapevine (*Vitis vinifera* L. cv. Albariño) conferred by the EgCCR promoter of *Eucalyptus gunnii*. *Plant Physiol. Biochem.* **2011**, *49*, 413–419. [CrossRef] [PubMed]

42. Sade, N.; Galle, A.; Flexas, J.; Lerner, S.; Peleg, G.; Yaaran, A.; Moshelion, M. Differential tissue-specific expression of *NtAQP1* in Arabidopsis thaliana reveals a role for this protein in stomatal and mesophyll conductance of CO2 under standard and salt-stress conditions. *Planta* **2013**, *239*, 357–366. [CrossRef]

43. Lu, L.; Dong, C.; Liu, R.; Zhou, B.; Wang, C.; Shou, H. Roles of soybean plasma membrane intrinsic protein GmPIP2;9 in drought tolerance and seed development. *Front. Plant Sci.* **2018**, *9*, 530. [CrossRef]

44. Forrest, K.L.; Bhave, M. The PIP and TIP aquaporins in wheat form a large and diverse family with unique gene structures and functionally important features. *Funct. Integr. Genom.* **2008**, *8*, 115e133. [CrossRef]

45. Siefritz, F.; Tyree, M.T.; Lovisolo, C.; Schubert, A.; Kaldenhoff, R. PIP1 plasma membrane aquaporins in tobacco: From cellular effects to function in plants. *Plant Cell* **2002**, *14*, 869–876. [CrossRef] [PubMed]

46. Mittova, V.; Tal, M.; Volokita, M.; Guy, M. Up-regulation of the leaf mitochondrial and peroxisomal antioxidative systems in response to salt-induced oxidative stress in the wild salt-tolerant tomato species *Lycopersicon pennellii*. *Plant Cell Environ.* **2003**, *26*, 845–856. [CrossRef]

47. Deng, L.; Chen, F.; Jiang, L.; Lam, H.M.; Xiao, G. Ectopic expression of *GmPAP3* enhances salt tolerance in rice by alleviating oxidative damage. *Plant Breed.* **2014**, *133*, 348–355. [CrossRef]

48. Hernández, I.; Alegrel, L.; Munné-Bosch, S. Drought-induced changes in flavonoids and other low molecular weight antioxidants in *Cistus clusii* grown under Mediterranean field conditions. *Tree Physiol.* **2004**, *24*, 1303–1311. [CrossRef]

49. Ben Amor, N.; Ben Hamed, K.; Debez, A.; Grignon, C.; Abdelly, C. Physiological and antioxidant responses of the perennial halophyte *Crithmum maritimum* to salinity. *Plant Sci.* **2005**, *168*, 889–899. [CrossRef]

50. Mandhania, S.; Madan, S.; Sawhney, V. Antioxidant defense mechanism under salt stress in wheat seedling. *Biol. Plant.* **2006**, *227*, 227–231. [CrossRef]

51. Koca, H.; Bor, M.; Ozdemir, F.; Turkan, I. Effect of salt stress on lipid peroxidation, antioxidative enzymes and proline content of sesame cultivars. *Environ. Exp. Bot.* **2007**, *60*, 344–351. [CrossRef]

52. Ellouzi, H.; Ben Hamed, K.; Cela, J.; Munne'-Bosch, S.; Abdelly, C. Early effects of salt stress on the physiological and oxidative status of *Cakile maritima* (halophyte) and *Arabidopsis thaliana* (glycophyte). *Physiol. Plant.* **2011**, *142*, 128–143. [CrossRef] [PubMed]

53. Hood, E.E.; Gelvin, S.B.; Melchers, S.; Hoekema, A. New Agrobacterium helper plasmids for gene transfer to plant. *Transgenic Res.* **1993**, *2*, 208–218. [CrossRef]

54. Chen, H.; Nelson, R.S.; Sherwood, J.L. Enhanced recovery of transformants of *Agrobacterium tumifaciens* after freeze–thaw transformation and drug selection. *Biotechniques* **1994**, *16*, 664–668. [PubMed]

55. Supartana, P.; Shimizu, T.; Nogawa, M.; Shioiri, H.; Nakajima, T.; Haramoto, N.; Nozue, M.; Kojima, M. Development of simple and efficient in-planta transformation method for wheat (*Triticum aestivum* L.) using *Agrobacterium tumefaciens*. *J. Biosci Bioeng.* **2006**, *102*, 162–170. [CrossRef] [PubMed]

56. Garfinkel, D.J.; Nester, E.W. *Agrobacterium tumefaciens* mutants affected in crown gall tumorigenesis and octopine catabolism. *J. Bacteriol.* **1980**, *144*, 732–774. [PubMed]

57. Ohira, K.; Ojima, K.; Fujiwara, A. Studies on the nutrition of rice cell culture 1: A simple defined medium for rapid growth in suspension culture. *Plant Cell Physiol.* **1973**, *14*, 1113–1121.

58. Livak, K.J.; Schmittgen, T.D. Analysis of relative gene expression data using real-time quantitative PCR and the 2 DDCT method. *Methods* **2001**, *25*, 402–408. [CrossRef]

59. Sinclair, T.R.; Zwieniecki, M.A.; Holbrook, N.M. Low leaf hydraulic conductance associated with drought tolerance in soybean. *Physiol. Plant.* **2008**, *132*, 446–451. [CrossRef]

60. Sinclair, T.R.; Ludlow, M.M. Influence of soil water supply on the plant water balance of four tropical grain legumes. *Aust. J. Plant Physiol.* **1986**, *13*, 329–341. [CrossRef]

61. Velikova, V.; Yordanov, I.; Edreva, A. Oxidative stress and some antioxidant system in acid rain treated bean plants: Protective role of exogenous polyamines. *Plant Sci.* **2000**, *151*, 59–66. [CrossRef]
62. Aebi, H. Catalase in vitro. *Method Enzymol.* **1984**, *105*, 121–126.
63. Marklund, S.; Marklund, G. Involvement of the superoxide anion radical in the autoxidation of pyrogallol and a convenient assay for superoxide dismutase. *Eur. J. Biochem.* **1974**, *47*, 469–474. [CrossRef] [PubMed]
64. Kelliher, T.; Starr, D.; Su, X.; Tang, G.; Chen, Z.; Carter, J.; Wittich, P.E.; Dong, S.; Green, J.; Burch, E.; et al. One-step genome editing of elite crop germplasm during haploid induction. *Nat. Biotechnol.* **2019**, *37*, 287–292. [CrossRef] [PubMed]

© 2019 by the authors. Licensee MDPI, Basel, Switzerland. This article is an open access article distributed under the terms and conditions of the Creative Commons Attribution (CC BY) license (http://creativecommons.org/licenses/by/4.0/).

International Journal of
Molecular Sciences

MDPI

Article

Genome-Wide Association Study Reveals Novel Genomic Regions Associated with 10 Grain Minerals in Synthetic Hexaploid Wheat

Madhav Bhatta [1], P. Stephen Baenziger [1], Brian M. Waters [1], Rachana Poudel [2], Vikas Belamkar [1], Jesse Poland [3] and Alexey Morgounov [4,*]

[1] Department of Agronomy and Horticulture, University of Nebraska, Lincoln, NE 68583-0915, USA; madhav.bhatta@huskers.unl.edu (M.B.); pbaenziger1@unl.edu (P.S.B.); bwaters2@unl.edu (B.M.W.); vikas.belamkar@unl.edu (V.B.)
[2] Food Science and Technology Department, University of Nebraska, Lincoln, NE 68588-6205, USA; rpoudel2@huskers.unl.edu
[3] Wheat Genetics Resource Center, Department of Plant Pathology, Kansas State University, Manhattan, KS 66506-5502, USA; jpoland@ksu.edu
[4] International Maize and Wheat Improvement Center (CIMMYT), Emek, 06511 Ankara, Turkey
* Correspondence: a.morgounov@cgiar.org; Tel.: +93-530-406-2822

Received: 20 September 2018; Accepted: 12 October 2018; Published: 19 October 2018

Abstract: Synthetic hexaploid wheat (SHW; *Triticum durum* L. × *Aegilops tauschii* Coss.) is a means of introducing novel genes/genomic regions into bread wheat (*T. aestivum* L.) and a potential genetic resource for improving grain mineral concentrations. We quantified 10 grain minerals (Ca, Cd, Cu, Co, Fe, Li, Mg, Mn, Ni, and Zn) using an inductively coupled mass spectrometer in 123 SHWs for a genome-wide association study (GWAS). A GWAS with 35,648 single nucleotide polymorphism (SNP) markers identified 92 marker-trait associations (MTAs), of which 60 were novel and 40 were within genes, and the genes underlying 20 MTAs had annotations suggesting a potential role in grain mineral concentration. Twenty-four MTAs on the D-genome were novel and showed the potential of *Ae. tauschii* for improving grain mineral concentrations such as Ca, Co, Cu, Li, Mg, Mn, and Ni. Interestingly, the large number of novel MTAs (36) identified on the AB genome of these SHWs indicated that there is a lot of variation yet to be explored and to be used in the A and B genome along with the D-genome. Regression analysis identified a positive correlation between a cumulative number of favorable alleles at MTA loci in a genotype and grain mineral concentration. Additionally, we identified multi-traits and stable MTAs and recommended 13 top 10% SHWs with a higher concentration of beneficial grain minerals (Cu, Fe, Mg, Mn, Ni, and Zn), a large number of favorable alleles compared to low ranking genotypes and checks that could be utilized in the breeding program for the genetic biofortification. This study will further enhance our understanding of the genetic architecture of grain minerals in wheat and related cereals.

Keywords: *Triticum durum*; *Aegilops tauschii*; *Triticum aestivum*; marker-trait associations; genes; bread wheat; genetic biofortification; favorable alleles

1. Introduction

The global population is increasing rapidly and is expected to reach 9.8 billion in 2050 [1]. With the increase in global population, the demand for staple crops will continue to increase. Wheat (*Triticum aestivum* L.) is one of the most important staple crops, and it feeds more than one-third of the world's population, providing carbohydrates, proteins, vitamins, antioxidants, fibers, and minerals [2]. In 2017/2018, wheat production was estimated at 756.8 million tons [3]. Despite the significant growth in wheat production, a large percentage of the population who rely on wheat as a staple crop suffer from deficiencies in minerals such as calcium (Ca), copper (Cu), iron (Fe), magnesium (Mg), and zinc (Zn) [4–6] because of the of low grain mineral concentrations [7]. Increased concentrations of essential minerals and decreased concentrations of toxic minerals such as cadmium (Cd) in wheat grain will have a significant impact on human health. One sustainable and cost-effective approach to increasing essential mineral concentration is through genetic biofortification, which requires identification of cultivars with useful genetic variability for grain minerals and understanding of the physiological and genetic architecture of these minerals in wheat [8].

Grain mineral concentration is dependent on several processes, including mineral absorption from the soil, uptake by the roots, translocation, assimilation, and remobilization to the seed [9]. The involvement of several processes for the accumulation of minerals in grain makes them complex traits, which are most likely controlled by many genes [8]. Quantitative trait loci (QTL) analysis or genome-wide association study (marker-trait associations; MTAs) approaches are widely used to dissect complex traits. In wheat, to date, 13 QTLs and 485 MTAs were identified for Ca [4,8,10], one QTL and 13 MTAs identified for Cd [11,12], 17 QTLs for Cu [8,10,13], 58 QTLs for Fe [5,8,10,13–20], three QTLs for Mg [8,10], 15 QTLs for manganese (Mn) [10,13], and 46 QTLs and 16 MTAs for Zn [5,8,10,13–21]. The identification of QTLs or MTAs for high concentrations of beneficial grain minerals such as Ca, Cu, cobalt (Co), Fe, lithium (Li), Mg, Mn, nickel (Ni), and Zn, and low Cd concentration will assist in genetic biofortification through marker-assisted selection and ultimately assist in ensuring nutritional security.

Improved wheat cultivars contain low concentrations of grain minerals [5] and have narrow genetic variation for grain minerals compared to wheat's wild relatives [22]. Synthetic hexaploid wheat (SHW; *Triticum turgidum* L. × *Aegilops tauschii* Coss.) is being used as a means of introducing novel genes/genetic variation into bread wheat [23,24] and it is a potential source of high grain mineral concentrations [25]. Thus, we selected a panel of 123 synthetic hexaploid wheat genotypes to (i) explore the genetic variation of 10 grain minerals (Ca, Cd, Co, Cu, Fe, Li, Mg, Mn, Ni, and Zn) and grain protein concentration (GPC); (ii) identify marker–trait associations using a genome-wide association study (GWAS) and (iii) candidate genes containing nucleotide variants influencing grain minerals. This report is the first for Cu, Co, Fe, Li, Mg, Mn, and Ni in wheat. Results of this study will facilitate the selection of SHWs for use in wheat improvement programs and in enhancing the nutritive value through the integration of valuable grain mineral favorable alleles from SHWs to meet current and future dietary needs.

2. Results and Discussion

2.1. Phenotypic Variation for Grain Protein Content and Grain Minerals

Genotypic variability for GPC and grain minerals was assessed in 123 SHWs across two years (2016 and 2017) field studies in Turkey. The analysis of variance (ANOVA) combined over these years revealed a significant effect of genotype for all traits, whereas a significant genotype × year effect was observed for GPC, Ca, Cu, Mg, Mn, and Ni (Table 1). Non-significant genotype × year interactions for Cd, Co, Fe, Li, and Zn indicate the genetic stability of these traits across years. A wide range of genotypic variation for GPC and minerals was observed among the 123 SHWs (Table 1). A wide range of genetic variation observed for grain yield (GY) and thousand kernel weight (TKW) in these SHWs was described previously [26]. Variation for GPC ranged from 130 g·kg^{-1} to 168 g·kg^{-1} with an average of 151 g·kg^{-1} in 2016 and from 116 g·kg^{-1} to 169 g·kg^{-1} with an average of 138 g·kg^{-1} in 2017 (Table 1). Similarly, variation for grain Fe concentration combined over two years ranged from 17 to 65 mg·kg^{-1} with an average of 39 mg·kg^{-1} and for grain Zn concentration ranged from 10 to 39 mg·kg^{-1} with an average of 23 mg·kg^{-1}. Some of these SHWs had higher grain Co, Cu, Fe, Li, and Mg concentrations, some had similar grain concentrations of Mn, Ni, and Zn, and some had lower grain Cd and Ca concentrations than previously reported in the Hard Winter Wheat Association Mapping Panel (HWWAMP) consisting of 299 diverse genotypes representing the USA

Great Plains [12]. The lower concentration of grain Ca in the SHWs than in bread wheat cultivars has been reported previously [25]. A previous study had reported much higher grain Cd concentration (up to 0.6 mg·kg^{-1}) in winter wheat [12] than our study. The Cd concentration in the SHWs in this study was < 0.1 mg·kg^{-1}, which is below the regulatory toxic level of 0.2 mg·kg^{-1}. However, the low Cd concentration in SHWs may be reflective of low Cd concentration in the soil, and unless they are grown in a high Cd site, we cannot ascertain whether these lines will provide low-Cd alleles for breeding [12]. Additionally, a previous study on genetic variation for grain Fe, Mn, and Zn concentrations in SHWs reported between 25–30% higher grain mineral concentrations of Fe, Mn, and Zn than bread wheat cultivars and the higher grain mineral concentrations in SHWs were not only due to lower GYs, but also due to a higher nutrient uptake efficiency [25]. This result indicated that the SHWs are potential sources of high grain mineral concentrations and could be used for genetic biofortification of wheat.

Broad-sense heritability estimated across the two years was high (H^2 > 0.60) for GPC, Cu, Fe, Mg, Mn, and Zn concentrations; moderate (>0.40 and <0.60) for Ca and Ni concentrations, and low (<0.40) for Cd, Co, and Li concentrations (Table 1). Higher broad sense heritability indicated that the trait was largely governed by the genotypic effect. These results showed potential for the improvement of GPC, Cu, Fe, Mg, Mn, and Zn concentrations through phenotypic selection within SHWs. Similar heritability for these traits has been reported in previous studies [4,5,8,12,20].

2.2. Principal Component Analysis and Phenotypic Correlation

To understand the association among GY, GPC, and 10 mineral concentrations, a factor analysis using the principal component (PC) method was performed in each year (Figure 1). The first three PCs explained from 74.6% to 75.8% of the total variation in the data in 2016 and 2017, respectively. In 2016, the first PC explained 53.1% of the variation in the data and the variables included were Ca, Cd, Co, Cu, Fe, Mg, Mn, Ni, and Zn; the second PC explained 13.2% of the variation in data and variables included were GY and GPC; and the third PC explained 8.3% of the variation in the data and variable included was Li. Similarly, in 2017, the first PC explained 54.2% of the variation in the data and variables included were Ca, Cd, Cu, Fe, Mg, Mn, and Zn; the second PC explained 13.2% of the total variance and variables included were GY and GPC; and the third PC explained 8.4% of the total variance and the variables included were Co, Li, and Ni. Most of the grain minerals in both years were included in the first PC with positive loadings, implying that the first PC is a measure of overall mineral accumulations in the grain, which was similar to the conclusions of Guttieri et al. [12]. The second PC showed a negative correlation between GY and GPC. The association observed in the factor analysis was supported by the significant positive correlations (r) among most of the grain minerals and the negative correlation of GY and GPC (Table 2).

A significant negative correlation between GY and GPC was reported in previous studies [12,27] and the negative correlation was mainly due to the dilution effect. As expected, the present study also identified a significant negative correlation between GY and GPC (Table 2). Additionally, GY was positively correlated with TKW ($r = 0.37$, $p < 0.0001$ in 2016 and $r = 0.35$, $p < 0.0001$ in 2017), similar to previous studies [26,28]. However, GY was not correlated with grain minerals in this study whereas the significant negative correlation of GY with most of the grain minerals was observed after controlling for TKW (Table 2). This result indicated that TKW masked the true association of GY with minerals and controlling for the effect of TKW is important. Furthermore, canonical correlation analysis between GY and overall mineral concentration identified negative correlation ($r = -0.37$ in 2016 and $r = 0.16$ in 2017) between them. Similarly, several previous studies have identified a negative correlation of GY with grain minerals, including Fe [8,29] and Zn [8,10,12,29], which were reported to be associated with a dilution effect [12]. In the present study, GPC was significantly positively correlated with Ca, Cd, Cu, Fe, Mg, Mn, Ni, and Zn, however, the correlation was not very strong ($0.51 \geq r \geq 0.19$) (Table 2). Additionally, canonical correlation analysis between GPC and overall grain minerals identified a positive correlation ($r = 0.44$ in 2016 and $r = 0.62$ in 2017) between them. Many studies have shown a significant positive correlation of GPC with Fe and Zn concentrations [12,15,16,21], indicating that these traits might have a similar genetic basis and could be improved simultaneously [7]. Additionally, most of the grain minerals had highly significant positive correlations ($p < 0.01$) among each other. For instance, a strong correlation ($r > 0.70$, $p < 0.0001$) between Fe and Zn was observed, and they were also strongly correlated ($r > 0.70$, $p < 0.0001$) with other minerals such as Cu, Mg, Mn, and Zn. Positive correlations among grain minerals have been reported previously. For instance, many studies have shown the significant correlation between Fe and Zn concentrations in wheat [5,12,14,16,20,29]. However, other studies have shown no correlation between Fe and Zn [7,30], indicating the genotypic and environmental influence on the relationship between these traits.

Cadmium is a toxic heavy metal that causes harm to human health. Reducing the grain Cd concentration is one of the important plant breeding objectives for creating healthier grains along with the enhancement of beneficial grain mineral concentrations [31]. The current study identified the significant positive correlation between grain Cd concentration with other minerals (Table 2). The previous study in HWWAMP (in this case, using 286 genotypes) also identified a significant positive correlation between grain Cd and Zn concentration, however, the correlation was not very strong ($r < 0.49$) [12]. However, independent genetic regulation of Cd and Zn has been reported [31], which may help explain this weak correlation. The current study identified a weak to moderate correlation ($r < 0.70$) of grain Cd concentration with other grain minerals, implying that enhancement of beneficial mineral concentration may be possible without further increasing grain Cd concentration.

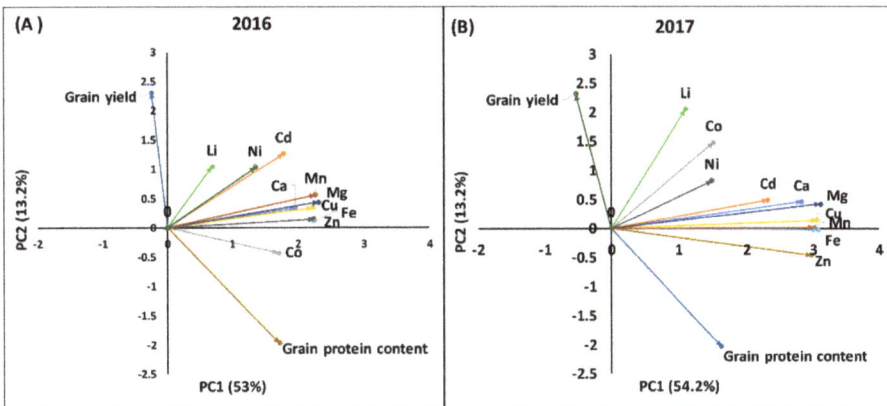

Figure 1. Factor analysis using the principal component method based on correlation matrices on grain yield, grain protein concentration, and 10 grain mineral concentrations in 123 synthetic hexaploid wheat lines grown in 2016 (**A**) and 2017 (**B**) in Konya, Turkey. PC1 = the first principal component analysis. PC2 = the second principal component analysis.

Table 1. Analysis of variance and phenotypic variation for 10 grain minerals, grain protein concentration, and grain yield with minimum (min), maximum (max), fold change (max/min), mean, and broad sense heritability (H^2) values of 123 synthetic hexaploid wheat (SHW) grown in 2016 and 2017 in Konya, Turkey.

Trait	2016				2017				Trials Combined			
	Min	Max	Fold	Mean	Min	Max	Fold	Mean	Year (Yr)	Genotype (G)	G x Yr	H^2
Ca (mg·kg⁻¹)	47.5	167.2	3.5	103.1	21.6	84.5	3.9	44.3	*	**	***	0.41
Cd (mg·kg⁻¹)	0.03	0.1	3.44	0.07	0.02	0.13	7.68	0.07	NS	*	NS	0.28
Co (mg·kg⁻¹)	0.01	0.06	6.53	0.03	0.01	0.04	6.86	0.02	***	*	NS	0.33
Cu (mg·kg⁻¹)	2.8	11.4	4.1	7.5	2.9	8.9	3.1	5.7	NS	***	***	0.63
Fe (mg·kg⁻¹)	17.7	61.8	3.5	40.2	15.4	67.7	4.4	38.5	NS	***	NS	0.78
Li (mg·kg⁻¹)	0.04	0.23	6.43	0.09	0.13	1.07	8.43	0.52	***	*	NS	0.35
Mg (mg·kg⁻¹)	617	2097	3	1391	659	2131	3	1458	NS	***	**	0.62
Mn (mg·kg⁻¹)	20.3	66.2	3.3	41.2	21.5	69.8	3.2	44.9	NS	***	**	0.67
Ni (mg·kg⁻¹)	0.21	2.22	10.81	0.91	0.13	1.16	8.75	0.48	NS	***	***	0.52
Zn (mg·kg⁻¹)	8.8	38.1	4.3	23.1	11.1	39.6	3.6	23	NS	***	NS	0.65
Grain protein (g·kg⁻¹)	129.8	167.6	1.3	151.2	116.4	168.9	1.5	137.8	***	***	**	0.68
Grain yield (g·m⁻²)	54.3	530	9.8	259	194.7	479.5	2.5	290.1	NS	*	*	0.44

*, **, and *** = significant at the 0.05, 0.01, and 0.001 probability level, respectively; NS = non-significant at the 0.05 probability level.

Table 2. Pearson's correlation coefficients of 10 grain minerals, grain protein content (GPC), grain yield (GY), and GY controlling for thousand kernel weight (GY_{pTKW}) in 123 synthetic hexaploid wheat grown in 2016 (upper triangle) and 2017 (lower triangle) in Konya, Turkey.

Trait	Ca	Cd	Co	Cu	Fe	Li	Mg	Mn	Ni	Zn	GPC	GY	GY_{pTKW}
Ca	1	0.63 ***	0.42 ***	0.64 ***	0.58 ***	0.31 ***	0.80 ***	0.79 ***	0.44 ***	0.60 ***	0.36 ***	−0.01	−0.11
Cd	0.64 ***	1	0.32 ***	0.68 ***	0.61 ***	0.38 ***	0.65 ***	0.67 ***	0.58 ***	0.63 ***	0.22 *	−0.03	−0.40 ***
Co	0.37 ***	0.34 ***	1	0.49 ***	0.63 ***	0.19 *	0.49 ***	0.49 ***	0.54 ***	0.53 ***	−0.05	0.13	0.01
Cu	0.82 ***	0.67 ***	0.42 ***	1	0.79 ***	0.27 **	0.84 ***	0.81 ***	0.43 ***	0.89 ***	0.22 *	0.04	−0.18 *
Fe	0.80 ***	0.67 ***	0.46 ***	0.89 ***	1	0.24 **	0.79 ***	0.82 ***	0.48 ***	0.84 ***	0.19 *	0.08	−0.13
Li	0.41 ***	0.33 ***	0.43 ***	0.90 ***	0.30 **	1	0.38 ***	0.26 **	0.29 **	0.17	0.09	−0.13	−0.14
Mg	0.87 ***	0.66 ***	0.47 ***	0.90 ***	0.89 ***	0.38 ***	1	0.88 ***	0.50 ***	0.83 ***	0.20 *	−0.01	−0.19 *
Mn	0.77 ***	0.64 ***	0.46 ***	0.87 ***	0.87 ***	0.26 **	0.91 ***	1	0.50 ***	0.83 ***	0.19 *	−0.08	−0.25 **
Ni	0.32 ***	0.27 **	0.37 ***	0.42 ***	0.38 ***	0.30 ***	0.46 ***	0.34 ***	1	0.38 ***	0.21 *	−0.02	−0.28 **
Zn	0.75 ***	0.65 ***	0.32 ***	0.85 ***	0.86 ***	0.14	0.87 ***	0.85 ***	0.31 ***	1	0.23 *	0.07	−0.19 *
GPC	0.31 ***	0.22 *	0.12	0.43 ***	0.48 ***	−0.08	0.40 ***	0.47 ***	0.19 *	0.51 ***	1	−0.37 ***	−0.36 ***
GY	0.03	0.04	0.02	−0.01	−0.07	0.13	0.05	0.02	−0.03	−0.07	−0.36 ***	1	NA
GY_{pTKW}	−0.11	−0.10	0.07	−0.19 *	0.22 *	0.06	−0.14	−0.15	−0.13	−0.24 **	−0.44 ***	NA	1

*, **, and *** = significant at the 0.05, 0.01, and 0.001 probability level, respectively. NA: Not Applicable.

2.3. Selection of Top-Ranking Genotypes

The 13 top 10% SHW lines were selected from two years of combined data that had higher amounts of GPC and beneficial grain mineral concentrations (Cu, Fe, Mg, Mn, Ni, and Zn) compared to checks and lower ranking genotypes and lower Cd concentration compared to lower ranking genotypes (Table S1). For instance, the Fe and Zn concentration in top ranking genotypes ranged from 49.5 to 56.0 mg·kg^{-1} and 29 to 35 mg·kg^{-1}, respectively, whereas Cd concentration ranged from 0.07 to 0.08 mg·kg^{-1}. This result indicated that these genotypes could be used in the breeding program as parents with a goal of increasing beneficial grain minerals for addressing global mineral deficiencies while decreasing toxic compounds such as Cd.

2.4. Population Structure and Genome-Wide Association Study

Population structure analysis of the 123 SHWs was performed using 35,648 high quality genotyping-by-sequencing (GBS)-derived single nucleotide polymorphisms (SNPs) (minor allele frequency; MAF > 0.05 and missing data < 20%) that were well distributed across 21 chromosomes [26]. Our previous study on genetic diversity and population structure analysis of 101 SHWs identified a large amount of novel genetic variation that could be utilized in broadening the genetic diversity of bread wheat germplasms [23]. The population structure analysis identified that the 123 SHWs can be divided into three subgroups as described previously [26].

The substantial genetic diversity in these SHWs and the dense SNP markers could be useful in identifying genetic factors underlying the variation for grain minerals using GWAS. A GWAS analysis performed using a multi-locus mixed linear model implemented in the FarmCPU algorithm with 35,648 GBS-derived SNPs for 10 grain minerals identified a total of 92 MTAs distributed across 20 chromosomes (Figure 2) with phenotypic variance explained (PVE) up to 25% (Table S2). Thirty-five MTAs were detected on the A genome, 32 MTAs on the B genome, and 25 MTAs on the D-genome of SHWs (Figure 2). The Manhattan and quantile–quantile (Q–Q) plots obtained from the GWAS were shown in Figure S1.

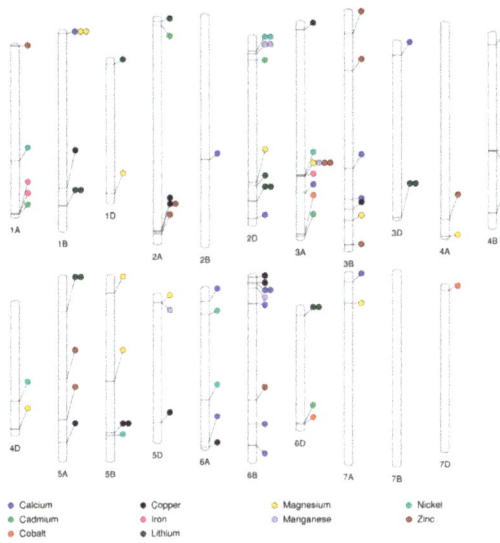

Figure 2. Significant marker–trait associations (MTAs) identified on each chromosome for 10 grain minerals from a genome-wide association study using 35,648 single nucleotide polymorphisms (SNPs) in 123 synthetic hexaploid wheat grown in 2016 and 2017 in Konya, Turkey.

2.4.1. Calcium

The 15 MTAs for Ca concentration were observed in 14 different genomic regions on chromosomes 1B, 2B, 2D, 3A, 3B, 3D, 6A, 6B, 7A (Figure 2) and the PVE by these MTAs ranged from 2.7% to 21.5% (Table S2), indicating a quantitative nature of inheritance for Ca concentration. Earlier studies have reported QTLs/MTAs for Ca on chromosomes 1A [8,10], 2A [10], 2D [10], 5A [4], 2B, 4A, 4B, 5B, 6A, and 7B [8] in wheat, indicating the involvement of these chromosomes in different mapping populations for Ca concentration. However, it is difficult to align our findings with earlier studies because of the employment of different marker systems (such as 90K SNP, short sequence repeat (SSR), and diversity arrays technology (DART)), the lack of precise location information in previous literature, or the utilization of a different version of the reference wheat genome other than the International Wheat Genome Sequencing Consortium (IWGSC) RefSeq v1.0 [26]. However, the associations identified on the same chromosomes as the previous study provided confidence in the reliability of these MTAs. The 11 MTAs identified in this study on chromosomes 1B, 3A, 3B, 3D, 6B, and 7A have not been reported and they are potentially novel MTAs controlling grain Ca concentration. Interestingly, no studies have identified a QTL in the D-genome.

2.4.2. Cadmium

The five MTAs for Cd concentration were observed in five different genomic regions on chromosomes 1A, 2A, 2D, 3A, and 6D (Figure 2) with PVE ranging from 1.8% to 14.4% (Table S2). A previous study on QTL analysis in durum wheat identified a major QTL on chromosome 5B [11]. A GWAS was conducted using 286 winter wheat association mapping populations and identified 12 MTAs for Cd on chromosome 5A [31]. All five MTAs identified in this study are potentially novel MTAs controlling grain Cd concentration. Our study did not find the QTLs identified in the earlier studies; this may be due to the complexity of the trait and different genotypes used in this study. The identification of novel MTAs in the D-genome (2D and 6D) clearly represent variation from the *Ae. tauschii* and show the potential of SHW for its utilization in a marker-assisted breeding program upon validation in an independent genetic background.

2.4.3. Cobalt, Lithium, and Nickel

The present study identified three MTAs on chromosomes 3A, 6D, and 7D for Co, 13 MTAs on chromosomes 1B, 1D, 2A, 2D, 3D, 5A, and 6D for Li, and eight MTAs on chromosomes 1A, 2D, 3A, 4D, 5B, and 6A for Ni (Figure 2 and Table S2). There is no previous report on QTL or GWAS analysis for Co, Li, and Ni in wheat. Therefore, all the MTAs identified for Co, Li, and Ni are potentially novel MTAs responsible for Co, Li, and Ni concentrations. Interestingly, our study identified several MTAs on the D-genome for Co, Li, and Ni, which showed the utility of SHWs for the improvement of these traits.

2.4.4. Copper

A total of 13 MTAs for Cu were identified on chromosomes 1B, 2A, 3A, 3B, 4B, 5A, 5B, 5D, 6A, and 6B (Figure 2) with PVE ranging from 1.2% to 17.1% (Table S2). Earlier studies have identified one QTL for Cu concentration on chromosome 5A in diploid wheat (*T. monococum*) [13], 10 QTLs on chromosomes 1A, 2A, 3B, 4A, 4B, 5A, 6A, 6B, 7A, and 7B in tetraploid wheat [8], and six QTLs on chromosomes 2A, 4A, 4D, 5A, 6A, and 7B in hexaploid wheat [10]. The five MTAs identified in this study on chromosomes 1B, 3A, 5B, and 5D have not been reported and they are potentially novel MTAs controlling grain Cu concentration.

2.4.5. Iron

A total of three MTAs for Fe concentration were identified on chromosomes 1A and 3A (Figure 2) with PVE ranging from 11.2% to 13.2% (Table S2). Earlier studies have identified 58

QTLs distributed on 16 chromosomes (1A, 1B, 2A, 2B, 2D, 3A, 3B, 3D, 4B, 4D, 5A, 5B, 6A, 6B, 6D, 7A, 7B, and 7D) [5,8,10,13–20].

2.4.6. Magnesium

A total of 13 MTAs for Mg concentration were identified on chromosomes 1B, 1D, 2D, 3A, 3B, 4A, 4B, 4D, 5B, 5D, and 7A (Figure 2) with PVE ranging from 1.4% to 14.6% (Table S2). Earlier studies have identified eight QTLs for Mg concentration on chromosomes 1B, 2A, 3A, 5B, 6A, 6B, 7A, and 7B in tetraploid wheat [8] and three QTLs on chromosomes 4A, 5A, and 6A in hexaploid wheat [10]. The six MTAs identified in this study on chromosomes 1D, 2D, 3B, 4B, 4D, and 5D have not been reported and they are potentially novel MTAs controlling grain Mg concentration.

2.4.7. Manganese

A total of six MTAs for Mn concentration were identified on chromosomes 2D, 3A, 4B, 5D, and 6B (Figure 2) with PVE ranging from 4.4% to 14.3% (Table S2). Earlier studies have identified one QTL on chromosome 5A in *T. monoccocum* [13], two QTLs for Mn concentration on chromosomes 2B and 7B in tetraploid wheat [8] and four QTLs on chromosomes 1A, 2B, 3B in hexaploid wheat [10]. All the six MTAs identified in this study on chromosomes 2D, 3A, 4B, 5D, and 6B have not been reported and they are potentially novel MTAs controlling grain Mn concentration.

2.4.8. Zinc

A total of 13 MTAs for Zn concentration were identified on chromosomes 1A, 2A, 3A, 3B, 4A, 4B, 5A, and 6B (Figure 2) with PVE ranging from 1.8% to 14.1% (Table S2). Earlier studies have identified 46 QTLs on 15 chromosomes 1A, 1B, 1D, 2A, 2B, 3A, 3D, 4A, 4B, 4D, 5A, 6A, 6B, 7A, and 7B for Zn concentration [5,8,10,13–21]. Additionally, a previous GWAS for Zn concentration identified 13 MTAs on chromosomes 1B, 3A, and 4B [31]. Three MTAs identified in this study on chromosome 3B have not been reported and they are potentially novel MTAs controlling grain Zn concentration.

2.5. Relationship Between Grain Mineral Concentrations and Number of Favorable Alleles

The number of favorable alleles in a genotype is the cumulative number of alleles from MTAs that increase the concentration of beneficial minerals while decreasing the Cd concentration. A linear relationship between grain mineral concentration and the number of favorable alleles per genotype was observed (Figure 3), implying that the addition of every favorable allele in a genotype contributed to increasing beneficial grain mineral concentrations while decreasing grain Cd concentration. The number of favorable alleles within 123 SHWs ranged from 9 to 37 alleles (Table S3) and variance explained (R^2) by favorable alleles on grain minerals ranged from 10% to 53% (Figure 3). The top-ranking 13 genotypes have a high number of favorable alleles, ranging from 23 to 27 alleles (Table S1). This result suggested that pyramiding these favorable alleles can enhance the grain mineral concentration and be used in a breeding program for genetic biofortification.

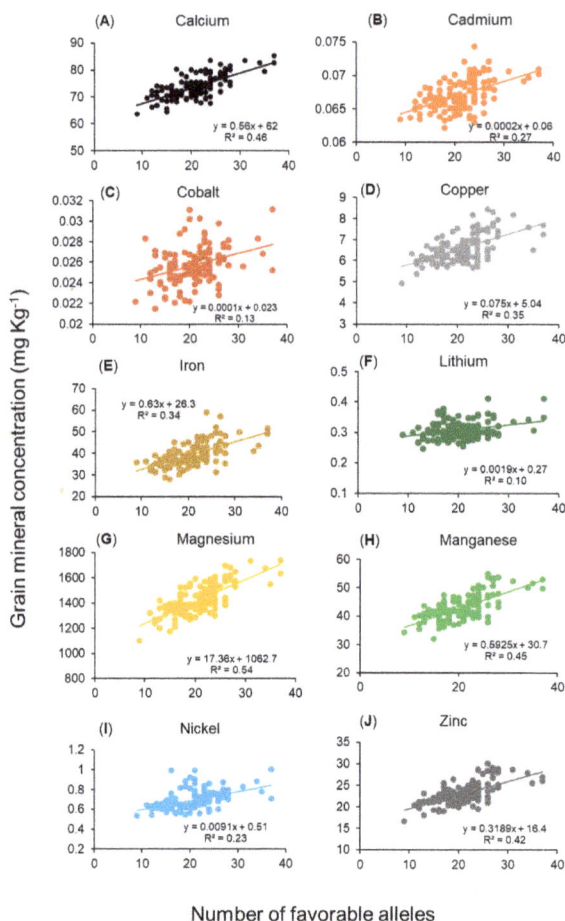

Figure 3. Regression analysis between the total number of favorable alleles per genotype and best linear unbiased predictor values of grain mineral concentrations obtained from two years (2016 and 2017) of experiments conducted in Konya, Turkey. The number of favorable alleles is the total number of alleles present in a genotype that increases the grain concentration of beneficial minerals such as calcium (**A**), cobalt (**C**), copper (**D**), iron (**E**), lithium (**F**), magnesium (**G**), manganese (**H**), nickel (**I**), and zinc (**J**), while decreasing grain cadmium (**B**) concentration.

2.6. Multi-Trait and Stable Marker-Trait Associations

The present study identified common regions associated with multiple traits on chromosomes 1B, 2A, 3A, and 5B (Figure 2 and Table S2). For instance, the MTA for Ca and Mg was identified on chromosome 1B at 6,867,825 bp, Cu and Zn on chromosomes 2A at 742,969,119 bp, Cu and Mg on chromosome 5B at 607,870,649 bp, and Mg, Mn, and Zn on chromosome 3A at 534,469,328 bp. The co-locations of MTAs for Ca, Cu, Mg, Mn, and Zn indicated the same genomic region controlling these traits, which was also supported by highly significant strong positive correlations among those minerals (Table 2). These results suggested that the relationship among these traits was at the molecular level, indicating a common genetic basis for these traits which could be improved simultaneously. The QTL co-localization for some of the minerals have previously been reported. The co-localization of grain Zn QTLs with grain Fe QTLs in tetraploid wheat [8] and hexaploid wheat [5,16,20,21], and QTLs for Mn co-located with Fe concentration in tetraploid wheat [8] have been observed. Co-localization may have occurred either by pleiotropy of the same gene involved in controlling mineral concentrations of different elements, or by the presence of different linked genes in the same regions controlling mineral concentrations of different elements independently. Although Cd was significantly associated with grain minerals, we did not find co-localization of grain Cd MTAs with the MTAs of other grain minerals, indicating that the grain Cd concentration may be governed by a different genetic mechanism, as described previously [31].

In this study, we identified a stable MTA for Ca on chromosome 6B (32,333,184 bp), Cu on 5B (607,870,649 bp), Mg on 1B (867,825 bp), Mn on 2D (58,740,285 bp), Ni on 2D (48,611,294 bp), Zn on 3A (534,469,328 bp), and five stable MTAs for Li on chromosomes 1B (606,491,241 bp), 2D (572,031,650 bp), 3D (610,567,350 bp), 5A (135,164,381 bp), and 6D (30,744,756 bp) (Table S2). The stable MTAs identified could be used for the genetic improvement of these traits.

2.7. Genes Underlying Marker-Trait Associations

The MTAs that were identified were searched against the IWGSC RefSeq v1.0 annotation to identify genes underlying the various MTAs identified in this study. Identification of underlying genes with annotations matching the trait function would provide further confidence for these MTAs. The 40 MTAs (Ca: 8 MTAs; Cd: 1 MTA; Co: 2 MTAs; Cu: 4 MTAs; Fe: 3 MTAs; Li: 4 MTAs; Mg: 5 MTAs; Mn: 3 MTAs; Ni: 3 MTAs; and Zn: 7 MTAs) for 10 grain minerals were found within genes distributed on chromosomes 1A, 1B, 2A, 2D, 3A, 3B, 4A, 4B, 4D, 5A, 6A, 6B, 6D, and 7A (Table S4). Of these, 28 MTAs were present in 19 genes whose annotations indicate they are associated with grain minerals (Table 3 and Table S4). For instance, MTAs for Fe were located in genes related to Fe concentration such as 2-oxoglutarate (2OG), Fe(II)-dependent oxygenase superfamily protein [32,33], ATP synthase gamma chain [34], F-box family protein domain [33], GDSL esterase/lipase [35], leucine-rich receptor-like protein kinase [36,37], Myb transcription factor [36,37], Na-translocating NADH–quinone reductase subunit A [38], no apical meristem (NAM) protein [39], protein DETOXIFICATION [37], ROP guanine nucleotide exchange factor 10 [33], and universal stress protein family [37]. Additional examples are provided in Table 3. This result provides further evidence for these MTAs and indicates that these genes could be important for grain minerals in wheat, however, functional characterization studies are needed to validate the function of these genes.

Furthermore, we identified several MTAs for the same or multiple traits located within genes that had the same gene annotations (Table 3 and Table S2). For instance, some of the MTAs for the same traits such as Mg on chromosomes 3B and 4D, and Zn on chromosomes 3B and 6B were within genes that were both annotated as F-box family protein domain. Similarly, some of the MTAs for multiple traits such as Ca (1 MTA) on chromosome 6B, Li (1 MTA) on 2D, Mg (1 MTA) on 4A, and Zn (1 MTA) on 3B were within genes annotated as leucine-rich repeat receptor-like protein kinase (Table 3), indicating that these genes may be important for improving multiple traits. Multiple MTAs for different traits within genes having the same gene annotation was also reported in our previous study on drought stress related-traits [26].

Table 3. Potential candidate genes containing/flanking marker–trait associations for improving grain minerals in SHWs.

Gene Annotation (Gene ID)	Trait in Our Study [a]	Chromosome	PVE (%) [b]	Traits Influenced Based on the Annotations	References for the Association of Annotations with Traits
2-oxoglutarate (2OG) and Fe (II)-dependent oxygenase superfamily protein (TraesCS2A01G519900-TraesCS2A01G520000)	Cu (1)	2A	5.3	Fe, Mg	[33,40]
AP2-like ethylene-responsive transcription factor (TraesCS2A01G514200)	Cu (1)	2A	3.1	As	
ATP synthase gamma chain (TraesCS6B01G117700)	Ca (1)	6B	19.9	Fe, Zn	[41]
Chaperone protein dnaJ (TraesCS4B01G187600)	Zn (1)	4B	14.1	Cd	[32]
F-box family protein domain (TraesCS6D01G360300, TraesCS3B01G479800, TraesCS3B01G111900, TraesCS6B01G268400, TraesCS6D01G064500, TraesCS2D01G106500, TraesCS4D01G333100)	Co (1), Li (1), Mg (2), Mn (1), Zn (2)	2D, 3B, 4D, 6B, 6D	1.8–25.2	Fe	[34]
GDSL esterase/lipase (TraesCS5A01G096300)	Li (1)	5A	4.4	Fe, Zn, Mn	[42]
Kinase family protein (TraesCS1B01G375400)	Li (1)	1B	13.5	Cd, Zn	[35,36]
Leucine rich receptor-like protein kinase (TraesCS2D01G466400, TraesCS6B01G384300-TraesCS6B01G384400, TraesCS4A01G490700, TraesCS3B01G192500)	Ca (1), Li (1), Mg, Zn	2D, 3B, 4A, 6B	1.8–12.6	Fe	[35,36,42]
MYB transcription factor (TraesCS6B01G053100)	Ca (1)	6B	9.9	Cd, Fe, Zn	[37]
Na-translocating NADH-quinone reductase subunit A (TraesCS1A01G432900)	Fe (1)	1A	11.2	Fe	[38]
No apical meristem (NAM) protein (TraesCS7A01G068200)	Ca (1)	7A	11.8	Fe, Zn, N	[42]
Peroxidase (TraesCS6A01G081700)	Ca (1)	6A	9	Cd	[43]
Phosphate translocator (TraesCS3B01G192400)	Zn (1)	3B	1.8	P	[44]
Potassium transporter (TraesCS2D01G106600)	Mn (1)	2D	8.7	K	[45]
Protein COBRA, putative (TraesCS4B01G187300)	Mn (1)	4B	13.4	Al	[46]
Protein DETOXIFICATION (TraesCS3A01G300400)	Mg (1)	3A	14.6	Fe	[46]
Protein ROOT HAIR DEFECTIVE 3 homolog (TraesCS1A01G003300-TraesCS1A01G003400)	Zn (1)	1A	3	Cd	[32]
ROP guanine nucleotide exchange factor 10 (TraesCS4D01G333000)	Mg (1)	4D	7.9	Fe	[36]
Universal stress protein family (TraesCS3B01G418000)	Ca (1)	3B	2.9	Fe, Zn	[37]

[a] The count of marker–trait associations (in the parenthesis) for either single or multiple traits located within genes that have the same gene annotation; [b] PVE, phenotypic variance explained by the MTA.

3. Materials and Methods

3.1. Plant Materials and Experimental Design

The detail of the experimental materials and design were described previously [26]. In brief, a diversity panel of 123 SHWs originating from the International Maize and Wheat Improvement Center (CIMMYT), Mexico and Kyoto University, Japan were used (Table S3). Grain samples from each plot were obtained from field trials conducted in 2016 and 2017 growing seasons at the research farm located at the Bahri Dagdas International Agricultural Research Institute in Konya, Turkey (37°51′15.894″ N, 32°34′3.936″ E; elevation = 1021 m). The mean monthly temperature in both growing seasons was similar [26], however, the total rainfall in the 2017 growing season (243 mm) was slightly higher than that observed in the 2016 growing season (222 mm) [26]. However, rainfall in both growing seasons was below the 25-year average (435 mm), suggesting the presence of drought-stressed environmental conditions [26]. The soil texture was clayey loam, with a mean pH of 7.7 in 2016 and 8.2 in 2017 (Table S5). Details on soil analysis are provided in Table S5. The experimental design in the 2016 growing season was an augmented design with replicated checks ("Gerek" and "Karahan") and modified alpha lattice design with replicated checks (Gerek and Karahan) and two replications in 2017 as described previously [26].

3.2. Grain Yield, Thousand Kernel Weight, Grain Protein Concentration, and Grain Mineral Analysis

Grain yield, TKW, and GPC were measured using previously reported protocols [26–28]. Whole grain mineral analysis was performed as described previously [12]. In brief, approximately 2 g of oven dried grains were digested with concentrated nitric acid (Optima, Fisher Chemical, Thermo Fisher Scientific Inc., Waltham, MA, USA) and hydrogen peroxide (30% H_2O_2, Fisher BioReagents, Thermo Fisher Scientific Inc., Waltham, MA, USA). Each digestion set of 50 samples included a reagent blank and 0.25 g of standard reference flour (standard reference material 1567a, National Bureau of Standards, MD). Grain mineral concentrations were determined in duplicate by inductively-coupled plasma mass spectrometry (ICP-MS; Agilent 7500cx, Agilent Technologies Inc., Santa Clara, CA, USA) with Ar carrier and a He collision cell at the University of Nebraska Redox Biology Center, Proteomics and Metabolomics Core. Mineral concentrations for Ca, Cd, Co, Cu, Fe, Li, Mg, Mn, Ni, and Zn were averaged over the duplicates and a reagent blank was subtracted. Mineral concentrations expressed as $mg \cdot kg^{-1}$ (dry weight basis) were used for further analysis [31].

3.3. Phenotypic Data Analysis

Combined over two years, individual year analyses of variance (ANOVA) were computed using a mixed linear model using PROC MIXED in SAS 9.4 [47]. This was performed to estimate the best linear unbiased predictors (BLUPs) and to determine whether significant variations exist among the genotype, year, and their interactions. The details of the mixed linear model used for the analysis was described previously [26]. In brief, for the combined ANOVA, year and check were assumed as fixed effects whereas genotype, genotype x year interaction, replication nested within a year, and incomplete block nested within replications were assumed as random effects. For augmented design in 2016, ANOVA was calculated by assuming check as a fixed effect whereas genotype and incomplete block were assumed as random effects. Incomplete blocks nested within replication, checks fitted into new variable (new variable: check was coded as 0 and entry was coded as 1, where genotype was taken as a new variable x entry), and replications were used to correct for spatial variation in the data. For modified alpha (α) lattice design in 2017, ANOVA was calculated by assuming check as a fixed effect and genotype, replication, and incomplete block nested within replication as random effects. Broad-sense heritability was calculated based on an entry mean basis using the following formula:

$$H^2 = \frac{\sigma^2 g}{\sigma^2 g + \frac{\sigma^2 yr}{n} + \frac{\sigma^2 gxyr}{nr}} \tag{1}$$

where, $\sigma^2{}_g$, $\sigma^2{}_{yr}$, and $\sigma^2{}_{gxyr}$ are the variance components for genotype, year, and genotype x year, respectively, and n and r are the number of years and replications, respectively.

The phenotypic correlation was computed using PROC CORR in SAS using BLUPs of each trait. To understand the association among grain minerals, GPC, and GY, a factor analysis using the principal component (PC) method and varimax rotation was performed on the correlation matrix in each year using the factoextra package in R software [48]. Furthermore, canonical correlation was performed between GY/GPC and mineral concentrations to determine the relationship between GY/GPC and overall mineral concentration.

3.4. Genotyping and SNP Discovery

Genotyping, SNP discovery, and SNP filtering procedures were described previously [23]. Briefly, DNA was extracted from fresh young leaves (approximately 14 days after sowing) using BioSprint 96 Plant Kits (Qiagen, Hombrechtikon, Switzerland). Genotyping was performed using a genotyping-by-sequencing approach [49]. SNP discovery was performed using TASSEL v. 5.2.40 GBS v2 Pipeline [50] with a physical alignment to the Chinese spring genome sequence (RefSeq v1.0) provided by the IWGSC [51]. The identified SNPs were filtered for MAF (>5%) and missing data (<20%) [23,52]. The GBS-derived SNP markers used in this study were provided previously [26].

3.5. Population Structure and Genome-Wide Association Analysis

The population structure analysis was described in our previous study [23]. In brief, the population structure of 123 genotypes was assessed using STRUCTURE v 2.3.4 [53] and the unweighted pair group method with arithmetic mean using TASSEL [54].

A GWAS was performed separately for each year (BLUPs from 2016 (BLUP16), BLUPs from 2017 (BLUP17), and BLUPs combined over years (CBLUP)) (Table S6) to identify MTAs for grain minerals using FarmCPU (fixed and random model circulating probability unification) with population structure (Q_{1-3}) as a fixed effect (covariate) and FarmCPU calculated kinship as a random effect [55] implemented in the MVP R software package (available online: https://github.com/XiaoleiLiuBio/MVP). The identified MTAs were tested against a Bonferroni correction at a 5% level of significance with a $p = 1.4026 \times 10^{-6}$ ($-\log_{10}p = 5.85$) for multiple testing correction. Regression analysis was performed between the cumulative number of favorable alleles in a genotype and the BLUPs of each trait. Functional annotations of genes were retrieved using the IWGSC RefSeq v1.0 annotations provided for Chinese spring [51]. The impact of nucleotide variants on predicted genes or proteins was investigated using SnpEff software (available online: http://snpeff.sourceforge.net/).

4. Conclusions

The SHWs under study are a valuable resource for the genetic improvement of wheat because they were reported to have large amounts of novel genetic diversity (including D-genome diversity [23]), are resistant to multiple stresses [24,26], and showed a weak correlation of GY with GPC and most of the minerals, indicating improvement of grain minerals and GPC without sacrificing yield could be possible. Further, the strong positive correlations observed among the most grain minerals suggested the simultaneous improvement of grain minerals could be possible. The top ranking 13 genotypes with higher concentrations of useful grain minerals and lower concentrations of Cd identified in this study have a large number of favorable alleles and these SHWs could be used as a donor parent in a wheat breeding program for genetic biofortification.

A GWAS identified 92 MTAs, of which 60 MTAs were novel (15 MTAs on the A genome, 21 MTAs on the B genome and 24 MTAs on the D genome). The large number of novel MTAs (36) identified in the AB genome of these SHWs indicated that there is a lot of variation yet to be explored and to be used in the A and B genome. Several MTAs identified in this study were within genes with potential roles in improving grain mineral concentrations based on information for their annotations in the literature, which provided further evidence for the reliability and usefulness of the MTAs identified.

However, further investigation on identified genomic regions could significantly assist in genetic biofortification program. Interestingly, this study identified several MTAs for grain minerals located in genes on different chromosomes that had the same gene annotation, suggesting that the same gene family may play a major role in affecting different grain mineral concentrations in SHWs. This study identified multi-trait (Ca and Mg; Cu and Mg; Mg and Cu; Mg, Mn, and Zn) MTAs on chromosomes 1B, 2A, 3A, and 5B which suggested a common genetic basis of these traits, showing the possibility of simultaneous improvement of these traits. Additionally, we identified several stable MTAs for Ca, Cu, Li, Mg, Mn, Ni, and Zn that could be used for the genetic improvement of grain minerals. In summary, a wide range of useful genetic variations for grain minerals and identification of several stable, co-localized, multi-trait, and novel genomic regions (especially on the D-genome) demonstrate the potential of SHWs in its utilization in the wheat breeding program for the genetic biofortification and this study also provided information towards further understanding the genetic complexity of grain mineral accumulation in wheat.

Supplementary Materials: Supplementary materials can be found at http://www.mdpi.com/1422-0067/19/10/3237/s1.

Author Contributions: Conceptualization, M.B., P.S.B., and A.M.; validation, M.B.; formal analysis, M.B.; methodology, M.B., P.S.B., R.P., V.B., and A.M.; investigation, M.B.; resources, M.B., P.S.B, B.M.W., J.P., and A.M.; data curation, M.B., R.P., and V.B.; writing—original draft preparation, M.B.; writing—review and editing, M.B., P.S.B., B.M.W., R.P., J.P., and A.M.; visualization, M.B.; supervision, P.S.B. and A.M.; project administration, P.S.B. and A.M.

Funding: This work was funded by Monsanto Beachell–Borlaug International Scholarship Program. This project is a collaborative effort between the University of Nebraska–Lincoln, Lincoln, USA, CIMMYT in Turkey (supported by CRP WHEAT; Ministry of Food, Agriculture and Livestock of Turkey; Bill and Melinda Gates Foundation, and UK Department for International Development, grant OPP1133199), and Omsk State Agrarian University (supported by the Russian Science Foundation Project No. 16-16-10005). Partial funding for P.S. Baenziger is from Hatch project NEB-22-328, AFRI/2011-68002-30029, the USDA National Institute of Food and Agriculture as part of the International Wheat Yield Partnership award number 2017-67007-25939, the CERES Trust Organic Research Initiative, and USDA under Agreement No. 59-0790-4-092 which is a cooperative project with the U.S. Wheat and Barley Scab Initiative. Any opinions, findings, conclusions, or recommendations expressed in this publication are those of the authors and do not necessarily reflect the view of the USDA. Cooperative investigations of the Nebraska Agriculture Research Division, University of Nebraska, and USDA-ARS.

Acknowledgments: We would like to acknowledge the Monsanto Beachell–Borlaug International Scholarship Program, the University of Nebraska, Lincoln, USA, and the International Maize and Wheat Improvement Center (CIMMYT) in Turkey for their support throughout the project; to the International Winter Wheat Improvement Program (IWWIP) for providing seeds of synthetic hexaploid wheats; to the CIMMYT–Turkey staff, especially Adem Urglu and Ibrahim Ozturk; to the Bahri Dagdas International Agricultural Research Institute staff, including Emel Özer, Fatih Özdemir, and Enes Yakişir; to Shuangye Wu for technical support with GBS and Sarah Blecha for assisting in DNA extraction and sending samples for genotyping; to the Holland Computing Center at the University of Nebraska, Lincoln, for providing the high performance computing resources; to Javier Seravalli for providing grain mineral data from the University of Nebraska Redox Biology Center Metabolomics and Proteomics Core Facility; to Mitch Montgomery and Greg Dorn for assisting in the greenhouse and field experiments; and to all staff and graduate students in the Baenziger laboratory for their valuable suggestions.

Conflicts of Interest: The authors declare no conflict of interest.

Abbreviations

BLUP	Best Linear Unbiased Predictor
Ca	Calcium
Cd	Cadmium
Co	Cobalt
Cu	Copper
Fe	Iron
GBS	Genotyping-By-Sequencing
GPC	Grain Protein Concentration
GWAS	Genome-Wide Association Study
GY	Grain Yield

Abbreviations

Li	Lithium
MAF	Minor Allele Frequency
Mg	Magnesium
Mn	Manganese
MTA	Marker-Trait Association
Ni	Nickel
PVE	Phenotypic Variance Explained
QTL	Quantitative Trait Loci
SHW	Synthetic Hexaploid Wheat
SNP	Single Nucleotide Polymorphism
TKW	Thousand Kernel Weight
Zn	Zinc

References

1. World Population Prospects. 2017. Available online: https://esa.un.org/unpd/wpp/Download/Standard/Population/ (accessed on 8 August 2018).
2. Poudel, R.; Bhatta, M. Review of nutraceuticals and functional properties of whole wheat. *J. Nutr. Food Sci.* **2017**, *7*. [CrossRef]
3. World food situation. Available online: http://www.fao.org/worldfoodsituation/csdb/en/ (accessed on 8 August 2018).
4. Alomari, D.Z.; Eggert, K.; von Wirén, N.; Pillen, K.; Röder, M.S. Genome-wide association study of calcium accumulation in grains of european wheat cultivars. *Front. Plant Sci.* **2017**, *8*, 1797. [CrossRef] [PubMed]
5. Velu, G.; Tutus, Y.; Gomez-Becerra, H.F.; Hao, Y.; Demir, L.; Kara, R.; Crespo-Herrera, L.A.; Orhan, S.; Yazici, A.; Singh, R.P.; et al. QTL mapping for grain zinc and iron concentrations and zinc efficiency in a tetraploid and hexaploid wheat mapping populations. *Plant Soil* **2017**, *411*, 81–99. [CrossRef]
6. White, P.J.; Broadley, M.R. Biofortification of crops with seven mineral elements often lacking in human diets–iron, zinc, copper, calcium, magnesium, selenium and iodine. *New Phytol.* **2009**, *182*, 49–84. [CrossRef] [PubMed]
7. Welch, R.M.; Graham, R.D. Breeding for micronutrients in staple food crops from a human nutrition perspective. *J. Exp. Bot.* **2004**, *55*, 353–364. [CrossRef] [PubMed]
8. Peleg, Z.; Cakmak, I.; Ozturk, L.; Yazici, A.; Jun, Y.; Budak, H.; Korol, A.B.; Fahima, T.; Saranga, Y. Quantitative trait loci conferring grain mineral nutrient concentrations in durum wheat × wild emmer wheat RIL population. *Theor. Appl. Genet.* **2009**, *119*, 353–369. [CrossRef] [PubMed]
9. Welch, R.M. Importance of seed mineral nutrient reserves in crop growth and development. In *Mineral Nutrition of Crops: Fundamental Mechanisms and Implications*; Rengel, Z., Ed.; CRC Press: Boca Raton, FL, USA, 1999; pp. 205–226. ISBN 978-1-56022-880-6.
10. Shi, R.; Tong, Y.; Jing, R.; Zhang, F.; Zou, C. Characterization of quantitative trait loci for grain minerals in hexaploid wheat (*Triticum aestivum* L.). *J. Integr. Agric.* **2013**, *12*, 1512–1521. [CrossRef]
11. Knox, R.E.; Pozniak, C.J.; Clarke, F.R.; Clarke, J.M.; Houshmand, S.; Singh, A.K. Chromosomal location of the cadmium uptake gene (*Cdu1*) in durum wheat. *Genome* **2009**, *52*, 741–747. [CrossRef] [PubMed]
12. Guttieri, M.J.; Baenziger, P.S.; Frels, K.; Carver, B.; Arnall, B.; Waters, B.M. Variation for Grain mineral concentration in a diversity panel of current and historical great plains hard winter wheat germplasm. *Crop Sci.* **2015**, *55*, 1035–1052. [CrossRef]
13. Ozkan, H.; Brandolini, A.; Torun, A.; AltIntas, S.; Eker, S.; Kilian, B.; Braun, H.J.; Salamini, F.; Cakmak, I. Natural Variation and identification of microelements content in seeds of einkorn wheat (*Triticum Monococcum*). In *Wheat Production in Stressed Environments*; Developments in Plant Breeding; Springer: Dordrecht, The Netherlands, 2007; pp. 455–462. ISBN 978-1-4020-5496-9.
14. Tiwari, V.K.; Rawat, N.; Chhuneja, P.; Neelam, K.; Aggarwal, R.; Randhawa, G.S.; Dhaliwal, H.S.; Keller, B.; Singh, K. Mapping of Quantitative trait loci for grain iron and zinc concentration in diploid A genome wheat. *J. Hered.* **2009**, *100*, 771–776. [CrossRef] [PubMed]

15. Genc, Y.; Verbyla, A.P.; Torun, A.A.; Cakmak, I.; Willsmore, K.; Wallwork, H.; McDonald, G.K. Quantitative trait loci analysis of zinc efficiency and grain zinc concentration in wheat using whole genome average interval mapping. *Plant Soil* **2009**, *314*, 49. [CrossRef]

16. Krishnappa, G.; Singh, A.M.; Chaudhary, S.; Ahlawat, A.K.; Singh, S.K.; Shukla, R.B.; Jaiswal, J.P.; Singh, G.P.; Solanki, I.S. Molecular mapping of the grain iron and zinc concentration, protein content and thousand kernel weight in wheat (*Triticum aestivum* L.). *PLoS ONE* **2017**, *12*, e0174972. [CrossRef] [PubMed]

17. Srinivasa, J.; Arun, B.; Mishra, V.K.; Singh, G.P.; Velu, G.; Babu, R.; Vasistha, N.K.; Joshi, A.K. Zinc and iron concentration QTL mapped in a *Triticum spelta* × *T. aestivum* cross. *Theor. Appl. Genet.* **2014**, *127*, 1643–1651. [CrossRef] [PubMed]

18. Roshanzamir, H.; Kordenaeej, A.; Bostani, A. Mapping QTLs related to Zn and Fe concentrations in bread wheat (*Triticum aestivum*) grain using microsatellite markers. *Iranian J. Genet. Plant Breed.* **2013**, *2*, 10–17.

19. Distelfeld, A.; Cakmak, I.; Peleg, Z.; Ozturk, L.; Yazici, A.M.; Budak, H.; Saranga, Y.; Fahima, T. Multiple QTL-effects of wheat Gpc-B1 locus on grain protein and micronutrient concentrations. *Physiol. Plant.* **2007**, *129*, 635–643. [CrossRef]

20. Crespo-Herrera, L.A.; Velu, G.; Singh, R.P. Quantitative trait loci mapping reveals pleiotropic effect for grain iron and zinc concentrations in wheat. *Ann. Appl. Biol.* **2016**, *169*, 27–35. [CrossRef]

21. Xu, Y.; An, D.; Liu, D.; Zhang, A.; Xu, H.; Li, B. Molecular mapping of QTLs for grain zinc, iron and protein concentration of wheat across two environments. *Field Crops Res.* **2012**, *138*, 57–62. [CrossRef]

22. Cakmak, I. Enrichment of cereal grains with zinc: Agronomic or genetic biofortification? *Plant Soil* **2008**, *302*, 1–17. [CrossRef]

23. Bhatta, M.; Morgounov, A.; Belamkar, V.; Poland, J.; Baenziger, P.S. Unlocking the novel genetic diversity and population structure of synthetic hexaploid wheat. *BMC Genom.* **2018**, *19*, 591. [CrossRef] [PubMed]

24. Morgounov, A.; Abugalieva, A.; Akan, K.; Akın, B.; Baenziger, S.; Bhatta, M.; Dababat, A.A.; Demir, L.; Dutbayev, Y.; Bouhssini, M.E.; et al. High-yielding winter synthetic hexaploid wheats resistant to multiple diseases and pests. *Plant Genet. Res.* **2018**, *16*, 273–278. [CrossRef]

25. Calderini, D.F.; Ortiz-Monasterio, I. Are synthetic hexaploids a means of increasing grain element concentrations in wheat? *Euphytica* **2003**, *134*, 169–178. [CrossRef]

26. Bhatta, M.; Morgounov, A.; Belamkar, V.; Baenziger, P. Genome-Wide Association Study Reveals Novel Genomic Regions for Grain Yield and Yield-Related Traits in Drought-Stressed Synthetic Hexaploid Wheat. *Int. J. Mol. Sci.* **2018**, *19*, 3011. [CrossRef] [PubMed]

27. Bhatta, M.; Regassa, T.; Rose, D.J.; Baenziger, P.S.; Eskridge, K.M.; Santra, D.K.; Poudel, R. Genotype, environment, seeding rate, and top-dressed nitrogen effects on end-use quality of modern Nebraska winter wheat. *J. Sci. Food Agric.* **2017**, *97*, 5311–5318. [CrossRef] [PubMed]

28. Bhatta, M.; Eskridge, K.M.; Rose, D.J.; Santra, D.K.; Baenziger, P.S.; Regassa, T. Seeding rate, genotype, and topdressed nitrogen effects on yield and agronomic characteristics of winter wheat. *Crop Sci.* **2017**, *57*, 951–963. [CrossRef]

29. Morgounov, A.; Gómez-Becerra, H.F.; Abugalieva, A.; Dzhunusova, M.; Yessimbekova, M.; Muminjanov, H.; Zelenskiy, Y.; Ozturk, L.; Cakmak, I. Iron and zinc grain density in common wheat grown in Central Asia. *Euphytica* **2007**, *155*, 193–203. [CrossRef]

30. Graham, R.; Senadhira, D.; Beebe, S.; Iglesias, C.; Monasterio, I. Breeding for micronutrient density in edible portions of staple food crops: Conventional approaches. *Field Crops Res.* **1999**, *60*, 57–80. [CrossRef]

31. Guttieri, M.J.; Baenziger, P.S.; Frels, K.; Carver, B.; Arnall, B.; Wang, S.; Akhunov, E.; Waters, B.M. Prospects for selecting wheat with increased zinc and decreased cadmium concentration in grain. *Crop Sci.* **2015**, *55*, 1712. [CrossRef]

32. Moran Lauter, A.N.; Peiffer, G.A.; Yin, T.; Whitham, S.A.; Cook, D.; Shoemaker, R.C.; Graham, M.A. Identification of candidate genes involved in early iron deficiency chlorosis signaling in soybean (*Glycine max*) roots and leaves. *BMC Genom.* **2014**, *15*, 702. [CrossRef] [PubMed]

33. Li, W.; Lan, P. Genome-wide analysis of overlapping genes regulated by iron deficiency and phosphate starvation reveals new interactions in Arabidopsis roots. *BMC Res. Notes* **2015**, *8*. [CrossRef] [PubMed]

34. Laganowsky, A.; Gómez, S.M.; Whitelegge, J.P.; Nishio, J.N. Hydroponics on a chip: Analysis of the Fe deficient *Arabidopsis* thylakoid membrane proteome. *J. Proteom.* **2009**, *72*, 397–415. [CrossRef]

35. Singh, S.P.; Jeet, R.; Kumar, J.; Shukla, V.; Srivastava, R.; Mantri, S.S.; Tuli, R. Comparative transcriptional profiling of two wheat genotypes, with contrasting levels of minerals in grains, shows expression differences during grain filling. *PLoS ONE* **2014**, *9*, e111718. [CrossRef] [PubMed]

36. Zheng, L.; Huang, F.; Narsai, R.; Wu, J.; Giraud, E.; He, F.; Cheng, L.; Wang, F.; Wu, P.; Whelan, J.; et al. Physiological and transcriptome analysis of iron and phosphorus interaction in rice seedlings. *Plant Physiol.* **2009**, *151*, 262–274. [CrossRef] [PubMed]

37. King, K.E.; Peiffer, G.A.; Reddy, M.; Lauter, N.; Lin, S.F.; Cianzio, S.; Shoemaker, R.C. Mapping of iron and zinc quantitative trait loci in soybean for association to iron deficiency chlorosis resistance. *J. Plant Nutr.* **2013**, *36*, 2132–2153. [CrossRef]

38. Elias, D.A.; Yang, F.; Mottaz, H.M.; Beliaev, A.S.; Lipton, M.S. Enrichment of functional redox reactive proteins and identification by mass spectrometry results in several terminal Fe(III)-reducing candidate proteins in *Shewanella oneidensis* MR-1. *J. Microbiol. Methods* **2007**, *68*, 367–375. [CrossRef] [PubMed]

39. Waters, B.M.; Uauy, C.; Dubcovsky, J.; Grusak, M.A. Wheat (*Triticum aestivum*) NAM proteins regulate the translocation of iron, zinc, and nitrogen compounds from vegetative tissues to grain. *J. Exp. Bot.* **2009**, *60*, 4263–4274. [CrossRef] [PubMed]

40. Liang, W.-W.; Huang, J.-H.; Li, C.-P.; Yang, L.-T.; Ye, X.; Lin, D.; Chen, L.-S. MicroRNA-mediated responses to long-term magnesium-deficiency in *Citrus sinensis* roots revealed by Illumina sequencing. *BMC Genom.* **2017**, *18*. [CrossRef] [PubMed]

41. Billard, V.; Maillard, A.; Garnica, M.; Cruz, F.; Garcia-Mina, J.-M.; Yvin, J.-C.; Ourry, A.; Etienne, P. Zn deficiency in *Brassica napus* induces Mo and Mn accumulation associated with chloroplast proteins variation without Zn remobilization. *Plant Physiol. Biochem.* **2015**, *86*, 66–71. [CrossRef] [PubMed]

42. Song, G.; Yuan, S.; Wen, X.; Xie, Z.; Lou, L.; Hu, B.; Cai, Q.; Xu, B. Transcriptome analysis of Cd-treated switchgrass root revealed novel transcripts and the importance of HSF/HSP network in switchgrass Cd tolerance. *Plant Cell Rep.* **2018**. [CrossRef] [PubMed]

43. Van De Mortel, J.E.; Schat, H.; Moerland, P.D.; Van Themaat, E.V.L.; Van Der Ent, S.; Blankestijn, H.; Ghandilyan, A.; Tsiatsiani, S.; Aarts, M.G.M. Expression differences for genes involved in lignin, glutathione and sulphate metabolism in response to cadmium in Arabidopsis thaliana and the related Zn/Cd-hyperaccumulator Thlaspi caerulescens. *Plant Cell Environ.* **2008**, *31*, 301–324. [CrossRef] [PubMed]

44. Wang, G.; Leonard, J.M.; von Zitzewitz, J.; James Peterson, C.; Ross, A.S.; Riera-Lizarazu, O. Marker–trait association analysis of kernel hardness and related agronomic traits in a core collection of wheat lines. *Mol. Breed.* **2014**. [CrossRef]

45. Waters, B.M.; Grusak, M.A. Quantitative trait locus mapping for seed mineral concentrations in two *Arabidopsis thaliana* recombinant inbred populations. *New Phytol.* **2008**, *179*, 1033–1047. [CrossRef] [PubMed]

46. Chandran, D.; Sharopova, N.; Ivashuta, S.; Gantt, J.S.; VandenBosch, K.A.; Samac, D.A. Transcriptome profiling identified novel genes associated with aluminum toxicity, resistance and tolerance in Medicago truncatula. *Planta* **2008**, *228*, 151–166. [CrossRef] [PubMed]

47. SAS 9.4 Product Documentation. Available online: https://support.sas.com/documentation/94/ (accessed on 16 August 2018).

48. R: The R Project for Statistical Computing. Available online: https://www.r-project.org/ (accessed on 16 August 2018).

49. Poland, J.A.; Brown, P.J.; Sorrells, M.E.; Jannink, J. Development of high-density genetic maps for barley and wheat using a novel two-enzyme genotyping-by- sequencing approach. *PLoS ONE* **2012**, *7*. [CrossRef] [PubMed]

50. Glaubitz, J.C.; Casstevens, T.M.; Lu, F.; Harriman, J.; Elshire, R.J.; Sun, Q.; Buckler, E.S. TASSEL-GBS: A high capacity genotyping by sequencing analysis pipeline. *PLoS ONE* **2014**, *9*, e90346. [CrossRef] [PubMed]

51. International Wheat Genome Sequencing Consortium (IWGSC). Shifting the limits in wheat research and breeding using a fully annotated reference genome. *Science* **2018**, *361*, eaar7191. [CrossRef] [PubMed]

52. Belamkar, V.; Guttieri, M.J.; Hussain, W.; Jarquín, D.; El-basyoni, I.; Poland, J.; Lorenz, A.J.; Baenziger, P.S. Genomic selection in preliminary yield trials in a winter wheat breeding program. *G3* **2018**, *8*, 2735–2747. [CrossRef] [PubMed]

53. Pritchard, J.K.; Stephens, M.; Donnelly, P. Inference of population structure using multilocus genotype data. *Genetics* **2000**, *155*, 945–959. [PubMed]

54. Bradbury, P.J.; Zhang, Z.; Kroon, D.E.; Casstevens, T.M.; Ramdoss, Y.; Buckler, E.S. TASSEL: Software for association mapping of complex traits in diverse samples. *Bioinformatics* **2007**, *23*, 2633–2635. [CrossRef] [PubMed]
55. Liu, X.; Huang, M.; Fan, B.; Buckler, E.S.; Zhang, Z. Iterative usage of fixed and random effect models for powerful and efficient genome-wide association studies. *PLoS Genet.* **2016**, *12*, e1005767. [CrossRef] [PubMed]

© 2018 by the authors. Licensee MDPI, Basel, Switzerland. This article is an open access article distributed under the terms and conditions of the Creative Commons Attribution (CC BY) license (http://creativecommons.org/licenses/by/4.0/).

International Journal of
Molecular Sciences

MDPI

Article

RNA Sequencing-Based Bulked Segregant Analysis Facilitates Efficient D-genome Marker Development for a Specific Chromosomal Region of Synthetic Hexaploid Wheat

Ryo Nishijima [1], Kentaro Yoshida [1,*], Kohei Sakaguchi [1], Shin-ichi Yoshimura [1], Kazuhiro Sato [2] and Shigeo Takumi [1,*]

[1]   Graduate School of Agricultural Science, Kobe University, Rokkodai 1-1, Nada, Kobe 657-8501, Japan; ryosnishijima@gmail.com (R.N.); k.sakaguchi@brown.plala.or.jp (K.S.); shin.82353@gmail.com (S.-i.Y.)
[2]   Institute of Plant Science and Resources, Okayama University, Kurashiki, Okayama 710-0046, Japan; kazsato@rib.okayama-u.ac.jp
*   Correspondence: kentaro.yoshida@port.kobe-u.ac.jp (K.Y.); takumi@kobe-u.ac.jp (S.T.); Tel.: +81-78-803-5858 (K.Y.); +81-78-803-5860 (S.T.)

Received: 2 November 2018; Accepted: 22 November 2018; Published: 26 November 2018

Abstract: Common wheat originated from interspecific hybridization between cultivated tetraploid wheat and its wild diploid relative *Aegilops tauschii* followed by amphidiploidization. This evolutionary process can be reproduced artificially, resulting in synthetic hexaploid wheat lines. Here we performed RNA sequencing (RNA-seq)-based bulked segregant analysis (BSA) using a bi-parental mapping population of two synthetic hexaploid wheat lines that shared identical A and B genomes but included with D-genomes of distinct origins. This analysis permitted identification of D-genome-specific polymorphisms around the *Net2* gene, a causative locus to hybrid necrosis. The resulting single nucleotide polymorphisms (SNPs) were classified into homoeologous polymorphisms and D-genome allelic variations, based on the RNA-seq results of a parental tetraploid and two *Ae. tauschii* accessions. The difference in allele frequency at the D-genome-specific SNP sites between the contrasting bulks (ΔSNP-index) was higher on the target chromosome than on the other chromosomes. Several SNPs with the highest ΔSNP-indices were converted into molecular markers and assigned to the *Net2* chromosomal region. These results indicated that RNA-seq-based BSA can be applied efficiently to a synthetic hexaploid wheat population to permit molecular marker development in a specific chromosomal region of the D genome.

Keywords: allohexaploid; homoeolog; hybrid necrosis; molecular marker; wheat

---

1. Introduction

Wild relatives of common wheat (*Triticum aestivum* L.), including *Aegilops* species, constitute important genetic resources for common wheat breeding. Notably, *Aegilops tauschii* Coss. is the D-genome progenitor of common wheat, and a limited number of *Ae. tauschii* strains appear to have contributed to the speciation of common wheat through interspecific crossing to the cultivated tetraploid wheat (*Triticum turgidum* L. ssp. *durum*) strains carrying the A and B genomes and subsequent amphidiploidization about 8000 years ago in the Fertile Crescent [1,2]. Artificial replication of this evolutionary process can be achieved by generation of synthetic hexaploid wheat derived from the interspecific crosses between cultivated tetraploid wheat and *Ae. tauschii* [3]. Thus, agriculturally useful alleles of *Ae. tauschii* are directly available for common wheat breeding via the transmission from synthetic hexaploid wheat to common wheat [4–10]. In common wheat, the D genome is known to harbor much lower genetic diversity than do the A and B genomes, whereas introgression of the

Ae. tauschii natural variation via synthetic hexaploid wheat has been used successfully to enlarge D-genome diversity [11–14].

Ae. tauschii is widely distributed, and is present from northern Syria and southeastern Turkey to western China [15,16]. Recent studies on the population structure of *Ae. tauschii* showed that only one (TauL1) of three major lineages has contributed to the wide species range [2,17,18]. The other two major lineages, TauL2 and TauL3, are restricted to the Transcaucasus/Middle East region. TauL2 includes both subspecies *tauschii* and subspecies *strangulata*, and the TauL3 accessions are limited in Georgia [18]. The *Ae. tauschii* strains associated with the origin of common wheat are assumed to be the TauL2 lineage [2,17], and only limited reproductive barriers are thought to exist between tetraploid wheat and many of the TauL2 accessions [19]. TauL1 accessions frequently carry a hybrid incompatibility gene, designated *Net2*, that triggers low-temperature-induced necrotic cell death upon interspecific hybridization with tetraploid wheat, impeding the generation of synthetic hexaploid wheat [20,21]. Thus, *Net2* is a major reproductive barrier for breeding use of the TauL1 accessions, and the development of markers closely linked to *Net2* is needed for efficient use of the TauL1 gene pool. A fine map for the *Net2* chromosomal region already has been constructed on the short arm of chromosome 2D [22].

Genomic approaches using next-generation sequencing (NGS) techniques have been applied to analysis of the genomes of the wild relatives of domesticated crops, expanding the genetic resources available for crop improvement [23]. In common wheat, the D-genome markers remain much less developed than those of the A and B genomes, whereas recent progress using the NGS technique has facilitated an increase in the number of D-genome markers [24]. RNA sequencing (RNA-seq) of the *Ae. tauschii* accessions has generated a huge number of genome-wide polymorphisms, including single nucleotide polymorphisms (SNPs) and insertions/deletions (indels); the genome-wide SNPs and indels can be efficiently anchored to the chromosomes of barley and *Ae. tauschii* [25–27]. The SNP- and indel-based markers are available for construction of linkage maps in the target chromosomal regions of not only *Ae. tauschii* but also the D genome of hexaploid wheat including synthetic wheat [26,27].

Bulked segregant analysis (BSA) combined with NGS allows efficient development of molecular markers linked to a genomic region associated with the target phenotype in cereals [28–31]. In maize, for example, an RNA-seq-based BSA approach has been used to construct high-density linkage maps and to screen among candidate genes for a target locus [32–34]. Hexaploid wheat has three closely related genomes (designated A, B, and D), each of which carries a set of highly similar genes (homoeology). Due to the genome complexity via allopolyploidy and the large proportion of repetitive DNA in polyploid wheat, whole-genome resequencing is still unviable and reduced-representation methods of NGS data have been employed in this species [35]. Recently, RNA-seq-based BSA was employed successfully for the development of molecular markers closely linked to target chromosomal genes such as a grain protein content gene (*GPC-B1*), a yellow rust resistance gene (*Yr15*), and a powdery mildew resistance gene (*Pm4b*) in tetraploid and common wheat [36–38]. However, a limited amount of information has been reported for the RNA-seq-based BSA approach in polyploid wheat. Here, we employed the RNA-seq-based BSA method to develop a molecular marker closely linked to *Net2*. This process used a mapping population generated from a cross of two synthetic hexaploid wheat lines that shared identical A and B genomes but contained diverse D genomes originating from two distinct pollen parents.

2. Results and Discussion

Two synthetic hexaploid wheat lines were derived from interspecies crosses of tetraploid wheat cultivar Langdon (Ldn) and two *Ae. tauschii* accessions (KU-2075 and KU-2025). Ldn/KU-2075 and Ldn/KU-2025, respectively, showed normal phenotype (wild-type) and type-II necrosis phenotype [20]. To obtain novel molecular markers tightly linked to *Net2*, RNA sequencing of four bulks of synthetic hexaploid wheat was performed (Figure 1). Each bulk was composed of ten *Net2*-homozygous individuals or non-carriers (*net2*-homozygous) selected from two F_5 populations between Ldn/KU-2075 and Ldn/KU-2025 (hereafter referred to as Segregating Populations SP1 and SP2). The two bulks of each SP (*Net2*-SP1, non-carrier-SP1, *Net2*-SP2, and non-carrier-SP2) were sequenced twice; the bulks were designated as follows: *Net2*-SP1-1st, *Net2*-SP1-2nd, non-carrier-SP1-1st, non-carrier-SP1-2nd, *Net2*-SP2-1st, *Net2*-SP2-2nd, non-carrier-SP2-1st, and non-carrier-SP2-2nd (Table 1). In each experiment, 4.06 million to 5.22 million paired-end reads were generated. After quality filtering, 2.80 million to 3.71 million high-quality reads were acquired. These reads were aligned with transcripts of the two parental synthetic hexaploid *Ae. tauschii* accessions, KU-2075 and KU-2025, transcriptomes that had been de novo assembled in our previous study [23]. In each experiment, 2.02 million to 2.76 million and 1.93 million to 2.68 million reads were aligned to the KU-2075 and KU-2025 transcriptomes, respectively. The alignment output files of the first and second runs were merged for each alignment pair of bulk and transcript. On average, 292,678 and 278,690 SNPs were detected on KU-2075 and KU-2025 transcripts, respectively (Table 2). Of these, 290,712 (99.33%) and 276,700 (99.29%) SNPs on the KU-2075 and KU-2025 transcripts (respectively) were anchored to the *Ae. tauschii* genome. SNP sites in the eight alignment pairs of bulk and transcript (e.g., *Net2*-SP1 vs. KU-2075), as shown in Table 3, were integrated with the genome sequence, yielding a total of 319,808 non-redundant (NR) SNP sites on the seven chromosomes (Table 3). This section may be divided by subheadings. It should provide a concise and precise description of the experimental results, their interpretation as well as the experimental conclusions that can be drawn.

To distinguish D-genome-specific variations from homoeologous polymorphisms between the tetraploid AB and diploid D genomes, the RNA-seq raw read data of Ldn, the tetraploid parental accession of the synthetic wheat lines, were processed as described above. Out of the total of 6.32 million read pairs obtained, 4.37 million read pairs passed the quality filtering (Table 1). To KU-2075 and KU-2025 transcripts, 2.97 million and 2.66 million reads were aligned, respectively.

Table 1. Summary of RNA sequencing results for four pairs of bulks of synthetic hexaploid wheat and for tetraploid wheat cv. Langdon.

Samples	Total Read Pairs	Filtered Read Pairs (%) [a]	Aligned to the *Ae. tauschii* Transcripts [b] (%) [c]	
			KU-2075	KU-2025
Synthetic hexaploids				
non-carrier-SP1-1st	4,202,114	2,799,202 (66.61%)	2,020,405 (72.18%)	1,930,400.5 (68.96%)
non-carrier-SP1-2nd	4,059,840	2,858,956 (70.42%)	2,062,291 (72.13%)	1,971,110.5 (68.95%)
non-carrier-SP2-1st	4,492,358	2,953,088 (65.74%)	2,164,656 (73.3%)	2,083,120 (70.54%)
non-carrier-SP2-2nd	4,115,352	2,864,271 (69.60%)	2,098,383 (73.26%)	2,020,752.5 (70.55%)
Net2-SP1-1st	4,710,499	3,148,652 (66.84%)	2,208,392 (70.14%)	2,110,452.5 (67.03%)
Net2-SP1-2nd	4,403,630	3,108,568 (70.59%)	2,178,403 (70.08%)	2,082,004 (66.98%)
Net2-SP2-1st	4,828,182	3,249,596 (67.30%)	2,420,056 (74.47%)	2,348,814 (72.28%)
Net2-SP2-2nd	5,216,082	3,709,478 (71.12%)	2,763,471 (74.5%)	2,684,019 (72.36%)
Tetraploid wheat				
cv. Langdon	6,316,174	4,372,660 (69.23%)	2,974,277 (68.02%)	2,661,487 (60.87%)

[a] The ratio of the filtered read pairs to the total read pairs. [b] Nishijima et al. [27]. [c] The ratio of the aligned reads to the filtered read pairs.

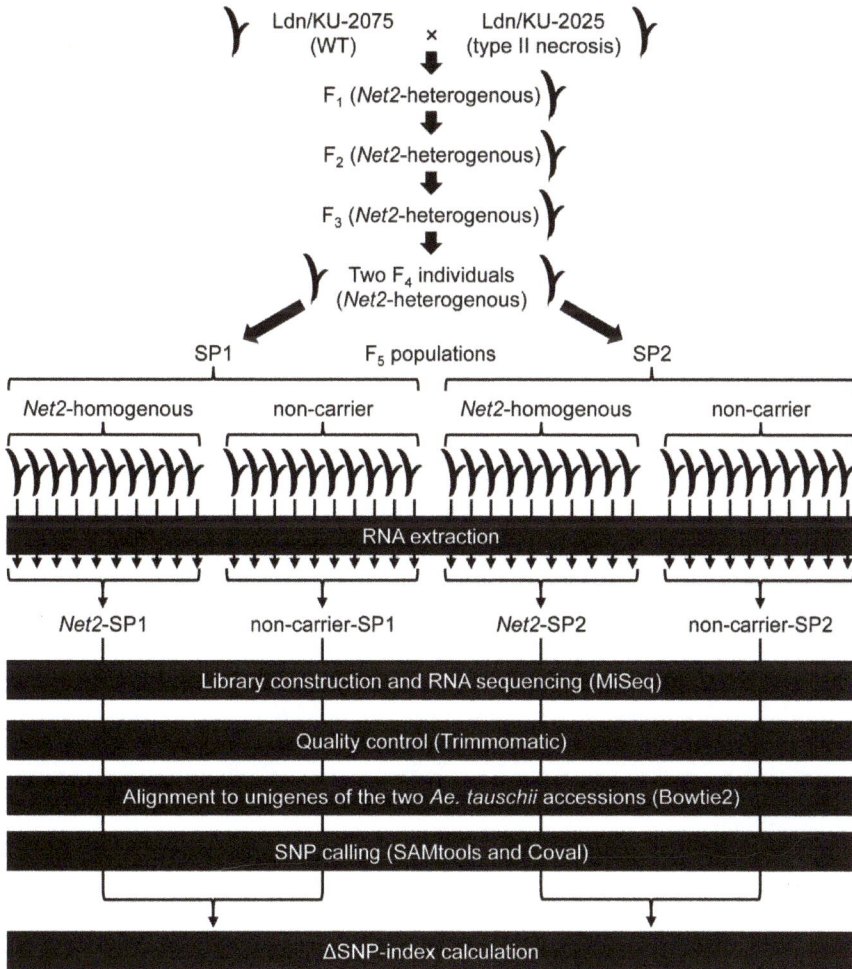

Figure 1. Workflow of the RNA sequencing analysis in this study.

Table 2. The number of single nucleotide polymorphisms (SNPs) detected in four bulks of synthetic hexaploid wheat and tetraploid wheat cv. Langdon compared to the two parental *Ae. tauschii* transcriptomes.

Transcripts [a]	KU-2075		KU-2025	
The Number of SNP	Total	Anchored to the Genome [b] (%)	Total	Anchored to the Genome [b] (%)
Synthetic hexaploids				
non-carrier-SP1	277,605	275,799 (99.35%)	262,966	261,128 (99.30%)
non-carrier-SP2	276,564	274,772 (99.35%)	269,175	267,249 (99.28%)
Net2-SP1	318,046	315,859 (99.31%)	296,819	294,739 (99.30%)
Net2-SP2	298,496	296,419 (99.30%)	285,798	283,684 (99.26%)
Tetraploid wheat				
cv. Langdon	429,346	421,957 (98.28%)	350,871	345,657 (98.51%)

[a] Nishijima et al. [27]. [b] Luo et al. [39].

Table 3. The number of SNPs classified into three categories, including D-genome-specific allelic variations, homoeologous polymorphisms, and unclassified (those falling into neither of the other two classes).

Chr.	D-genome-Specific	Homoeologous	Unclassified	Total
1D	1674	29,307	11,760	42,741
2D	2295	33,822	13,975	50,092
3D	1611	31,932	15,532	49,075
4D	1698	28,781	11,776	42,255
5D	2936	34,961	14,336	52,233
6D	3730	24,534	10,966	39,230
7D	3983	28,593	11,606	44,182
Total	17,927	211,930	89,951	319,808

Out of the 429,346 and 350,871 SNPs identified on KU-2075 and KU-2025 transcripts (respectively), 421,957 (98.28%) and 345,657 SNPs (98.51%) (respectively) were anchored to the *Ae. tauschii* genome (Table 2). Based on these SNPs derived from tetraploid wheat reads, and on the SNPs between the two *Ae. tauschii* accessions reported in our previous study [27], NR SNPs identified in synthetic hexaploid bulks were classified into homoeologous polymorphisms and D-genome-specific allelic variation. Out of the 319,808 SNPs, 17,927 were assigned to the allelic variation on the D genome, and 211,930 were grouped into the homoeologous polymorphisms, while the remained 89,951 did not fall into either of these two classes ("unclassified"; Table 3).

SNP-index values were computed for each of the eight alignment pairs, and then ΔSNP-index values were calculated for the comparisons between *Net2*-homogenous bulks and non-carrier bulks, yielding indices for *Net2*-SP1 minus non-carrier-SP1 and *Net2*-SP2 minus non-carrier-SP2. The average of the four ΔSNP-index values at each SNP site was mapped to the *Ae. tauschii* genome (Figure 2). The ΔSNP-index distribution on the chromosomes indicated that SNPs with high ΔSNP-index values tended to be located on the short arm of chromosome 2D, where the *Net2* locus resides (Figure 2). Comparison of ΔSNP-index values among D-genome-specific SNPs showed that the D-genome-specific SNPs on chromosome 2D possessed significantly higher (Steel-Dwass test, $p < 0.001$) ΔSNP-index values than those on the other six chromosomes (Figure 3). Theoretically, homozygous chromosomal regions account for 87.5% of the total genome in an F_4 individual derived from two parents. D-genome-specific SNP sites with low ΔSNP-index values appeared to already have been fixed as homozygous for the KU-2075 or KU-2025 allele by the F_4 generation of the synthetic hexaploid populations. The ΔSNP-index values of D-genome-specific SNPs on chromosome 2D also were significantly higher (Steel-Dwass test, $p < 0.001$) than those of homoeologous and unclassified SNPs (Figure 3). Moreover, most of the homoeologous and unclassified polymorphisms showed ΔSNP-index values of approximately zero (Figure 3), indicating that homoeologous polymorphisms had been efficiently removed from candidate SNPs for development of molecular markers, since such candidate SNPs should have possessed high ΔSNP-index values. Low ΔSNP-index values at homoeologous polymorphic sites were likely due to the similar allele frequency of A- and B-genome-derived reads between the *Net2*-homogenous and non-carrier bulks, which thereby offset the SNP-index values of each other.

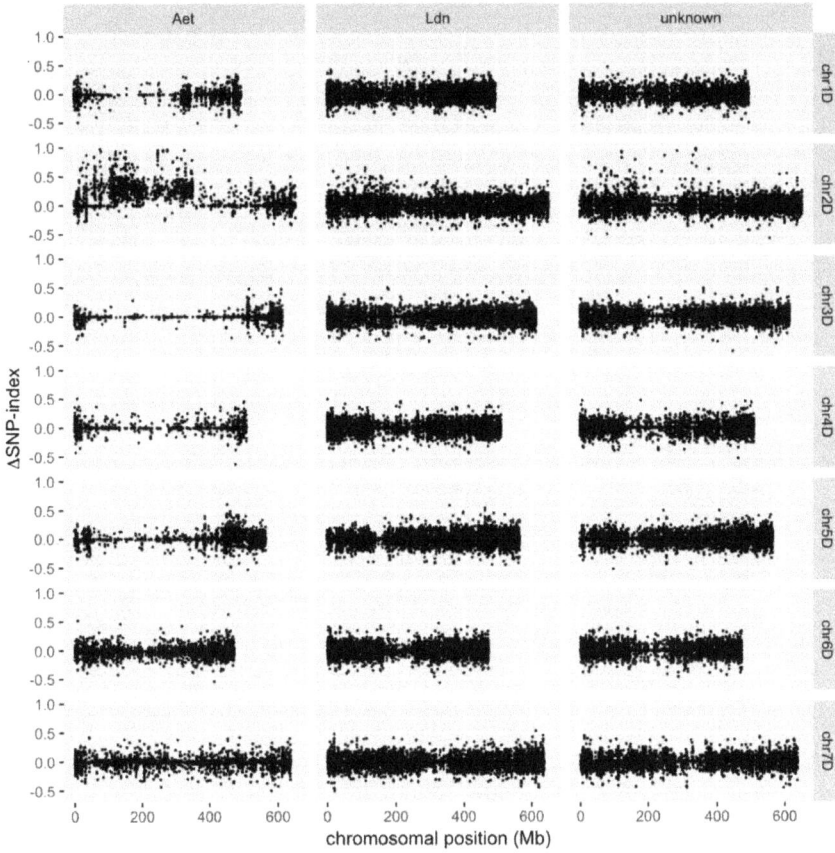

Figure 2. Distribution of ΔSNP-index values along the *Ae. tauschii* chromosomes. The three categories of SNPs (D-genome-specific allelic variations, homoeologous polymorphisms, and unclassified SNPs) are designated as "Aet", "Ldn", and "Unknown", respectively.

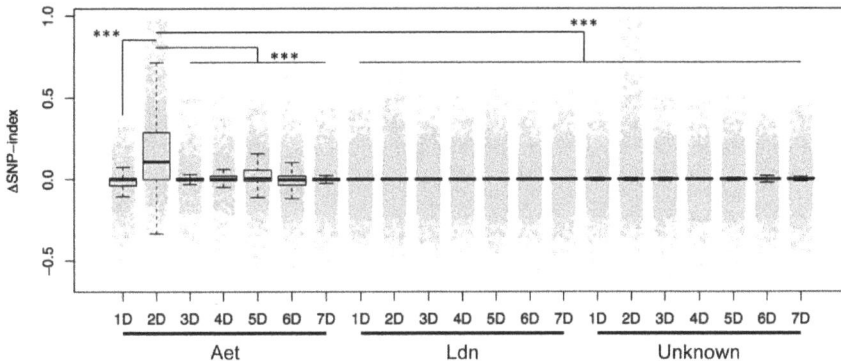

Figure 3. Box and dot plots for ΔSNP-index values based on the seven chromosomes and on the SNP classifications in Figure 2. D-genome-specific SNPs on chromosome 2D had significantly higher ΔSNP-index values than SNPs on the other six chromosomes, homoeologous SNPs, and unclassified SNPs. *** $p < 0.001$ by the Steel-Dwass test.

To assess whether SNPs with high ΔSNP-index values can be used as molecular markers for the *Net2* locus, derived cleaved amplified polymorphic sequence (dCAPS) markers were designed from these SNPs. Within the physical region from 81.8 Mb to 83.3 Mb on chromosome 2D that was defined by the two markers noted in our previous study [22], 208 SNPs were detected in the present study, including 10 D-genome specific, 144 homoeologous, and 54 unclassified SNPs. The ΔSNP-index values of this subset of SNPs ranged from −0.4025 to 0.5425. Four of the D-genome-specific SNPs with ΔSNP-index values higher than 0.38 were converted into dCAPS markers (Table 4), and three of these markers (designated *bsa1*, *bsa2*, and *bsa4*) were successfully genotyped and mapped in the Ldn/KU-2025//Ldn/KU-2075 population (Figure 4). All three of these markers were anchored within 2.5 cM distal to the *Net2* locus, and two of these markers were closer to *Net2* than to *scaf52*, the most closely linked markers in our previous study [22]. This result indicates that SNPs derived from RNA-seq-based BSA of synthetic hexaploid wheat can be used for molecular marker development in a specific chromosomal region of the D genome.

Table 4. List of the derived cleaved amplified polymorphic sequence (dCAPS) markers developed in this study.

Marker Name	Primer Sequence (5′ to 3′)	Restriction Enzyme
bsa1	TCATGACCTGCTGGTTTGTT GATTCCAATGTTATTTCTGAACCCT	*Sty*I
bsa2	TCACAACATTCGCAGGTCAT TGGTTCTGTTGATCTCACTGCC	*Hpa*II
bsa4	ACAAGTCGGATATCGCCAAA CAGCTAAAAACTGTTTGCTTGAGA	*Hinf*I

The distribution of ΔSNP-index values was not continuous (Figure 2), even in a single transcript. The discontinuous pattern was due to the homoeologous and nonpolymorphic RNA-seq reads, which could be derived from homoeologous regions in the A and B genomes and overlap with the reads containing the D-genome-specific SNPs. For complete masking of the overlapping reads from the A and B genomes, precise alignment to all three genomes of hexaploid wheat would be needed. In the present study, although transcripts of the parental diploid species were used as the reference sequences, our strategy succeeded in efficiently detecting candidate SNPs on a specific chromosomal region of synthetic hexaploid wheat. Therefore, this strategy of RNA-seq-based BSA is expected to be applicable to other synthetic polyploids derived from crosses among tetraploid and/or diploid wheat relatives.

Taken together, these data demonstrate that RNA-seq-based BSA can be applied to synthetic hexaploid wheat for the development of molecular markers in specific chromosomal regions. The use of a single tetraploid wheat cultivar as a parental line for synthetic hexaploid wheat effectively cancelled the increase in ΔSNP-index values at homoeologous polymorphic sites. To date, a large number of synthetic wheat hexaploids have been produced as part of efforts to enlarge the D-genome genetic diversity [5,6,40]. Epistatic interactions can occur between the parental genomes in the newly synthesized allopolyploids [41], and several phenotypes specific to the synthetic wheat hexaploids have been reported [20,42]. RNA-seq and subsequent de novo transcriptome assembly can be performed even in wild diploid wheat relatives with no reference sequences. This strategy would facilitate molecular marker development in diverse wheat synthetics with various genome constitutions. Thus, RNA-seq-based BSA is expected to serve as a rapid and efficient approach for genetic evaluation of target traits from wild wheat relatives with allopolyploid backgrounds.

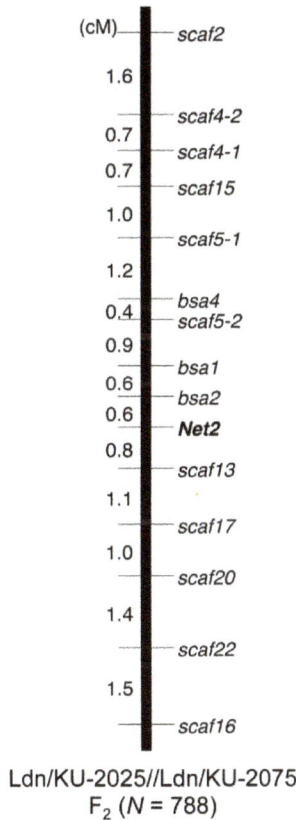

Figure 4. Genetic map of the short arm of chromosome 2D in the Ldn/KU-2025//Ldn/KU-2075 population. The three newly developed markers are designated with "*bsa*" prefixes.

3. Materials and Methods

3.1. Plant Materials

Two synthetic hexaploid wheat lines derived from interspecies crosses of tetraploid wheat cultivar Langdon (Ldn) and two *Ae. tauschii* accessions (KU-2075 and KU-2025) were used in this study. These lines were produced in our previous study [21]. A synthetic hexaploid line (Ldn/KU-2075) showed normal growth, whereas another synthetic hexaploid line (Ldn/KU-2025) exhibited the type-II necrosis phenotype when grown at low temperature [21]. An F_2 mapping population was generated through a cross of the two synthetic hexaploid lines [22]. Subsequently, the F_3 and F_4 generations were developed from selfing F_2 and F_3 individuals (respectively) heterozygous for the chromosomal region of *Net2*, a type-II necrosis-causing gene [22]. Two F_5 populations (SP1 and SP2) were generated through selfing of two independent F_4 individuals heterozygous for the *Net2* region (Figure 1).

3.2. Library Construction and RNA Sequencing

Ten *Net2*-homozygous individuals and ten non-carriers (*net2*-homozygous) each were selected from the SP1 and SP2 populations. Total RNA was extracted from leaves of each of the forty selected plants at the seedling stage using a Plant Total RNA Extraction Miniprep System (Viogene, Taipei Hsien, Taiwan, ROC). The resulting RNA preparations were pooled independently into four bulk preparations according to the genotype of the *Net2* region and to the SP source population (*Net2*-SP1, non-carrier-SP1,

Net2-SP2, and non-carrier-SP2). A total of 8 µg of each bulk preparation of RNA was used to construct paired-end libraries using TruSeq RNA Library Preparation Kit v2 (Illumina, San Diego, CA, USA). The libraries were sequenced twice for 300 cycles × 2 on an Illumina MiSeq sequencer with 300-bp paired-end reads. The read data was deposited to the DDBJ Sequence Read Archive under the accession number DRA007501. RNA sequencing data of Ldn, which was obtained in our previous report [43] and deposited under the accession number DRA007097, also was used for the subsequent analysis.

3.3. Alignment of RNA-seq Reads to de Novo Assembled Transcripts of the Parental Ae. tauschii Accessions

Low-quality bases (average Phred quality score per 4 bp < 30), adapter sequences, and reads < 100 bp were removed from the short reads using the Trimmomatic version 0.32 tool [44]. RNA-seq analysis of KU-2075 and KU-2025 for de novo transcriptome assembly was described in our previous study [27], and the corresponding sequencing data was obtained from the DDBJ Sequence Read Archive DRA004604. The quality-filtered reads of each bulk and Ldn were aligned to the transcripts of KU-2075 and KU-2025 using the Bowtie2 tool [45] with local alignment, obtaining bam outputs.

3.4. Identification of D-genome Specific SNPs and Calculation of ΔSNP-Index

After merging the two bam files of each bulk library using SAMtools software (version 1.9; La Jolla, CA, USA) with the command "samtools merge", SNPs were called from the alignment files using SAMtools and Coval software (version 1.5; Iwate, Japan) with option "-freq 0.1 -m 1000000 -n 10" [46,47]. The bam files of the synthetic hexaploid bulks were supposed to hold homoeologous polymorphisms and allelic variations on the D genome, whereas the alignment outputs of Ldn would contain homoeologous polymorphisms only. We assumed that, of the SNPs derived from the bulks, SNPs also detected in the pairwise comparison of KU-2075 and KU-2025 would represent D-genome specific polymorphisms, and SNPs also found in the alignment files of Ldn would represent homoeologous polymorphisms. The transcript sequences of the two *Ae. tauschii* accessions were mapped to the reference genome of *Ae. tauschii* [39] using GMAP version 2013-03-31 software (South San Francisco, CA, USA) [48,49], and the identified SNPs were anchored to the *Ae. tauschii* genome based on the GMAP outputs. The allele frequency at the SNP sites, a value designated as the SNP-index [28,29], was calculated in each of the eight respective alignment pairs, and the difference in the SNP-index values of the contrasting bulks (*Net2*-homogenous *minus* non-carrier bulk alignment, e.g., *Net2*-SP1-KU-2075 *minus* non-carrier-SP1-KU-2075) was defined as the ΔSNP-index by analogy to the work of Takagi et al. [29]. RStudio ver. 0.99.902 [50] with R software ver. 3.3.1 [51] was used for statistical analyses and ΔSNP-index plotting of the *Ae. tauschii* chromosomes.

3.5. Molecular Marker Development and Genotyping

In our previous study [22], the *Net2* locus was fine-mapped within a 0.6-cM region on the short arm of chromosome 2D. The two most closely linked markers were assigned to the *Ae. tauschii* genome [39] using the BLASTN search function of the BLAST+ software (Bethesda, MD, USA) [52] to define the physical region where the *Net2* locus was assumed to reside. D-genome-specific SNPs of high ΔSNP-index values within this chromosomal region were converted into derived cleaved amplified polymorphic sequence (dCAPS) markers (Table 4), and these novel markers were used to genotype 788 individuals of the Ldn/KU-2025//Ldn/KU-2075 population, as described in our previous report [22]. The genotyped markers were assigned to the genetic map around *Net2* using the MAPMAKER/EXP version 3.0 package (Cambridge, MA, USA) [53].

Author Contributions: Conceptualization, S.T. and K.Y.; methodology, R.N. and K.Y.; formal analysis, R.N., K.S., S.-i.Y. and K.S.; resources, S.T.; writing—original draft preparation, R.N. and S.T.; writing—review and editing, K.Y., K.S. and S.T.; project administration, S.T.; funding acquisition, S.T., R.N. and K.S.

Funding: This work was supported by grants from the Ministry of Education, Culture, Sports, Science and Technology of Japan (MEXT; Grant-in-Aid for Scientific Research (B) No. 16H04862 and Grant-in-Aid for Scientific Research on Innovative Areas No. 17H05842) to ST; by a Research Fellowship from the Japan Society for the Promotion of Science for Young Scientists (No. 16J03477) to RN; and by MEXT as part of a Joint Research Program implemented at the Institute of Plant Science and Resources, Okayama University, Japan. KY was supported by the Japan Science and Technology Agency, PRESTO (No. JPMJPR15QB).

Acknowledgments: The *Ae. tauschii* seeds used in this study were supplied by the National BioResource Project-Wheat (http://shigen.nig.ac.jp/wheat/komugi/).

Conflicts of Interest: The authors declare no conflict of interest. The funders had no role in the design of the study; in the collection, analyses, or interpretation of data; in the writing of the manuscript, or in the decision to publish the results.

Abbreviations

RNA-seq	RNA sequencing
NGS	next-generation sequencing
BSA	bulked segregant analysis
SNP	single nucleotide polymorphism
Indel	insertion/deletion
NR	non-redundant
SP	Segregating Population
dCAPS	derived cleaved amplified polymorphic sequence
Ldn	Langdon

References

1. Matsuoka, Y. Evolution of polyploid *Triticum* wheats under cultivation: The role of domestication, natural hybridization and allopolyploid speciation in their diversification. *Plant Cell Physiol.* **2011**, *52*, 750–764. [CrossRef] [PubMed]

2. Wang, J.; Luo, M.C.; Chen, Z.; You, F.M.; Wei, Y.; Zheng, Y.; Dvorak, J. *Aegilops tauschii* single nucleotide polymorphisms shed light on the origins of wheat D-genome genetic diversity and pinpoint the geographic origin of hexaploid wheat. *New Phytol.* **2013**, *198*, 925–937. [CrossRef] [PubMed]

3. Matsuoka, Y.; Nasuda, S. Durum wheat as a candidate for the unknown female progenitor of bread wheat: An empirical study with a highly fertile F_1 hybrid with *Aegilops tauschii* Coss. *Theor. Appl. Genet.* **2004**, *109*, 1710–1717. [CrossRef] [PubMed]

4. Zohary, D.; Harlan, J.R.; Vardi, A. The wild diploid progenitors of wheat and their breeding value. *Euphytica* **1969**, *18*, 58–65. [CrossRef]

5. Mujeeb-Kazi, A.; Rosas, V.; Roldan, S. Conservation of the genetic variation of *Triticum tauschii* (Coss.) Schmalh. (*Aegilops squarrosa* auct. non L.) in synthetic hexaploid wheats (*T. turgidum* L. s.lat. x *T. tauschii*; 2n=6x=42, AABBDD) and its potential utilization for wheat improvement. *Genet. Resour. Crop Evol.* **1996**, *43*, 129–134. [CrossRef]

6. Jones, H.; Gosman, N.; Horsnell, R.; Rose, G.A.; Everst, L.A.; Bentley, A.R.; Tha, S.; Uauy, C.; Kowalski, A.; Novoselovic, D.; et al. Strategy for exploiting exotic germplasm using genetic, morphological, and environmental diversity: The *Aegilops tauschii* Coss. example. *Theor. Appl. Genet.* **2013**, *126*, 1793–1808. [CrossRef] [PubMed]

7. Bhatta, M.; Morgounov, A.; Belamkar, V.; Yorgancilar, A.; Baenziger, P.S. Genome-wide association study reveals favorable alleles associated with common bunt resistance in synthetic hexaploid wheat. *Euphytica* **2018**, *214*, 200. [CrossRef]

8. Bhatta, M.; Morgounov, A.; Belamkar, V.; Baenziger, P.S. Genome-wide association study reveals novel genomic regions for grain yield and yield-related traits in drought-stressed synthetic hexaploid wheat. *Int. J. Mol. Sci.* **2018**, *19*, 3011. [CrossRef] [PubMed]

9. Bhatta, M.; Baenziger, P.S.; Waters, B.M.; Poudel, R.; Belamkar, V.; Poland, J.; Morgounov, A. Genome-wide association study reveals novel genomic regions associated with 10 grain minerals in synthetic hexaploid wheat. *Int. J. Mol. Sci.* **2018**, *19*, 3237. [CrossRef] [PubMed]

10. Gorafi, Y.S.A.; Kim, J.S.; Elbashir, A.A.E.; Tsujimoto, H. A population of wheat multiple synthetic derivatives: An effective platform to explore, harness and utilize genetic diversity of *Aegilops tauschii* for wheat improvement. *Theor. Appl. Genet.* **2018**, *131*, 1615–1626. [CrossRef] [PubMed]

11. Zhang, P.; Dreisigacker, S.; Melchinger, A.E.; Reif, J.C.; Mujeeb-Kazi, A.; Van Ginkel, M.; Hoisington, D.; Warburton, M.L. Quantifying novel sequence variation and selective advantage in synthetic hexaploid wheats and their backcross-derived lines using SSR markers. *Mol. Breed.* **2005**, *15*, 1–10. [CrossRef]

12. Jafarzadeh, J.; Bonnett, D.; Jannink, J.L.; Akdemir, D.; Dreisigacker, S.; Sorrells, M.E. Breeding value of primary synthetic wheat genotypes for grain yield. *PLoS ONE* **2016**, *11*, e0162860. [CrossRef] [PubMed]

13. Bhatta, M.; Morgounov, A.; Belamkar, V.; Poland, J.; Baenziger, P.S. Unlocking the novel genetic diversity and population structure of synthetic hexaploid wheat. *BMC Genom.* **2018**, *19*, 591. [CrossRef] [PubMed]

14. Rasheed, A.; Mujeeb-Kazi, A.; Ogbonnaya, F.C.; He, Z.; Rajaram, S. Wheat genetic resources in the post-genomics era: Promise and challenges. *Ann. Bot.* **2018**, *121*, 603–616. [CrossRef] [PubMed]

15. Van Slageren, M.W. *Wild Wheats: A Monograph of Aegilops L. and Amblyopyrum (Jaub. & Spach) Eig (Poaceae)*; Wageningen Agricultural University: Wageningen, The Netherlands, 1994; pp. 326–344, ISBN 90-6754-377-2.

16. Matsuoka, Y.; Takumi, S.; Kawahara, T. Flowering time diversification and dispersal in central Eurasian wild wheat *Aegilops tauschii* Coss.: Genealogical and ecological framework. *PLoS ONE* **2008**, *3*, e3138. [CrossRef] [PubMed]

17. Mizuno, N.; Yamasaki, M.; Matsuoka, Y.; Kawahara, T.; Takumi, S. Population structure of wild wheat D-genome progenitor *Aegilops tauschii* Coss.: Implications for intraspecific lineage diversification and evolution of common wheat. *Mol. Ecol.* **2010**, *19*, 999–1013. [CrossRef] [PubMed]

18. Matsuoka, Y.; Kawahara, T.; Takumi, S. Intraspecific lineage divergence and its association with reproductive trait change during species range expansion in central Eurasian wild wheat *Aegilops tauschii* Coss. (Poaceae). *BMC Evol. Biol.* **2015**, *15*, 213. [CrossRef] [PubMed]

19. Matsuoka, Y.; Takumi, S. The role of reproductive isolation in allohexaploid speciation pattern: Empirical insights from the progenitors of common wheat. *Sci. Rep.* **2017**, *7*, 16004. [CrossRef] [PubMed]

20. Mizuno, N.; Hosogi, N.; Park, P.; Takumi, S. Hypersensitive response-like reaction is associated with hybrid necrosis in interspecific crosses between tetraploid wheat and *Aegilops tauschii* Coss. *PLoS ONE* **2010**, *5*, e11326. [CrossRef] [PubMed]

21. Mizuno, N.; Shitsukawa, N.; Hosogi, N.; Park, P.; Takumi, S. Autoimmune response and repression of mitotic cell division occur in inter-specific crosses between tetraploid wheat and *Aegilops tauschii* Coss. that show low temperature-induced hybrid necrosis. *Plant J.* **2011**, *68*, 114–128. [CrossRef] [PubMed]

22. Sakaguchi, K.; Nishijima, R.; Iehisa, J.C.M.; Takumi, S. Fine mapping and genetic association analysis of *Net2*, the causative D-genome locus of low temperature-induced hybrid necrosis in interspecific crosses between tetraploid wheat and *Aegilops tauschii*. *Genetica* **2016**, *144*, 523–533. [CrossRef] [PubMed]

23. Brozynska, M.; Furtado, A.; Henry, R.J. Genomics of crop wild relatives: Expanding the gene pool for crop improvement. *Plant Biotechnol. J.* **2016**, *14*, 1070–1085. [CrossRef] [PubMed]

24. Ishikawa, G.; Saito, M.; Tanaka, T.; Katayose, Y.; Kanamori, H.; Kurita, K.; Nakamura, T. An efficient approach for the development of genome-specific markers in allohexaploid wheat (*Triticum aestivum* L.) and its application in the construction of high-density linkage maps of the D genome. *DNA Res.* **2018**, *25*, 317–326. [CrossRef] [PubMed]

25. Iehisa, J.C.M.; Shimizu, A.; Sato, K.; Nasuda, S.; Takumi, S. Discovery of high-confidence single nucleotide polymorphisms from large-scale de novo analysis of leaf transcripts of *Aegilops tauschii*, a wild wheat progenitor. *DNA Res.* **2012**, *19*, 487–497. [CrossRef] [PubMed]

26. Iehisa, J.C.M.; Shimizu, A.; Sato, K.; Nishijima, R.; Sakaguchi, K.; Matsuda, R.; Nasuda, S.; Takumi, S. Genome-wide marker development for the wheat D genome based on single nucleotide polymorphisms identified from transcripts in the wild wheat progenitor *Aegilops tauschii*. *Theor. Appl. Genet.* **2014**, *127*, 261–271. [CrossRef] [PubMed]

27. Nishijima, R.; Yoshida, K.; Motoi, Y.; Sato, K.; Takumi, S. Genome-wide identification of novel genetic markers from RNA sequencing assembly of diverse *Aegilops tauschii* accessions. *Mol. Genet. Genom.* **2016**, *291*, 1681–1694. [CrossRef] [PubMed]

28. Abe, A.; Kosugi, S.; Yoshida, K.; Natsume, S.; Takagi, H.; Kanzaki, H.; Matsumura, H.; Yoshida, K.; Mitsuoka, C.; Tamiru, M.; et al. Genome sequencing reveals agronomically important loci in rice using MutMap. *Nat. Biotechnol.* **2012**, *30*, 174–178. [CrossRef] [PubMed]

29. Takagi, H.; Abe, A.; Yoshida, K.; Kosugi, S.; Natsume, S.; Mitsuoka, C.; Uemura, A.; Utsushi, H.; Tamiru, M.; Takuno, S.; et al. QTL-seq: Rapid mapping of quantitative trait loci in rice by whole genome resequencing of DNA from two bulked populations. *Plant J.* **2013**, *74*, 174–183. [CrossRef] [PubMed]

30. Schneeberge, K. Using next-generation sequencing to isolate mutant genes from forward genetic screens. *Nat. Rev. Genet.* **2014**, *15*, 662–676. [CrossRef] [PubMed]

31. Zou, C.; Wang, P.; Xu, Y. Bulked sample analysis in genetics, genomics and crop improvement. *Plant Biotechnol. J.* **2016**, *14*, 1941–1955. [CrossRef] [PubMed]

32. Liu, S.; Yeh, C.T.; Tang, H.M.; Nettleton, D.; Schnable, P.S. Gene mapping via bulked segregant RNA-Seq (BSR-Seq). *PLoS ONE* **2012**, *7*, e36406. [CrossRef] [PubMed]

33. Su, A.; Song, W.; Xing, J.; Zhao, Y.; Zhang, R.; Li, C.; Duan, M.; Luo, M.; Shi, Z.; Zhao, J. Identification of genes potentially associated with the fertility instability of S-type cytoplasmic male sterility in maize via bulked segregant RNA-Seq. *PLoS ONE* **2016**, *11*, e0163489. [CrossRef] [PubMed]

34. Du, H.; Zhu, J.; Su, H.; Huang, M.; Wang, H.; Ding, S.; Zhang, B.; Luo, A.; Wei, S.; Tian, X.; et al. Bulked segregant RNA-seq revealed differential expression and SNPs of candidate genes associated with waterlogging tolerance in maize. *Front. Plant Sci.* **2017**, *8*, 1022. [CrossRef] [PubMed]

35. Borrill, P.; Adamski, N.; Uauy, C. Genomics as the key to unlocking the polyploid potential of wheat. *New Phytol.* **2015**, *208*, 1008–1022. [CrossRef] [PubMed]

36. Trick, M.; Adamski, N.M.; Mugford, S.G.; Jiang, C.C.; Febrer, M.; Uauy, C. Combining SNP discovery from next-generation sequencing data with bulked segregant analysis (BSA) to fine-map genes in polyploid wheat. *BMC Plant Biol.* **2012**, *12*, 14. [CrossRef] [PubMed]

37. Ramirez-Gonzalez, R.H.; Segovia, V.; Bird, N.; Fenwick, P.; Holdgate, S.; Bery, S.; Jack, P.; Caccamo, M.; Uauy, C. RNA-Seq bulked segregant analysis enables the identification of high-resolution genetic markers for breeding in hexaploid wheat. *Plant Biotechnol. J.* **2015**, *13*, 613–624. [CrossRef] [PubMed]

38. Wu, P.; Xie, J.; Hu, J.; Qiu, D.; Liu, Z.; Li, J.; Li, M.; Zhang, H.; Yang, L.; Liu, H.; et al. Development of molecular markers linked to powdery mildew resistance gene *Pm4b* by combining SNP discovery from transcriptome sequencing data with bulked segregant analysis (BSR-Seq) in wheat. *Front. Plant Sci.* **2018**, *9*, 95. [CrossRef] [PubMed]

39. Luo, M.-C.; Gu, Y.Q.; Puiu, D.; Wang, H.; Ywardziok, S.O.; Deal, K.R.; Huo, N.; Zhu, T.; Wang, L.; Wang, Y.; et al. Genome sequence of the progenitor of the wheat D genome *Aegilops tauschii*. *Nature* **2017**, *551*, 498–502. [CrossRef] [PubMed]

40. Comai, L. The advantages and disadvantages of being polyploid. *Nat. Rev. Genet.* **2005**, *6*, 836–846. [CrossRef] [PubMed]

41. Okada, M.; Yoshida, K.; Takumi, S. Hybrid incompatibilities in interspecific crosses between tetraploid wheat and its wild relative *Aegilops umbellulata*. *Plant Mol. Biol.* **2017**, *95*, 625–645. [CrossRef] [PubMed]

42. Miki, Y.; Yoshida, K.; Mizuno, N.; Nasuda, S.; Sato, K.; Takumi, S. Origin of the wheat B-genome chromosomes conferred by RNA sequencing analysis of leaf transcripts in the section Sitopsis species of *Aegilops*. *DNA Res.* **2018**, under review.

43. Bolger, A.M.; Lohse, M.; Usadel, B. Trimmomatic: A flexible trimmer for Illumina sequence data. *Bioinformatics* **2014**, *30*, 2114–2120. [CrossRef] [PubMed]

44. Langmead, B.; Salzberg, S.L. Fast gapped-read alignment with Bowtie 2. *Nat. Methods* **2012**, *9*, 357–359. [CrossRef] [PubMed]

45. Li, H.; Handsaker, B.; Wysoker, A.; Fennell, T.; Ruan, J.; Homer, N.; Marth, G.; Abecasis, G.; Durbin, R. Subgroup 1000 Genome Project Data Processing. The sequence alignment/map format and SAMtools. *Bioinformatics* **2009**, *25*, 2078–2079. [CrossRef] [PubMed]

46. Kosugi, S.; Natsume, S.; Yoshida, K.; MacLean, D.; Cano, L.; Kamoun, S.; Terauchi, R. Coval: Improving alignment quality and variant calling accuracy for next-generation sequencing data. *PLoS ONE* **2013**, *8*, e75402. [CrossRef] [PubMed]

47. Kajimura, T.; Murai, K.; Takumi, S. Distinct genetic regulation of flowering time and grain-filling period based on empirical study of D genome diversity in synthetic hexaploid wheat lines. *Breed. Sci.* **2011**, *61*, 130–141. [CrossRef]

48. Wu, T.D.; Watanabe, C.K. GMAP: A genomic mapping and alignment program for mRNA and EST sequences. *Bioinformatics* **2005**, *21*, 1859–1875. [CrossRef] [PubMed]

49. Quinlan, A.R.; Hall, I.M. BEDTools: A flexible suite of utilities for comparing genomic features. *Bioinformatics* **2010**, *26*, 841–842. [CrossRef] [PubMed]

50. RStudio Team. *RStudio: Integrated Development for R*; RStudio, Inc.: Boston, MA, USA, 2016; Available online: http://www.rstudio.com/ (accessed on 1 November 2016).

51. R Core Team. *R: A Language and Environment for Statistical Computing*; R Foundation for Statistical Computing: Vienna, Austria; Available online: https://www.R-project.org/ (accessed on 1 November 2016).

52. Altschul, S.F.; Gish, W.; Miller, W.; Myers, E.W.; Lipman, D.J. Basic local alignment search tool. *J. Mol. Biol.* **1990**, *215*, 403–410. [CrossRef]

53. Lander, E.S.; Green, P.; Abrahamson, J.; Barlow, A.; Daly, M.J.; Lincoln, S.E.; Newburg, L. MAPMAKER: An interactive computer package for constructing primary genetic linkage maps of experimental and natural populations. *Genomics* **1987**, *1*, 174–181. [CrossRef]

© 2018 by the authors. Licensee MDPI, Basel, Switzerland. This article is an open access article distributed under the terms and conditions of the Creative Commons Attribution (CC BY) license (http://creativecommons.org/licenses/by/4.0/).

International Journal of
Molecular Sciences

MDPI

Article

Development of SNP, KASP, and SSR Markers by BSR-Seq Technology for Saturation of Genetic Linkage Map and Efficient Detection of Wheat Powdery Mildew Resistance Gene *Pm61*

Jinghuang Hu [1], Jingting Li [2,*], Peipei Wu [1], Yahui Li [1], Dan Qiu [1], Yunfeng Qu [3], Jingzhong Xie [4], Hongjun Zhang [1], Li Yang [1], Tiantian Fu [2], Yawei Yu [2], Mengjuan Li [2], Hongwei Liu [1], Tongquan Zhu [5], Yang Zhou [1], Zhiyong Liu [4,*] and Hongjie Li [1,*]

[1] The National Engineering Laboratory of Crop Molecular Breeding, Institute of Crop Sciences, Chinese Academy of Agricultural Sciences, Beijing 100081, China; hujh2016@163.com (J.H.); ml3051378331_2@163.com (P.W.); 13522357314@163.com (Y.L.); smart_qd@163.com (D.Q.); zhanghongjun01@caas.cn (H.Z.); yangli02@caas.cn (L.Y.); liuhongwei@caas.cn (H.L.); zhouyang@caas.cn (Y.Z.)
[2] College of Chemistry and Environment Engineering, Pingdingshan University, Pingdingshan 467000, China; 15936057197@163.com (T.F.); 776117360@139.com (Y.Y.); mengjuan9768@163.com (M.L.)
[3] College of Life Science and Technology, Harbin Normal University, Harbin 150080, China; quyunfeng1993@163.com
[4] Institute of Genetics and Developmental Biology, Chinese Academy of Sciences, Beijing 100101, China; jzxie@genetics.ac.cn
[5] Zhumadian Academy of Agricultural Sciences, Zhumadian 463000, China; zmdwheat@163.com
[*] Correspondence: lijingting@pdsu.edu.cn (J.L.); zyliu@genetics.ac.cn (Z.L.); lihongjie@caas.cn (H.L.)

Received: 24 December 2018; Accepted: 29 January 2019; Published: 11 February 2019

Abstract: The gene *Pm61* that confers powdery mildew resistance has been previously identified on chromosome arm 4AL in Chinese wheat landrace Xuxusanyuehuang (XXSYH). To facilitate the use of *Pm61* in breeding practices, the bulked segregant analysis-RNA-Seq (BSR-Seq) analysis, in combination with the information on the Chinese Spring reference genome sequence, was performed in the $F_{2:3}$ mapping population of XXSYH × Zhongzuo 9504. Two single nucleotide polymorphism (SNP), two Kompetitive Allele Specific PCR (KASP), and six simple sequence repeat (SSR) markers, together with previously identified polymorphic markers, saturated the genetic linkage map for *Pm61*, especially in the proximal side of the target gene that was short of gene-linked markers. In the newly established genetic linkage map, *Pm61* was located in a 0.71 cM genetic interval and can be detected in a high throughput scale by the KASP markers *Xicsk8* and *Xicsk13* or by the standard PCR-based markers *Xicscx497* and *Xicsx538*. The newly saturated genetic linkage map will be useful in molecular marker assisted-selection of *Pm61* in breeding for disease resistant cultivar and in its map-based cloning.

Keywords: *Triticum aestivum*; Landrace; Powdery mildew; Bulked segregant analysis-RNA-Seq (BSR-Seq); Single nucleotide polymorphism (SNP); Kompetitive Allele Specific PCR (KASP)

---

1. Introduction

Powdery mildew is one of the most widely epidemic diseases in wheat (*Triticum aestivum* L.) grown in the temperate and humid regions of the world. The causal agent of powdery mildew, *Blumeria graminis* f. sp. *tritici* (*Bgt*), is an obligate biotrophic fungus, which usually colonizes wheat leaves and develops white pustule symptoms on leaf blades. The penalty in wheat yield caused by infection of powdery mildew has been reported to be from 5% to 40% in various countries, depending on the disease severity [1]. The impact of powdery mildew on grain quality, such as test weight and protein content, has been reported [2]. Changes in grain proteome and composition and grain starch and protein contents caused by the disease were observed using the proteomics analysis [3].

In China, powdery mildew has become an economically important disease since the 1970s. In recent years, the annual areas of powdery mildew infected wheat fields ranged from 6 to 8 million hectares in most winter wheat and parts of spring wheat fields throughout the country (available online: https://www.natesc.org.cn/sites/cb/). Management of wheat powdery mildew relies on growing disease resistant cultivars in accompany with application of fungicides such as triadimefon when necessary.

Breeding for powdery mildew resistant cultivars necessitates the availability of powdery mildew resistance genes (*Pm* genes). Sources of *Pm* genes include common wheat and its close or distant relative species. Wheat landraces from China are a class of historically grown and maintained common wheat cultivars, which have provided quite a number of *Pm* genes. A group of landraces carry allelic genes in the *Pm5* locus on chromosome 7BL, for example, *Pm5d* (IGV1-455) [4], *Pm5e* (Fuzhuang 30) [5], *PmH* (Hongquanmang) [6], *PmTm4* (Tangmai 4) [7], *Mlmz* (Mazhamai) [8], *Mlxbd* (Xiaobaidong) [9], *pmHYM* (Hongyoumai) [10], *PmBYYT* (Baiyouyantiao) [11], and *PmSGD* (Shangeda) [12]. In the *Pm24* locus, there are two alleles, *Pm24a* in Chiyacao [13] and *Pm24b* in Baihulu [14]. The designated genes *Pm2c* on 5DS [15], *Pm45* on 6DS [16], and *Pm47* on 7BS [17] were identified in Niaomai, Wuzhaomai, and Hongyanglazi, respectively. *PmX* in Xiaohongpi [18] and *MlHLT* in Hulutou [19] were detected on chromosomes 2AL and 1DS, respectively. Recently, a group of scientists in the USA characterized three *Pm* genes in wheat landraces: two from Afghanistan, *Pm223899* on 1AS in PI 223,899 [20] and *Pm59* on 7AL in PI 181,356 [21], and the other one from Iran, *Pm63* on 2BL in PI 628,024 [22].

Landraces of wheat are traditionally grown in agriculture until they have been replaced by more productive cultivars since the initiation of modern crop breeding in the mid-20th century. At present, landraces can only be grown in certain marginal lands with less productivity [23]. They are no longer adapting to most of the improved agricultural environments in spite of possessing desirable genes. There is a need to introgress the useful genes, for example, *Pm* genes, into modern improved genetic backgrounds in order to be used efficiently in modern breeding practices. Molecular approaches facilitate identification and transfer of wheat genes for disease resistance [24]. In fact, many recently characterized *Pm* genes are identified with the aid of various classes of molecular markers, such as SSR (microsatellite) markers and STS (sequence-tagged site) markers, which are useful in marker-assisted selection (MAS) of target genes.

The breeder-friendly molecular markers associated with resistance genes are useful in the breeding programs during development of disease resistant cultivars. The PCR-based markers are affordable in most wheat breeding programs, so this type of molecular markers can be routinely used in breeding practice. Moreover, high-throughput genotyping is needed in a large scale of population study. The newly improved next-generation sequencing techniques allow the discovery of numerous single nucleotide polymorphism (SNP) markers. The Chinese Spring wheat reference genome sequence has been updated [25], which facilitates the identification of gene-linked molecular markers and map-based cloning of disease resistance genes. The abundance of SNP markers is far greater than the traditional PCR-based markers. The SNP markers can be visualized by converting them into Kompetitive Allele Specific PCR (KASP) markers for establishing high-throughput genotyping platform for MAS of the target genes [26]. They have also been used to detect disease resistance genes, such as *Sr26* for resistance to stem rust (caused by *Puccinia graminis* f. sp. *tritici*) [27], *Yr34* and *Yr48* for resistance to stripe rust (caused by *Puccinia striiformis* f. sp. *tritici*) [28].

BSR-seq technique, which integrates bulked segregant analysis and RNA-seq [29], has proven to be a rapid and efficient strategy to identify gene-linked molecular markers. It provides a fast and high-throughput method to localize resistance genes in crops with large genome, e.g., wheat.

This technique has been used in the molecular characterization of wheat disease resistance genes, such as *Yr15* [30], *YrZH22* [31], *YrMM58* and *YrHY1* [32], *Yr26* [33], *Pm4b* [34], and *PmSGD* [12].

Wheat landrace Xuxusanyuehuang (XXSYH) was resistant to several *Bgt* isolates from China, and a recessive gene, *Pm61*, was located on chromosome 4AL [35]. In the genetic linkage map that was developed based on the mapping population of XXSYH × Mingxian 169, *Pm61* was mapped in a 0.46 cM genetic interval on 4AL. However, only two linked markers were identified in the proximal side of *Pm61*. Taking the advantage of BSR-seq and the Chinese Spring reference genome sequence, this study was conducted to (1) saturate genetic linkage map for *Pm61*, and (2) develop PCR-based markers for breeder-friendly use and KASP markers for large scale and high-throughput detection of *Pm61* during its MAS.

2. Results

2.1. Evaluation of Resistance to Bgt Isolates in XXSYH

Thirty *Bgt* isolates collected from wheat fields in Shandong, Shanxi, Beijing, Hebei, and Sichuan provinces were used to test response of XXSYH to powdery mildew. Twenty isolates were avirulent on XXSYH with ITs 0, 0; 1 or 2. Ten isolates were virulent on XXSYH with ITs 3 or 4 (Table 1). All the *Bgt* isolates tested were virulent on Zhongzuo 9504, the susceptible control.

Table 1. Infection types of Xuxusanyuehuang (XXSYH) and Zhongzuo 9504 to 30 *Blumeria graminis* f. sp. *tritici* (*Bgt*) isolates from different provinces of China.

Bgt Isolate	XXSYH	Zhongzuo 9504	Province
1	1	4	Shandong
2	1	3	Shandong
3	1	3	Shandong
4	2	3	Shandong
5	1	3	Shandong
6	2	3	Shandong
7	3	3	Shandong
8	3	4	Shanxi
9	2	3	Shanxi
10	0;	4	Shanxi
11	3	3	Shanxi
12	2	3	Shanxi
13	1	3	Beijing
14	1	3	Beijing
15	1	3	Beijing
16	3	4	Beijing
17	1	3	Hebei
18	3	3	Hebei
19	0;	3	Hebei
20	3	4	Hebei
21	3	3	Hebei
22	1	3	Hebei
23	1	3	Hebei
24	3	3	Hebei
25	1	3	Hebei
26	2	4	Hebei
27	3	3	Hebei
28	2	4	Hebei
29	3	3	Sichuan
30	1	4	Sichuan
31	2	3	Sichuan

Note: the infection type on leaves was rated on a 0–4 scale for determine the response of wheat genotypes to powdery mildew, where 0 = immune, no symptom, 0; = hypersensitive necrotic flecks, 1 = highly resistant, necrosis with low sporulation, 2 = moderately resistant, necrosis with moderate sporulation, 3 = moderately susceptible, moderate to high sporulation, and 4 = highly susceptible, no necrosis with full sporulation.

2.2. Genetic Analysis of Powdery Mildew Resistance in XXSYH

When inoculated with isolate *Bgt1* from Shandong, province, XXSYH was resistant with an IT 1, while Zhongzuo 9504 was susceptible with an IT 3. Therefore, this *Bgt* isolate was able to differentiate the phenotypes of the two parents that were crossed to develop the populations for genetic analysis. The IT of the 15 F_1 plants from the XXSYH × Zhongzuo 9504 resembled the susceptible parent Zhongzuo 9504 (Figure 1). The 211 $F_{2:3}$ lines produced 51 homozygous resistant, 115 heterozygous, and 45 homozygous susceptible lines, which agrees to the 1:2:1 segregating ratio ($\chi^2 = 2.05$, $P = 0.3584$). This indicates that the resistance of XXSYH to isolate *Bgt1* was in accordance with the single recessive mode of inheritance.

Xuxusanyuehuang

Xuxusanyuehuang × Zhongzuo 9504 F_1

Zhongzuo 9504

Figure 1. The phenotypic reactions of resistant parent Xuxusanyuehuang, susceptible parent Zhongzuo 9504, and their F_1 plants to isolate *Bgt1*.

2.3. BSR-Seq Analysis of the Bulked RNA Pools with Distinct Phenotypes to Isolate Bgt1

RNA-seq analysis generated a total of 40.0 Gb of raw data from the resistant and susceptible RNA samples and 38.3 Gb (95.77%) of clean data were obtained after quality control. Among the 62,760,214 and 67,764,274 high-quality reads for the two bulks, 58,213,973 (92.76%) and 62,295,839 (91.93%) were uniquely mapped to the Chinese Spring reference genome sequence, respectively (Supplementary Table S1). With the criteria of $P < 1 \times 10^{-10}$ and AFD > 0.6, 134 SNP variants, potentially associated with the target powdery mildew resistance gene, were identified. Further analysis indicated that 80 (59.7%) candidate SNP were distributed on chromosome 4AL (Figure 2A) and corresponded to a 31 Mb interval in the terminal region of 4AL in the reference genome (Figure 2B). This is consistent with previous study in which *Pm61* was mapped in a 1.3 Mb genomic region (717,963,176–719,260,469) on chromosome 4AL [35].

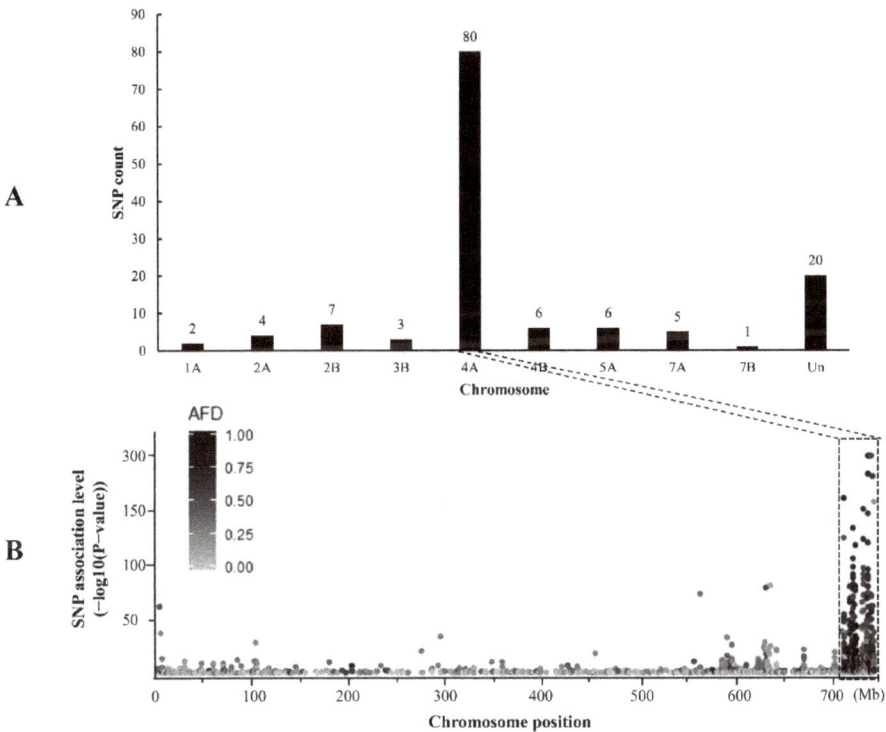

Figure 2. Number of polymorphic single nucleotide polymorphism (SNP) distributed on different wheat chromosomes (**A**) and distribution of SNP variants on chromosome 4A (**B**).

2.4. Validation of the Candidate SNP and Development of SNP Markers

Using the sequences flanking the 80 SNP that were anchored on chromosome 4AL as the queries, Blast analysis against the Chinese spring whole genome assembly (available online: https://urgi. versailles.inra.fr/) produced 24 homologous scaffolds. The 3 kb sequences containing the candidate SNP and corresponding to the above homologous scaffold were used as templates to design 28 pairs of SNP primers on the GSP website (Supplementary Table S2). The amplified products from XXSYH, Zhongzuo 9504, and the resistant and susceptible DNA bulks were sequenced for polymorphism analysis, and seven SNP markers were polymorphic. Based on the linkage analysis with 16 randomly selected F$_{2:3}$ lines, *Xicsn1*, *Xicsn2*, and *Xicsn3* were potentially mapped on one side of *Pm61*, and *Xicsn4*, *Xicsn5*, *Xicsn6*, and *Xicsn7*, on the other side of target gene. Sequencing analysis of *Xicsn2* and *Xicsn4* was carried out in 211 F$_{2:3}$ lines (Figure 3). *Pm61* was localized in a 4.5 cM genetic interval between the SNP markers *Xicsn2* and *Xicsn4* corresponding to a 5.3 Mb physical region (713,523,186–718,866,838) on the distal end of chromosome 4AL. The other five SNP markers were not used in genotyping the mapping population because of poor clustering of the fluorescence signals.

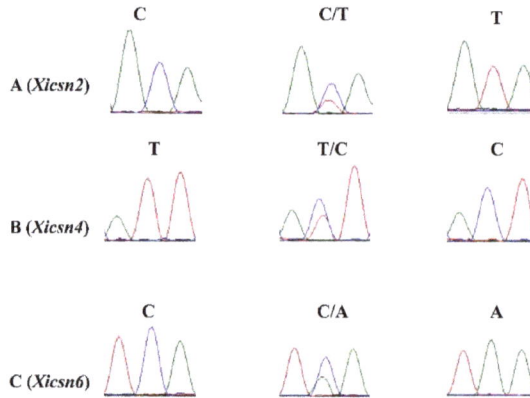

Figure 3. Sanger sequencing profiles of SNP markers *Xicsn2* (**A**), *Xicsn4* (**B**) and *Xicsn6* (**C**) in the homozygously resistant (R), homozygously susceptible (S), and heterozygous $F_{2:3}$ lines (H) from the mapping population of the Xuxusanyuehuang × Zhongzuo 9504 cross. Blue, green, and red lines represent bases of cytosine (C), adenine (A), and thymine (T), respectively.

2.5. Development of KASP Markers

To cost-effectively detect *Pm61* in MAS, 15 SNP generated from BSR-Seq analysis between the SNP markers *Xicsn1* and *Xicsn7* were converted into 13 KASP primer pairs (Supplementary Table S3). They were subjected to polymorphism analysis on the parental cultivars and 15 $F_{2:3}$ lines (including five homozygous resistant, heterozygous, and homozygous susceptible lines each). Six polymorphic KASP markers, i.e., *Xicsk4*, *Xicsk5*, *Xicsk7*, *Xicsk8*, *Xicsk9*, and *Xicsk13*, were identified. Moreover, the amplicons of 31 SSR primer pairs, which were previously designed based on the genomic sequence corresponding to the genetic interval between markers *Xicsn2* and *Xicsn4*, were sequenced to detect the SNP variants that differed between the two parents. Seven SNP variants were identified in the amplicons of 7 SSR primer pairs and were converted into KASP primer pairs (Supplementary Table S3). The KASP markers *Xicsk14* and *Xicsk15* showed clear polymorphism between the two parents and 15 $F_{2:3}$ lines by sequencing analysis. By genotyping the $F_{2:3}$ mapping population, KASP markers *Xicsk8* and *Xicsk13* were linked to *Pm61* (Figure 4A,B).

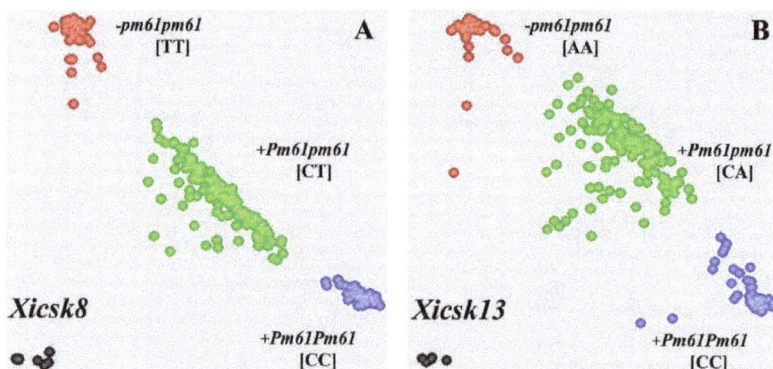

Figure 4. Genotyping results of *Xicsk8* (**A**) and *Xicsk13* (**B**) by Kompetitive Allele Specific PCR (KASP) assay. The scatter plot with the axes *x* and *y* represents the allelic discrimination of *Xicsk8* or *Xicsk13* genotypes. The red, green and blue dots represent the homozygously susceptible, heterozygous, and homozygously resistant $F_{2:3}$ lines from the mapping population of the Xuxusanyuehuang × Zhongzuo 9504 cross, respectively.

2.6. Development of SSR Markers

The polymorphism of the eleven SSR markers, previously linked to *Pm61* using the mapping population of XXSYH × Mingxian 169 (Figure 5A), was analyzed against the mapping population of XXSYH × Zhongzuo 9504. Seven SSR markers, *Xicsx29*, *Xicsx65*, *Xicsx73*, *Xicsx511*, *Xicsx520*, *Xicsx530*, and *Xicsx538*, were polymorphic. Linkage analysis indicated that all of these polymorphic markers were located on the distal side of *Pm61* (Figure 5B). Markers *Xicsx79* and *Xicsx436* previously located in the proximal side of *Pm61* were not polymorphic in the XXSYH × Zhongzuo 9504 mapping population. To develop more gene-linked markers in the proximal side of *Pm61*, the 5.3 Mb sequences (713,523,186–718,866,838) of the reference genome corresponding to the *Pm61* flanking markers *Xicsn2* and *Xicsn4* were used as templates to design SSR primer pairs. The 1.5 Mb (713,528,439–715,057,737) genomic sequences extended from marker *Xicsn2* toward *Pm61* was used to design 347 SSR primer pairs (Figure 5C, Supplementary Table S4). The 2.5 Mb (718,854,012–716,351,892) sequences extended from marker *Xicsn4* to *Pm61* were used to design 617 SSR primer pairs (Figure 5C, Supplementary Table S5). Five co-dominant SSR markers, *Xicscx305*, *Xicscx497*, *Xicscx543*, *Xicscx741*, and *Xicscx834* and a dominant SSR marker *Xicscx848*, were polymorphic between the two parents and two DNA bulks, which indicates their possible linkage to *Pm61*.

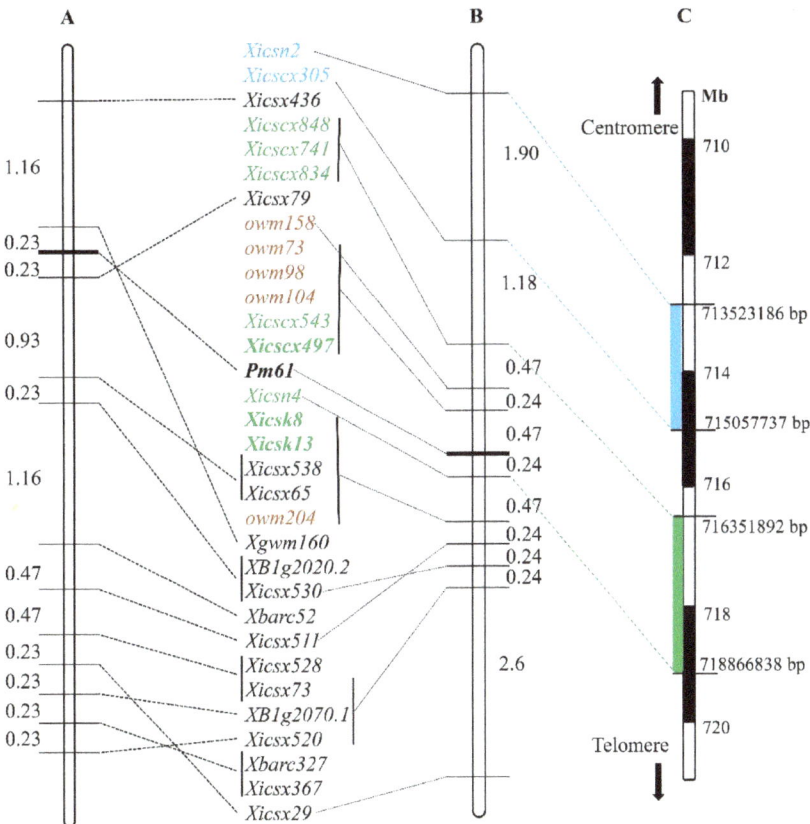

Figure 5. Genetic linkage maps of *Pm61* in previous study [35] (**A**) and newly developed in the present study (**B**). The positions of the *Pm61*-linked molecular markers are indicated on a scale bar based on the Chinese Spring genome sequence (**C**).

2.7. Construction of the Genetic Linkage Map for Pm61

The polymorphic markers developed, including two SNP markers (*Xicsn2* and *Xicsn4*), two KASP markers (*Xicsk8* and *Xicsk13*), and six SSR markers (*Xicscx305*, *Xicscx497*, *Xicscx543*, *Xicscx741*, *Xicscx834*, and *Xicscx848*), together with seven polymorphic *Pm61*-linked markers developed in the previous study [35], were used to construct the genetic linkage map after genotyping the $F_{2:3}$ mapping population of XXSYH × Zhongzuo 9504. In this linkage map, *Pm61* was placed in a 0.71 cM genetic interval that corresponded to 0.61 Mb genomic interval (718,257,529–718,866,730) of the genomic region in the reference genome sequence of Chinese Spring. The SSR markers *Xicscx543* and *Xicscx497* were located in the same locus at the proximal side of *Pm61* with genetic distance of 0.47 cM. The SNP marker *Xicsn4* was located in the distal side of *Pm61* with genetic distance of 0.24 cM. The KASP markers *Xicsk8* and *Xicsk13* (Figure 4A,B) and SSR markers *Xicscx497* and *Xicsx538* (Figure 6A,B) produced clear banding patterns and were able to differentiate individuals of the mapping population with distinct phenotypes.

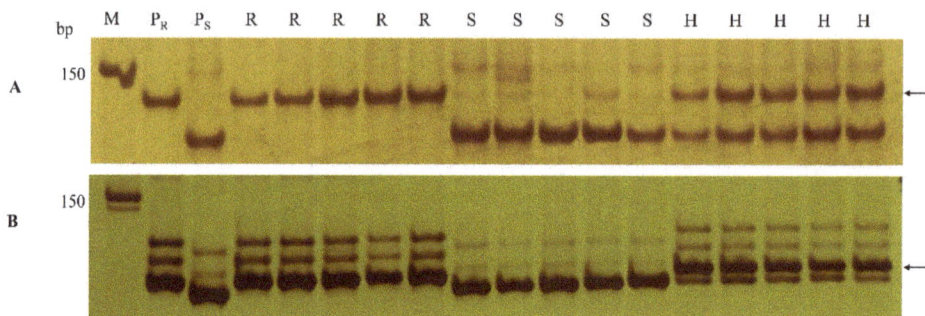

Figure 6. The banding patterns of *Pm61*-linked simple sequence repeat (SSR) markers *Xicscx497* (**A**) and *Xicsx538* (**B**) in the parents and the selected $F_{2:3}$ lines of the Xuxusanyuehuang × Zhongzuo 9504 cross. Lane M, 50 bp DNA ladder; P_R, Xuxusanyuehuang; P_S, Zhongzuo 9504; R, homozygously resistant $F_{2:3}$ lines; S, homozygously susceptible $F_{2:3}$ lines; and H, heterozygous $F_{2:3}$ lines. Arrows indicate the polymorphic bands specific for *Pm61*.

2.8. Physical Locations of Pm61, QPm.tut-4A and MlIW30 on Chromosome Arm 4AL

In addition to *Pm61*, a dominant gene *MlIW30* from wild emmer wheat [36] and a QTL *QPm.tut-4A* from *T. militinae* Zhuk. et Migusch. [37] were mapped on chromosome arm 4AL. Two *MlIW30*-linked markers, *XB1g2020.2* and *XB1g2070.1*, were linked to *Pm61* using the mapping population of XXSYH × Mingxian 169 [35]. However, only *XB1g2070.1* was linked to *Pm61* with genetic distance of 1.43 cM in the mapping population of XXSYH × Zhongzuo 9504. Polymorphism of 73 markers linked to *QPm.tut-4A* [37] were examined against the parents and the contrasting DNA bulks of the XXSYH × Zhongzuo 9504 cross, resulting in 5 polymorphic markers *owm73*, *owm98*, *owm104*, *owm158*, and *owm204* that were linked to *Pm61*. The closest *QPm.tut-4A*–linked marker *owm158* to *Pm61* was 0.95 cM from *Pm61*. Based on the physical position on the Chinese Spring reference genome sequence, *Pm61* was located in 0.61 Mb physical interval (718,257,529–718,866,730) between *MlIW30* (21 kb, 732,769,506–732,790,522) and *QPm.tut-4A* (1.54 Mb, 715,294,437–716,829,606 bp) (Figure 7).

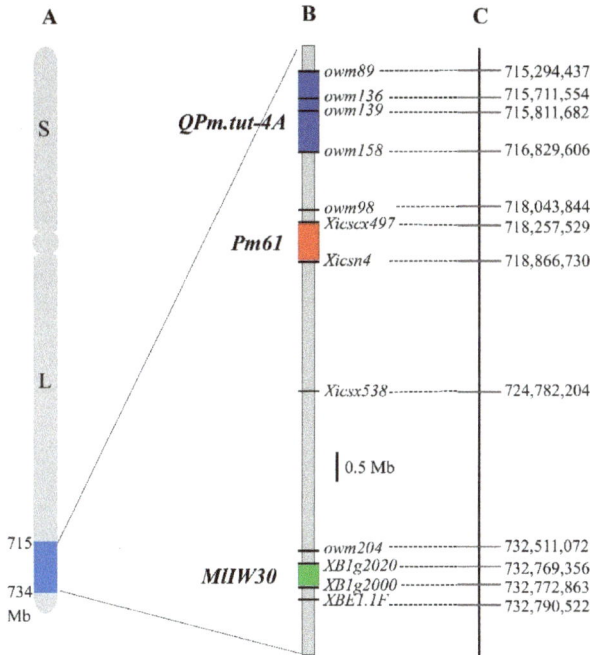

Figure 7. A sketch physical map of chromosome 4A (**A**), comparative analysis of the three *Pm* genes or QTL on chromosome arm 4AL (**B**), and corresponding physical locations in chromosome 4AL in the Chinese Spring reference genome (**C**).

3. Discussion

Using the strategy of BSR-Seq analysis that was performed on the mapping population of XXSYH × Zhongzuo 9504, two *Pm61*-linked SNP markers (*Xicsn2* and *Xicsn4*) and two KASP markers (*Xicsk8* and *Xicsk13*) were developed. Based on the genomic sequences of the Chinese Spring reference genome that flanked the *Pm61* locus, six SSR markers were linked to the target gene. Using these molecular markers, together with previously developed gene-linked markers [35,37], a new saturated genetic linkage map was established, which placed *Pm61* in a 0.71 cM genetic interval corresponding to a 0.61 Mb physical interval (718,257,529–718,866,730) on the terminal region of chromosome 4AL. The closest flanking markers of *Pm61* were *Xicscx497/Xicscx543/owm73/owm98/owm104* and *Xicsn4* with genetic distances of 0.47 and 0.24 cM, respectively. *Pm61* can be detected with the KASP markers *Xicsk8* and *Xicsk13* and SSR markers *Xicscx497* and *Xicsx538*. Compared to the previous study [35], the newly developed genetic linkage map was saturated with more molecular markers especially in the proximal side of *Pm61*. The physical location of *Pm61* in the Chinese Spring reference genome between the two genetic linkage maps with different mapping populations is comparable.

Previously identified *Pm61*-linked markers, *Xicsx79* and *Xicsx436* [35], the only two markers in the proximal side of the target gene, were not polymorphic between the parents XXSYH and Zhongzuo 9504. Therefore, BSR-seq analysis was performed to find more polymorphic markers that flanked *Pm61*. Two SNP markers *Xicsn2* and *Xicsn4* were located on the opposite sides of *Pm61*. The fragments of genomic sequences from the Chinese Spring reference genome corresponding to the two SNP markers were used to develop SSR markers, which produced 6 markers that were linked to *Pm61*. After genetic linkage analysis, all these markers were located in the proximal side of *Pm61*. The newly developed *Pm61*-linked markers are being used to genotype large scale of F_2 and $F_{2:3}$ population that was derived from XXSYH × Zhongzuo 9504 cross. The saturation of the genetic linkage map for *Pm61* facilitates its fine mapping and ultimately map-based cloning.

The race-specific *Pm* genes inherit either in a dominant mode or in a recessive mode. Many Chinese landraces were identified to possess recessive *Pm* genes, for example, several genes in the *Pm5* locus [4–6,8–12], *Pm47* [17], and *PmX* [18]. *Pm61* in XXSYH is also a recessive gene [35]. A *Pm* gene with such a recessive mode of inheritance needs additional generation to allow the expression of resistant phenotype when the target gene is homozygous.

The establishment of the MAS technique for *Pm61* will facilitate its application in developing disease resistant wheat cultivars. The KASP markers *Xicsk8* and *Xicsk13* are able to identify *Pm61* in a high throughput scale. KASP assay is also known as tolerance to DNA quality, low cost and high specificity [38]. However, KASP assay may produce a certain proportion of calling errors (0.7~1.6%) [38] and (8.8% ± 5.5%) missing data [39]. The SSR markers *Xicscx497* and *Xicsx538* are useful to detect *Pm61* using a standard DNA amplification method. Many breeders are able to use PCR-based molecular markers in their routine breeding practices, since such type of molecular markers is cost-effective. The combination KASP assay and PCR analysis can provide a more accurate identification of the target gene during the process of MAS.

Three *Pm* genes or QTL, i.e., *MlIW30*, *Pm61*, and *QPm.tut-4A*, have been detected on a 17.50 Mb genomic region on the terminal part of chromosome arm 4AL (Figure 7). Based on their positions in the Chinese Spring reference genome, *MlIW30* was located in a 21 kb (732,769,506–732,790,522) genomic interval [36]. *QPm.tut-4A* was identified in a 1.54 Mb (715,294,437–716,829,606) genomic interval. *MlIW30* and *QPm.tut-4A* were transferred into common wheat from the wild emmer (*T. turgidum* ssp. *dicoccoides*) and *T. militinae*, respectively [36,37]. *Pm61* is derived from a Chinese landrace and is located in a 0.61 Mb (718,257,529–718,866,730) genomic interval between *MlIW30* and *QPm.uga-4A*. It appears that *Pm61* was located on different genomic intervals as *MlIW30* and *QPm.tut-4A*. Isolation and functional analysis of these *Pm* genes will ultimately understand their relationship.

In summary, BSR-seq analysis, in combination with the Chinese Spring reference genome sequence, identified 10 SNP, KASP, and SSR markers, which saturated the genetic linkage map of *Pm61*, especially in the proximal side of the target gene. The development of KASP markers, *Xicsk8* and *Xicsk13*, and SSR markers, *Xicscx497* and *Xicsx538*, allows the detection of *Pm61* in different scales and platforms. Results from this study will facilitate the fine mapping and ultimate map-based cloning, as well as application in breeding and agriculture, of *Pm61* gene.

4. Materials and Methods

4.1. Plant Materials

Xuxusanyuehuang as the maternal parent was crossed to powdery mildew susceptible winter wheat Zhongzuo 9504 to generate F_1, F_2 and $F_{2:3}$ populations for analyzing the inheritance mode of the resistance gene, determining polymorphism of molecular markers, and establishing linkage relationships between polymorphic markers and *Pm61*. Zhongzuo 9504 also served as the susceptible control in the powdery mildew tests and provided the host plants to maintain and increase *Bgt* isolates.

4.2. Assessments of Resistance to Powdery Mildew

Thirty *Bgt* isolates collected from Shandong, Shanxi, Beijing, Hebei, and Sichuan wheat producing provinces were used to test responses of XXSYH to powdery mildew. These isolates were subjected to three rounds of single-pustule culture on Zhongzuo 9504 plants prior to inoculation. *Bgt1* from Shandong province was used in phenotyping the $F_{2:3}$ mapping population. At least 15 seeds of each $F_{2:3}$ line and the parents were planted in 8-cm-diameter plastic pots. Inoculation of *Bgt* isolates was conducted when wheat seedlings were at 2-leaf stage. Seedlings were dusted with freshly increased conidiophores, incubated in a dew chamber with 90% relative humidity, and grown in a greenhouse to allow development of powdery mildew symptoms. The conditions of plant growth were set at

20–22 °C/14 °C (day/night) with 16 h light/8 h dark photoperiod. When the disease symptoms were fully developed on the susceptible control plants 15 day after inoculation, symptom scoring was conducted by determining the infection type (IT) of each plant on a 0–4 scale as described previously [40]. Based on the scores of ITs, plants with ITs 0–2 were categorized into the resistant group, and those with ITs 3–4 into the susceptible group.

4.3. BSR-Seq Analysis

Based on the phenotypic evaluations, 30 homozygous resistant (IT 1) and 30 homozygous susceptible (IT 4) lines in the $F_{2:3}$ mapping populations of XXSYH × Zhongzuo 9504 were selected to construct the phenotypically contrasting bulks. Each line was represented by a single plant that was free of *Bgt* inoculation. The leaf segments approximately 3 cm long from each plant two-week old were sampled and pooled as the resistant and susceptible bulks for isolating total RNA following the TRIzol protocol (Invitrogen, Carlsbad, CA, USA). RNA-seq analysis was performed in the platform of Illumina HiSeq 2500 in Beijing Novogene Bioinformatics Technology Co. Ltd. (Beijing, China). To remove the adapter sequences and low-quality sequences, raw reeds of generated were subjected to quality control using the software Trimmomatic v0.36 (available online: http://www.sadellab.org/cms/index.php?page=trimmomatic) [41]. The high-quality reads were aligned to the Chinese Spring reference genome sequence v1.0 and its annotation files [25], which was carried using the software STAR v2.5.1b (available online: http://code.google.com/p/rna-star/.) [42], with the mismatch rate of less than 5%. After removing PCR optical duplicates and spliting the mapped reads spanning introns, the unique and confident alignments were used to call SNP variants using "HaplotypeCaller" module in the software GATK v3.6 (available online: http://www.broadinstitute.org/gsa/wiki/index.php/The_Genome_Analysis_Toolkit) [43]. A Fish Exact Test (FET) and allele frequency for each variant and allele frequency difference (AFD) between the resistant and susceptible bulks were used to identify SNP variants. The SNP variants with $P < 1 \times 10^{-10}$ and AFD > 0.6 were regarded as candidate SNP linked to the target gene and used as templates for developing SNP markers.

4.4. DNA Isolation, Amplification and Electrophoresis

The leaf tissues of each $F_{2:3}$ line were used for DNA extraction after disease resistance test following the CTAB protocol [44]. DNA concentration was determined using the Napdrop One (Thermo Fisher Scientific Inc, Madison, WI, USA) and was adjusted to 50 ng·mL^{-1}. DNA bulks with resistant or susceptible phenotype were constructed by pooling equal amounts of DNA from 10 resistant and susceptible $F_{2:3}$ lines each. DNA was amplified in a Biometra T3000 Thermocycler (ABI, New York, NY, USA). Each reaction mixture (10 μL) was composed of 5 μL mixture (including *Taq* polymerase, dNTPs, and 10× PCR buffer and Mg^{2+}), 2 μL ddH$_2$O, 1 μL DNA, and 1 μL 10 μM each of the forward and reverse primers. The profile of DNA amplification was set at 98 °C for 3 min; 35 cycles of 98 °C for 10 s, 55 °C–60 °C (depending on the specific primers) for 10 s and 72 °C for 25 s; 72 °C for 10 min. Products amplified were separated by 2% agarose gel or 8% non-denaturing polyacrylamide gel (Acr:Bis = 29:1).

4.5. Development and Validation of SNP Markers

Approximately 3 kb sequences extracted from the RefSeqv1.0 Chinese Spring genome sequence [25] containing the candidate SNP linked to the target gene were used as queries in searching the Chinese Spring whole genome assembly (available online: https://urgi.versailles.inra.fr/) to acquire the homologous scaffolds. The sequences containing candidate SNP that corresponded to the above homologous scaffolds were used as templates to design SNP primers on the GSP website (available online: https://probes.pw.usda.gov/GSP/) [45]. The primers designed contained at least one variant site at the 3' ends and were anticipated to amplify products in a range of 300~1000 bp in length. The polymorphism of SNP primers was validated in the parents and the two DNA bulks by analyzing the sequences of the amplicons (Invitrogen Trading Co., Ltd., Shanghai, China), and the polymorphic SNP primers were used to genotype the $F_{2:3}$ mapping population.

4.6. Conversion of SNP Markers to KASP Markers

The polymorphic SNP primers were converted to KASP markers using the PolyMarker software (available online: http://polymarker.tgac.ac.uk) [46]. Each KASP reaction was carried out using a 5 µL reaction mixture consisting of 2.2 µL DNA, 2.5 µL 2× KASP master mix, 0.056 µL primer mix (12 mM of each allele-specific primer and 30 mM of the common primer), 0.039 µL Mg^{2+}, and 0.205 µL ddH$_2$O. Amplification was performed in the ABI 7500 device (Applied Biosystems, Foster City, CA, USA) with the program of 94 °C for 15 min; 35 cycles of 94 °C for 20 s and 60 °C for 1 min. The FLUOstar Omega microplate reader (BMG Labtech, Durham, NC, USA) was used to read the green (521 nm) and pink (556 nm) fluorescence signals at 25 °C for 2 min after the reactions were completed. The fluorescence signals were transformed into different genotypes, i.e., FAM homozygotes, HEX homozygotes, and FAM/HEX heterozygotes using the Klustering Caller software (available online: http://www.lgcgroup.com/).

4.7. Development of SSR Markers

The sequences of *Pm61*-linked SNP markers designed were used to search for the Chinese Spring reference genome sequence v1.0 [25]. The corresponding genomic sequences were used as templates to design SSR markers with BatchPrimer3 software (available online: https://wheat.pw.usda.gov/demos/batchprimer3). Polymorphism of SSR markers were examined using the parents and the contrasting DNA bulks. DNA amplification and visualization of banding patterns were carried out as described previously [35]. The polymorphic markers were used to establish a genetic linkage map of *Pm61* with the $F_{2:3}$ mapping population.

4.8. Construction of High-Density Genetic Linkage Map

The *Pm61*-linked markers, including previously developed SSR markers and the SNP markers, KASP markers, and SSR markers developed in present studies [35,37], were used to construct the high-density genetic linkage map of *Pm61* using the software Mapdraw v.2.1 (Huazhong Agricultural Sciences, Wuhan, China) [47]. The genetic distance was measured by the Kosambi function. Linkage relationship between markers and *Pm61* was established with the software Mapmaker/Exp Version 3.0b and a logarithm of the odd ratio (LOD) score threshold of 3.0 [48].

Supplementary Materials: Supplementary materials can be found at http://www.mdpi.com/1422-0067/20/3/750/s1. Table S1 Sequencing quality assessment details; Table S2 A list of SNP marker primers used in this study; Table S3 A list of KASP marker primers used in this study; Table S4 A list of 347 SSR primer pairs based on the 1.5 Mb sequences extended from marker *Xicsn2* toward *Pm61*; Table S5 A list of 617 SSR primer pairs based on the 2.5 Mb sequences extended from marker *Xicsn4* toward *Pm61*.

Author Contributions: H.J.L., J.L., and Z.L. conceived and designed the study. J.H., P.W. and D.Q. conducted the experiments. J.H., Y.L., Y.Q., H.Z., T.Z., and L.Y. participated the disease tests and sample preparation. Y.Y., M.L. and T.F. helped with electrophoresis. J.H., J.L. and J.X. analyzed data. J.H., J.L. and H.J.L. wrote the manuscript with contributions of Z.L. and Y.Z. All authors read and approved the final manuscript.

Acknowledgments: Financial support of this research by the National Natural Science Foundation of China (31501310 and 31471491), the National Key Research and Development Program of China (2017YFD0101000), Scientific and Technological Research Project of Henan Province of China (172102110110), the CAAS Innovation Team, Henan Province Young College Key Teacher Subsidy Program (2017GGJS177) are gratefully appreciated.

Conflicts of Interest: The authors declare no conflict of interest.

References

1. Singh, R.P.; Singh, P.K.; Rutkoski, J.; Hodson, D.P.; He, X.; Jørgenssen, L.N.; Huertaespino, J. Disease impact on wheat yield potential and prospects of genetic control. *Annu. Rev. Phytopathol.* **2016**, *54*, 303–322. [CrossRef] [PubMed]
2. Samobor, V.; Vukobratović, M.; Jošt, M. Effect of powdery mildew attack on quality parameters and experimental bread baking of wheat. *Acta Agric. Slov.* **2006**, *87*, 381–391.
3. Li, J.; Liu, X.H.; Yang, X.W.; Li, Y.C.; Wang, C.Y.; He, D.X. Proteomic analysis of the impacts of powdery mildew on wheat grain. *Food Chem.* **2018**, *261*, 30–35. [CrossRef] [PubMed]
4. Hsam, S.L.K.; Huang, X.Q.; Zeller, F.J. Chromosomal location of genes for resistance to powdery mildew in common wheat (*Triticum aestivum* L. em. Thell.). 6. Alleles at the *Pm5* locus. *Theor. Appl. Genet.* **2001**, *102*, 127–133. [CrossRef]
5. Huang, X.Q.; Wang, L.X.; Xu, M.X.; Röder, M.S. Microsatellite mapping of the powdery mildew resistance gene *Pm5e* in common wheat (*Triticum aestivum* L.). *Theor. Appl. Genet.* **2003**, *106*, 858–865. [CrossRef] [PubMed]
6. Zhou, R.H.; Zhu, Z.D.; Kong, X.Y.; Huo, N.X.; Tian, Q.Z.; Li, P.; Jin, C.Y.; Dong, Y.C.; Jia, J.Z. Development of wheat near-isogenic lines for powdery mildew resistance. *Theor. Appl. Genet.* **2005**, *110*, 640–648. [CrossRef] [PubMed]
7. Hu, T.Z.; Li, H.J.; Xie, C.J.; You, M.S.; Yang, Z.M.; Sun, Q.X.; Liu, Z.Y. Molecular mapping and chromosomal location of powdery mildew resistance gene in wheat cultivar Tangmai 4. *Acta Agron. Sin.* **2008**, *34*, 1193–1198.
8. Zhai, W.W.; Duan, X.Y.; Zhou, Y.L.; Ma, H.Q. Inheritance of resistance to powdery mildew in four Chinese landraces. *Plant Protect* **2008**, *34*, 37–40.
9. Xue, F.; Zhai, W.W.; Duan, X.Y.; Zhou, Y.L.; Ji, W.Q. Microsatellite mapping of powdery mildew resistance gene in wheat landrace Xiaobaidong. *Acta Agron. Sin.* **2009**, *35*, 1806–1811. [CrossRef]
10. Fu, B.S.; Zhang, Z.L.; Zhang, Q.F.; Wu, X.Y.; Wu, J.Z.; Cai, S.B. Identification and mapping of a new powdery mildew resistance allele in the Chinese wheat landrace Hongyoumai. *Mol. Breed.* **2017**, *37*, 133. [CrossRef]
11. Xu, X.D.; Jing, F.; Fan, J.R.; Liu, Z.Y.; Li, Q.; Zhou, Y.L.; Ma, Z.H. Identification of the resistance gene to powdery mildew in Chinese wheat landrace Baiyouyantiao. *J. Integr. Agric.* **2018**, *17*, 37–45. [CrossRef]
12. Xu, X.D.; Li, Q.; Ma, Z.H.; Fan, J.R.; Zhou, Y.L. Molecular mapping of powdery mildew resistance gene *PmSGD* in Chinese wheat landrace Shangeda using RNA-seq with bulk segregant analysis. *Mol. Breed.* **2018**, *38*, 23. [CrossRef]
13. Huang, X.Q.; Hsam, S.L.K.; Zeller, F.J.; Wenzel, G.; Mohler, V. Molecular mapping of the wheat powdery mildew resistance gene *Pm24* and marker validation for molecular breeding. *Theor. Appl. Genet.* **2000**, *101*, 407–414. [CrossRef]
14. Xue, F.; Wang, C.Y.; Li, C.; Duan, X.Y.; Zhou, Y.L.; Zhao, N.J.; Wang, Y.J.; Ji, W.Q. Molecular mapping of a powdery mildew resistance gene in common wheat landrace Baihulu and its allelism with *Pm24*. *Theor. Appl. Genet.* **2012**, *125*, 1425–1432. [CrossRef] [PubMed]
15. Xu, H.X.; Yi, Y.J.; Ma, P.T.; Qie, Y.M.; Fu, X.Y.; Xu, Y.F.; Zhang, X.T.; An, D.G. Molecular tagging of a new broad-spectrum powdery mildew resistance allele *Pm2c* in Chinese wheat landrace Niaomai. *Theor. Appl. Genet.* **2015**, *128*, 2077–2084. [CrossRef]
16. Ma, H.Q.; Kong, Z.X.; Fu, B.S.; Li, N.; Zhang, L.X.; Jia, H.Y.; Ma, Z.Q. Identification and mapping of a new powdery mildew resistance gene on chromosome 6D of common wheat. *Theor. Appl. Genet.* **2011**, *123*, 1099–1106. [CrossRef] [PubMed]
17. Xiao, M.G.; Song, F.J.; Jiao, J.F.; Wang, X.M.; Xu, H.X.; Li, H.J. Identification of the gene *Pm47* on chromosome 7BS conferring resistance to powdery mildew in the Chinese wheat landrace Hongyanglazi. *Theor. Appl. Genet.* **2013**, *126*, 1397–1403. [CrossRef]
18. Fu, B.S.; Chen, Y.; Li, N.; Ma, H.Q.; Kong, Z.X.; Zhang, L.X.; Jia, H.Y.; Ma, Z.Q. *pmX*: A recessive powdery mildew resistance gene at the *Pm4* locus identified in wheat landrace Xiaohongpi. *Theor. Appl. Genet.* **2013**, *126*, 913–921. [CrossRef]
19. Wang, Z.Z.; Li, H.W.; Zhang, D.Y.; Guo, L.; Chen, J.J.; Chen, Y.X.; Wu, Q.H.; Xie, J.Z.; Zhang, Y.; Sun, Q.X.; et al. Genetic and physical mapping of powdery mildew resistance gene *MlHLT* in Chinese wheat landrace Hulutou. *Theor. Appl. Genet.* **2015**, *128*, 365–373. [CrossRef]

20. Li, G.; Carver, B.F.; Cowger, C.; Bai, G.; Xu, X. *Pm223899*, a new recessive powdery mildew resistance gene identified in Afghanistan landrace PI 223899. *Theor. Appl. Genet.* **2018**, *131*, 2775–2783. [CrossRef]
21. Tan, C.C.; Li, G.Q.; Cowger, C.; Carver, B.F.; Xu, X.Y. Characterization of *Pm63*, a powdery mildew resistance gene in Iranian landrace PI 628024. *Theor. Appl. Genet.* **2018**. [CrossRef] [PubMed]
22. Tan, C.C.; Li, G.Q.; Cowger, C.; Carver, B.F.; Xu, X.Y. Characterization of *Pm59*, a novel powdery mildew resistance gene in Afghanistan wheat landrace PI 181356. *Theor. Appl. Genet.* **2018**, *131*, 1145–1152. [CrossRef] [PubMed]
23. Newton, A.C.; Akar, T.; Baresel, J.P.; Bebeli, P.J.; Bettencourt, E.; Blandenopoulos, K.V.; Czembor, J.H.; Fasoula, D.A.; Katsiotis, A.; Koutis, K.; et al. Cereal landraces for sustainable agriculture. A review. *Agron. Sustain. Dev.* **2010**, *20*, 237–269. [CrossRef]
24. Kaur, N.; Street, K.; Mackay, M.; Yahiaoui, N.; Keller, B. Molecular approaches for characterization and use of natural disease resistance in wheat. *Eur. J. Plant Pathol.* **2008**, *121*, 387–397. [CrossRef]
25. International Wheat Genome Sequencing Consortium (IWGSC). Shifting the limits in wheat research and breeding using a fully annotated reference genome. *Science* **2018**, *361*, eaar7191. [CrossRef] [PubMed]
26. Rasheed, A.; Wen, W.; Gao, F.; Zhai, S.; Jin, H.; Liu, J.; Guo, Q.; Zhang, Y.; Dreisigacker, S.; Xia, X.; et al. Development and validation of KASP assays for genes underpinning key economic traits in bread wheat. *Theor. Appl. Genet.* **2016**, *129*, 1843–1860. [CrossRef] [PubMed]
27. Qureshi, N.; Kandiah, P.; Gessese, M.K.; Nsabiyera, V.; Wells, V.; Babu, P.; Wong, D.; Hayden, M.J.; Bariana, H.; Bansal, U. Development of co-dominant KASP markers co-segregating with Ug99 effective stem rust resistance gene *Sr26* in wheat. *Mol. Breed.* **2018**, *38*, 97. [CrossRef]
28. Qureshi, N.; Bariana, H.S.; Zhang, P.; McIntosh, R.; Bansal, U.K.; Wong, D.; Hayden, M.J.; Dubcovsky, J.; Shankar, M. Genetic relationship of stripe rust resistance genes *Yr34* and *Yr48* in wheat and identification of linked KASP markers. *Plant Dis.* **2018**, *102*, 413–420. [CrossRef]
29. Trick, M.; Adamski, N.M.; Mugford, S.G.; Jiang, C.; Febrer, M.; Uauy, C. Combining SNP discovery from next-generation sequencing data with bulked segregant analysis (BSA) to fine-map genes in polyploid wheat. *BMC Plant Biol.* **2012**, *12*, 14. [CrossRef]
30. Ramirez-Gonzalez, R.H.; Segovia, V.; Bird, N.; Fenwick, P.; Holdgate, S.; Berry, S.; Jack, P.; Caccamo, M.; Uauy, C. RNA-Seq bulked segregant analysis enables the identification of high-resolution genetic markers for breeding in hexaploid wheat. *Plant Biotechnol. J.* **2015**, *13*, 613–624. [CrossRef]
31. Wang, Y.; Xie, J.Z.; Zhang, H.Z.; Guo, B.M.; Ning, S.Z.; Chen, Y.X.; Lu, P.; Wu, Q.H.; Li, M.M.; Zhang, D.Y.; et al. Mapping stripe rust resistance gene *YrZH22* in Chinese wheat cultivar Zhoumai 22 by bulked segregant RNA-Seq (BSR-Seq) and comparative genomics analyses. *Theor. Appl. Genet.* **2017**, *130*, 2191–2201. [CrossRef] [PubMed]
32. Wang, Y.; Zhang, H.Z.; Xie, J.Z.; Guo, B.M.; Chen, Y.X.; Zhang, H.Y.; Lu, P.; Wu, Q.H.; Li, M.M.; Zhang, D.Y.; et al. Mapping stripe rust resistance genes by BSR-Seq, *YrMM58* and *YrHY1* on chromosome 2AS in Chinese wheat lines Mengmai 58 and Huaiyang 1 are *Yr17*. *Crop J.* **2018**, *6*, 91–98. [CrossRef]
33. Wu, J.H.; Zeng, Q.D.; Wang, Q.L.; Liu, S.J.; Yu, S.Z.; Mu, J.M.; Huang, S.; Sela, H.; Distelfeld, A.; Huang, L.L.; et al. SNP-based pool genotyping and haplotype analysis accelerate fine-mapping of the wheat genomic region containing stripe rust resistance gene *Yr26*. *Theor. Appl. Genet.* **2018**, *131*, 1481–1496. [CrossRef] [PubMed]
34. Wu, P.P.; Xie, J.Z.; Hu, J.H.; Qiu, D.; Liu, Z.Y.; Li, J.T.; Li, M.M.; Zhang, H.J.; Yang, L.; Liu, H.W.; et al. Development of molecular markers linked to powdery mildew resistance gene *Pm4b* by combining SNP discovery from transcriptome sequencing data with bulked segregant analysis (BSR-Seq) in wheat. *Front. Plant Sci.* **2018**, *9*, 95. [CrossRef] [PubMed]
35. Sun, H.G.; Hu, J.H.; Song, W.; Qiu, D.; Cui, L.; Wu, P.P.; Zhang, H.J.; Liu, H.W.; Yang, L.; Qu, Y.F.; et al. *Pm61*: A recessive gene for resistance to powdery mildew in wheat landrace Xuxusanyuehuang identified by comparative genomics analysis. *Theor. Appl. Genet.* **2018**, *131*, 2085–2097. [CrossRef] [PubMed]
36. Geng, M.M.; Zhang, J.; Peng, F.X.; Liu, X.; Lv, X.D.; Mi, Y.Y.; Li, Y.H.; Li, F.; Xie, C.J.; Sun, Q.X. Identification and mapping of *MlIW30*, a novel powdery mildew resistance gene derived from wild emmer wheat. *Mol. Breed.* **2016**, *36*, 130. [CrossRef]

37. Janáková, E.; Jakobson, I.; Peusha, H.; Abrouk, M.; Škopová, M.; Šimková, H.; Šafář, J.; Jan Vrána, J.; Doležel, J.; Järve, K.; et al. Divergence between bread wheat and *Triticum militinae* in the powdery mildew resistance *QPm.tut-4A* locus and its implications for cloning of the resistance gene. *Theor. Appl. Genet.* **2018**. [CrossRef]

38. Semagn, K.; Babu, R.; Hearne, S.; Olsen, M. Single nucleotide polymorphism genotyping using Kompetitive Allele Specific PCR (KASP): Overview of the technology and its application in crop improvement. *Mol. Breed.* **2014**, *33*, 1–14. [CrossRef]

39. Allen, A.M.; Barker, G.L.; Wilkinson, P.; Burridge, A.; Winfield, M.; Coghill, J.; Uauy, C.; Griffiths, S.; Jack, P.; Berry, S.; et al. Discovery and development of exome-based, co-dominant single nucleotide polymorphism markers in hexaploid wheat (*Triticum aestivum* L.). *Plant Biotechnol. J.* **2013**, *11*, 279–295. [CrossRef]

40. Liu, Z.; Sun, Q.; Ni, Z.; Yang, T.; McIntosh, R.A. Development of SCAR markers linked to the *Pm21* gene conferring resistance to powdery mildew in common wheat. *Plant Breed.* **1999**, *118*, 215–219. [CrossRef]

41. Bolger, A.M.; Lohse, M.; Usadel, B. Trimmomatic: A flexible trimmer for Illumina sequence data. *Bioinformatics* **2014**, *30*, 2114–2120. [CrossRef] [PubMed]

42. Dobin, A.; Davis, C.A.; Schlesinger, F.; Drenkow, J.; Zaleski, C.; Jha, S.; Batut, P.; Chaisson, M.; Gingeras, T.R. STAR: Ultrafast universal RNA-seq aligner. *Bioinformatics* **2013**, *29*, 15–21. [CrossRef] [PubMed]

43. McKenna, A.; Hanna, M.; Banks, E.; Sivachenko, A.; Cibulskis, K.; Kernytsky, A.; Garimella, K.; Altshuler, D.; Gabriel, S.; Daly, M.; et al. The Genome Analysis Toolkit: A MapReduce framework for analyzing next-generation DNA sequencing data. *Genome Res.* **2010**, *20*, 1297–1303. [CrossRef] [PubMed]

44. Saghai-Maroof, M.A.; Soliman, K.M.; Jorgensen, R.A.; Allard, R.W. Ribosomal DNA spacer-length polymorphisms in barley: Mendelian inheritance, chromosomal location, and population dynamics. *Proc. Natl. Acad. Sci. USA* **1984**, *81*, 8014–8018. [CrossRef] [PubMed]

45. Wang, Y.; Tiwari, V.K.; Rawat, N.; Gill, B.S.; Huo, N.X.; You, F.M.; Coleman-Derr, D.; Gu, Y.Q. GSP: A web-based platform for designing genome-specific primers in polyploids. *Bioinformatics* **2016**, *32*, 2382–2383. [CrossRef] [PubMed]

46. Ramirez-Gonzalez, R.H.; Uauy, C.; Caccamo, M. PolyMarker: A fast polyploid primer design pipeline. *Bioinformatics* **2015**, *31*, 2038–2039. [CrossRef] [PubMed]

47. Liu, R.H.; Meng, J.L. MapDraw: A Microsoft excel macro for drawing genetic linkage maps based on given genetic linkage data. *Hereditas* **2003**, *25*, 317–321. [PubMed]

48. Lincoln, S.E.; Daly, M.J.; Lander, E.S. *Constructing Genetic Linkage Maps with MAPMAKER/EXP Version 3.0: A Tutorial and Reference Mannual*, 3rd ed.; Whitehead Institute for Medical Research: Cambridge, MA, USA, 1993.

© 2019 by the authors. Licensee MDPI, Basel, Switzerland. This article is an open access article distributed under the terms and conditions of the Creative Commons Attribution (CC BY) license (http://creativecommons.org/licenses/by/4.0/).

International Journal of
Molecular Sciences

MDPI

Article

Fine Mapping of the Wheat Leaf Rust Resistance Gene *Lr42*

Harsimardeep S. Gill [1,†], Chunxin Li [2,†], Jagdeep S. Sidhu [1], Wenxuan Liu [3], Duane Wilson [4], Guihua Bai [5], Bikram S. Gill [4] and Sunish K. Sehgal [1,*]

[1] Department of Agronomy, Horticulture & Plant Science, South Dakota State University, Brookings, SD 57006, USA; harsimardeep.gill@sdstate.edu (H.S.G.); jagdeeproots@gmail.com (J.S.S.)
[2] National Engineering Laboratory of Wheat, Wheat Research Institute, Henan Academy of Agricultural Sciences, Zhengzhou 450002, China; licx82@yahoo.com
[3] The State Key Laboratory of Wheat and Maize Crop Science, College of Life Sciences, Henan Agricultural University, Zhengzhou 450002, China; wxliu2003@gmail.com
[4] Wheat Genetics Resource Center, Department of Plant Pathology, Kansas State University, Manhattan, KS 66506, USA; dlwil@ksu.edu (D.W.); bsgill@ksu.edu (B.S.G.)
[5] USDA-ARS, Hard Winter Wheat Genetic Research Unit, Manhattan, KS 66506, USA; guihua.bai@ars.usda.gov
*   Correspondence: sunish.sehgal@sdstate.edu
†   These authors contributed equally to this work.

Received: 22 March 2019; Accepted: 14 May 2019; Published: 17 May 2019

Abstract: Leaf rust caused by *Puccinia triticina* Eriks is one of the most problematic diseases of wheat throughout the world. The gene *Lr42* confers effective resistance against leaf rust at both seedling and adult plant stages. Previous studies had reported *Lr42* to be both recessive and dominant in hexaploid wheat; however, in diploid *Aegilops tauschii* (TA2450), we found *Lr42* to be dominant by studying segregation in two independent F_2 and their $F_{2:3}$ populations. We further fine-mapped *Lr42* in hexaploid wheat using a KS93U50/Morocco F_5 recombinant inbred line (RIL) population to a 3.7 cM genetic interval flanked by markers *TC387992* and *WMC432*. The 3.7 cM *Lr42* region physically corresponds to a 3.16 Mb genomic region on chromosome 1DS based on the Chinese Spring reference genome (RefSeq v.1.1) and a 3.5 Mb genomic interval on chromosome 1 in the *Ae. tauschii* reference genome. This region includes nine nucleotide-binding domain leucine-rich repeat (NLR) genes in wheat and seven in *Ae. tauschii*, respectively, and these are the likely candidates for *Lr42*. Furthermore, we developed two kompetitive allele-specific polymorphism (KASP) markers (*SNP113325* and *TC387992*) flanking *Lr42* to facilitate marker-assisted selection for rust resistance in wheat breeding programs.

Keywords: wheat; *Aegilops tauschii*; *Lr42*; disease resistance; molecular mapping; KASP markers; marker-assisted selection

---

1. Introduction

Wheat is one of the leading staple foods worldwide, providing one-fifth of the calories and protein to more than 4.5 billion people [1]. Wheat production is constrained not only due to changing climate, but to a great extent by the emergence of new and more virulent races of economically important pathogens. Several diseases and insect pests, including leaf rust (caused by *Puccinia triticina* Eriks), threaten sustainable wheat production in the major wheat-growing areas of the world [2]. In Kansas alone, the leaf rust epidemic of 2007 caused yield losses of 13.9% in winter wheat [3]. Yield losses are attributed to fewer kernels, aggregated by lower kernel weight [2], and losses can be severe if wheat is infected early in development and may reach epidemic proportions in susceptible cultivars under favorable conditions [4]. Diving into history, one can find a reference to leaf rust in the Bible and literature of classical Greece and Rome [5]. Also, the existence of prevalence of this disease from that era until today indicates that this pathogen has evolved along with wheat or other grass species and there has been no permanent solution to control this disease, and likewise for other rusts.

Breeding for rust resistance is considered as one of the most economical approaches to manage rust diseases, and wheat breeding programs throughout the world are deploying rust resistance genes in commercial cultivars. The populations of *P. triticina* are reported to be highly variable in North America [6], with many different virulence pathotypes or races detected annually. Therefore, a more durable approach for long-lasting resistance is the pyramiding of different race-specific and race non-specific genes in single cultivars [7]. However, combining different resistance genes using phenotypic selection is a challenging and cumbersome process. Molecular markers can be very useful for simultaneous stacking of different resistance genes. Thus, the availability of tightly linked molecular markers for different genes is essential for facilitating the stacking of these genes into a single genetic background.

To date, around 80 leaf rust resistance (*Lr*) genes have been formally reported in wheat and its wild relatives [8]. *Aegilops tauschii* Coss, the D-genome donor of wheat, has been a rich source of resistance genes [9,10] and agronomic traits [11]. Several leaf rust resistance genes (*Lr21* (1D), *Lr32* (3D), *Lr39* (2D)) have been transferred into bread wheat from *Ae. tauschii*, including *Lr42* [12]. *Lr42* was introgressed into wheat through a direct cross with *Ae. tauschii* (accession TA 2450) and released as KS91WGRC11 (Century*/TA2450) for further utilization in hexaploid wheat breeding. KS91WGRC11 (carrying *Lr42*) has been successfully used by breeders in several breeding programs [13,14]. Studying near-isogenic lines (NILs) for the *Lr42*, Martin et al. [15] reported that *Lr42* plays an important role in increasing yield, test weight, and kernel weight in wheat. *Lr42* still is one of the highly effective genes, conferring resistance at both seedling and adult plant stages and being used in CIMMYT lines for breeding against leaf rust.

Cox et al. [16] first reported *Lr42* on chromosome 1DS using monosomic analysis in wheat, and it was found to be closely linked to *Lr21*. In their study, Cox et al. [16] reported *Lr42* to be a partially dominant race-specific gene. However, Czembor et al. [17] used Diversity Arrays Technology (DArT) markers to map *Lr42* gene on chromosome 3D and reported that *Lr42* behaved as a dominant gene. By analyzing a set of near-isogenic lines (NILs), Sun et al. [18] mapped *Lr42* on the distal end of chromosome arm 1DS by employing simple sequence repeat (SSR) markers. Using a segregating population of NILs (for *Lr42*) and evaluating for rust infection at both seedling and adult plant stages, they identified three molecular markers—*WMC432*, *CFD15*, and *GDM33*—closely linked to *Lr42*. *WMC432*, about 0.8 cM from *Lr42*, was found to be the closest marker; however, no flanking markers were reported. Further, Liu et al. [19] analyzed an F$_2$ population derived from a cross of KS93U50 (*Lr42*)/Morocco to map *Lr42* on chromosome 1DS using six SSR loci with the quantitative calculation method and reported *Lr42* as a recessive gene. *Lr42* was flanked by *WMC432* and *GDM33* onto a 17 cM region at the distil end of chromosome 1DS with the closest proximal marker (*WMC432*) around 4 cM away. Although the previous studies were able to map *Lr42* to wheat chromosome 1DS, the gene lies in a very gene-rich and recombination-hotspot region at the terminal tip of wheat chromosome 1DS; therefore, SSR markers flanking 17 cM regions are not suitable for marker-assisted selection.

The objectives of this study were to (1) determine the genetic and physical location of *Lr42* on the chromosome 1D and (2) develop kompetitive allele-specific polymorphism (KASP) markers to facilitate marker-assisted selection of *Lr42* in wheat breeding programs. This work will lay the foundation for further cloning of the gene.

2. Results

2.1. Genetic Analysis of Lr42 in Ae. tauschii

The parental lines TA2433 and TA10132 (AL8/78) showed a highly susceptible response to the leaf rust isolate PNMRJ, with an infection type (IT) score of 3, whereas the *Lr42*-carrying accession

TA2450 showed highly resistant response with an IT score. Of the 66 F_2 plants screened from the TA2450/TA2433 population, 50 were resistant and 16 were susceptible, fitting a 3:1 ratio ($\chi^2 = 0.20$, $p = 0.89$) for a single dominant gene (*Lr42*) segregation in this population (Table 1). Further, the 100 $F_{2:3}$ families also exhibited a good fit for the expected 1:2:1 (resistant:segregating:susceptible) ratio ($\chi^2 = 1.92$, $p = 0.38$).

Table 1. Segregation of *Lr42* in *Ae. tauschii* and hexaploid wheat populations against leaf rust race PNMRJ. The observed and expected ratios correspond to the resistance:susceptible in F_2 generations and homozygous resistant:segregating:homozygous susceptible in $F_{2:3}$ generations.

Sr No.	Species	Population	Generation	Lines Evaluated	Observed Ratio	Expected Ratio	χ^2	*p*-Value*
1	*Ae. tauschii*	TA2450/TA2433	F_2	66	50:16	3:1	0.20	0.89
			$F_{2:3}$	100	27:54:19	1:2:1	1.92	0.38
2	*Ae. tauschii*	TA2450/TA10132 (AL8/78)	F_2	67	53:14	3:1	0.60	0.44
			$F_{2:3}$	100	33:53:14	1:2:1	7.58	0.02
3	*T. aestivum*	KS93U50/Morocco	F_5 RIL	234	99:135	1:1	5.54	0.02

* $\alpha = 0.01$.

The second F_2 population derived from the cross TA2450/TA10132 showed 53 resistant and 14 susceptible individuals, fitting a 3:1 ratio ($\chi^2 = 0.60$, $p = 0.44$) and confirming that *Lr42* behaves as a dominant gene (Table 1). Similarly, 100 $F_{2:3}$ families (TA2450/TA10132) evaluated also fit the 1:2:1 (resistant:segregating:susceptible) ratio ($\chi^2 = 7.58$, $p = 0.02$), with skewing toward resistant families. Our results from segregation of resistance and susceptibility in these two populations suggest that *Lr42* shows dominant inheritance in *Ae. tauschii* backgrounds.

2.2. Phenotypic Evaluation of KS93U50/Morocco RIL Population

The hexaploid wheat line KS93U50 carrying *Lr42* showed resistance reaction against isolate PNMRJ producing an infection type (IT) score of 2, whereas Morocco, the susceptible parent of the RIL population, exhibited an IT score of 3+ as expected. The individual plants of 234 F_5 RILs from the KS93U50/Morocco population were evaluated for responses to PNMRJ and 99 RILs were found to be rust-susceptible, while the other 135 RILs showed a resistant response. The 1S:1R segregation ratio ($\chi^2 = 5.54$) suggests the presence of a single resistance gene *Lr42* in the RIL population (Table 1).

2.3. Marker Discovery and Molecular Mapping

Numerous genomic resources were employed to develop new markers to saturate the target *Lr42* region. The flanking markers for *Lr42*, namely *GDM33*, *WMC432*, and *CFD15* [19], were amplified from Chinese Spring (CS) chromosomes 1D, 4D, and 6D bacterial artificial chromosome (BAC) pools. These pools were developed from a BAC library contructed from fraction-I chromosomes (1D, 4D, and 6D) obtained through flow cytometry separation of CS DNA. *CFD15* was physically mapped to two BAC clones (146DhC878D17 and 146DhC799I02), whereas *WMC432* was mapped to four BAC clones (146DhB488K16, 146DhC808O03, 146DhB488K16, and 146DhB458C07). Both these markers were mapped to the single BAC contig ctg1768; however, the distal marker *GWM33* could not be mapped to a unique BAC contig. Our BAC-based physical map of CS chromosome 1D is anchored to *Ae. tauschii* 10K Infinium SNP-based genetic maps, and we identified BAC contigs proximal and distal to ctg1768 in a 1D physical map (https://urgi.versailles.inra.fr/gb2/gbrowse/wheat_phys_1D_v1/). Six BAC contigs were identified spanning this very terminal region of chromosome 1DS. The 24 BACs in these six contigs were end-sequenced to identify five new SSR markers.

We further employed comparative genomic analysis with chromosome 1H of barley [20] for the development of new molecular markers. The collinear interval on chromosome 1H was determined, and 24 genes were predicted to carry plant defense-related domains. Wheat expressed sequence tag (EST) sequences collinear to barley genes were used to develop 19 new EST markers for saturation of the *Lr42* region. In addition, we identified 44 SNPs from *Ae. tauschii* 10K Infinium SNPs mapped on the terminal end of chromosome 1D in a *Prelude* (TA2988)/synthetic wheat (TA8051) RIL population (Figure 1) and also in an *Ae. tauschii* AL8/78 (TA10132)/AS75 F_2 population [21]. These SNPs were anchored on a CS 1D physical map. Thus, a total of five SSR markers, 19 EST markers, and 44 KASP markers were designed to enrich the candidate region. Of the 68 markers, 11 were polymorphic between KS93U50 and Morocco and used for mapping of *Lr42* (Figure 1, Table 2). We were able to narrow *Lr42* to a 3.7 cM interval flanked by markers *TC387992* and *WMC432* as against the previously reported 17 cM interval on the terminal end of the short arm of chromosome 1D in wheat (Figure 1).

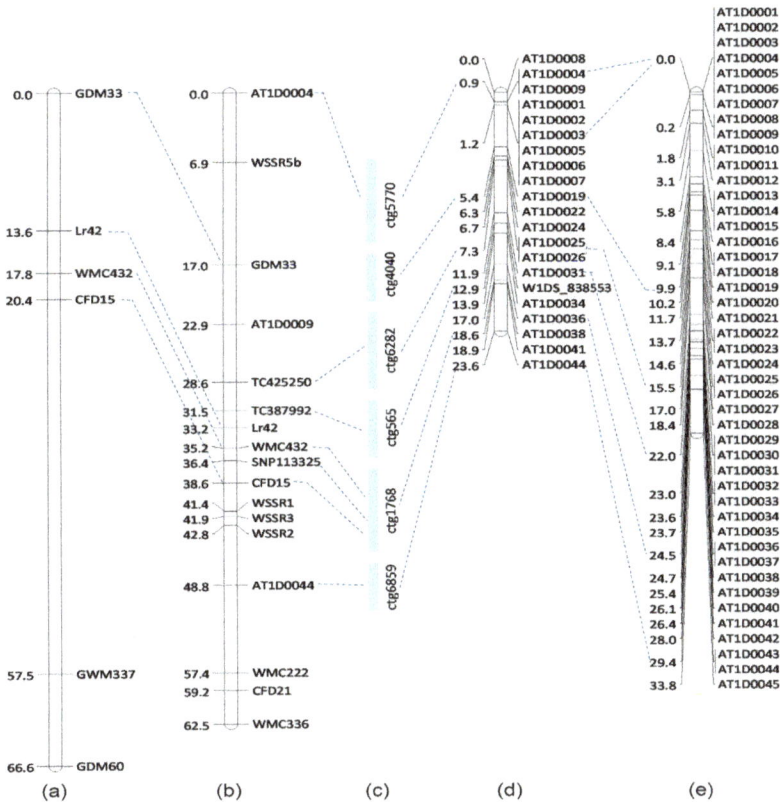

Figure 1. Comparative genetic and physical map of the leaf rust resistance gene *Lr42*: (**a**) genetic map in a KS93U50/Morocco F_2 population [21]; (**b**) genetic map in a KS93U50/Morocco F_5 recombinant inbred line (RIL) population (current study); (**c**) physical mapping of the *Lr42* region on Chinese Spring 1D bacterial artificial chromosome (BAC) contigs (current study), and (**d**) *Prelude* (TA2988)/synthetic wheat (TA8051) RIL population (current study), and (**e**) an *Ae. tauschii* AL8/78 (TA10132)/AS75 F_2 population [19].

Table 2. Simple sequence repeat (SSR) and KASP markers developed and mapped on the KS93U50/Morocco RIL population.

Sr. No.	Primer Name	Assay Type	Sequence(s)
1	WSSR1	SSR	ACGACGTTGTAAAACGACTGGAGACAGACGAACGCATA TGCATGCATACACACACCAG
2	WSSR2	SSR	ACGACGTTGTAAAACGACAGCAATGCAGTTGCAAAGAG GCAAAGATGGACAGATGGCT
3	WSSR3	SSR	ACGACGTTGTAAAACGACAAGATCAGCTCCGACAGCTC CGAAGTCAGCACAAACCAAA
4	WSSR5	SSR	ACGACGTTGTAAAACGACTGGTGAATCTTGCACCACAT CTGGACACCGTTCGTTAGGT
5	AT1D004	KASP	GGTACCATGTTGTTTCGCATGTCTAT GTACCATGTTGTTTCGCATGTCTAC GGAGGCAGAGACAATAAGTTTATGTTACAA
6	AT1D0009	KASP	GGAGATCTTTATATTTGTGGTTTGCCA GAGATCTTTATATTTGTGGTTTGCCG CCAGGTCACAGGCTGTGATGTTTAA
7	TC425250	KASP	GCACTACTTTTATTGATGTTGTGTAACC AAGCACTACTTTTATTGATGTTGTGTAACT CAGAGGGAAGAAAACAACACTGAACAAAA
8	TC387992	KASP	TTGGATCTGCATTCCTTCTCCCA GGATCTGCATTCCTTCTCCCG CTTTGGGATGTTGCTGCTGGAGAT
9	SNP113325	KASP	GGTGTTTGGCAGCATCATCACG GGTGTTTGGCAGCATCATCACC GACAACTTGAGACACTAGATATCAGAGAT

Further, an additional KASP marker, SNP113325, developed from comparative analysis with barley, was mapped 3.2 cM proximal to *Lr42*, but physically mapped to the same BAC contig as WMC432 and CFD15. The two KASP markers, SNP113325 and TC387992, could be very useful in marker-assisted selection for *Lr42* (Figure 2).

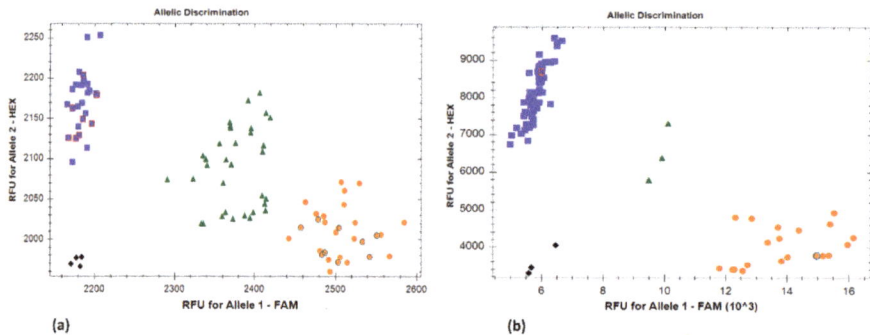

Figure 2. Kompetitive allele-specific polymorphism (KASP) marker *SNP113325* identifying the resistant and susceptible parents and selected progenies of (**a**) an *Ae. tauschii* (TA2450/TA2433) F$_2$ population; (**b**) a hexaploid wheat (KS93U50/Morocco) F$_5$ RIL population. Blue: susceptible homozygotes; green: heterozygotes; orange: resistant homozygotes; black: non-template controls. The parental resistant (orange with a blue border) and susceptible (blue with a red border) lines have been highlighted in the respective figures.

2.4. Candidate Genes in the Lr42 Region in Wheat and Ae. tauschii

The flanking markers (*TC42520* and *CFD15*) and BAC end sequences from the region flanking *Lr42* were BLASTN searched against CS Wheat RefSeq v1.1 [22]. We identified a corresponding physical segment of 3.16 Mb (6,327,249 bp to 9,490,443 bp) on the tip of the short arm of chromosome 1D of CS

Wheat. On the other hand, we identified a 3.5 Mb syntenic *Lr42* region in *Ae. tauschii* chromosome 1. There are 109 high-confidence genes in the 3.16 Mb *Lr42* region based on CS Wheat RefSeq v1.1, of which 23 genes were associated with disease resistance function and three were annotated as serine/threonine protein kinase genes (Table S1). Among the 23 disease resistance genes, 19 genes had NLR domains (associated with most of the rust resistance genes cloned in wheat to date) and another four genes had wall-associated kinase (WAK) domains. Further analysis of these 19 NLR genes showed that 10 genes carry pseudo-NLRs (Table S2); therefore, only nine carry functional NLR domains in the *Lr42* region (Figure 3).

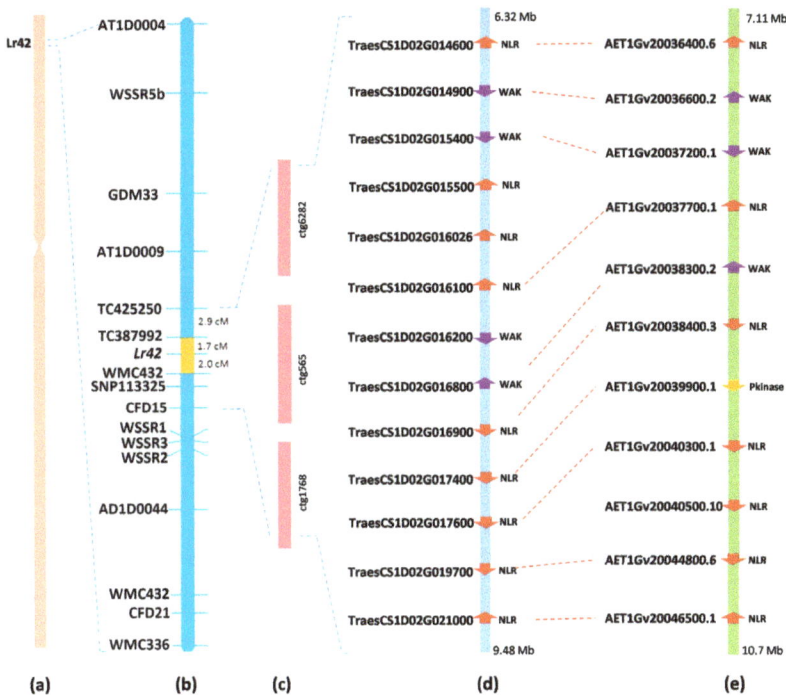

Figure 3. Candidate genes in the *Lr42* region in wheat and *Ae. tauschii*. (**a**) Physical location of *Lr42* on chromosome 1DS of wheat; (**b**) genetic location of *Lr42* in KS93U50/Morocco F5 RIL; (**c**) BAC contigs spanning the *Lr42* physical region; (**d**) annotated genes in the *Lr42* region in CS RefSeq v1.1 [22]; (**e**) annotated genes in the *Lr42* region in the *Ae. tauschii* chromosome 1 sequence [21]. NLR: nucleotide-binding leucine-rich repeats; WAK: wall-associated kinase; Pkinase: protein kinase.

In *Ae. tauschii*, we identified 98 genes in the 3.5 Mb *Lr42* region, with 21 genes encoding plant disease defense-related proteins (Table S3), using the PFAM database [23]. Among the 21 genes, three have putative wall-associated kinase (WAK) domains, seven genes had serine/threonine protein kinase (Pkinase) domains, and 11 genes had NB-ARC domains (Figure 3). Comparative analysis of wheat and *Ae. tauschii* genes in the *Lr42* region showed that four of the 11 NLR genes in *Ae. tauschii* are orthologues of pseudo-NLRs in wheat. Furthermore, the comparison of genes from two species showed that six NLRs and two WAKs from wheat have high sequence similarity (>96%) and are orthologues in *Ae. tauschii*. One NLR gene from wheat could be an orthologue of a protein kinase (Pkinase) gene from *Ae. tauschii* as it shared a sequence similarity of 90%. Apart from these genes, no orthologues were found in *Ae. tauschii* for one WAK and two NLR genes that were present in wheat. By contrast, one NLR gene was present in *Ae. tauschii* but absent in wheat (Figure 3).

3. Discussion

Leaf rust can cause severe losses in wheat yield and grain quality. Host resistance is a key component in managing leaf rust, and thus, molecular genetic characterization of resistance and identification of tightly linked molecular markers can help achieve better understanding of the mechanism of leaf rust resistance and facilitate marker-assisted breeding and gene pyramiding. Liu et al. [19] mapped *Lr42* on the short arm of wheat chromosome 1D after analyzing F_2 and F_3 generations of KS93U50/Morocco population. In the current study, we advanced the population to F_5 RILs. Segregation observed in F_2 and F_3 generations of the KS93U50/Morocco population as studied by Liu et al. [19] suggested that *Lr42* was recessive. In the current study, we evaluated two independent *Ae. tauschii* F_2 populations (TA2450/TA2433 and TA2450/TA10132) for a response to leaf rust using the same isolate, PNMRJ. In both the populations, the majority of plants were resistant, indicating a dominant monogenic control of the resistance reaction. Further segregation pattern in the $F_{2:3}$ generation in the two *Ae. tauschii* populations was consistent with F_2, suggesting that *Lr42* behaves as a dominant gene in *Ae. tauschii*. *Ae. tauschii* (the D-genome donor of wheat) is the source of *Lr42*, however, the segregation behavior of *Lr42* had not been studied earlier in *Ae. tauschii*. There are several conflicting reports regarding the segregation behavior of the *Lr42* gene in hexaploid wheat. Cox et al. [16] reported that *Lr42* was partially dominant, whereas Czembor et al. [17] reported it as being a dominant gene in wheat. However, these studies used several different genetic backgrounds or different leaf rust isolates, which could have resulted in the differences, as demonstrated by Kolmer and Dyck [24].

Applying BAC-based physical mapping, BAC-end sequencing, and comparative genomic analysis with barley [20], we identified 68 new markers and further mapped *Lr42* to a 3.7 cM interval. Liu et al. [19] reported *Lr42* as being located on the distal tip of chromosome 1DS flanked by a 17 cM interval. In the current study, we significantly reduced the size of the *Lr42* flanking segment by mapping *TC387792* and *WMC432,* as being 1.7 cM and 2.0 cM away from *Lr42* on the distal and proximal regions, respectively. Additionally, *SNP113325* lies 3.2 cM distal to *Lr42* and is also tightly linked to *Lr42* in the RIL population. The two KASP markers *SNP113325* and *TC387792* have been used for screening of hard winter wheat in regional nurseries (see 2018 SRPN, Table 10 from https://www.ars.usda.gov/ARSUserFiles/30421000/HardWinterWheatRegionalNurseryProgram/ 2018%20SRPN%20021519.xlsx) in marker-assisted selection for *Lr42* in wheat breeding programs.

We further physically delimited the *Lr42* region to a 3.16 Mb (6,327,249 bp to 9,490,443 bp) region in chromosome 1D using the Chinese Spring reference genome RefSeq v1.1 [22]. As expected, *Lr42* is present in a high-recombination region with a higher genetic-to-physical map ratio, making the search for the candidate gene relatively easy compared to the centromeric region. Though we identified 109 genes in this region, 23 of these genes have kinase domains mostly associated with disease-related genes in wheat. Besides the 23 genes, three genes were annotated as protein kinases (Pkinases); however, Pkinase genes have not been associated with wheat rust resistance genes to date. Out of the 23 genes, 10 genes were found to be pseudo-NLRs, leaving only 13 candidate genes. Of the final 13 candidate genes, nine genes have NLR domains and four genes have WAK domains. In *Ae. tauschii*, we identified seven genes with NLR domains that could be candidates for *Lr42*. Currently, several rust resistance genes have been cloned in wheat, including *Lr10* [25], *Lr21* [26], *Lr1* [27], *Lr22a* [28], *Sr33* [29], *Sr35* [30], *Sr45* [31], and *Yr10* [32], and all encode an NLR-type protein. Though wall-associated kinase domains (WAKs) have been reported to confer resistance to fungi in wheat, such as *Snn1* against Septoria nodorum blotch [33] and *Stb6* against *Septoria tritici* blotch [34], none of them showed resistance to wheat rusts. Therefore, the nine NLR-type genes in the *Lr42* region are the most probable candidates for the *Lr42* gene. Nonetheless, it is possible that *Lr42* has a regulatory mechanism different from NBS-LRR type genes. Molecular cloning of *Lr42* is thus required to reveal the complete regulation mechanism, which can be facilitated using novel cloning techniques such as MutRenSeq [31] or TACCA [28]. These techniques have been recently used to clone several disease-related genes in wheat, such as

Lr22a [28] and *Pm2* [35]. Thus, we have identified an *Lr42* susceptible mutant by EMS mutagenesis of *Ae. tauschii* accession TA2450, and future efforts will be made to further characterize *Lr42*.

Lr42 is physically located on the distal region of wheat chromosome 1DS, where a number of disease-related genes have been mapped to, such as *Lr21* [26], *Sr33* [29], *Sr45* [31], *Pm24* [36], and several other genes of agronomic importance. In wheat, the terminal gene-rich regions have been found to be positively correlated with recombination frequency [37]. Identification of tightly linked flanking markers could facilitate marker-assisted selection of *Lr42* in wheat breeding programs. The KASP markers developed in our study closely flank *Lr42* and could be efficiently used for marker-assisted selection of *Lr42* and help in stacking of the *Lr42* gene with other race non-specific disease resistance genes to develop wheat cultivars with more durable resistance against leaf rust.

4. Materials and Methods

4.1. Plant Materials and Rust Resistance Evaluation

Lr42 was identified from *Ae. tauschii* accession TA2450 [16]. We screened several *Ae. tauschii* accessions with leaf rust race PNMRJ and identified two *Ae. tauschii* accessions, TA2433, and TA10132 (AL8/78), that were susceptible to PNMRJ (PNMRJ is avirulent to *Lr42*). TA2450 was then crossed to TA2433 and TA10132 to develop two F_2 populations (TA2450/TA2433 and TA2450/TA10132). A total of 100 F_2 plants were derived from each cross (TA2450/TA2433 and TA2450/TA10132), of which a sample of around 70 plants were artificially inoculated with leaf rust race PNMRJ in a greenhouse following the method described by Liu et al. [19]. Briefly, the F_2 plants from each of the two crosses, along with the parental lines, were planted in plastic trays. The plants were artificially inoculated with PNMRJ at the two-leaf stage. For artificial inoculation, the PNMRJ spores were suspended in Soltrol 170 mineral oil (CPChemicals LLC, Garland, TX, USA). The suspension was sprayed on the seedlings using a pressure sprayer, followed by incubation at 20 °C for 24 h in a humid chamber. Following incubation, the F_2 plants were grown in a greenhouse at 20–24 °C for the establishment of infection. At ten days after inoculation, the seedlings were scored for rust infection type (IT) on a 1 to 4 scale (; = hypersensitive flecks, 1 = small uredinia with necrosis, 2 = moderate size pustules with chlorosis, 3 = moderate-large size uredinia without necrosis or chlorosis, and 4 = large uredinia lacking necrosis or chlorosis) [38,39]. The scoring was repeated after two days for confirmation. For further study, $F_{2:3}$ progenies for each cross were grown and evaluated for leaf rust to identify homozygous non-segregating families. About 20 seeds for each family were grown and evaluated using the same procedure used for the F_2 populations. Based on the reactions of parental lines to PNMRJ, the plants with IT ≤ 2 were considered resistant and those with IT > 2 susceptible.

In addition to the two *Ae. tauschii* populations, we used the F_5 RIL population of 234 progenies developed from the wheat cross KS93U50/Morocco. The RIL population, along with parents KS93U50 and Morocco, were grown and evaluated for leaf rust resistance under controlled greenhouse conditions with three replications. The F_5 RILs were advanced lines from the F_2 population analyzed by Liu et al. [19] to map *Lr42*. The inoculation and scoring method was the same as described for *Ae. tauschii* populations. The data from the three populations were used to deduce the inheritance pattern of *Lr42*.

4.2. Marker Discovery and Saturation in the Candidate Region

Previously reported *Lr42* flanking markers [19] were mapped on to the physical map of chromosome 1D of Chinese Spring wheat (https://urgi.versailles.inra.fr/gb2/gbrowse/wheat_phys_1D_v1/) by identifying the BACs associated with the candidate region using three-dimensional BAC pools. Identified BACs spanning this region were end-sequenced to generate five new PCR-based SSR markers using the DesignPrimer tool (https://kofler.or.at/bioinformatics/SciRoKo/DesignPrimer.html). Further, the genomic information from the BACs and comparative analysis with chromosome 1H of barley [20] was conducted to develop an additional 19 EST markers. In addition, KASP markers were developed from *Ae. tauschii* 10K Infinium single nucleotide polymorphisms (SNPs) [19] that were mapped onto the BACs of this region. A total of 68 new molecular markers, along with previously reported markers, were used to genotype the F$_5$ RILs along with the parents. The polymorphic markers were used for the construction of a genetic map for the *Lr42* region. The EST markers that were polymorphic between parents were later converted into KASP-based markers for easy and reliable genotyping and to select *Lr42* in breeding programs.

4.3. DNA Extraction and Genotyping of the RIL Population

Ten-day-old uninfected seedlings of RILs, along with parental lines, were used for DNA extraction using the cetyl trimethyl ammonium bromide (CTAB) method [40]. Extracted DNA samples were quantified using a Nanodrop ND1000 spectrophotometer (Nanodrop Technologies, US) for normalization of DNA concentration. For SSR and EST genotyping, polymerase chain reaction (PCR) was carried out in a GeneAmp PCR System 9700 (Applied Biosystems, CA, USA) using 20 µL of reaction mixture containing 100 ng genomic DNA, 25 ng each of forward and reverse primers, and 10 µL of 2× PCR Mastermix. Thermocycling profile was set as follows: initial denaturation at 95 °C for 5 min; 30 cycles each at 95 °C for 30 sec, 50–61 °C (depending upon the annealing temperature of particular primers) for 30 sec, 72 °C for 1 min; and a final extension step at 72 °C for 7 min. The EST and SSR PCR products were visualized using 3% high-resolution agarose gel (GeneMate, ISC Bioexpress, Inc, Kaysville, UT, USA). The kompetitive allele-specific polymorphism (KASP) genotyping was carried out using 8 µL of total reaction mixture containing 3 µL of 20–25 ng/µL genomic DNA, 5 µL of KASP Mastermix (LGC Genomics, Teddington, UK) consisting of a FAM and HEX specific FRET cassette, Taq polymerase and optimized buffer; and 0.07 µL of KASP assay mix consisting of two allele-specific primers and one common primer. For accurate distribution of the small volume of assay mix, KASP Mastermix and assay mix were combined before dispensing. PCR was carried out using a Bio-Rad CFX96 Real-time PCR system (Bio-Rad Laboratories, Hercules, CA, USA). The PCR profile was designed as follows: 94 °C for 20 min (hot start activation); 10 touchdown cycles each at 94 °C for 20 sec, 61–55 °C for 60 sec (dropping by 0.6 °C per cycle); followed by 35 cycles at 94 °C for 20 sec and 55 °C for 60 sec. Bio-Rad CFX Manager (Bio-Rad Laboratories, USA) was used for reading the plates after PCR.

4.4. Statistical Analysis and Genetic Mapping

Pearson's chi-squared analysis was used to test the goodness of fit for observed frequencies to the expected genetic frequencies. The chi-squared analysis was performed in R ver 3.4 [41] using the function 'chisq.test'. CarthaGene v1.3 was employed for the construction of genetic maps and automatical marker ordering to obtain a multipoint maximum likelihood map [42]. Firstly, a logarithm of odds (LOD) score of 3.0 was used as a threshold value for identifying the linkage groups, followed by ordering of the markers using the Kosambi mapping function [43]. The genetic linkage map was improved using the 'simulated annealing' algorithm and 'verification algorithms' in Carthagene [42]. MapChart version 2.2 [44] was used to draw and combine different genetic maps.

4.5. Physical Mapping of the Lr42 Region on Chromosome 1DS

Using BAC-end sequences and the sequences of EST markers flanking *Lr42*, we identified the *Lr42* region in the hexaploid wheat Chinese Spring (CS) chromosome 1D RefSeq v1.1 [22] and the *Ae. tauschii* reference genome sequence [45] using BLASTN [46]. Gene annotation of the *Lr42* genomic interval on CS chromosome 1D of wheat [22] and *Ae. tauschii* chromosome 1 [45] was obtained to identify candidate disease-resistance genes in the region.

5. Conclusions

In the current study, we fine-mapped the leaf rust resistance gene *Lr42* to a 3.7 cM region from the previously reported 17 cM region on chromosome 1DS in wheat. We further physically mapped the *Lr42* region on Chinese Spring RefSeq v1.1 and *Ae. tauschii* reference genomes and identified genes with defense-related functions as possible candidates for *Lr42*. The KASP markers flanking *Lr42* developed in the current study will facilitate marker-assisted selection of *Lr42* and pyramiding of the gene with other adult plant resistance genes for effective management of leaf rust in wheat.

Supplementary Materials: Supplementary materials can be found at http://www.mdpi.com/1422-0067/20/10/2445/s1.

Author Contributions: S.K.S. and B.S.G. conceptualized the experiment and designed the methodology; S.K.S., H.S.G., and J.S.S. performed data curation and formal analysis; C.L., D.W., H.S.G., J.S.S., W.L., and S.K.S. performed the investigation; G.B., D.W., and S.K.S. developed the mapping populations; H.S.G. and S.K.S. performed the software analysis; S.K.S. supervised the experiment; H.S.G. and S.K.S. wrote the original manuscript; B.S.G., G.B., D.W., W.L., and J.S.S. contributed to the interpretation of results and revision of the manuscript. All authors approved the manuscript.

Funding: This project was collectively funded by the USDA hatch projects SD00H538-15 and the Agriculture and Food Research Initiative Competitive Grants 2011-68002-30029 (Triticeae-CAP), 2017-67007-25939 (Wheat-CAP) and 2019-67013-29015 from the USDA National Institute of Food and Agriculture and South Dakota Wheat Commission grant 3X9267. The funders had no role in the study design, data collection and analysis, decision to publish, or preparation of the manuscript.

Acknowledgments: The authors would like to thank the South Dakota Agriculture Experimental Station (Brookings, SD, USA) and Kansas State University Agriculture Experimental Station (Manhattan KS) for providing the resources to conduct the experiments.

Conflicts of Interest: The authors declare that the research was conducted in the absence of any commercial or financial relationships that could be construed as a potential conflict of interest.

Abbreviations

MAS	Marker-assisted selection
KASP	Kompetitive allele-specific PCR
BAC	Bacterial artificial chromosome
EST	Expressed sequence tag
SSR	Simple sequence repeat

References

1. Shiferaw, B.; Smale, M.; Braun, H.J.; Duveiller, E.; Reynolds, M.; Muricho, G. Crops that feed the world 10. Past successes and future challenges to the role played by wheat in global food security. *Food Sec.* **2013**, *5*, 291–317. [CrossRef]

2. Bolton, M.D.; Kolmer, J.A.; Garvin, D.F. Wheat Leaf Rust Caused by *Puccinia Triticina*. *Mol. Plant Pathol.* **2008**, *9*, 563–575. [CrossRef] [PubMed]

3. Appel, J.A.; Dewolf, E.; Todd, T.; Bockus, W.W. *Preliminary 2015 Kansas Wheat Disease Loss Estimates*; Kansas Cooperative Plant Disease Survey Report; Kansas Department of Agriculture: Manhattan, KS, USA, 2015. Available online: http://agriculture.ks.gov/docs/default-source/PP-Disease-Reports-2014/2014-ks-wheat-disease-loss-estimates.pdf (accessed on 18 March 2019).

4. Marasas, C.N.; Melinda, S.; Singh, R.P. *The Economic Impact in Developing Countries of Leaf Rust Resistance Breeding in CIMMYT-Related Spring Bread Wheat*; CIMMYT (International Maize and Wheat Improvement Center): Texcoco, Mexico, 2004.

5. Chester, K.S. *The Nature and Prevention of the Cereal Rusts as Exemplified in the Leaf Rust of Wheat*; Chronica Botanica Company: Walthan, MA, USA, 1946; p. 269.

6. Kolmer, J.A.; Hughes, M.E. Physiologic Specialization of Puccinia Triticina on Wheat in the United States in 2012. *Plant Dis.* **2014**, *98*, 1145–1150. [CrossRef]

7. Kolmer, J.A.; Chen, X.; Jin, Y. Diseases Which Challenge Global Wheat Production the Wheat Rusts. In *Wheat: Science and Trade*; Carver, B.F., Ed.; Wiley-Blackwell: Ames, IA, USA, 2008.

8. Qureshi, N.; Bariana, H.; Kumran, V.V.; Muruga, S.; Forrest, K.L.; Hayden, M.J.; Bansal, U. A New Leaf Rust Resistance Gene Lr79 Mapped in Chromosome 3BL from the Durum Wheat Landrace Aus26582. *Theor. Appl. Genet.* **2018**, *131*, 1091–1098. [CrossRef] [PubMed]

9. Gill, B.S.; Raupp, W.J.; Sharma, H.C.; Browder, L.E.; Hatchett, J.H.; Harvey, T.L.; Moseman, J.G.; Waines, J.G. Resistance in Aegilops Squarrosa to Wheat Leaf Rust, Wheat Powdery Mildew, Greenbug, and Hessian Fly. *Plant Dis.* **1986**, *70*, 553–556. [CrossRef]

10. Dhaliwal, H.S.; Singh, H.; Gupta, S.; Bagga, P.S.; Gill, K.S. Evaluation of Aegilops and wild Triticum species for resistance to leaf rust (Puccinia recondita f. Sp. Tritici) of Wheat. *Int. J. Trop. Agric.* **1991**, *9*, 118–122.

11. Okamoto, Y.; Nguyen, A.T.; Yoshioka, M.; Iehisa, J.C.M.; Takumi, S. Identification of Quantitative Trait Loci Controlling Grain Size and Shape in the D Genome of Synthetic Hexaploid Wheat Lines. *Breed. Sci.* **2013**, *63*, 423–429. [CrossRef]

12. Gill, B.S.; Friebe, B.; Raupp, W.J.; Wilson, D.L.; Cox, T.S.; Sears, R.G.; Brown-Guedira, G.L.; Fritz, A.K. Wheat genetics resource center: The first 25 years. *Adv. Agron.* **2006**, *1*, 73–136. [CrossRef]

13. Bacon, R.K.; Kelly, J.T.; Milus, E.A.; Parsons, C.E. Registration of Soft Wheat Germplasm AR93005 Resistant to Leaf Rust. *Crop Sci.* **2006**, *46*, 1398. [CrossRef]

14. Singh, R.P.; Huerta-Espino, J.; Sharma, R.; Joshi, A.K.; Trethowan, R. High Yielding Spring Bread Wheat Germplasm for Global Irrigated and Rainfed Production Systems. *Euphytica* **2007**, *157*, 351–363. [CrossRef]

15. Martin, J.N.; Carver, B.F.; Hunger, R.M.; Cox, T.S. Contributions of Leaf Rust Resistance and Awns to Agronomic and Grain Quality Performance in Winter Wheat. *Crop Sci.* **2003**, *43*, 1712. [CrossRef]

16. Cox, T.S.; Raupp, W.J.; Gill, B.S. Leaf Rust-Resistance Genes *Lr41*, *Lr42*, and *Lr43* Transferred from Triticum Tauschii to Common Wheat. *Crop Sci.* **1994**, *34*, 339. [CrossRef]

17. Czembor, P.C.; Radecka-Janusik, M.; Pietrusińska, A.; Czembor, H.J. *Proceedings of the 11th International Wheat Genetics Symposium, Brisbane, QLD, Australia, 24–29 August 2008*; Appel, R., Eastwood, R., Lagudah, E., Langridge, P., Mackay, M., McIntyre, L., Sharp, P., Eds.; Sydney University Press: Sidney, Australia, 2008; pp. 739–740.

18. Sun, X.; Bai, G.; Carver, B.F.; Bowden, R. Molecular mapping of wheat leaf rust resistance gene *Lr42*. *Crop Sci.* **2010**, *50*, 59–66. [CrossRef]

19. Liu, Z.; Bowden, R.L.; Bai, G. Molecular markers for leaf rust resistance gene *Lr42* in Wheat. *Crop Sci.* **2013**, *53*, 1566–1570. [CrossRef]

20. International Barley Genome Sequencing Consortium; Mayer, K.F.; Waugh, R.; Brown, J.W.; Schulman, A.; Langridge, P.; Platzer, M.; Fincher, G.B.; Muehlbauer, G.J.; Sato, K.; et al. A physical, genetic and functional sequence assembly of the barley genome. *Nature* **2012**, *491*, 711–716. [CrossRef] [PubMed]

21. Luo, M.-C.; Gu, Y.Q.; You, F.M.; Deal, K.R.; Ma, Y.; Hu, Y.; Huo, N.; Wang, Y.; Wang, J.; Chen, S.; et al. A 4-gigabase physical map unlocks the structure and evolution of the complex genome of *Aegilops tauschii*, the wheat D-genome progenitor. *Proc. Natl. Acad. Sci. USA* **2013**, *110*, 7940–7945. [CrossRef]

22. International Wheat Genome Sequencing Consortium (IWGSC); IWGSC RefSeq principal investigators; Appels, R.; Eversole, K.; Feuillet, C.; Keller, B.; Rogers, J.; Stein, N.; IWGSC whole-genome assembly principal investigators; Pozniak, C.J.; et al. Shifting the Limits in Wheat Research and Breeding Using a Fully Annotated Reference Genome. *Science* **2018**, *361*, eaar7191. [CrossRef]

23. El-Gebali, S.; Mistry, J.; Bateman, A.; Eddy, S.R.; Luciani, A.; Potter, S.C.; Qureshi, M.; Richardson, L.J.; Salazar, G.A.; Smart, A.; et al. The Pfam Protein Families Database in 2019. *Nucleic Acids Res.* **2019**, *47*, D427–D432. [CrossRef] [PubMed]

24. Kolmer, J.; Phytopathology, P.D. Gene Expression in the Triticum Aestivum-Puccinia Recondita f. Sp. tritici gene-for-gene system. *Phytopathology* **1994**, *84*, 437–440. [CrossRef]

25. Feuillet, C.; Travella, S.; Stein, N.; Albar, L.; Nublat, A.; Keller, B. Map-Based Isolation of the Leaf Rust Disease Resistance Gene Lr10 from the Hexaploid Wheat (*Triticum aestivum*, L.) Genome. *Proc. Natl. Acad. Sci. USA* **2003**, *100*, 15253–15258. [CrossRef]

26. Huang, L.; Brooks, S.A.; Li, W.; Fellers, J.P.; Trick, H.N.; Gill, B.S. Map-Based cloning of leaf rust resistance gene lr21 from the large and polyploid genome of bread wheat. *Genetics* **2003**, *164*, 655–664. [PubMed]

27. Cloutier, S.; McCallum, B.D.; Loutre, C.; Banks, T.W.; Wicker, T.; Feuillet, C.; Keller, B.; Jordan, M.C. Leaf rust resistance gene Lr1, isolated from bread wheat (*Triticum aestivum*, L.) is a member of the large psr567 gene family. *Plant Mol. Biol.* **2007**, *65*, 93–106. [CrossRef] [PubMed]

28. Thind, A.K.; Wicker, T.; Šimková, H.; Fossati, D.; Moullet, O.; Brabant, C.; Vrána, J.; Doležel, J.; Krattinger, S.G. Rapid Cloning of Genes in Hexaploid Wheat Using Cultivar-Specific Long-Range Chromosome Assembly. *Nat. Biotechnol.* **2017**, *35*, 793–796. [CrossRef] [PubMed]

29. Periyannan, S.; Moore, J.; Ayliffe, M.; Bansal, U.; Wang, X.; Huang, L.; Deal, K.; Luo, M.; Kong, X.; Bariana, H.; et al. the gene Sr33, an ortholog of barley mla genes, encodes resistance to wheat stem rust race Ug99. *Science* **2013**, *341*, 786–788. [CrossRef]

30. Saintenac, C.; Zhang, W.; Salcedo, A.; Rouse, M.N.; Trick, H.N.; Akhunov, E.; Dubcovsky, J. Identification of wheat gene Sr35 that confers resistance to Ug99 stem rust race group. *Science* **2013**, *341*, 783–786. [CrossRef]

31. Steuernagel, B.; Periyannan, S.K.; Hernández-Pinzón, I.; Witek, K.; Rouse, M.N.; Yu, G.; Hatta, A.; Ayliffe, M.; Bariana, H.; Jones, J.D.G.; et al. Rapid cloning of disease-resistance genes in plants using mutagenesis and sequence capture. *Nat. Biotechnol.* **2016**, *34*, 652–655. [CrossRef]

32. Liu, W.; Frick, M.; Huel, R.; Nykiforuk, C.L.; Wang, X.; Gaudet, D.A.; Eudes, F.; Conner, R.L.; Kuzyk, A.; Chen, Q.; et al. The stripe rust resistance gene Yr10 encodes an evolutionary-conserved and unique CC–NBS–LRR sequence in wheat. *Mol. Plant* **2014**, *7*, 1740–1755. [CrossRef]

33. Shi, G.; Zhang, Z.; Friesen, T.L.; Raats, D.; Fahima, T.; Brueggeman, R.S.; Lu, S.; Trick, H.N.; Liu, Z.; Chao, W.; et al. The hijacking of a receptor kinase–driven pathway by a wheat fungal pathogen leads to disease. *Sci. Adv.* **2016**, *2*, e1600822. [CrossRef] [PubMed]

34. Saintenac, C.; Lee, W.-S.; Cambon, F.; Rudd, J.J.; King, R.C.; Marande, W.; Powers, S.J.; Bergès, H.; Phillips, A.L.; Uauy, C.; et al. Wheat receptor-kinase-like protein stb6 controls gene-for-gene resistance to fungal pathogen zymoseptoria tritici. *Nat. Genet.* **2018**, *50*, 368–374. [CrossRef] [PubMed]

35. Sánchez-Martín, J.; Steuernagel, B.; Ghosh, S.; Herren, G.; Hurni, S.; Adamski, N.; Vrána, J.; Kubaláková, M.; Krattinger, S.G.; Wicker, T.; et al. Rapid gene isolation in barley and wheat by mutant chromosome sequencing. *Genome Biol.* **2016**, *17*, 221. [CrossRef]

36. Huang, X.-Q.; Röder, M.S. High-density genetic and physical bin mapping of wheat chromosome 1D reveals that the powdery mildew resistance gene Pm24 is located in a highly recombinogenic region. *Genetica* **2011**, *139*, 1179–1187. [CrossRef]

37. Gill, K.S.; Gill, B.S.; Endo, T.R.; Taylor, T. Identification and high-density mapping of gene-rich regions in chromosome group 1 of wheat. *Genetics* **1996**, *144*, 1001–1012.

38. Stakman, E.C.; Stewart, D.M.; Loegering, W.Q. Identification of physiologic races of Puccinia graminis var. tritici. *Can. J. Plant Pathol.* **1981**, *3*, 33–39.

39. Roelfs, A.P.; Bushnell, W.R.; San, O.; New, D.; London, Y.; Montreal, T.; Tokyo, S. *The Cereal Rusts: Diseases, Distribution, Epidemiology, and Control*; Elsevier: Amsterdam, The Netherlands, 1985.

40. Saghai-Maroof, M.A.; Soliman, K.M.; Jorgensen, R.A.; Allard, R.W. Ribosomal DNA Spacer-Length Polymorphisms in Barley: Mendelian Inheritance, Chromosomal Location, and Population Dynamics. *Proc. Natl. Acad. Sci. USA* **1984**, *81*, 8014–8018. [CrossRef]

41. The R Development Core Team. *R: A Language and Environment for Statistical Computing*; R Foundation for Statistical Computing: Vienna, Austria, 2017. Available online: https://www.R-project.org (accessed on 18 March 2019).

42. de Givry, S.; Bouchez, M.; Chabrier, P.; Milan, D.; Schiex, T. CARHTA GENE: Multipopulation integrated genetic and radiation hybrid mapping. *Bioinformatics* **2005**, *21*, 1703–1704. [CrossRef] [PubMed]

43. Kosambi, D.D. The estimation of map distances from recombination values. *Ann. Eugen.* **1943**, *12*, 172–175. [CrossRef]

44. Voorrips, R.E. MapChart: Software for the graphical presentation of linkage maps and QTLs. *J. Hered.* **2002**, *93*, 77–78. [CrossRef]

45. Luo, M.-C.; Gu, Y.Q.; Puiu, D.; Wang, H.; Twardziok, S.O.; Deal, K.R.; Huo, N.; Zhu, T.; Wang, L.; Wang, Y.; et al. Genome sequence of the progenitor of the wheat d genome aegilops tauschii. *Nature* **2017**, *551*, 498. [CrossRef]

46. Altschul, S.F.; Gish, W.; Miller, W.; Myers, E.W.; Lipman, D.J. Basic local alignment search tool. *J. Mol. Biol.* **1990**, *215*, 403–410. [CrossRef]

© 2019 by the authors. Licensee MDPI, Basel, Switzerland. This article is an open access article distributed under the terms and conditions of the Creative Commons Attribution (CC BY) license (http://creativecommons.org/licenses/by/4.0/).

International Journal of
Molecular Sciences

MDPI

Article

Instability of Alien Chromosome Introgressions in Wheat Associated with Improper Positioning in the Nucleus

Kateřina Perničková [1], Veronika Koláčková [1], Adam J. Lukaszewski [2], Chaolan Fan [2], Jan Vrána [1], Martin Duchoslav [3], Glyn Jenkins [4], Dylan Phillips [4], Olga Šamajová [5], Michaela Sedlářová [3], Jozef Šamaj [5], Jaroslav Doležel [1] and David Kopecký [1,*]

[1] Institute of Experimental Botany, Centre of the Region Haná for Biotechnological and Agricultural Research, Šlechtitelů 31, 78371 Olomouc, Czech Republic; pernickova@ueb.cas.cz (K.P.); kolackova@ueb.cas.cz (V.K.); vrana@ueb.cas.cz (J.V.); dolezel@ueb.cas.cz (J.D.)
[2] Department of Botany and Plant Sciences, University of California, Riverside, CA 92521, USA; adam.lukaszewski@ucr.edu (A.J.L.); chaolanf@ucr.edu (C.F.)
[3] Department of Botany, Faculty of Science, Palacký University Olomouc, Šlechtitelů 27, 783 71 Olomouc, Czech Republic; martin.duchoslav@upol.cz (M.D.); michaela.sedlarova@upol.cz (M.S.)
[4] Institute of Biological, Environmental and Rural Sciences, Aberystwyth University, Aberystwyth, Ceredigion, Wales SY23 3DA, UK; gmj@aber.ac.uk (G.J.); dwp@aber.ac.uk (D.P.)
[5] Department of Cell Biology, Centre of the Region Haná for Biotechnological and Agricultural Research, Faculty of Science, Palacký University Olomouc, Šlechtitelů 27, 783 71 Olomouc, Czech Republic; olga.samajova@upol.cz (O.Š.); jozef.samaj@upol.cz (J.Š.)
* Correspondence: kopecky@ueb.cas.cz; Tel.: +42-058-523-8723

Received: 27 February 2019; Accepted: 19 March 2019; Published: 22 March 2019

Abstract: Alien introgressions introduce beneficial alleles into existing crops and hence, are widely used in plant breeding. Generally, introgressed alien chromosomes show reduced meiotic pairing relative to the host genome, and may be eliminated over generations. Reduced pairing appears to result from a failure of some telomeres of alien chromosomes to incorporate into the leptotene bouquet at the onset of meiosis, thereby preventing chiasmate pairing. In this study, we analysed somatic nuclei of rye introgressions in wheat using 3D-FISH and found that while introgressed rye chromosomes or chromosome arms occupied discrete positions in the Rabl's orientation similar to chromosomes of the wheat host, their telomeres frequently occupied positions away from the nuclear periphery. The frequencies of such abnormal telomere positioning were similar to the frequencies of out-of-bouquet telomere positioning at leptotene, and of pairing failure at metaphase I. This study indicates that improper positioning of alien chromosomes that leads to reduced pairing is not a strictly meiotic event but rather a consequence of a more systemic problem. Improper positioning in the nuclei probably impacts the ability of introgressed chromosomes to migrate into the telomere bouquet at the onset of meiosis, preventing synapsis and chiasma establishment, and leading to their gradual elimination over generations.

Keywords: chromatin; 3D-FISH; nucleus; introgression; rye; hybrid; wheat; genome stability

1. Introduction

During interphase, chromosomes are decondensed and occupy distinct regions of the nucleus, called chromosome domains or chromosome territories [1,2]. In flowering plants, chromosome territories show two predominant configurations, known as the Rabl's orientation and Rosette [3,4] with the former being more frequent. The Rabl's configuration [5] reflects the orientation of chromosomes from the preceding anaphase: all centromeres are grouped at or close to the nuclear periphery at one pole of the nucleus while telomeres are dispersed toward the opposite pole [6]. Such an orientation presumably simplifies homologue search and the initiation of chromosome pairing in early meiosis. Chromosome pairing is initiated during leptotene (or telomere) bouquet, where all telomeres are located close to each other at one pole of the nucleus. The association of the telomeres into the bouquet at the onset of the first meiotic division facilitates the initiation of synapsis of homologous chromosomes, which in turn is a prerequisite for crossing over and chiasmate pairing at metaphase I [7], both of which are critical for the success of meiosis. It has been clearly demonstrated that misalignment of telomeres of homologues during the bouquet formation restricts synapsis and drastically reduces metaphase I pairing [8,9].

Many crops are allopolyploids, that is, products of wide hybridization, a natural process commandeered frequently by plant breeders to widen the gene pool of a crop by introduction of agronomically important alleles. Such introgressions may take the form of new amphiploids, such as triticale (X Triticosecale Wittmack), introgressions of single chromosomes, chromosome arms or even smaller chromosome segments [10]. One of the most successful intergeneric introgressions is the 1RS.1BL centric wheat-rye translocation in wheat (*Triticum aestivum* L.), where the short arm of wheat chromosome 1B is replaced by its counterpart from the rye genome (*Secale cereale* L.). Many other introgressions have been released in wheat breeding and research programs involving rye, barley (*Hordeum* sp.), *Aegilops*, *Agropyron* and *Haynaldia* species [11,12].

As a general rule, early generations of interspecific hybrids suffer irregular chromosome pairing which may affect different genomes in various ways. This hampers wider commercial utilization of amphiploids and introgression lines in agriculture. Nishiyama [13] described reduction of chromosome number in successive generations of synthetic decaploid interspecific hybrids of oats (*Avena*) from 2n = 70 to 2n = 42−58. Similarly, reversion to bread wheat (loss of the entire rye genome) has been reported in octoploid triticale (amphiploids of bread wheat with rye) [14]. In tetraploid and hexaploid triticales, rye chromosomes fail to pair more frequently than wheat chromosomes [15]. Among disomic additions and substitutions of individual rye chromosomes in bread wheat, Orellana et al. [16] found reduced metaphase I pairing of rye chromosomes and significantly higher numbers of univalents compared to wheat chromosomes.

The problem of reduced chromosome pairing in amphiploids has been a subject of much debate over the decades. In wheat-rye hybrids, five different concepts have been presented, but none has been proven to be correct (reviewed in [17]). After several recent studies, a relationship between the behaviour of telomeres and the success of chromosome pairing has gained credence. Murphy and Bass [18] have shown that the desynaptic (*dy*) mutant of maize displays multiple defects in telomere-nuclear envelope interactions, homologous chromosome synapsis, recombination and chromosome segregation. Similarly, Naranjo [19] reported that reduced pairing of rye chromosomes in wheat appears to be a consequence of disturbed migration of rye telomeres into the leptotene bouquet. Telomere positioning and migration are preferentially studied in meiosis. However, our previous study suggested that the problem may be systemic in nature, and aberrant arrangement of telomeres in pollen mother cells (PMCs) is only an extension of their erratic behaviour in other (somatic) tissues. We observed that the frequency of out-of-bouquet rye telomere position at leptotene was virtually identical to that in the nuclei of somatic cells, and was similar to the frequency of synapsis of the normal and inverted chromosome arms in a heterozygote for an inversion of a rye chromosome arm in wheat [20].

In this study, we analysed wheat lines with introgressed chromatin from rye, involving disomic whole chromosome substitutions, ditelosomic line and centric (whole arm) translocations in the 3D space of wheat nuclear volume of somatic tissue in order further investigate the possible link between telomere positioning in somatic tissues, chromosome pairing and stability of introgressed alien chromatin. Special attention was paid to the distribution of the telomeres and centromeres.

2. Results

2.1. Morphometrical Characteristics of G1 Interphase Nuclei of Wheat-Rye Introgression Lines

In total, we analysed 315 nuclei with most of the parameters described in Material and Methods. The morphology of G1 nuclei flow sorted into a polyacrylamide gel ranged from spherical and ellipsoidal to irregular shapes with varying degrees of contortion. Only nuclei with spherical and slightly ellipsoidal shapes were selected for analyses. Nuclear volumes ranged from 753 to 2996 μm^3 (mean $\pm$ SD 1677 $\pm$ 506 μm^3). 3D-GISH analysis showed that the chromosome territories (CTs) of rye chromosome arms appeared as compact structures of regular shapes spanning the entire nucleus and arranged in a typical Rabl's orientation (Figure 1). Centromeres were generally close to each other and located on the nuclear periphery at one pole, while telomeres were at the periphery of the opposite pole. In a majority of the analysed nuclei, centromeres of rye chromosome arms were closer to each other than their telomeres (Table 1).

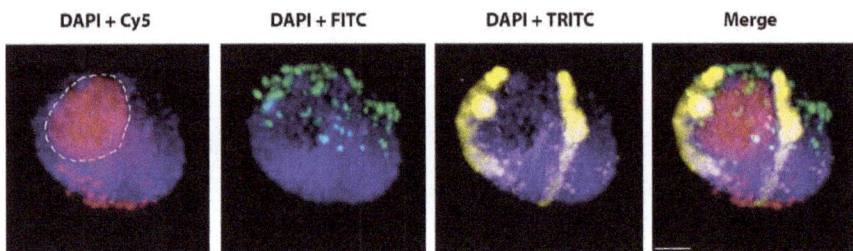

Figure 1. Rabl's orientation of rye chromosomes. Nucleus with a pair of homologous rye del1RS.1RL chromosomes in the proper Rabl's orientation. Total genomic DNA of rye was labelled with TRITC using Nick translation (yellow color), centromeres of both wheat and rye chromosomes were visualized using oligonucleotide probe (red color), and telomere-specific sequence was PCR-labelled with FITC (green color). Nuclear DNA was counterstained with DAPI (blue color). Note the difference in signal intensity along the chromosome arms with subtelomeric heterochromatin labelled dark yellow and the remaining rye arm labelled light yellow. Nucleoli are indicated by white dashed lines. Scale bar 5 μm.

Table 1. Morphometrical characteristics of rye chromosome arms in wheat-rye introgression lines. The values (Mean $\pm$ SD) of arm length, distances between centromere to centromere (C-C) and telomere to telomere (T-T) (given in μm) were normalized according to the volume of the nucleus (absolute distance in μm/nuclear volume $\times$ 1000).

Introgression	Nuclear Volume (μm^3)	Arm Length (μm)	C-C (μm)	T-T (μm)
1AS.1RL	1545 $\pm$ 224	7.73 $\pm$ 1.48	4.01 $\pm$ 1.98	4.88 $\pm$ 1.71
1RS.1BL	1501 $\pm$ 362	8.21 $\pm$ 1.69	3.49 $\pm$ 1.50	4.18 $\pm$ 1.66
1RS.1DL	1998 $\pm$ 285	6.86 $\pm$ 1.31	2.83 $\pm$ 1.15	3.71 $\pm$ 1.74
2RS.2BL	1597 $\pm$ 490	8.51 $\pm$ 2.58	3.38 $\pm$ 1.70	4.09 $\pm$ 1.95
2BS.2RL	1510 $\pm$ 357	8.96 $\pm$ 1.62	4.05 $\pm$ 1.80	4.99 $\pm$ 2.25
5RS.5BL	1656 $\pm$ 329	6.95 $\pm$ 1.54	3.06 $\pm$ 1.29	3.97 $\pm$ 1.48
del1RS.1RL	1986 $\pm$ 390	S [1]: 4.04 $\pm$ 1.03	2.88 $\pm$ 1.37	3.86 $\pm$ 1.79
		L [2]: 7.56 $\pm$ 1.67		3.56 $\pm$ 1.87
1RS. del1RL	2006 $\pm$ 500	S [1]: 6.18 $\pm$ 1.58	3.57 $\pm$ 1.69	4.20 $\pm$ 1.92
		L [2]: 5.21 $\pm$ 1.41		4.12 $\pm$ 1.53

[1] S: short arm [2] L: long arm.

2.2. Positions of Rye Telomeres and Centromeres Relative to the Nuclear Envelope and Positioning of Rye Telomeres in the Telomere Cluster

In the majority of nuclei, telomeres and centromeres of rye chromosomes/chromosome arms were positioned at the nuclear periphery (NP). As we did not perform any tests for the attachment of telomeres to the inner nuclear envelope, we cannot be certain that the positioning of telomeres or centromeres at the nuclear periphery reflects their attachment to the nuclear envelope. However, it is a safe assumption that the positioning of a telomere away from the nuclear periphery indicates an absence of such attachment. It also needs to be stressed that the position of the nuclear envelope was inferred from the edge of the DAPI-stained chromatin. Therefore, in the absence of definitive proof, all telomeres located away from the edge of nuclear periphery are assumed to be unattached to the nuclear envelope; those at the nuclear periphery are labelled below as "in contact with NP".

In a small proportion of nuclei across most of the genotypes, a centromere or a telomere of one rye chromosome arm was positioned away from the NP (Table 2). In control diploid rye, essentially all chromosomes were in Rabl's orientation with all centromeres positioned at the centromere pole and 99.3% of telomeres located at the telomere pole. Similarly, 94.3% of the telomeres were in contact with the NP. Rye chromosome arms in the introgression lines differed in the proportions of the telomere (log-linear models using likelihood-ratio chi-square test, $\chi^2 = 54.36$, DF = 12, $p < 0.001$) and centromere ($\chi^2 = 25.61$, DF = 12, $p = 0.012$) positions relative to the NP, while only minor differences were observed among genotypes in proportions of telomeres positioned at the telomere pole ($\chi^2 = 22.08$, DF = 12, $p = 0.036$).

Table 2. Frequencies of proper positioning of telomeres of rye chromosome arms (at the telomere pole) and rye telomeres and centromeres in contact with the nuclear periphery (NP).

Introgression	Rye Chromosome arm Length (Mb) [3]	Chromosome Length (Mb)	Arm Ratio	Number of Nuclei	Telomere Proper Positioning (%)	Telomere in Contact with the NP (%)	Centromere in Contact with the NP (%)
1AS.1RL	626	902	2.27	25	98.00	98.00	84.00
1RS.1BL	423	959	1.27	40	98.75	98.75	100.00
1RS.1DL	423	804	0.90	40	97.50	95.00	98.75
2RS.2BL	595	1102	0.85	40	98.75	96.25	98.75
2BS.2RL	693	1116	1.64	40	98.75	100.00	100.00
5RS.5BL	346	928	1.68	40	93.75	87.50	96.25
del 1RS.1RL	short arm: 271 [1] long arm: 626	897	2.31	25	1RS: 82.00 1RL: 100.00	1RS: 46.00 1RL: 100.00	90.00
1RS.del 1RL	short arm: 423 long arm: 438 [2]	861	1.04	21	1RS: 92.90; 1RL: 100.00	1RS: 90.50 1RL: 97.60	88.10
1R(1A)	short arm: 423 long arm: 626	1049	1.48	22	1RS: 100.00 1RL: 100.00	1RS: 93.20; 1RL: 100.00	90.90
t1RS	423	423	-	22	100.00	93.20	79.50

[1] deletion of about 36% of 1RS arm (proximal part); [2] deletion of about 30% of 1RL arm (proximal part); [3] estimated values of chromosome and chromosome arm length and arm ratio have been calculated from karyotypes of Schlegel et al. [21] and Naranjo [22] for rye and Gill et al. [23] for wheat and genome size estimations [24].

In the majority of cases, centromeres were tightly clustered at one pole of the nucleus; telomeres were located in the about one third of the nuclear volume opposite to the centromere pole (see measurements of C-C and T-T in Table 1). Rarely, a telomere was located away from the telomere pole (Figure 2, Figure A2 and Video S1) and/or away from the nuclear envelope (NE) (Figure 3; Figures A1 and A2 and Video S2). The frequencies of telomere proper positioning and telomere contact with the nuclear periphery (NP) were correlated (Spearman correlation coefficient, $r_s = 0.57$, $p = 0.043$), and they appeared to correlate with the length of a chromosome arm (telomere proper positioning: $r_s = 0.57$, $p = 0.040$, telomere in contact with NP: $r_s = 0.53$, $p = 0.063$): shorter arms were less likely to be in contact with the NP and more frequently were out-of-position (Table 2). Chromosome length and arm ratio had no effect on the proper positioning of telomeres in Rabl's orientation and the proportions of centromeres and telomeres in contact with the NP. Shortening of 1RS by a deletion reduced the in-contact with the NP frequency from 93.2% (in 1R) to 46.0% (in del1RS.1RL) and proper positioning from 100% to 82%. In all cases, telomeres of long arms were in contact with the NP much more frequently than those of the short arms. Even the reduction in length of 1RL did not change this pattern. Telomeres of the 1RS arm were less likely to be in contact with the NP relative to those of the $_{del}$1RL arm (90.5% vs. 97.6%, respectively). Telocentric chromosome t1RS had its telomere at the nuclear periphery with the same frequency as the short arm of 1R (93.2%) and was always properly arranged in the Rabl's configuration (the same as 1RS in normal 1R). Interestingly, the telomere located at the centromere of t1RS was positioned at the centromere pole but less frequently in contact with the NP than the centromere of 1R (79.5% vs. 90.9%).

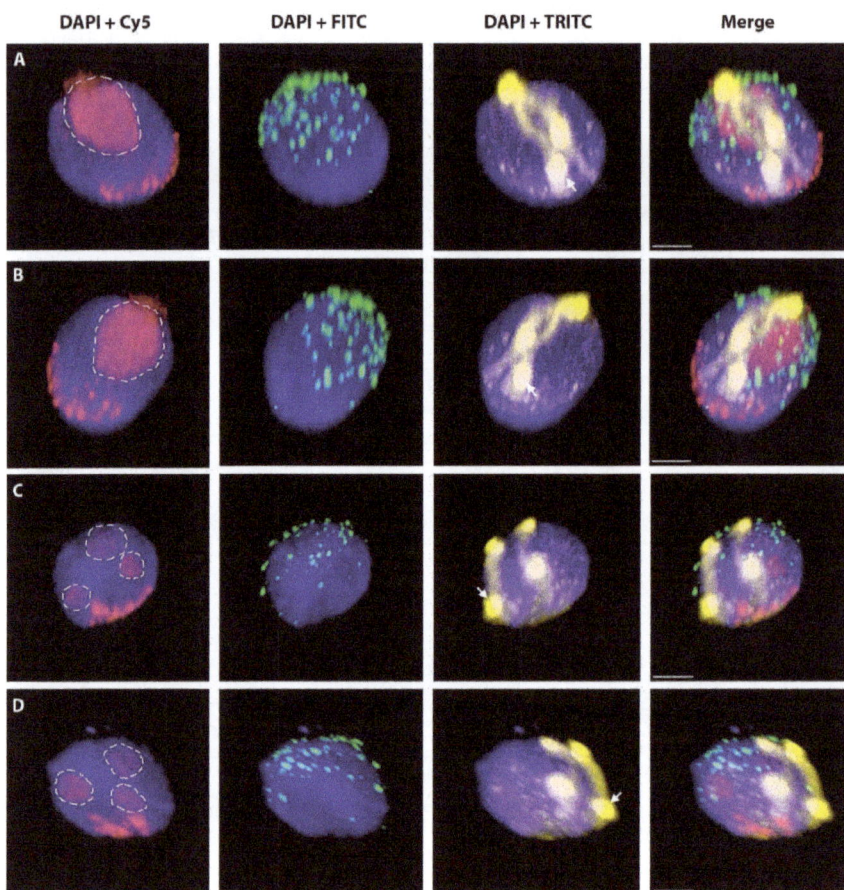

Figure 2. Telomere out-positioning of rye chromosome arms. Two nclei ((**A,B**) is the same nucleus from different angle; the same for (**C,D**)) with a pair of homologous rye $_{del}$1RS.1RL chromosomes after 3D-FISH. Total genomic DNA of rye was labelled with TRITC using Nick translation (yellow colour), centromeres of both wheat and rye chromosomes were visualized using oligonucleotide probe (red colour), and telomere-specific sequence was PCR-labelled with FITC (green colour). Nuclear DNA was counterstained with DAPI (blue colour). Rye telomeres positioned out of the telomere pole are indicated by arrows. Nucleoli are indicated by white dashed lines. Scale bar 5 μm.

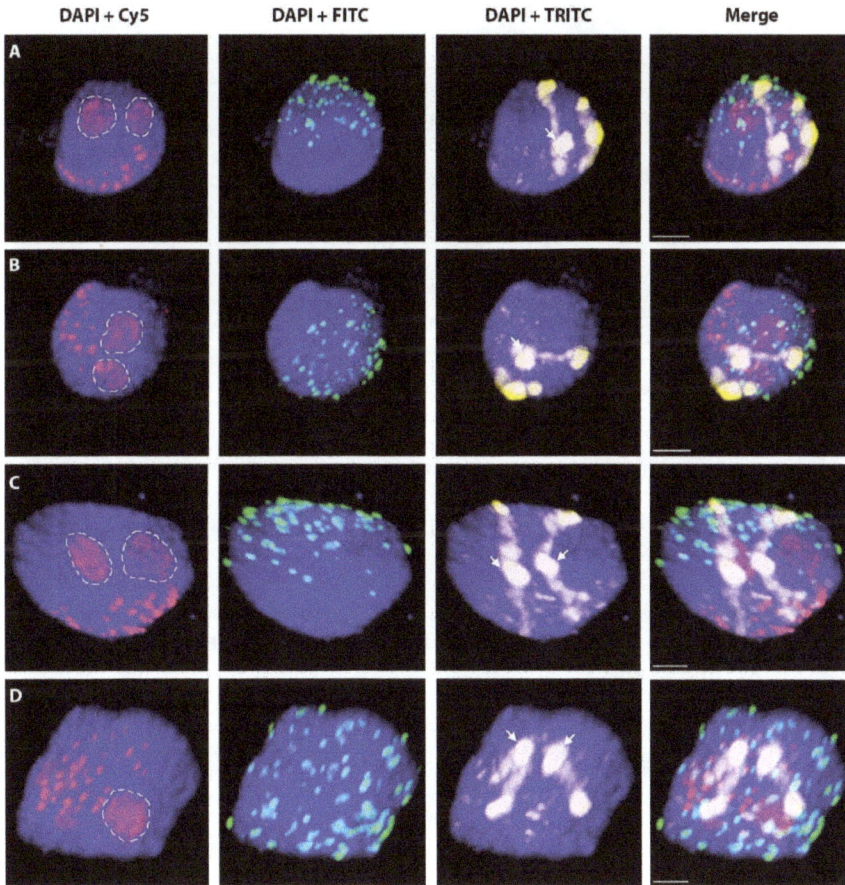

Figure 3. Non-attachment of rye telomeres. Two nuclei ((**A,B**) is the same nucleus from different angles; same for (**C,D**)) with a pair of homologous rye $_{del}$1RS.1RL chromosomes after 3D-FISH. Total genomic DNA of rye was labelled with TRITC using Nick translation (yellow colour), centromeres of both wheat and rye chromosomes were visualized using oligonucleotide probe (red colour), and telomere-specific sequence was PCR-labelled with FITC (green colour). Nuclear DNA was counterstained with DAPI (blue color). Rye telomeres without visual contact to nuclear envelope are indicated by arrows. Nucleoli are indicated by white dashed lines. Scale bar 5 μm.

2.3. Chromosome Pairing and Transmission Rate

The pairing frequencies of rye chromosome arms, whether in chromosome 1R or in centric translocations, were high, but varied for different chromosome arms and even for the same arms but in different translocations or configurations. In centric translocations 1RS.1BL and 2BS.2RL, rye chromosome arms paired less frequently than the wheat arms present (90.0% vs. 94.0% and 93.9% vs. 95.9%, respectively). In the disomic substitution and deletion lines of chromosome 1R, the short arm (1RS) always paired less frequently than the long one. Deletion of a large portion of an arm, whether in the short or the long arm, did not have any major effect on pairing frequency of the arm with the deletion, or the other arm of the chromosome (Figure 4).

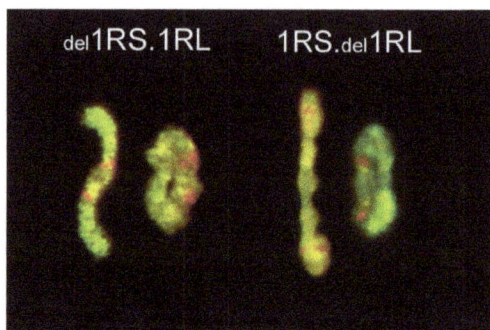

Figure 4. Metaphase I pairing of chromosomes 1R with deletions. From left to right: chromosome del1RS.1RL paired as a rod in the short arm and as a ring and chromosome 1RS. del1RL paired as a rod in the long arm and as a ring. Short arms are marked by non-hybridized (red) bands of the NOR region.

The transmission rate of rye chromosomes or chromosome arms in wheat-rye centric translocation was also high, but with some variation among the lines (Table 3). Centric translocations were highly stable with the transmission rate in most cases at 100%. The only exception was 5RS.5BL with a transmission rate of 98.6% (one out of 71 plants was nullisomic for this translocation). Similarly, high stability was observed in 1R (99.4%; one monosomic among 80 plants) and del1RS.1RL (98.7%; two monosomics among 80 plants). On the other hand, much lower transmission rates were observed for 1RS.del1RL (91.9%; 13 monosomics among 80 plants) and ditelosomic 1RS (87.2%; two nullisomics and 15 monosomics among 74 plants).

Table 3. Pairing of rye chromosome arms during metaphase I and their transmission to the next generation.

Introgression	Number of PMC	Chromosome Pairing	Number of Progeny	Transmission (%)
1AS.1RL			63	100.0
1RS.1BL	50	1RS: 90.0; 1BL: 94.0	32	100.0
1RS.1DL			30	100.0
2RS.2BL			38	100.0
2BS.2RL	49	2BS: 95.9; 2RL: 93.9	76	100.0
5RS.5BL			71	98.6
del1RS.1RL	60	1RS: 90.0; 1RL: 100.0	80	98.7
1RS.del1RL	59	1RS: 89.8; 1RL: 94.9	80	91.9
1R(1A)	289 [1]	1RS: 80.8; 1RL: 94.4	80	99.4
t1RS			74	87.2

[1] [25].

3. Discussion

Interspecific hybrids and allopolyploids frequently display genome instability. This instability may take different forms, such as parental chromosome competition. In *Lolium* × *Festuca* hybrids, the *Lolium* genome predominates: chromosomes of *Festuca* are slowly, but continuously replaced by those of *Lolium* in consecutive generations [26,27]. The mechanism underlying this phenomenon is not yet clear. Alternatively, genome instability may manifest itself as reduced meiotic pairing of one of the parental genomes, causing the higher univalency rate for the low pairing genome, and its gradual elimination from the hybrid. Wheat-rye hybrids are perhaps the best studied in this respect, with known cases of complete elimination of the rye genome, especially from octoploid triticales [14]. This phenomenon is also well documented in single chromosome wheat-rye addition lines [16]; as a general rule rye chromosomes pair poorly and are eliminated over time, which requires careful control of such lines to maintain their status. It has been argued that meiotic instability of wheat-rye hybrids is a consequence of mis-matched chromosome pairing control systems of the two parental species, and, especially, the effect of the *Ph1* system of wheat [17]. However, the *Ph1* system in wheat appears to impose strict stringency levels for crossover formation, and rye homologues in established triticale lines and wheat-rye addition lines are virtually the same. In stocks analysed in this study, rye chromosome arms always paired less frequently than their wheat counterparts. This fits well with the general pattern of rye chromosome behaviour when introgressed into wheat. As shown earlier by [16], rye chromosomes introgressed into wheat as disomic substitution or addition lines generate univalents in frequencies ranging from 1 to 19% (6% on average; Figure 5). Similarly, rye univalents are twice as frequent as those of wheat in tetraploid triticale, despite equal proportions of wheat and rye genomes [15].

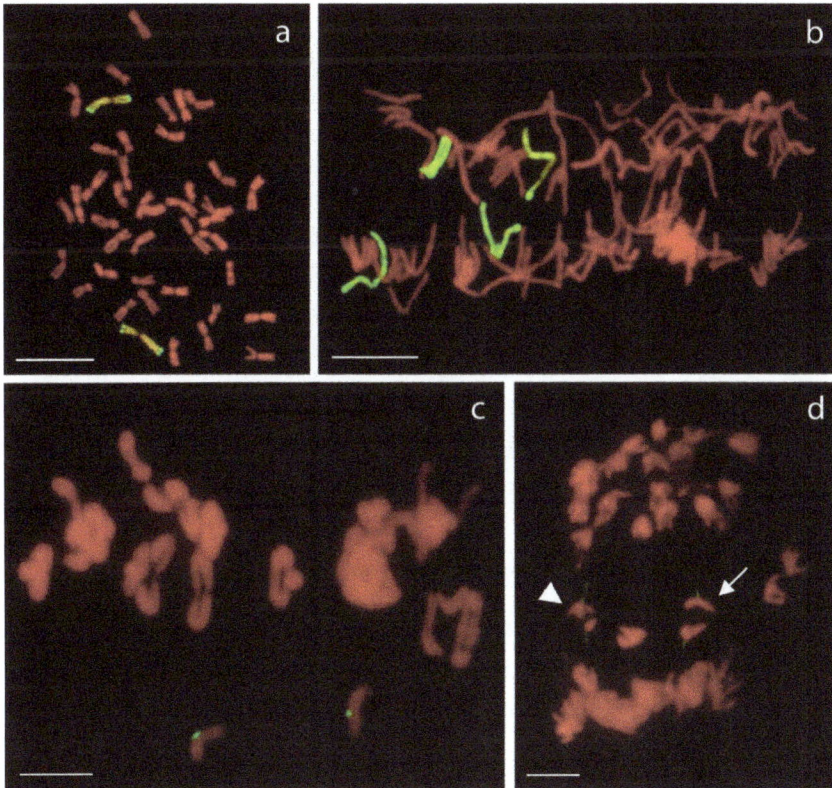

Figure 5. Mitotic (**a**,**b**) and meiotic (**c**,**d**) behavior of rye chromosome pair introgressed into wheat background. Rye chromosomes usually behaves perfectly normal during mitotic metaphase (**a**) and anaphase (**b**), while reduced pairing was observed during meiotic metaphase I (**c**). This may result in the separation of sister chromatids (arrow) or misdivision (in this case one arm of one chromatid is being separated from the rest of the chromosome; arrowhead) during anaphase I (**d**). Rye mitotic chromosomes were visualized using labelled genomic DNA of rye (labelled with FITC; green color), while rye-specific centromeric probe was used for meiotic preparations (labelled with FITC; green color). Scale bar 10 μm.

Naranjo [22] indicated that reduced metaphase I pairing of rye chromosomes in wheat might be a consequence of a lower probability of rye telomeres clustering into the leptotene bouquet at the onset of meiosis. Proper positioning of telomeres in the bouquet is believed to be prerequisite for the initiation of synapsis, and thus, regular chromosome pairing [7]. In our previous work [20], we observed a surprising agreement between the frequency of out-of-position telomeres at leptotene and in somatic nuclei on the one hand, and the failure of metaphase I pairing on the other. However, the situation is probably much more complex. There is chromosome movement at the onset of meiosis mediated by protein complexes of the inner nuclear envelope [28]. Such movement may be less frequent, or perhaps impossible, if the telomeres are not attached to the nuclear envelope (NE) already in the premeiotic interphase. The frequencies of non-attached rye telomeres observed in this study (0–12.5% for normal length chromosome arms) are again surprisingly similar to the frequencies of their out-of-bouquet positioning at the leptotene-zygotene transition as shown by Naranjo [19,22] and the frequencies of their pairing failure ([16] and our results). Interestingly, the transmission rate of rye introgressions in disomic centric wheat-rye translocations was high. This is because wheat chromosome arms in such translocations pair with normal frequencies assuring normal disjunction of the translocated chromosomes, their regular inclusion into the products of meiosis and then normal transmission of rye chromatin into subsequent generations. This is despite reduced (relative to diploid rye) pairing of rye chromosome arms. On the other hand, general instability of ditelosomics (such as t1RS analysed here) appears to be associated with lower frequencies of telomeres attached to the NE (here detected as telomeres in contact with NP). Correlation also exists between the out-positioning of rye telomeres in diploid rye nuclei observed here (0.66%) and average pairing failure per arm in a population of rye (0.21%) [29,30]. Thus, what appears to emerge from rather fragmentary data is that there is a direct link between somatic arrangement of chromosome arms, the leptotene bouquet and metaphase I pairing success.

The study of Naranjo [19] and our results indicate that there may be a relationship between chromosome arm length and the positioning of that arm's telomere at the telomere cluster/pole. Naranjo found that only 83.3% and 73.5% of telomeres of the short arms of chromosomes 1R and 6R, respectively, were located in the telomere cluster at the leptotene-zygotene transition, while none of 1RL telomeres and only 0.5% of 6RL telomeres were out of the telomere clusters [19]. Similarly, the lowest frequency of rye telomere clustering was for chromosome arm 5RS, the shortest rye chromosome arm, and the highest observed frequencies were for the arm 6RL, the longest in the karyotype [19,22,31]. This correlates well with our results, where telomeres of shorter chromosome arms were attached to the NE (or positioned at the telomere cluster/pole) less frequently than the telomeres of longer chromosome arms and significantly higher frequencies of pairing of long arms compared to short arms of chromosomes 1R, 2R and 5R in disomic addition lines (84.5 vs. 91.0, 64.8 vs. 96.8% and 31.0 vs. 79.0, respectively) [16]. Similarly, reduction in length of 1RS arm by deletion in the 1R introgression line caused reduction of its telomere attachment from 93.2% (in regular 1R introgression) to 46%.

4. Materials and Methods

4.1. Plant Material

The plant material consisted of a set of lines of hexaploid bread wheat (*Triticum aestivum* L., 2n = 6× = 42) cv. 'Pavon 76' with disomic (homozygous) introgressions of rye chromosomes or chromosome arms: substitution of rye chromosome 1R for wheat chromosome 1A [1R(1A)], ditelosomic addition line 1RS, a deletion line $_{del}$1RS.1RL where ca. proximal 36% of the short arm is missing, a deletion line 1RS.$_{del}$1RL where proximal ca. 30% of the long arm is missing (Figure 6), and centric wheat-rye chromosome translocations 1RS.1BL, 1RS.1DL, 1AS.1RL, 2RS.2BL, 2BS.2RL and 5RS.5BL. The telosomic line and all centric translocation lines were created by centric misdivision of complete normal chromosomes of rye and their wheat homoeologues; the deletion chromosomes were identified during selection of centric translocations [32]. As a control, we used a population of rye (*Secale cereale* L.) cv. Dankowskie Zlote. About 25 seedlings were used for each line.

Note on terminology: this manuscript uses the original terminology of Bridges for chromosome aberrations where "deficiency" is a loss of a terminal chromosome segment and "deletion" indicates a loss of an intercalary segment [33].

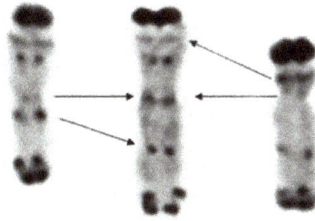

Figure 6. Three 1R chromosomes differing in the length of their arms. From left to right: 1RS.$_{del}$1RL with deletion of about 30% of the 1RL arm (proximal part), regular 1R chromosome and $_{del}$1RS.1RL with deletion of about 36% of 1RS arm (proximal part).

4.2. Isolation of Nuclei and Flow Sorting

Seeds of the introgression lines were germinated in Petri dishes on moist filter paper at 25 °C in the dark. Root tips from young seedlings were collected and fixed in 2% (*v*/*v*) freshly prepared formaldehyde in meiocyte buffer A (15 mM PIPES, pH 6.8, 80 mM KCl, 20 mM NaCl, 0.5 mM EGTA, 2 mM EDTA, 0.15 mM spermine tetra-HCl, 0.05 mM spermidine tri-HCl, 1 mM dithiothreitol, 0.32 M sorbitol) for 20 min at 5 °C. After fixation, root tips were washed three times for 15 min at room temperature in meiocyte buffer A. Meristem tissue of root tips was cut and transferred into a 5 mL sample tube containing 400 µL of meiocytes buffer A, and homogenized using a Polytron PT1200 homogenizer (20,000 rpm/13 sec). The homogenate was filtered through 20 µm nylon mesh into a 5 mL polystyrene tube and stored on ice until used. The nuclear suspension was stained with 2 µg/mL DAPI (4′,6-diamidino-2-phenylindole). Nuclei in G1 phase of the cell cycle were identified and sorted using a FACSAria II SORP flow cytometer (BD Biosciences, San Jose, CA, USA) into a sample tube containing 10 µl of meiocyte buffer A [34]. About 50,000 nuclei at G1 were obtained from one sample prepared from 50 root tips. Using flow cytometry enabled us to have a sample of the nuclei exclusively from one phase of cell cycle (G1), rather than a mixture of cells from different phases, which could bias the results from different lines.

4.3. Probe Preparation and 3D-FISH

Total genomic DNA of *S. cereale* L. was labelled with Texas Red or TRITC using a Nick Translation Kit (Roche Applied Science, Penzberg, Germany) according to manufacturer's instructions, and applied as a probe. Total genomic DNA of rye labelled the rye chromosome arms yellow, and their subtelomeric heterochromatin dark yellow. Both rye and wheat centromeres were visualized by an oligonucleotide probe based on the sequence of clone pHind258 [35] and directly labelled with Cy5. The telomeric probe was prepared using PCR and FITC-directly labelled nucleotides. Total genomic DNA of wheat was sheared to 200–500 bp fragments by boiling and used as blocking DNA at a ratio of 1:150 (probe/blocking DNA). As a control, we used rye nuclei with probes specific for telomeres and centromeres.

3D-FISH experiments were performed according to [36,37] with minor modifications. The suspension of flow-sorted nuclei was dripped onto a 22 × 22 mm coverslip and mixed with acrylamide solution (30% Acrylamide/Bis-acrylamide Mix Solution, ratio 29:1, Sigma-Aldrich, St. Louis, USA) in meiocyte buffer A, ammonium persulfate (20%) and sodium sulphate anhydrous (20%). The drop of solidifying nuclear suspension was covered with another coverslip and the "sandwich" was placed in an oven at 37 °C for 1 h for polymerization. Thereafter, the coverslip "sandwich" was gently separated by a razor blade. The coverslip with polymerized nuclei was washed in meiocyte buffer A in a small Petri dish for 3 × 5 min at room temperature. The coverslip was then placed on a new slide with a drop of hybridization mixture for FISH. The FISH mixture included probes, blocking DNA, 70% formamide and 2× saline sodium citrate (SSC). Hybridization was conducted according to [37]. The nuclei were counterstained with 1.5 µg/mL DAPI in Vectashield antifade mounting medium (Vector Laboratories, Burlingame, USA).

4.4. Image Acquisition and Analysis

Probed nuclei were optically sectioned using an inverted laser spinning disk microscope (Axio Observer Z1, ZEISS) and ZEN Blue 2012 software and an inverted motorized microscope Olympus IX81 equipped with a Fluoview FV1000 confocal system (Olympus, Tokyo, Japan) and FV10-ASW software. Fluorescein (binding to telomere-specific repeats) was excited by a 488 nm line of an argon laser, while Texas Red and TRITC (total rye genomic DNA) were excited by He-Ne laser at 543 nm or 561 nm. Cy5 (labelling both rye and wheat centromeres) was excited at 639 nm. The excitation of DAPI (to visualize nuclear DNA) was performed by a 405 nm diode laser.

For each nucleus, 80–120 optical sections in 160 – 200 nm steps were taken and merged into a 3D model. Subsequent analyses were performed using Imaris 9.2 software (Bitplane, Oxford Instruments, Zurich, Switzerland). Imaris applications 'Contour Surface', 'Spot Detection' and '3D Measurement' were used for manual analysis of each nucleus. The volume and the centre of the nucleus (CN) were determined from the rendering of primary intensity of DAPI staining using the function 'Surfaces'. The 'Spot' function was used to mark the positions of centromeres (C) and telomeres (T). Distances between the centromeres of rye homologues (C-C) and between their telomeres (T-T) were measured using the 'Line' function.

Special attention was paid to the positioning of rye telomeres and centromeres relative to the nuclear envelope and to the (wheat) telomere cluster. No additional steps were taken to visualize the nuclear envelope: its position was inferred from the edge of the DAPI-stained chromatin. 'Display Adjustment' was used to adjust the channel contrast and thus, to improve the visualization of all analysed objects. Between 21 and 40 nuclei were analysed per genotype.

4.5. Chromosome Pairing and Transmission Rate

Meiotic pairing of selected chromosomes was analysed using standard protocols for material collection, fixation and genomic in situ hybridization (GISH) described in detail previously [34]. Briefly, from each sampled flower, a portion of one anther was removed, fresh-squashed in a drop of acetocarmine and if the desired meiotic stage was present, the remaining anthers from the flower were fixed in a mixture of absolute ethyl alcohol and glacial acetic acid in proportion 3:1 at 37 °C for a week and stored at −20 °C until used. Chromosome pairing was analysed on squashed preparations using GISH. Rye chromosomes and chromosome arms were visualized with DIG-labelled total genomic rye DNA and anti-DIG FITC, with wheat DNA sheared to ca. 200–500 bp fragments as a blocking DNA at 1:150 ratio (probe: blocking DNA). Following hybridization, preparations were counterstained with 1.5% propidium iodide (PI) in VectaShield antifade (BioRad, Hercules, CA, USA), mounted and observed under a microscope. Individual arms of rye chromosome 1R were identified by the presence of an unlabelled NOR band on the short arm. Pairing frequencies of individual arms of chromosome 1R were taken from a previous study [25]. The same protocol was used on root-tip meristems to estimate transmission rates to further generations. Seeds were germinated on wet filter paper in Petri dishes, root tips were collected to ice water for 26–30 h and fixed in a mixture of absolute alcohol and glacial acetic acid (3:1) at 37 °C for seven days. GISH was done the same way as for meiotic preparations with a probe prepared from total genomic DNA of rye and unlabelled genomic DNA of wheat as a blocking DNA.

5. Conclusions

To conclude, our study indicates that reduced chromosome pairing of introgressed alien chromosomes (or chromosome arms) may be predetermined already in somatic cells, and is more systemic in nature: telomeres of rye chromosomes in wheat often fail to attach to the nuclear envelope, and less frequently assume proper positions within the telomere cluster. This may hamper their migration to the telomere bouquet at the onset of meiosis resulting in reduced synapsis and reduced metaphase I pairing. Consequently, reduced metaphase I pairing lowers the transmission rate of such chromosomes to successive generations, and thus, destabilizes the integrity of a hybrid genome.

Supplementary Materials: Supplementary materials can be found at http://www.mdpi.com/1422-0067/20/6/1448/s1. **Video S1 (.mp4)**. Visualization of nucleus of disomic deletion line 1RS.$_{del}$1RL after 3D-FISH. Total genomic DNA of rye was labelled with TRITC using Nick translation (yellow color), centromeres of both wheat and rye chromosomes were visualized using oligonucleotide probe (red color), and telomere-specific sequence was PCR-labelled with FITC (green color). Nuclear DNA was counterstained with DAPI (blue color). Out-positioning of one rye chromosome arm is indicated by arrow. **Video S2 (.mp4)**. Visualization of nucleus of disomic deletion line $_{del}$1RS. 1RL after 3D-FISH. Total genomic DNA of rye was labelled with TRITC using Nick translation (yellow color), centromeres of both wheat and rye chromosomes were visualized using oligonucleotide probe (red color), and telomere-specific sequence was PCR-labelled with FITC (green color). Nuclear DNA was counterstained with DAPI (blue color). Non-attachment of telomeres of both rye $_{del}$1RS chromosome arms is indicated by arrows.

Author Contributions: Conceptualization, A.J.L. and D.K.; Methodology, K.P., V.K., A.J.L. and D.K.; Formal Analysis, K.P., V.K., J.V., M.D., D.P., O.Š. and D.K.; Investigation, K.P., V.K., A.J.L., C.F. and D.K.; Resources, A.J.L., J.V., M.S., G.J., O.Š., J.Š. and J.D.; Data Curation, M.D. and D.K.; Writing—Original Draft Preparation, K.P., V.K., A.J.L. and D.K.; Writing— Review & Editing, G.J., J.Š., M.S., O.Š. and J.D.; Visualization, K.P., A.J.L. and D.K.; Supervision, D.K.; Project Administration, D.K.; Funding Acquisition, J.Š., J.D. and D.K.

Funding: This research was funded by the Czech Science Foundation (grant award 17-13853S) and by the European Regional Development Fund OPVVV project "Plants as a tool for sustainable development" number CZ.02.1.01/0.0/16_019/0000827 supporting Excellent Research at CRH.

Acknowledgments: We would like to thank Radka Tušková for excellent technical support.

Conflicts of Interest: The authors declare no conflict of interest.

Abbreviations

3D-FISH	Three-dimensional fluorescent in situ hybridization
CN	Centre of the nucleus
CT	Chromosome territory
GISH	Genomic in situ hybridization
NE	Nuclear envelope
NOR	Nucleolar organizing region
NP	Nuclear periphery
PMC	Pollen mother cell

Appendix A

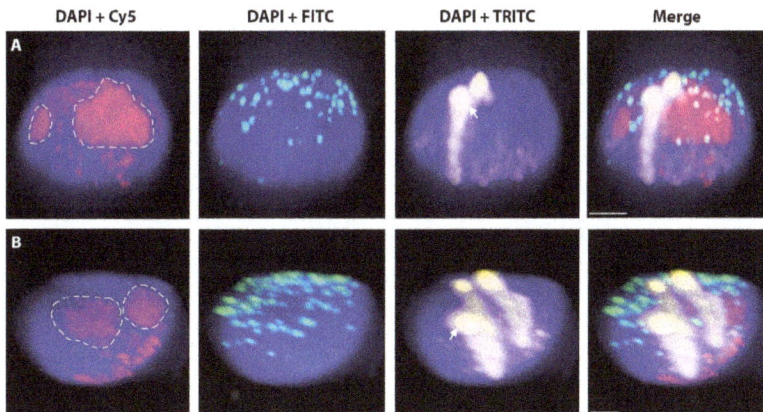

Figure A1. Non-attachment of rye telomeres. Nuclei with a pair of telocentric 1RS rye chromosomes (**A**) and a pair of 1R rye chromosomes (**B**) after 3D-FISH. Total genomic DNA of rye was labelled with TRITC using Nick translation (yellow color), centromeres of both wheat and rye chromosomes were visualized using oligonucleotide probe (red color), and telomere-specific sequence was PCR-labelled with FITC (green color). Nuclear DNA was counterstained with DAPI (blue color). Rye telomeres without visual contact to nuclear envelope are indicated by arrows. Nucleoli are indicated by white dashed lines. Scale bar 5 μm.

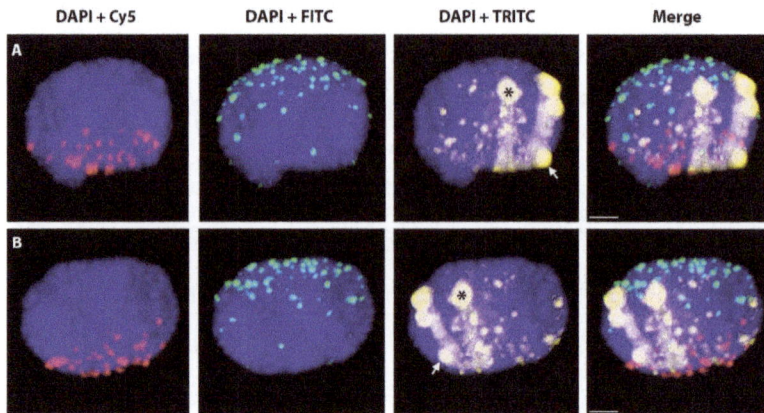

Figure A2. Combination of non-attachment and out-positioning of rye telomeres. A nucleus ((**A,B**) is the same nucleus from different angles) with a pair of homologous rye 1RS. $_{del}$1RL chromosomes after 3D-FISH. Total genomic DNA of rye was labelled with TRITC using Nick translation (yellow color), centromeres of both wheat and rye chromosomes were visualized using oligonucleotide probe (red color), and telomere-specific sequence was PCR-labelled with FITC (green color). Nuclear DNA was counterstained with DAPI (blue color). Rye telomere without visual contact to nuclear envelope is indicated by asterisk, rye telomere positioned out of the telomere pole (at the centromere pole) is indicated by arrow. Nucleoli are indicated by white dashed lines. Scale bar 5 µm.

References

1. Cremer, T.; Cremer, C. Chromosome territories, nuclear architecture and gene regulation in mammalian cells. *Nat. Rev. Genet.* **2001**, *2*, 292–301. [CrossRef] [PubMed]
2. Fritz, A.J.; Barutcu, A.R.; Martin-Buley, L.; van Wijnen, A.J.; Zaidi, S.K.; Imbalzano, A.N.; Lian, J.B.; Stein, J.L. Chromosomes at Work: Organization of Chromosome Territories in the Interphase Nucleus. *J. Cell. Biochem.* **2016**, *117*, 9–19. [CrossRef] [PubMed]
3. Fransz, P.; de Jong, J.H.; Lysak, M.; Castiglione, M.R.; Schubert, I. Interphase chromosomes in Arabidopsis are organized as well defined chromocenters from which euchromatin loops emanate. *Proc. Natl. Acad. Sci. USA* **2002**, *9*, 14584–14589. [CrossRef] [PubMed]
4. Tiang, C.L.; He, Y.; Pawlowski, W.P. Chromosome Organization and Dynamics during Interphase, Mitosis, and Meiosis in Plants. *Plant Physiol.* **2012**, *158*, 26–34. [CrossRef]
5. Rabl, C. Über Zellteilung. *Morph. Jahrb.* **1885**, *10*, 214–330.
6. Dong, F.G.; Jiang, J.M. Non-Rabl patterns of centromere and telomere distribution in the interphase nuclei of plant cells. *Chromosome Res.* **1998**, *6*, 551–558. [CrossRef] [PubMed]
7. Dawe, R.K. Meiotic chromosome organization and segregation in plants. *Annu. Rev. Plant Physiol. Plant Mol. Biol.* **1998**, *49*, 371–395. [CrossRef] [PubMed]
8. Moens, P.B.; Bernei-Moens, C.; Spyropoulos, B. Chromosome core attachment to the meiotic nuclear envelope regulates synapsis in Chloealtis (Orthoptera). *Genome* **1989**, *32*, 601–610. [CrossRef]
9. Curtis, C.A.; Lukaszewski, A.J.; Chrzastek, M. Metaphase I pairing of deficient chromosomes and genetic mapping of deficiency breakpoints in common wheat. *Genome* **1991**, *34*, 553–560. [CrossRef]
10. Mujeeb-Kazi, A. Intergeneric crosses: Hybrid production and utilization. In *Utilizing Wild Grass Biodiversity in Wheat Improvement, 15 Years of Wide cross Research at CIMMYT*; Mujeeb-Kazi, A., Hettel, G.P., Eds.; CIMMYT: Texcoco, Mexico, 1995; p. 140.
11. Friebe, B.; Jiang, J.; Raupp, J.W.; McIntosh, R.A.; Gill, B.S. Characterization of wheat-alien translocations conferring resistance to diseases and pests: Current status. *Euphytica* **1996**, *91*, 59–87. [CrossRef]
12. Molnár-Láng, M. The crossability of wheat with rye and other related species. In *Alien Introgression in Wheat: Cytogenetics, Molecular Biology, and Genomics*; Molnár-Láng, M., Ceoloni, C., Doležel, J., Eds.; Springer International Publishing: Chem, Switzerland, 2015; pp. 103–120.

13. Nishiyama, I. Cytogenetic studies in Avena, IX New synthetic oats in the progenies of induced decaploid interspecific hybrids. *Jpn. J. Genet.* **1962**, *37*, 118–130. [CrossRef]

14. Tsunewaki, K. Genetic studies on a 6x derivative from an 8x triticale. *Can. J. Genet. Cytol.* **1964**, *6*, 1–11. [CrossRef]

15. Lukaszewski, A.J.; Apolinarska, B.; Gustafson, J.P.; Krolow, K.D. Chromosome-pairing and aneuploidy in tetraploid triticale 1. Stabilized karyotypes. *Genome* **1987**, *29*, 554–561. [CrossRef]

16. Orellana, J.; Cermeno, M.C.; Lacadena, J.R. Meiotic pairing in wheat-rye addition and substitution lines. *Can. J. Genet. Cytol.* **1984**, *26*, 25–33. [CrossRef]

17. Lukaszewski, A.J.; Gustafson, J.P. Cytogenetics of Triticale. *Plant Breed. Rev.* **1987**, *5*, 41–93.

18. Murphy, S.P.; Bass, H.W. The maize (*Zea mays*) *desynaptic* (*dy*) mutation defines a pathway for meiotic chromosome segregation, linking nuclear morphology, telomere distribution and synapsis. *J. Cell Sci.* **2012**, *125*, 3681–3690. [CrossRef] [PubMed]

19. Naranjo, T. Dynamics of Rye Telomeres in a Wheat Background during Early Meiosis. *Cytogenet. Genome Res.* **2014**, *143*, 60–68. [CrossRef]

20. Pernickova, K.; Linc, G.; Gaal, E.; Kopecky, D.; Samajova, O.; Lukaszewski, A.J. Out-of-position telomeres in meiotic leptotene appear responsible for chiasmate pairing in an inversion heterozygote in wheat (*Triticum aestivum* L.). *Chromosoma* **2019**, *128*, 31–39. [CrossRef]

21. Schlegel, R.; Melz, G.; Nestrowicz, R. A universal reference karyotype in rye, *Secale cereale* L. *Theor. Appl. Genet.* **1987**, *74*, 820–826. [CrossRef]

22. Naranjo, T. Variable Patterning of Chromatin Remodeling, Telomere Positioning, Synapsis, and Chiasma Formation of Individual Rye Chromosomes in Meiosis of Wheat-Rye Additions. *Front. Plant Sci.* **2018**, *9*. [CrossRef]

23. Gill, B.S.; Friebe, B.; Endo, T.R. Standard karyotype and nomenclature system for description of chromosome bands and structural abberations in wheat (*Triticum aestivum*). *Genome* **1991**, *34*, 830–839. [CrossRef]

24. Doležel, J.; Greilhuber, J.; Lucretti, S.; Meister, A.; Lysák, M.A.; Nardi, L.; Obermayer, R. Plant genome size estimation by flow cytometry: Inter-laboratory comparison. *Ann. Bot.* **1998**, *82*, 17–26. [CrossRef]

25. Lukaszewski, A.J. Unexpected behavior of an inverted rye chromosome arm in wheat. *Chromosoma* **2008**, *117*, 569–578. [CrossRef]

26. Kopecky, D.; Loureiro, J.; Zwierzykowski, Z.; Ghesquiere, M.; Dolezel, J. Genome constitution and evolution in *Lolium x Festuca* hybrid cultivars (Festulolium). *Theor. Appl. Genet.* **2006**, *113*, 731–742. [CrossRef]

27. Zwierzykowski, Z.; Kosmala, A.; Zwierzykowska, E.; Jones, N.; Joks, W.; Bocianowski, J. Genome balance in six successive generations of the allotetraploid *Festuca pratensis x Lolium perenne. Theor. Appl. Genet.* **2006**, *113*, 539–547. [CrossRef] [PubMed]

28. Alleva, B.; Smolikove, S. Moving and stopping: Regulation of chromosome movement to promote meiotic chromosome pairing and synapsis. *Cell* **2017**, *8*, 613–624. [CrossRef] [PubMed]

29. Schlegel, R.; Mettin, D. Studies on intraindividual and interindividual variation of chromosome-pairing in diploid and tetraploid populations. 2. interindividual variation. *Biol. Zbl.* **1975**, *94*, 703–715.

30. Schlegel, R. *Rye: Genetics, Breeding, and Cultivation*; CRC Press: Boca Raton, FL, USA, 2014; p. 359.

31. Naranjo, T.; Valenzuela, N.T.; Perera, E. Chiasma Frequency Is Region Specific and Chromosome Conformation Dependent in a Rye Chromosome Added to Wheat. *Cytogenet. Genome Res.* **2010**, *129*, 133–143. [CrossRef] [PubMed]

32. Lukaszewski, A.J. Behavior of centromeres in univalents and centric misdivision in wheat. *Cytogenet. Genome Res.* **2010**, *129*, 97–109. [CrossRef]

33. Bridges, C.B. Deficiency. *Genetics* **1917**, *2*, 445–465.

34. Vrana, J.; Simkova, H.; Kubalakova, M.; Cihalikova, J.; Dolezel, J. Flow cytometric chromosome sorting in plants: The next generation. *Methods* **2012**, *57*, 331–337. [CrossRef] [PubMed]

35. Ito, H.; Nasuda, S.; Endo, T.R. A direct repeat sequence associated with the centromeric retrotransposons in wheat. *Genome* **2004**, *47*, 747–756. [CrossRef] [PubMed]

36. Phillips, D.; Nibau, C.; Ramsay, L.; Waugh, R.; Jenkins, G. Development of a Molecular Cytogenetic Recombination Assay for Barley. *Cytogenet. Genome Res.* **2010**, *129*, 154–161. [CrossRef] [PubMed]

37. Howe, E.S.; Murphy, S.P.; Bass, H.W. Three-Dimensional Acrylamide Fluorescence in Situ Hybridization for Plant Cells. In *Plant Meiosis. Methods and Protocols*; Pawlowski, W., Grelon, M., Armstrong, S., Eds.; Springer International Publishing AG: Dordrecht, The Netherlands, 2014; pp. 53–66.

© 2019 by the authors. Licensee MDPI, Basel, Switzerland. This article is an open access article distributed under the terms and conditions of the Creative Commons Attribution (CC BY) license (http://creativecommons.org/licenses/by/4.0/).

International Journal of
Molecular Sciences

MDPI

Article

Using the 6RL^{Ku} Minichromosome of Rye (*Secale cereale* L.) to Create Wheat-Rye 6D/6RL^{Ku} Small Segment Translocation Lines with Powdery Mildew Resistance

Haimei Du [1,†]**, Zongxiang Tang** [1,2,†]**, Qiong Duan** [1,2]**, Shuyao Tang** [1,2] **and Shulan Fu** [1,*]

[1] College of Agronomy, Sichuan Agricultural University, Wenjiang, Chengdu 611130, China; 18314441196@163.com (H.D.); zxtang@sicau.edu.cn (Z.T.); 18380443767@163.com (Q.D.); tangshuyao705708@sina.com (S.T.)
[2] Institute of Ecological Agriculture, Sichuan Agricultural University, Wenjiang, Chengdu 611130, China
* Correspondence: fushulan@sicau.edu.cn
† These authors contributed equally to this work.

Received: 18 October 2018; Accepted: 4 December 2018; Published: 7 December 2018

Abstract: Long arms of rye (*Secale cereale* L.) chromosome 6 (6RL) carry powdery mildew resistance genes. However, these sources of resistance have not yet been successfully used in commercial wheat cultivars. The development of small segment translocation chromosomes carrying resistance may result in lines carrying the 6R chromosome becoming more commercially acceptable. However, no wheat-rye 6RL small segment translocation line with powdery mildew resistance has been reported. In this study, a wheat-rye 6RL^{Ku} minichromosome addition line with powdery mildew resistance was identified, and this minichromosome was derived from the segment between L2.5 and L2.8 of the 6RL^{Ku} chromosome arm. Following irradiation, the 6RL^{Ku} minichromosome divided into two smaller segments, named 6RL^{Kumi200} and 6RL^{Kumi119}, and these fragments participated in the formation of wheat-rye small segment translocation chromosomes 6DS/6RL^{Kumi200} and 6DL/6RL^{Kumi119}, respectively. The powdery mildew resistance gene was found to be located on the 6RL^{Kumi119} segment. Sixteen 6RL^{Kumi119}-specific markers were developed, and their products were cloned and sequenced. Nucleotide BLAST searches indicated that 14 of the 16 sequences had 91–100% similarity with nine scaffolds derived from 6R chromosome of *S. cereale* L. Lo7. The small segment translocation chromosome 6DL/6RL^{Kumi119} makes the practical utilization in agriculture of powdery mildew resistance gene on 6RL^{Ku} more likely. The nine scaffolds are useful for further studying the structure and function of this small segment.

Keywords: wheat; rye; 6R; small segment translocation; powdery mildew

---

1. Introduction

It has already been reported that the long arms of rye (*Secale cereale* L.) chromosome 6 (6RL) carry powdery mildew resistance gene *Pm20* [1], and this gene was introduced into wheat background in the form of a 6BS/6RL translocation chromosome [1]. The gene *Pm20* still has a broad spectrum of resistance to *Blumeria graminis* f. sp. *tritici* (Bgt) isolates [2,3]. The 6RL chromosome arm that carries *Pm20* was derived from *S. cereale* L. cv. Prolific [1]. Recently, some reports indicated that 6RL arms derived from *S. cereale* cv. Jingzhouheimai, *S. cereale* cv. German White, and *S. cereale* cv. Kustro also carried powdery mildew resistance genes [3–6]. It has already been established that the powdery mildew resistance gene on 6RL of German White was different from the gene *Pm20* [3], and this indicates that different 6RL arms may also display genetic diversity for powdery mildew resistance genes. However, the powdery mildew resistance genes on 6RL arms have not been successfully used in commercial wheat cultivars because of agronomic disadvantages, possibly caused by non-compensation and linkage drag of the 6RL arm. The development of small segment translocations between wheat chromosomes and 6RL, which causes minimal loss of indispensable wheat genes, may resolve this problem [3,4,7,8]. Only one translocation chromosome carrying a small segment of a 6RL arm, Ti4AS.4AL-6RL-4AL, has been reported. In this case, the 6RL segment was derived from the telomeric region and carried the Hessian fly-resistant gene *H25* [9,10]. So far, no wheat-rye 6RL small segment translocation lines with powdery mildew resistance have been reported. In this study, a wheat-rye 6RL^{Ku} minichromosome addition line was developed, and 6D/6RL^{Ku} small segment translocation lines were identified from the irradiated seeds of this minichromosome addition line.

2. Results

2.1. Obtaining Wheat-Rye 6RL^{Ku} Minichromosome Addition Line

A 6RL^{Ku} minichromosome addition line, MiA6RL^{Ku}, was found among the self-pollinated progeny of wheat-rye 6RL^{Ku} monotelosomic addition line MTA6RL^{Ku}. Line MiA6RL^{Ku} contained one minichromosome derived from the 6RL^{Ku} arm (Figure 1). According to the FISH map of 6RL^{Ku} arm constructed based on the signal patterns of probes Oligo-pSc200, Oligo-pSc250, and Oligo-pSc119.2-1 [6], the 6RL^{Ku} minichromosome was derived from the segment between L2.5 and L2.8 (Figure 1). Through measuring the fraction length of the 6RL^{Ku} arm, combined with the fraction length standard of 6RL built by Mukai et al. [11], it can be deduced that the 6RL^{Ku} minichromosome comprised about 11% of the original 6RL^{Ku} length.

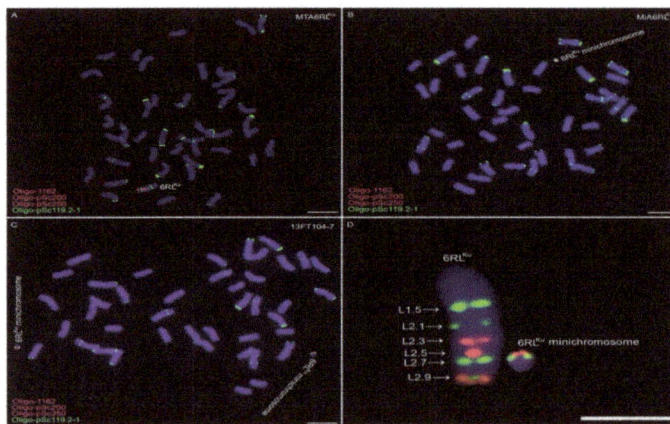

Figure 1. ND-FISH analysis of MTA6RL^{Ku} and MiA6RL^{Ku}. (**A,B,C**) Oligo-1162 (**red**), Oligo-pSc200 (**red**), Oligo-pSc250 (**red**), and Oligo-pSc119.2-1 (**green**) were used as probes. (**D**) Cut and pasted 6RL^{Ku} from MTA6RL^{Ku} and 6RL^{Ku} minichromosome from 13FT104-7. The FISH map of 6RL^{Ku} is the same as the one reported by Li et al. [6]. Chromosomes were counterstained with 4′-6-diamidino-2-phenyllindole (DAPI) (**blue**). Scale bar 10 μm.

2.2. Transmission of 6RL^{Ku} Minichromosome

Thirty-four seeds were randomly selected from the self-fertilized progeny of line MiA6RL^{Ku} for ND-FISH analysis. Among the 34 plants, 24 had no 6RL^{Ku} minichromosome; nine contained one 6RL^{Ku} minichromosome; and one plant, 13FT104-7, contained a pair of this minichromosome (Figure 1C). From the self-fertilized progeny of line 13FT104-7, 100 seeds were randomly selected for ND-FISH analysis; 25 plants contained two 6RL^{Ku} minichromosomes, 62 plants contained one 6RL^{Ku} minichromosomes, and the remaining 13 plants had none of this minichromosome. The progeny of

13FT104-7 were named 15T154, and some of these plants were used for developing additional 6RL^{Ku} minichromosome-specific markers and producing wheat-rye small segment translocations.

2.3. Development of 6D/6RL^{Ku} Small Segment Translocation Lines

Some seeds that were derived from line 14T154-35 with two 6RL^{Ku} minichromosomes were exposed to ^{60}Co-γ rays. A total of 1428 M1 seeds were analyzed using ND-FISH, and ten wheat-rye 6D/6RL^{Ku} small segment translocation lines were detected and named 16T379 or 16T380 (Table 1, Figure 2). In these 6D/6RL^{Ku} small segment translocation lines, the 6RL^{Ku} minichromosome divided into two smaller segments. One small segment, with the Oligo-pSc200 and Oligo-pSc250 signals and a strong Oligo-pSc119.2-1 signal, was named 6RL^{Kumi119}. The other small segment, with the Oligo-pSc200 and Oligo-pSc250 signals and a weak Oligo-pSc119.2-1 signal, was named 6RL^{Kumi200}. The segment 6RL^{Kumi200} had been translocated onto the 6DS arm, while the segment 6RL^{Kumi119} was translocated onto the 6DL arm (Figure 2). The small segment translocation chromosome fused with 6DS was named 6DS/6RL^{Kumi200}, and the other translocation attached with 6DL was named 6DL/6RL^{Kumi119} (Figure 2).

Table 1. Wheat-rye 6D/6RL^{Ku} small segment translocation lines with 6DS/6RL^{Kumi200} and 6DL/6RL^{Kumi119} translocation chromosomes.

Small Segment Translocation Lines	Small Segment Translocation Chromosomes
16T379-1	one 6DS/6RL^{Kumi200} and one 6DL/6RL^{Kumi119}
16T379-4	two 6DS/6RL^{Kumi200} and two 6DL/6RL^{Kumi119}
16T379-6	one 6DS/6RL^{Kumi200} and one 6DL/6RL^{Kumi119}
16T379-8	two 6DS/6RL^{Kumi200} and two 6DL/6RL^{Kumi119}
16T379-9	one 6DS/6RL^{Kumi200} and one 6DL/6RL^{Kumi119}
16T379-11	one 6DS/6RL^{Kumi200} and one 6DL/6RL^{Kumi119}
16T379-13	one 6DS/6RL^{Kumi200} and one 6DL/6RL^{Kumi119}
16T379-14	two 6DS/6RL^{Kumi200} and two 6DL/6RL^{Kumi119}
16T380-2	one 6DS/6RL^{Kumi200} and one 6DL/6RL^{Kumi119}
16T380-3	one 6DS/6RL^{Kumi200} and one 6DL/6RL^{Kumi119}

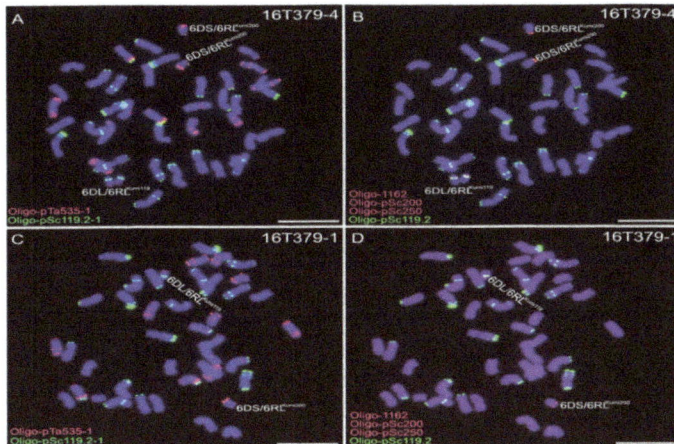

Figure 2. Oligo-pTa535-1 (**red**), Oligo-pSc119.2-1 (**green**), Oligo-1162 (**red**), Oligo-pSc200 (**red**), and Oligo-pSc250 (**red**) were used as probes for ND-FISH analysis of wheat-rye 6D/6RL^{Ku} small segment translocation lines. (**A,B**) Line 16T379-4 represents lines with two 6DS/6RL^{Kumi200} and two 6DL/6RL^{Kumi119} chromosomes. (**A**) and (**B**) are the same cell. (**C,D**) Line 16T379-1 represents lines with one 6DS/6RL^{Kumi200} and one 6DL/6RL^{Kumi119} chromosome. (**C**) and (**D**) are the same cell. Chromosomes were counterstained with DAPI (**blue**). Scale bar 10 μm.

2.4. Transmission Rates of 6DS/6RL^{Kumi200} and 6DL/6RL^{Kumi119}

Among 100 randomly selected seeds from the progeny of 16T379-1, 21 derived seedlings contained no translocation chromosomes, 16 seedlings contained two 6DS/6RL^{Kumi200} and two 6DL/6RL^{Kumi119} chromosomes, 42 seedlings contained one 6DS/6RL^{Kumi200} and one 6DL/6RL^{Kumi119} chromosome, eight contained two 6DS/6RL^{Kumi200} and one 6DL/6RL^{Kumi119} chromosome (Figure 3A), three plants contained one 6DS/6RL^{Kumi200} and two 6DL/6RL^{Kumi119} chromosomes (Figure 3B), three plants contained only one 6DS/6RL^{Kumi200} chromosome (Figure 3C), and seven plants contained only one 6DL/6RL^{Kumi119} chromosome (Figure 3D). The progeny of 16T379-1 were named 17T256, and some of these plants were used for further marker development. All of the randomly selected 100 seeds from the progeny of 16T379-4 contained two 6DS/6RL^{Kumi200} and two 6DL/6RL^{Kumi119} chromosomes. These results indicate that the 6DS/6RL^{Kumi200} and 6DL/6RL^{Kumi119} chromosomes show a high frequency of transmission to progeny.

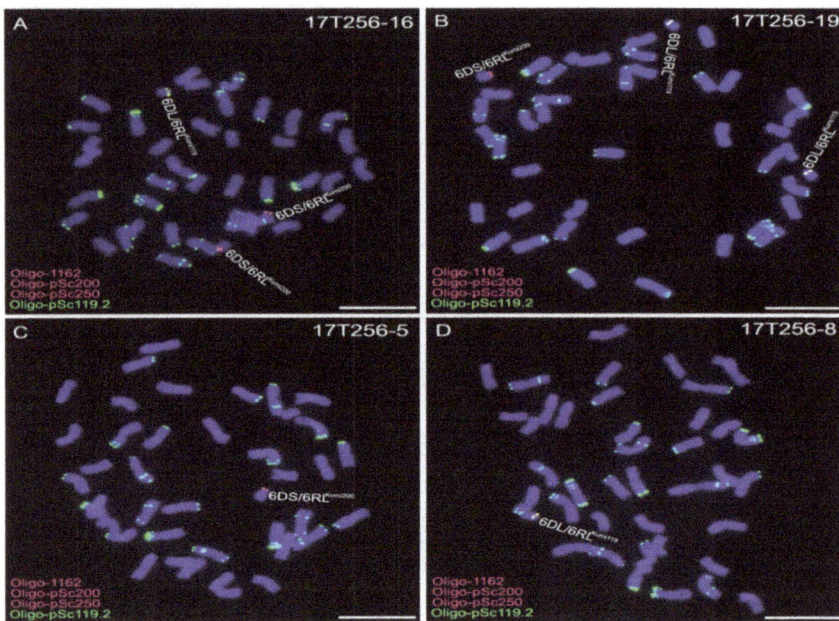

Figure 3. Oligo-pSc119.2-1 (**green**), Oligo-1162 (**red**), Oligo-pSc200 (**red**), and Oligo-pSc250 (**red**) were used as probes for ND-FISH analysis of the progeny of line 16T379-1. (**A**) Line 17T256-16 carried two 6DS/6RL^{Kumi200} chromosomes and one 6DL/6RL^{Kumi119} chromosome. (**B**) Line 17T256-19 carried one 6DS/6RL^{Kumi200} and two 6DL/6RL^{Kumi119} chromosomes. (**C**) Line 17T256-5 possessed only one 6DS/6RL^{Kumi200} chromosome. (**D**) Line 17T256-8 had only one 6DL/6RL^{Kumi119} chromosome. Chromosomes were counterstained with DAPI (**blue**). Scale bar 10μm.

2.5. Development of 6RL^{Ku} Minichromosome-Specific Markers

Common wheat Chinese Spring (CS), Mianyang 11 (MY11), rye Kustro, line MTA6RL^{Ku}, line 14T154-35 with two 6RL^{Ku} minichromosomes, and two lines (14T154-31 and 14T154-41) without the minichromosome were used to develop 6RL^{Ku} minichromosome-specific markers. One thousand and eighty-six pairs of primers were designed according to the 6RL^{Ku} minichromosome-specific pair-end reads. Sixteen of the 1086 primer pairs amplified their target bands from the genomic DNAs of rye Kustro, line MTA6RL^{Ku}, and line 14T154-35, and not from the genomic DNA of CS, MY11, and lines 14T154-31 and 14T154-41 (Figure 4). Therefore, the 16 primer pairs were tentatively regarded as 6RL^{Ku} minichromosome-specific markers and are listed in Table 2. From the progeny of 16T379-1, three lines (17T256-7, 17T256-10, and 17T256-17) without 6D/6RL^{Ku} translocation chromosomes, three lines (17T256-8, 17T256-12, and 17T256-14) that only contained one 6DL/6RL^{Kumi119} chromosome, and three lines (17T256-5, 17T256-9, and 17T256-11) that only contained one 6DS/6RL^{Kumi200} chromosome, were used to validate the 16 primer pairs. The 16 primer pairs only amplified target bands from lines 17T256-8, 17T256-12, and 17T256-14 (Figure 5). This indicated that all the 16 6RL^{Ku} minichromosome-specific markers were located on the segment 6RL^{Kumi119}. Therefore, the 16 markers were 6RL^{Kumi119}-specific.

Table 2. 6RL^{Ku} minichromosome-specific markers and the similarity between the sequences amplified by the markers and the *S. cereale* Lo7 scaffolds.

Marker	Forward (5′-3′)	Reverse (5′-3′)	GenBank Accession Number of Amplified Sequence	Similarity of Amplified Sequence with the *S. cereale* Lo7 Scaffolds
6RL-M8	CAACCTATTCG GACCAGAGC	GATTAAACCGCT GGTGAGAAAC	MK051036	99% similarity with 7794–8206 bp of Lo7_v2_scaffold_453717 6R
6RL-M11	GGGGGAACTTT GAGTATGCTT	GATCGGATCGGT TGAGTTGT	MK051037	99% similarity with 464–1239 bp of Lo7_v2_scaffold_651086 0R
6RL-M55	TGATGCAAGTT CGTTGGTGT	CGTTGACTCCCTT CCGTTAG	MK051038	91% similarity with 1–108 bp of Lo7_v2_scaffold_457844 6R
6RL-M63	TCGAAATGCAT CGGACAAT	TCCATGGTCTCCT CGAGTGT	MK051039	100% similarity with 144–422 bp of Lo7_v2_scaffold_492428 6R
6RL-M102	CGGGAGAGGA CTGGTTCTT	CATATGTACAAC AGAGGCATCTTC	MK051040	98% similarity with 26957–27168 bp, 27841-27881 bp and 27923-28047 bp of Lo7_v2_scaffold_445202 6R
6RL-M118	TCCCCCTTCTA GGGTTTCAT	ATAGCCCCATCTG CAAACAC	MK051041	100% similarity with 488–896 bp of Lo7_v2_scaffold_484582 6R
6RL-M149	AATGGCTGCAA TTTCTTGGA	AAAAAGCCACA AAACACTGC	MK051042	100% similarity with 34805–34422 bp of Lo7_v2_scaffold_445202 6R
6RL-M220	GCACAAGTCCA TGTCCTTCA	GATCCATCTGGCT GTGTGTG	MK051043	99% similarity with 4112–4449 bp of Lo7_v2_scaffold_448816 6R
6RL-M221	CGCTATATGCA ATGCAGGTG	CTTGCTTGCAAC ACCAAAAA	MK051044	98% similarity with 41511–41912 bp of Lo7_v2_scaffold_445202 6R
6RL-M255	CCTTATGACCA CCCATGCTC	TTCATAGCTGCCT CTTTAGGTG	MK051045	99% similarity with 31812–32230 bp of Lo7_v2_scaffold_445202 6R
6RL-M710	CAAACTCACAC GAAGCCAAA	CTGATCCAAATTT GCCCAGT	MK051046	92% similarity with 3600–3681 bp of Lo7_v2_scaffold_457146 6R
6RL-M828	TTTGTCGAGAG CAACAATGG	CCCGCTTCTAAGT TCAATCG	MK051047	100% similarity with 39318–39668 bp of Lo7_v2_scaffold_445202 6R
6RL-M869	GGGTCAACCCA TCTTGTTTC	CCTCTTCCACTGC AGAGCTT	MK051048	99% similarity with 589–962 bp of Lo7_v2_scaffold_451612 6R
6RL-M896	GACGAAACAC AACAAATCATTCA	GGGAAAATCGAA AACTGCAA	MK051049	100% similarity with 10161–10344 bp of Lo7_v2_scaffold_620512 0R
6RL-M1074	AAAGCCGATG AAAAATGGTG	GAAGAAGAAGA AGATGGGGTGTT	MK051050	100% similarity with 9159–9365 bp of Lo7_v2_scaffold_445253 6R
6RL-M1081	TTGCATGCTCG CTTTAGTTG	CCACTTGACGTT GCCCTATT	MK051051	100% similarity with 8940–9193 bp of Lo7_v2_scaffold_445253 6R

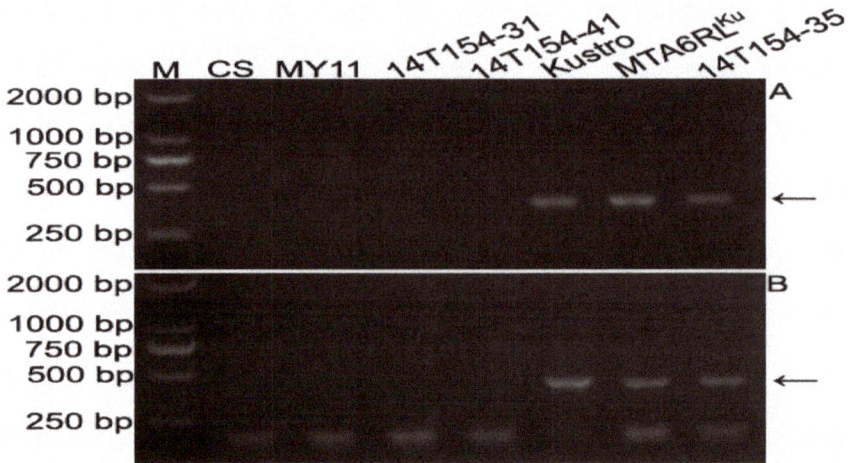

Figure 4. Representative results of developing 6RL^{Ku} minichromosome-specific markers. (**A**) Products amplified marker 6RL-M55. (**B**) Products amplified marker 6RL-M63. M, DNA marker. Arrows indicate 6RL^{Ku} minichromosome-specific bands.

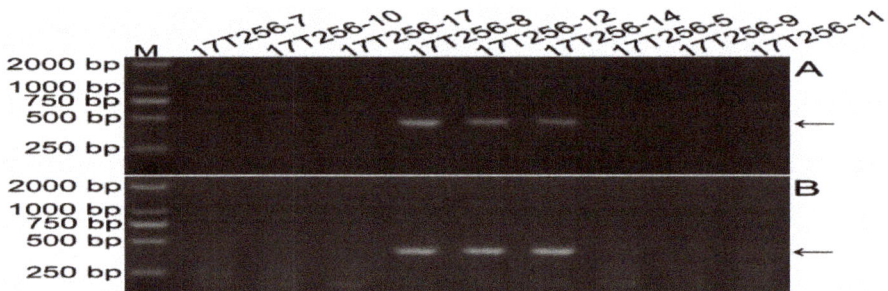

Figure 5. Representative results of locating 6RL^{Ku} minichromosome-specific markers on 6RL^{Kumi119} segment. (**A**) Products amplified by marker 6RL-M11. (**B**) Products amplified by marker 6RL-M102. M, DNA marker. Arrows indicate 6RL^{Kumi119}-specific bands.

2.6. Sequence Characteristics of the Products Amplified by the 16 Markers

The sequences amplified by the 16 6RL^{Kumi119}-specific markers were isolated. These sequences have been deposited in the GenBank Database (GenBank accession numbers: MK051036–MK051051). Sequence alignment using the BLAST tool in NCBI and gene annotation indicated that all the 16 sequences did not belong to any gene sequence; therefore, they are probably not involved in active genes. Analysis using RepeatMasker software (http://repeatmasker.org/) also indicated that these sequences were not involved in repetitive DNA sequences. Nucleotide BLAST searches indicated that 14 of the 16 sequences had 91–100% similarity with the scaffolds from the 6R chromosome of *S. cereale* L. Lo7, and the other two sequences had 99–100% similarity with the scaffolds derived from 0R of *S. cereale* L. Lo7 scaffolds (Table 2). It can be noted that all the five sequences (MK051040, MK051042, MK051044, MK051045, and MK051047) were located in the scaffold Lo7_v2_scaffold_445202 6R, and both MK051050 and MK051051 were located in the scaffold Lo7_v2_scaffold_445253 6R (Table 2). According to the gene annotation of these scaffolds published by Bauer et al. [12], they did not contain genes; however, these scaffolds were still useful for the future study on the structure and function of the 6RL^{Kumi119} segment.

2.7. Powdery Mildew Resistance

Powdery mildew resistance testing indicated that parental wheat MY11, lines without 6RL^{Ku} minichromosome, and lines with only one 6DS/6RL^{Kumi200} chromosome were highly susceptible to powdery mildew (Figure 6). Lines with both the 6DS/6RL^{Kumi200} and 6DL/6RL^{Kumi119} chromosomes and lines with only one 6DL/6RL^{Kumi119} chromosome displayed high resistance to powdery mildew (Figure 6). These results indicated that the powdery mildew resistance gene on 6RL^{Ku} was located on the small segment 6RL^{Kumi119}.

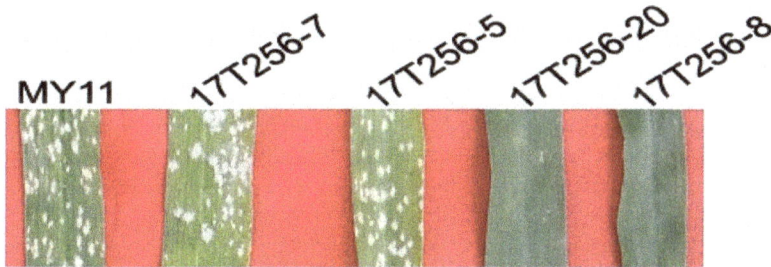

Figure 6. Identification of resistance to powdery mildew. Parental wheat and lines without 6D/6RL^{Ku} translocation chromosomes and lines with only one 6DS/6RL^{Kumi200} chromosome are highly susceptible to powdery mildew. Lines with one 6DS/6RL^{Kumi200} and one 6DL/6RL^{Kumi119} chromosome and lines with only one 6DL/6RL^{Kumi119} chromosome are highly resistant to powdery mildew.

3. Discussion

3.1. Extending Genetic Basis of Powdery Mildew Resistance Genes

Several reports have highlighted the current narrowness of genetic diversity of powdery mildew resistance genes in wheat breeding programs in China [13–15]. For example, the wheat cultivars (lines) from the Yangtze River region mainly contained genes *Pm4a* and *Pm21* [13]. The gene *Pm21* is widely used in wheat cultivars from Sichuan and Guizhou provinces [14], while the wheat cultivars (lines) from Henan province mainly carry genes *Pm2*, *Pm4*, *Pm21* and unknown gene(s) located on a new 1RS.1BL translocation chromosome [15]. It should be noted that gene *Pm21* has played an important role in wheat powdery mildew resistance breeding programs in China, but there is the risk of pathogen directional selection caused by the high frequency of *Pm21* gene in wheat breeding programs [15]. Therefore, extending the genetic basis of powdery mildew resistance genes in wheat breeding programs in China must be an urgent priority. 6R chromosomes of rye (*S. cereale* L.) carry powdery mildew resistance genes, and still display effective resistance to powdery mildew pathotypes in China [1–6,16]. In addition, the powdery mildew resistance genes on 6R chromosomes have genetic diversity [3,16], which may provide recipient wheat lines with resistance to future virulent powdery mildew races that might defeat existing genes. However, these 6R-derived powdery mildew resistance genes have not been successfully used in commercial wheat cultivars because of the loss of indispensable wheat genes and alien chromosome-associated linkage drag. Therefore, the development of new wheat-rye 6RL small segment translocation chromosomes with powdery mildew resistance gene(s) is of pressing concern [3,4,8]. It has been reported that few recombinants (sometime none) can be recovered between 6R and wheat chromosomes, even in a *ph1b* mutant background. Furthermore, it is difficult to produce suitable wheat-rye 6RL small segment translocations [17]. In this study, a wheat-rye small segment translocation chromosome 6DL/6RL^{Kumi119} with powdery mildew resistance was developed using a 6RL^{Ku} minichromosome. According to the fraction length standard of the 6RL arm constructed by Mukai et al. [11], the size of the 6RL^{Kumi119} segment might be slightly larger than the small 6RL segment in the Ti4AS.4AL-6RL-4AL translocation chromosome, in which the 6RL segment occupied 10% of the total length of 6RL [11]. Therefore, the occurrence of possible deleterious genes on the 6RL^{Ku} arm may have been greatly reduced. Additionally, the 6DS/6RL^{Kumi200} and 6DL/6RL^{Kumi119} chromosomes showed high transmission to progeny. These characteristics make the successful utilization of the powdery mildew resistance gene on 6RL^{Ku} more likely, although the agronomic traits of the lines with 6DS/6RL^{Kumi200} and 6DL/6RL^{Kumi119} chromosomes have not been investigated.

3.2. 6RL^{Ku} Minichromosome Specific Markers

For the effective utilization of elite genes in wild germplasm, it is necessary to increase our understanding of the molecular basis of target alien chromosome regions [18,19]. In this study, the small segment translocation chromosome 6DL/6RL^{Kumi119} provides an opportunity to understand the molecular basis of the small segment 6RL^{Kumi119} carrying powdery mildew resistance. Additionally, 16 6RL^{Ku} minichromosome-specific markers were developed, and these markers were subsequently located onto the segment 6RL^{Kumi119} with powdery mildew resistance. These markers represent convenient tools to use the 6DL/6RL^{Kumi119} translocation chromosome in wheat breeding programs. Using the sequences amplified by the 16 markers, 11 scaffolds derived from winter rye inbred line *S. cereale* L. Lo7 [12] were located on the segment 6RL^{Kumi119}. These scaffolds are useful for the further detailed studying of this segment. However, published data indicated that no active genes could be found among the 11 scaffolds [12]. There are two possible explanations for this phenomenon. First, the 6RL^{Ku} minichromosome-specific markers may be insufficient in number. Second, large gaps may exist in these scaffolds. Therefore, a greater number of 6RL^{Ku} minichromosome-specific markers are required and more complete rye genomic sequences are needed.

4. Materials and Methods

4.1. Plant Materials

A monotelosomic addition line MTA6RL^{Ku} was developed according to the methods described by Qiu et al. [20]. From the self-pollinated progeny of MTA6RL^{Ku}, a 6RL^{Ku} minichromosome addition line MiA6RL^{Ku} was identified. Some seeds selected from the progeny of MiA6RL^{Ku} were irradiated with ^{60}Co-γ rays at the Biotechnology and Nuclear Technology Research Institute, Sichuan Academy of Agricultural Sciences, China. Common wheat (*Triticum aestivum* L.) Mianyang 11 (MY11) and Chinese Spring (CS) were used as controls.

4.2. Cytological Analysis

The root-tip metaphase cells were analyzed using non-denaturing fluorescence *in situ* hybridization (ND-FISH) technology. Oligonucleotide (oligo) probes Oligo-1162, Oligo-pSc200, Oligo-pSc250, Oligo-pSc119.2-1, and Oligo-pTa535-1 [21,22] were used. These oligo probes were 5′-end-labelled with 6-carboxyfluorescein (6-FAM) or 6-carboxytetramethylrhodamine (Tamra). Root-tip metaphase chromosomes were prepared following the methods described by Han et al. [23]. ND-FISH analysis was performed following the methods described by Fu et al. [22] with the following minor modification. When dropped onto cell spreads, the probe mixture containing Oligo-1162 around the slides was kept above 28 °C, and the slides were immediately placed into a moist box that was incubated at 42 °C in advance for 1–2 h. Then, the slides were washed 15–20 s in 2 × SSC at 42 °C.

4.3. Development of 6RL^{Ku} Minichromosome-Specific PCR-Based Markers

Genomic DNAs of *S. cereale* L. Kustro and MiA6RL^{Ku} were sequenced by using the Specific Length Amplified Fragment Sequencing (SLAF-seq) technique (Biomarker, Beijing, China). The sequencing procedure followed the methods described by Duan et al. [24]. The pair-end reads derived from Kustro and MiA6RL^{Ku} were compared with the whole genome shotgun assembly sequences (IWGSC WGA v0.4) of common wheat *cv.* Chinese Spring (*Triticum aestivum* L.) (http://www.wheatgenome.org) using SOAP software (Beijing Genomics Institute (BGI), Beijing, China) [25].

Int. J. Mol. Sci. **2018**, *19*, 3933

The pair-end reads with high wheat homology were discarded. At last, after comparing specific pair-end reads of Kustro and MiA6RL^{Ku}, the 6RL^{Ku} minichromosome-specific pair-end reads were obtained. Primers were designed according to the 6RL^{Ku} minichromosome-specific pair-end reads using the software Primer 3 (version 4.0). The optimal melting temperature and size values of primers were set following the methods described by Duan et al. [24].

4.4. PCR Analysis and Sequence Cloning

The PCR amplifications were carried out according to the procedure described by Li et al. [6]. Agarose gel (2%) electrophoresis was used to detect the amplicons in 1 × TAE buffer. The products amplified by the 6RL^{Kumi119}-specific primers were recovered using Universal DNA Purification Kit (Tiangen Biotech Co., Ltd, Beijing, China) and cloned using pClone007 Vector Kit (TsingKe Biotech Co., Ltd, Beijing, China). Inserts were sequenced by TsingKe Biotech (Beijing) Co., Ltd. These sequences were deposited in the GenBank Database. Finally, these sequences were used for Nucleotide BLAST searches against the *S. cereale* L. Lo7 scaffolds database using the BLAST tool in GrainGenes (https://wheat.pw.usda.gov/cgi-bin/seqserve/blast_rye.cgi) and gene annotation was carried out according to the data sets related to the publication by Bauer et al. [12]. Additionally, RepeatMasker software (http://repeatmasker.org/) was used to identify whether these sequences amplified by the 6RL^{Kumi119}-specific primers are repetitive DNA.

4.5. Powdery Mildew Resistance Test

The resistance of line MiA6RL^{Ku}, the progeny of MiA6RL^{Ku}, 6D/6RL^{Ku} small segment translocation lines, and parental wheat MY11 to powdery mildew were evaluated. Plants were grown in Qionglai, Sichuan, China. The materials were naturally inoculated by locally occurring field derived powdery mildew, and infection types (IT) were scored according to the standard described by Fu et al. [5].

5. Conclusions

In conclusion, a new wheat-rye small segment translocation chromosome 6DL/6RL^{Kumi119} with powdery mildew resistance was produced. Sixteen 6RL^{Kumi119} specific markers were developed, and 11 *S. cereale* L. Lo7 scaffolds were located on the small rye segment 6RL^{Kumi119}. The small segment translocation chromosome 6DL/6RL^{Kumi119} makes the practical utilization of the powdery mildew resistance gene on 6RL^{Ku} more likely. The 6RL^{Kumi119}-specific markers and the *S. cereale* L. Lo7 scaffolds that were located on the 6RL^{Kumi119} segment may find application in future detailed studies of the structure and function of this small segment.

Author Contributions: S.F. and Z.T. created the materials, designed the study, analyzed the data, and wrote the manuscript. H.D. and Q.D. did the ND-FISH analysis, the PCR experiments, and cloned the sequences. S.T. did the BLAST searches. S.F. and Z.T. did the powdery mildew resistance test.

Funding: This research was funded by the National Natural Science Foundation of China (31770373 and 31470409).

Acknowledgments: We gratefully acknowledge Ian Dundas, The University of Adelaide, Australia and Zujun Yang, School of Life Science and Technology, University of Electronic Science and Technology of China for discussion and revision of this manuscript.

Conflicts of Interest: This research has no conflict of interest. The funding sponsors had no role in the experiment design, material creation, data analysis, manuscript writing, and decision to publish the results.

References

1. Friebe, B.; Heun, M.; Tuleen, N.; Zeller, F.J.; Gill, B.S. Cytogenetically monitored transfer of powdery mildew resistance from rye into wheat. *Crop Sci.* **1994**, *34*, 621–625. [CrossRef]
2. Zhao, Z.H.; Sun, H.G.; Song, W.; Lu, M.; Huang, J.; Wu, L.F.; Wang, X.M.; Li, H.J. Genetic analysis and detection of the gene MlLX99 on chromosome 2BL conferring resistance to powdery mildew in the wheat cultivar Liangxing 99. *Theor. Appl. Genet.* **2013**, *126*, 3081–3089. [CrossRef]

3. An, D.; Zheng, Q.; Luo, Q.; Ma, P.; Zhang, H.; Li, L.; Han, F.; Xu, H.; Zhang, X.; Zhou, Y. Molecular cytogenetic identification of a new wheat-rye 6R chromosome disomic addition line with powdery mildew resistance. *PLoS ONE* **2015**, *10*, e0134534. [CrossRef] [PubMed]

4. Wang, D.; Zhuang, L.F.; Sun, L.; Feng, Y.G.; Pei, Z.Y.; Qi, Z.J. Allocation of a powdery mildew resistance locus to the chromosome arm 6RL of *Secale cereale* L. cv. 'Jingzhouheimai'. *Euphytica* **2010**, *176*, 157–166. [CrossRef]

5. Fu, S.; Ren, Z.; Chen, X.; Yan, B.; Tan, F.; Fu, T.; Tang, Z. New wheat-rye 5DS-4RS.4RL and 4RS-5DS.5DL translocation lines with powdery mildew resistance. *J. Plant Res.* **2014**, *127*, 743–753. [CrossRef] [PubMed]

6. Li, M.; Tang, Z.X.; Qiu, L.; Wang, Y.Y.; Tang, S.Y.; Fu, S.L. Identification and physical mapping of new PCR-based markers specific for the long arm of rye (*Secale cereale* L.) chromosome 6. *J. Genet. Genomics* **2016**, *43*, 199–206. [CrossRef] [PubMed]

7. Fu, S.; Tang, Z.; Ren, Z.; Zhang, H. 2010 Transfer to wheat (*Triticum aestivum*) of small chromosome segments from rye (*Secale cereale*) carrying disease resistance genes. *J. Appl. Genet.* **2010**, *51*, 115–121. [CrossRef] [PubMed]

8. Lukaszewski, A.J. *Alien Introgression in Wheat*, 1st ed.; Molnár-Láng, M., Ceoloni, C., Doležel, J., Eds.; Springer: Cham, Switzerland, 2015; Chaprter 7; pp. 163–189.

9. Friebe, B.; Hatchett, J.H.; Gill, B.S.; Mukai, Y.; Sebeata, E.E. Transfer of Hessian fly resistance from rye to wheat via radiation-induced terminal and intercalary chromosomal translocations. *Theor. Appl. Genet.* **1991**, *83*, 33–40. [CrossRef]

10. Friebe, B.; Jiang, J.; Raupp, W.J.; McIntosh, R.A.; Gill, B.S. Characterization of wheat-alien translocations conferring resistance to diseases and pests: current status. *Euphytica* **1996**, *91*, 59–87. [CrossRef]

11. Mukai, Y.; Friebe, B.; Hatchett, J.H.; Yamamoto, M.; Gill, B.S. Molecular cytogenetic analysis of radiation-induced wheat-rye terminal and intercalary chromosomal translocations and the detection of rye chromatin specifying resistance to Hessian fly. *Chromosoma* **1993**, *102*, 88–95. [CrossRef]

12. Bauer, E.; Schmutzer, T.; Barilar, I.; Mascher, M.; Gundlach, H.; Martis, M.M.; Twardziok, S.O.; Hackauf, B.; Gordillo, A.; Wilde, P.; et al. Towards a whole-genome sequence for rye (*Secale cereale* L.). *Plant J.* **2017**, *89*, 853–869. [CrossRef] [PubMed]

13. Li, H.J.; Wang, X.M.; Song, F.J.; Wu, C.P.; Wu, X.F.; Zhang, N.; Zhou, Y.; Zhang, X.Y. Response to powdery mildew and detection of resistance genes in wheat cultivars from China. *Acta Agron. Sin.* **2011**, *37*, 943–954, (in Chinese with English abstract). [CrossRef]

14. Jiang, Z.; Wang, Q.L.; Wu, J.H.; Xue, W.B.; Zeng, Q.D.; Huang, L.L.; Kang, Z.S.; Han, D.J. Distribution of powdery mildew resistance gene *Pm21* in Chinese winter wheat cultivars and breeding lines based on gene-specific marker. *Scientia Agri. Sin.* **2014**, *47*, 2078–2087, (in Chinese with English abstract). [CrossRef]

15. Wu, Q.H.; Chen, Y.X.; Li, D.; Wang, Z.Z.; Zhang, Y.; Yuan, C.G.; Wang, X.C.; Zhao, H.; Cao, T.J.; Liu, Z.Y. Large scale detection of powdery mildew resistance genes in wheat via SNP and bulked segregate analysis. *Acta Agron. Sin.* **2018**, *44*, 1–14. [CrossRef]

16. Hao, M.; Luo, J.; Fan, C.; Yi, Y.; Zhang, L.; Yuan, Z.; Ning, S.; Zheng, Y.; Liu, D. Introgression of powdery mildew resistance gene *Pm56* on rye chromosome arm 6RS into wheat. *Front. Plant Sci.* **2018**, *9*, 1040. [CrossRef] [PubMed]

17. Dundas, I.S.; Frappell, D.E.; Crack, D.M.; Fisher, J.M. Deletion mapping of a nematode resistance gene on rye chromosome 6R in wheat. *Crop Sci.* **2001**, 1771–1778. [CrossRef]

18. Feuillet, C.; Langridge, P.; Waugh, R. Cereal breeding takes a walk on the wild side. *Trends Genet.* **2008**, *24*, 24–32. [CrossRef]

19. Tester, M.; Langridge, P. Breeding technologies to increase crop production in a changing world. *Science* **2010**, *327*, 818–822. [CrossRef]

20. Qiu, L.; Tang, Z.X.; Li, M.; Fu, S.L. Development of new PCR-based markers specific for chromosome arms of rye (*Secale cereal* L.). *Genome* **2016**, *59*, 159–165. [CrossRef]

21. Tang, Z.X.; Yang, Z.J.; Fu, S.L. Oligonucleotides replacing the roles of repetitive sequences pAs1, pSc119.2, pTa-535, pTa71, CCS1, and pAWRC.1 for FISH analysis. *J. Appl. Genet.* **2014**, *55*, 313–318. [CrossRef]

22. Fu, S.L.; Chen, L.; Wang, Y.Y.; Li, M.; Yang, Z.J.; Qiu, L.; Yan, B.J.; Ren, Z.L.; Tang, Z.X. Oligonucleotide probes for ND-FISH analysis to identify rye and wheat chromosomes. *Sci. Rep.* **2015**, *5*, 10552. [CrossRef] [PubMed]

23. Han, F.P.; Lamb, J.C.; Birchler, A. High frequency of centromere inactivation resulting in stable dicentric chromosomes of maize. *Proc. Natl. Acad. Sci. USA* **2006**, *103*, 3238–3243. [CrossRef] [PubMed]

24. Duan, Q.; Wang, Y.Y.; Qiu, L.; Ren, T.H.; Li, Z.; Fu, S.L.; Tang, Z.X. Physical location of new PCR-based markers and powdery mildew resistance gene(s) on rye (*Secale cereale* L.) chromosome 4 using 4R dissection lines. *Front. Plant Sci.* **2017**, *8*, 1716. [CrossRef] [PubMed]
25. Li, R.; Yu, C.; Li, Y.; Lam, T.W.; Yiu, S.M.; Kristiansen, K.; Wang, J. SOAP2: an improved ultrafast tool for short read alignment. *Bioinformatics* **2009**, *25*, 1966–1967. [CrossRef] [PubMed]

© 2018 by the authors. Licensee MDPI, Basel, Switzerland. This article is an open access article distributed under the terms and conditions of the Creative Commons Attribution (CC BY) license (http://creativecommons.org/licenses/by/4.0/).

International Journal of
Molecular Sciences

MDPI

Article

Comparative Proteomic Analysis of Wheat Carrying *Pm40* Response to *Blumeria graminis* f. sp. *tritici* Using Two-Dimensional Electrophoresis

Yinping Liang [1,2], Ye Xia [2], Xiaoli Chang [1,3], Guoshu Gong [1], Jizhi Yang [1], Yuting Hu [1], Madison Cahill [2], Liya Luo [1], Tao Li [1], Lu He [1] and Min Zhang [1,*]

[1] College of Agronomy & Key Laboratory for Major Crop Diseases, Sichuan Agricultural University, Chengdu 611130, China; liangyinping3@163.com (Y.L.); xl_changkit@126.com (X.C.); guoshugong@126.com (G.G.); yjzgxy@126.com (J.Y.); yutinghu116sicau@163.com (Y.H.); scmlsjc@163.com (L.L.); 13187874590@126.com (T.L.); hl415676745@163.com (L.H.)

[2] Department of Plant Pathology, The Ohio State University, Columbus, OH 43202, USA; xia.374@osu.edu (Y.X.); madisontaylor313@hotmail.com (M.C.)

[3] Institute of Ecological Agriculture, Sichuan Agricultural University, Chengdu 611130, China

* Correspondence: yalanmin@126.com; Tel.: +86-028-8629-3015

Received: 16 December 2018; Accepted: 16 February 2019; Published: 21 February 2019

Abstract: Wheat powdery mildew caused by *Blumeria graminis* f. sp. *tritici* (*Bgt*) is considered a major wheat leaf disease in the main wheat producing regions of the world. Although many resistant wheat cultivars to this disease have been developed, little is known about their resistance mechanisms. *Pm40* is a broad, effective resistance gene against powdery mildew in wheat line *L699*. The aim of this study was to investigate the resistance proteins after *Bgt* inoculation in wheat lines *L699*, *Neimai836*, and *Chuannong26*. *Neimai836* with *Pm21* was used as the resistant control, and *Chuannong26* without any effective *Pm* genes was the susceptible control. Proteins were extracted from wheat leaves sampled 2, 4, 8, 12, and 24 h after *Bgt* inoculation, separated by two-dimensional electrophoresis, and stained with Coomassie brilliant blue G-250. The results showed that different proteins were upregulated and downregulated in three wheat cultivars at different time points. For the wheat cultivar *L699*, a total of 62 proteins were upregulated and 71 proteins were downregulated after *Bgt* inoculation. Among these, 46 upregulated proteins were identified by mass spectrometry analysis using the NCBI nr database of *Triticum*. The identified proteins were predicted to be associated with the defense response, photosynthesis, signal transduction, carbohydrate metabolism, energy pathway, protein turnover, and cell structure functions. It is inferred that the proteins are not only involved in defense response, but also other physiological and cellular processes to confer wheat resistance against *Bgt*. Therefore, the resistance products potentially mediate the immune response and coordinate other physiological and cellular processes during the resistance response to *Bgt*. The lipoxygenase, glucan exohydrolase, glucose adenylyltransferasesmall, phosphoribulokinase, and phosphoglucomutase are first reported to be involved in the interactions of wheat-*Bgt* at early stage. The further study of these proteins will deepen our understanding of their detailed functions and potentially develop more efficient disease control strategies.

Keywords: *Blumeria graminis* f. sp. *tritici*; protein two-dimensional electrophoresis; mass spectrometry; *Pm40*

---

1. Introduction

Wheat powdery mildew caused by the obligate fungus *Blumeria graminis* f. sp. *tritici* (*Bgt*) is a major wheat leaf disease in the main wheat producing regions of world, leading to significant yield loss each year [1]. Agricultural and chemical methods are widely used to combat the disease. Wheat powdery mildew is an airborne disease and the chemical control methods for *Bgt* seriously pollute environments. Therefore, planting resistant cultivars is the most economical, most effective, and safest method to control wheat powdery mildew [2]. To date, approximately 90 formally designated powdery mildew resistance genes (*Pm* genes) are catalogued at 58 loci (*Pm1–Pm62*, *Pm18* = *Pm1c*, *Pm22* = *Pm1e*, *Pm23* = *Pm4c*, *Pm31* = *Pm21*) with the loci of *Pm1*, *Pm3*, *Pm4*, *Pm5*, and *Pm24* having 5, 17, 4, 5, and 2 alleles, respectively [3–13]. However, resistance genes often become ineffective due to the enrichment and variation of virulent races, particularly when a single resistance gene is used in large areas for long periods of time. Therefore, it is very important to identify effective resistance genes and develop multiple resistance cultivars in wheat breeding [14].

The resistant mechanisms of wheat cultivars against *Bgt* are not well-known. Bread wheat (*Triticum aestivum* L.) is a hexaploid ($2n = 42$; AABBDD) with a 17-gigabase genome that contains 124 201 genes [15]. Due to this complexity, cloning wheat genes by the standard map-based cloning strategy remains challenging. Although many powdery mildew resistance genes were identified and mapped in wheat, to date, only five genes, *Pm2*, *Pm3*, *Pm8*, *Pm21*, and *Pm60* have been cloned [9,16–20]. The resistance gene *Pm40* was transferred from *Elytrigia intermedium* into wheat line *GRY19* and mapped on chromosome arm 7BS [21]. *Pm40* is flanked by *Xwmc335* and *BF291338* at genetic distances of 0.58 cM and 0.26 cM, respectively, in deletion bin C-7BS-1-0.27 [22]. Wheat line *L699*, which is the high generation of wheat line *GRY19*, carries the resistance gene *Pm40* and confers resistance to all available isolates of *Bgt* in China [23].

Proteins are not only the final executant of life functions but also the key to understanding physiological, pathological, and pharmacological functions of plants [24]. Therefore, it is difficult to thoroughly explain the powdery mildew resistance mechanism using genomic and transcriptomic methods. Proteomic approaches have been extensively applied in plant pathology research [25,26]. However, only a few studies examined the changes of plant proteome in response to *Bgt*. Wheat cultivars Bainong/W2132 (*Pm21*), JD8/JD8-*Pm30*, N8038 (*PmG25*), N9134 (*PmAS846*), and Xinong979 (without effective *Pm* genes) were used to analyze the effect of *Bgt* on wheat protein expression. These studies showed that most of the upregulated proteins were involved in stress responses and primary metabolic pathways [24,27–30]. However, there is no such study investigating the differences of protein expressions in the period before *Bgt* haustoria formation, which is very critical for us to better understand the interactions of this pathogen with different wheat cultivars at early stage. To understand the molecular recognition of wheat-*Bgt* during the contact period and penetration period, we identified a set of proteins in wheat inoculated with *Bgt* using two-dimensional electrophoresis (2-DE). The possible roles of the identified proteins in the defense response at early interaction stage were discussed according to their functional implications. This study deepens and extends our knowledge on the interactions of wheat with *Bgt* and allows us to further understand the wheat immune systems against *Bgt*. All these will facilitate the development of more efficient strategies to control this devastating pathogen for enhancing wheat production, which can also potentially provide insights for the control of different plant diseases caused by diverse powdery mildew pathogens.

2. Results

2.1. Phenotypic Differences of Leaves Affected by Bgt

The bioassay revealed differences in resistance to *Bgt* among *L699*, *Chuannong26* and *Neimai836* (Figure 1). The susceptible cultivar *Chuannong26* was covered by a high number of sori and had the white powdery appearance due to the abundant conidia and conidiophores production on the leaf surface after 6 days of *Bgt* infection, with the infection type (IT) = 9 (Figure 1a). Meanwhile, the resistant wheat lines *L699* and *Neimai836* were observed to be healthy without any epidermal cell necrosis, chlorotic patches, and powdery appearance on the leaf surface, with the IT = 0 (*L699*: Figure 1b, *Neimai836*: Figure 1c) [23].

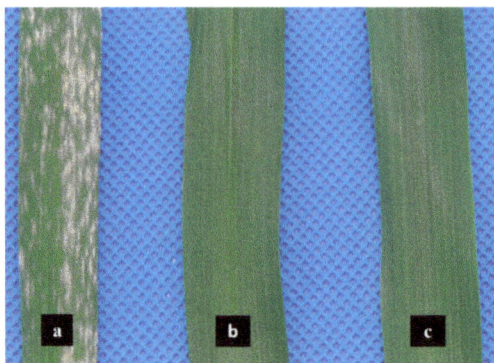

Figure 1. Different responses of wheat leaves to *Blumeria graminis* f. sp. *tritici (Bgt)* infection after six days. (**a**) Susceptible wheat cultivar *Chuannong26*. (**b**) Resistant wheat cultivar *L699*. (**c**) Resistant wheat cultivar *Neimai836*.

2.2. Estimation of Wheat-Bgt Interactions

To examine the development of *Bgt* and immune responses of wheat at 2, 4, 8, 12, and 24 h post-inoculation (hpi), the cytological observations of wheat samples were carried out. *Bgt* conidia successively formed primary germ tubes, appressorium germ tubes, appressoria, penetration pegs, and haustoria at 2, 4, 8, 12, and 24 hpi in susceptible wheat cultivar *Chuannong26*. However, in resistant wheat cultivars *L699* and *Neimai836*, only a small number of conidia successfully penetrated the epidermal cells at 24 hpi, and the hypersensitive reaction (HR) and formation of papilla (PA) effectively suppressed the development of haustoria and hyphae [31]. In addition, the appressoria of some conidia sprouted another lobe and stopped growing because of the lack of nutrition at 24 hpi (Figure 2).

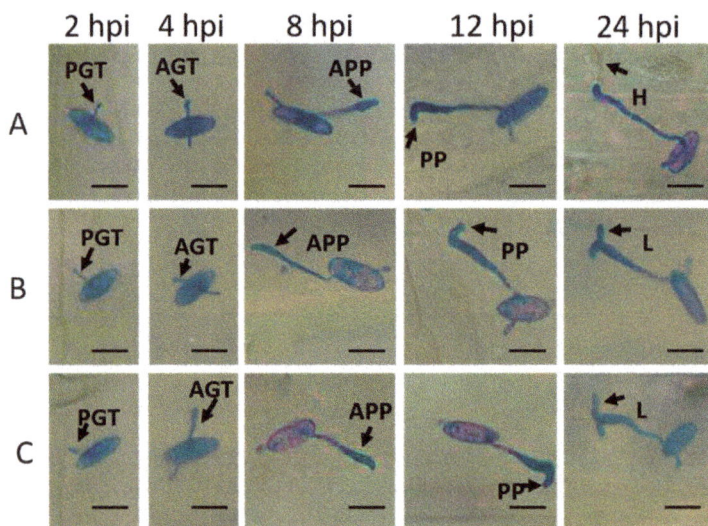

Figure 2. Microscopic observations of wheat-*Bgt* interactions on the leaf surface. The development of *Bgt* at 2, 4, 8, 12, and 24 hpi in wheat cultivars *Chuannong26* (**A**), *L699* (**B**), and *Neimai836* (**C**). PGT: primary germ tube, AGT: appressorium germ tube, APP: appressorium, PP: penetration peg, H: haustorium and L: lobe. Scale bar: 20 μm.

2.3. Detection of Differential Proteins by 2-DE

Approximately 500 protein spots were detected in all gels in this study. Using a twofold change cutoff, we found wheat cultivars *L699*, *Neimai836*, and *Chuannong26* all had upregulated and downregulated proteins affected by *Bgt*. The numbers of upregulated proteins were seven (spot 01, 02, 24, 27, 45, 46, 50), five (spot 12, 17, 36, 37, 44), 18 (spot 11, 13, 29, 30, 32–35, 38–46, 55), 26 (spot 02–06, 08–12, 14–16, 18–28, 48, 56), and 18 (spot 07, 31, 47–62). The numbers of downregulated proteins were four (spot 68–71), 12 (spot 45, 48, 58–67), 12 (spot 09, 12, 14, 49–57), 10 (spot 39–48), and 38 (spot 01–38) in wheat cultivar *L699* at 2, 4, 8, 12, 24 hpi, respectively (Figures 3 and 4). The numbers of upregulated proteins were 10 (spot 14, 19, 20, 22, 58–63), 13 (spot 32, 46–57), 15 (spot 31–45), 23 (spot 1–23), and nine (spot 19, 21, 24–30). The numbers of downregulated proteins were 11 (spot 01, 36, 43, 52, 69–75), 11 (spot 23, 28, 31, 50, 62–68), 11 (spot 50, 52–61), 35 (spot 1–35), and 16 (spot 36–51) in wheat cultivar *Neimai836* at 2, 4, 8, 12, 24 hpi, respectively (Figures 3 and 5). The numbers of upregulated proteins were three (spot 71, 109–110), nine (spot 33, 101–108), seven (spot 33, 34, 66, 97–100), 25 (spot 23, 34, 59, 75–96), and 74 (spot 1–74). The numbers of downregulated proteins were four (spot 42–45), 13 (spot 04, 25, 31–41), nine (spot 16, 19, 24–30), 7 (spot 01, 04, 19–23), and 18 (spot 1–18) in wheat cultivar *Chuannong26* at 2, 4, 8, 12, 24 hpi, respectively (Figures 3 and 6).

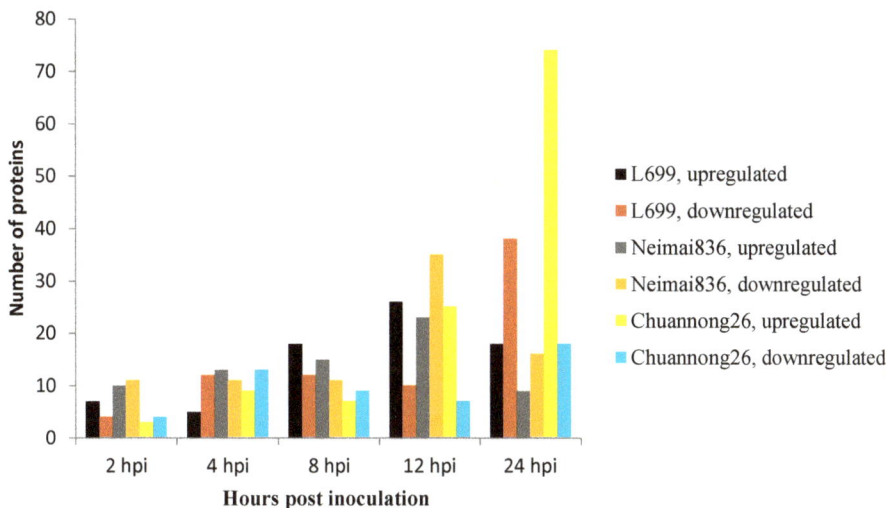

Figure 3. Number of proteins differentially expressed after powdery mildew infection at different time points.

2.4. Protein Identification

Sixty-two upregulated proteins in *L699* at five different inoculation time points were eluted from representative 2-DE gels for identification, and 46 were successfully identified. Bioinformatics analysis of the identified proteins revealed that these proteins were putatively involved in diverse biological processes including stress and disease resistance, photosynthesis, signal transduction, carbohydrate metabolism, energy pathway, gene expression, protein turnover, and cell structure (Table 1). Five proteins (approximately 16.5% of the total differentially expressed proteins (DEPs)) were unnamed or hypothetical proteins. The largest category of these upregulated proteins was protein turnover (28%, thirteen), followed by carbohydrate metabolism (22%, ten), stress and disease resistance (13%, six), photosynthesis (13%, six), energy pathway (6.5%, three), signal transduction (2%, one), gene expression (2%, one), and cell structure (2%, one).

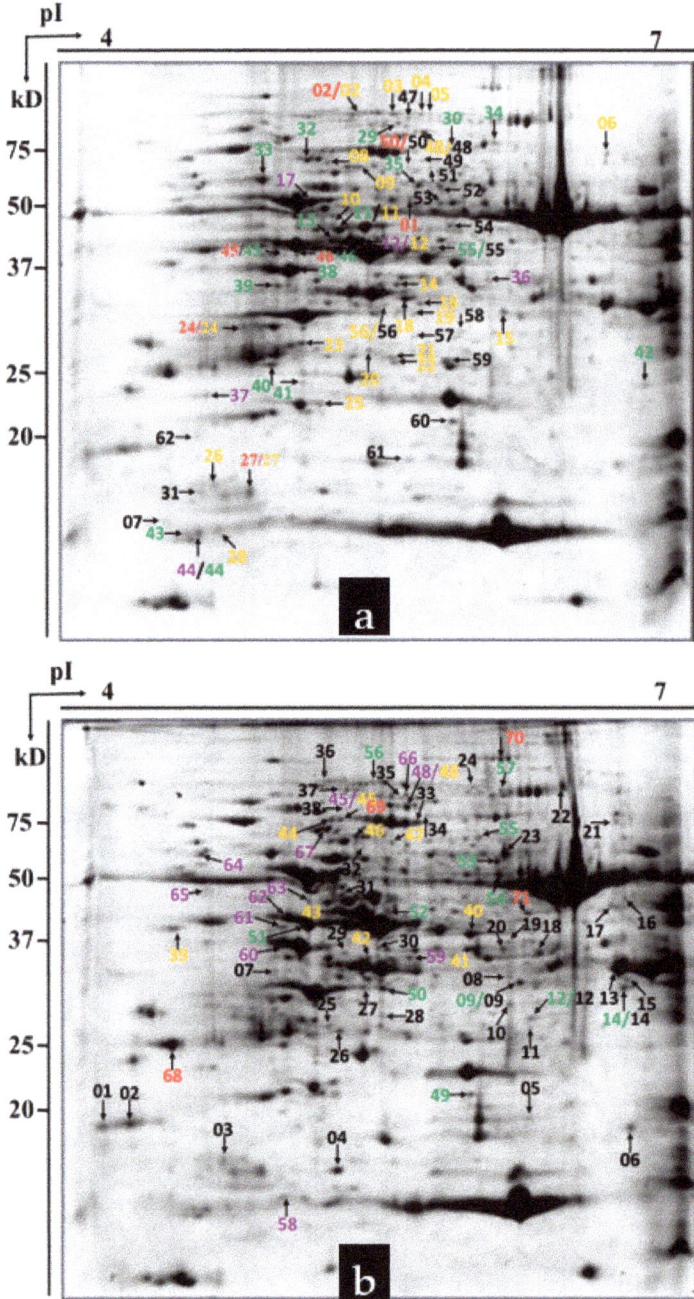

Figure 4. Differences in protein expression between resistant wheat cultivar *L699* wheat with and without powdery mildew infection. (**a**) Upregulated proteins are labeled in the representative 2-DE gel of *Bgt*-inoculated *L699* wheat at 24 hpi. (**b**) Downregulated proteins are labeled in the representative 2-DE gel of mock-inoculated *L699* wheat at 24 hpi. Red, 2 hpi; purple, 4 hpi; green, 8 hpi; yellow, 12 hpi; black, 24 hpi.

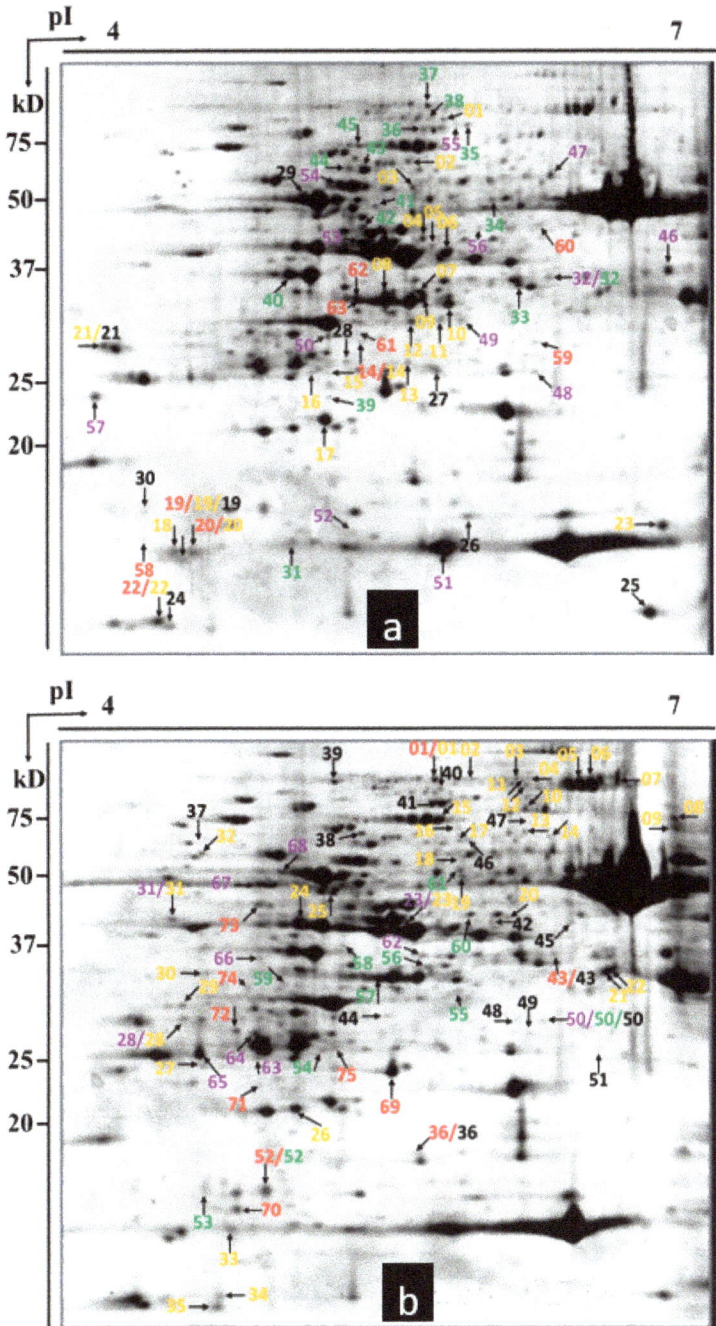

Figure 5. Differences in protein expression between resistant wheat cultivar *Neimai836* wheat with and without powdery mildew infection. (**a**) Upregulated proteins are labeled in the representative 2-DE gel of *Bgt*-inoculated *Neimai836* wheat at 12 hpi. (**b**) Downregulated proteins are labeled in the representative 2-DE gel of mock-inoculated *Neimai836* wheat at 12 hpi. Red, 2 hpi; purple, 4 hpi; green, 8 hpi; yellow, 12 hpi; black, 24 hpi.

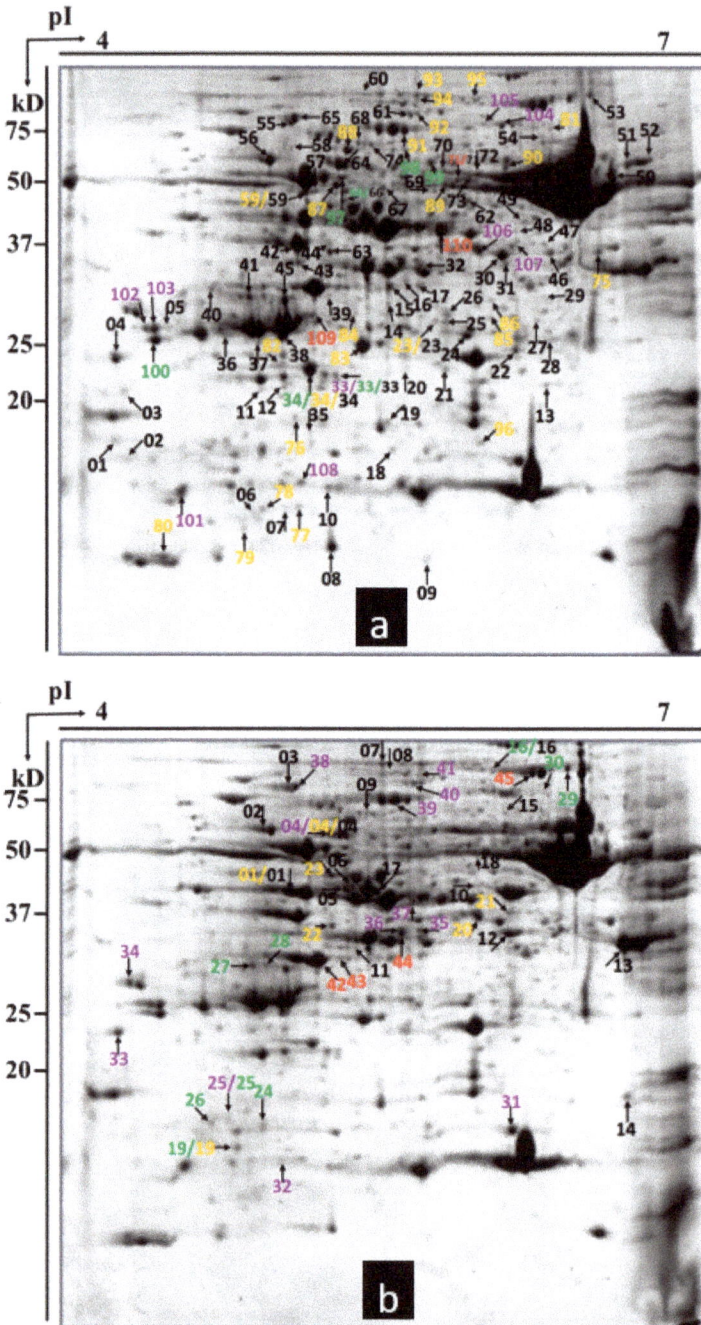

Figure 6. Differences in protein expression between susceptible wheat cultivar *Chuannong26* wheat with and without powdery mildew infection. (**a**) Upregulated proteins are labeled in the representative 2-DE gel of *Bgt*-inoculated *Chuannong26* wheat at 24 hpi. (**b**) Downregulated proteins are labeled in the representative 2-DE gel of mock-inoculated *Chuangnong26* wheat at 24 hpi. Red, 2 hpi; purple, 4 hpi; green, 8 hpi; yellow, 12 hpi; black, 24 hpi.

Table 1. Identification of differentially upregulated proteins in wheat resistance cultivar *L699* by MALDI-TOF-MS.

Spot	Protein Name	Accession	Matched Peptides	MW/PI	Score	Time (h)
	Proteins involved in disease defense response					
3	Lipoxygenase 2.1, chloroplastic	gi|473948122	21	105625.33/5.70	185	12
4	Lipoxygenase 2.1, chloroplastic	gi|473948122	41	105625.33/5.70	591	12
5	Lipoxygenase 2.1, chloroplastic	gi|473948122	17	105625.33/5.70	196	12
29	Lipoxygenase 1	gi|474399175	16	96333.65/5.91	299	8
32	Heat shock cognate 70 kDa protein 1	gi|474012573	37	71123.58/5.06	641	8
61	Germin-like protein 8-14	gi|473963025	4	21939.25/5.36	174	24
	Photosynthesis-related proteins					
11	Ribulose-1,5-bisphosphate carboxylase activase, partial	gi|37783283	10	22336.08/4.98	309	8, 12
28	Ribulose bisphosphate carboxylase small chain, chloroplastic	gi|473882355	14	18526.35/8.65	208	12
33	RuBisCO large subunit-binding protein subunit alpha, chloroplastic	gi|474113969	34	65380.60/5.17	864	8
36	Ribulose-1,5-bisphosphate carboxylase activase, partial	gi|37783283	10	22336.08/4.98	330	4
37	Photosystem II cytochrome b559 alpha subunit (chloroplast)	gi|699976019	6	9444.60/4.64	196	4
44	Ribulose bisphosphate carboxylase small chain PWS4.3, chloroplastic	gi|1132087	2	19417.36/8.99	92	4, 8
	Proteins involved in Signal transduction					
24	14-3-3 protein	gi|431822520	16	29264.88/4.83	434	12
	Carbohydrate metabolism-related proteins					
6	Beta-D-glucan exohydrolase	gi|20259685	14	67301.15/6.87	74	12
10	Glucose-1-phosphate adenylyltransferasesmall subunit, chloroplastic/amyloplastic	gi|474108293	23	64723.14/7.91	266	12
13	Phosphoglycerate kinase	gi|3293043	16	49839.53/6.57	580	8, 12
14	Glycerophosphodiester phosphodiesterase GDE1	gi|473847956	13	52899.68/5.69	43	12
18	Fructose-bisphosphate aldolase, chloroplastic	gi|473848356	15	42002.99/5.94	358	12
38	Phosphoribulokinase	gi|5924030	22	45141.39/5.72	587	8
50	Phosphoglucomutase, cytoplasmic	gi|473763033	18	63499.68/5.14	302	2, 24
51	Phosphoglucomutase, partial	gi|18076790	15	62789.15/5.66	218	24
54	6-phosphogluconate dehydrogenase, decarboxylating	gi|474379872	23	81169.95/8.56	608	24
55	Cytosolic 3-phosphoglycerate kinase, partial	gi|28172911	16	31334.35/4.98	291	8, 24
	Proteins involved in energy pathway					
9	Vacuolar proton-ATPase subunit A	gi|90025017	37	68454.90/5.23	583	12
17	ATP synthase CF1 beta subunit (chloroplast)	gi|66766997	33	53857.48/5.06	1200	4
39	Ferredoxin-NADP(H) oxidoreductase	gi|20302473	10	40232.03/6.92	120	8
	Proteins involved in gene expression and DNA remodeling					
15	Guanine nucleotide-binding protein subunit beta-like	gi|473957859	6	27150.69/6.29	211	12

Table 1. *Cont.*

Spot	Protein Name	Accession	Matched Peptides	MW/PI	Score	Time (h)
	Proteins involved in protein turnover					
8	ATP-dependent zinc metalloprotease FTSH 1, chloroplastic	gi|474350516	29	54477.49/5.58	673	12
27	50S Ribosomal protein L12-2, chloroplastic	gi|475532245	10	21837.90/5.35	452	12
30	Tyrosine phosphorylation protein A	gi|548319365	25	74252.07/6.61	434	8
34	5-methyltetrahydropteroyltriglutamate-Homocysteine methyltransferase	gi|473993302	14	84552.49/5.74	423	8
35	5, 10-methylene-tetrahydrofolate reductase	gi|115589742	12	64875.07/5.86	83	8
40	20 kDa chaperonin, chloroplastic	gi|474407512	10	29710.03/6.76	154	8
47	Putative alanyl-tRNA synthetase, chloroplastic	gi|474142555	12	111648.20/5.62	198	24
48	ATP-dependent Clp protease ATP-binding subunit ClpA-like protein CD4B, chloroplastic	gi|474241774	33	82735.21/5.16	513	12, 24
49	Lysyl-tRNA synthetase	gi|474147702	8	132545.46/6.28	92	24
52	Putative mitochondrial-processing peptidase subunit beta	gi|474142281	30	43290.34/5.41	486	24
53	Adenosylhomocysteinase	gi|474154141	8	45700.84/6.48	56	24
56	Cysteine synthase	gi|474315986	13	35583.27/5.82	216	12, 24
60	Ribosome-recycling factor, chloroplastic	gi|474043078	15	24770.60/8.92	504	24
	Cell structure-related proteins					
12	Actin-3	gi|474259583	18	44367.62/5.26	376	8, 12
	Proteins of unknow function					
2	Hypothetical protein TRIUR3_05354	gi|473755342	27	104676.25/5.87	368	2, 12
26	Unnamed protein product	gi|669029445	4	18152.74/5.60	244	12
58	Hypothetical protein TRIUR3_21449	gi|474384687	14	32942.26/9.31	129	24
59	Unnamed protein product	gi|669027704	13	26818.76/5.57	335	24
63	Unnamed protein product	gi|669029445	5	18152.74/5.60	269	24

2.5. Validation of Upregulated Proteins by qRT-PCR

To confirm the changes in protein abundance, qRT-PCR was used to analyze the mRNA expression levels of protein-coding genes after inoculation with *Bgt* in wheat cultivars *L699*, *Neimai836*, and *Chuannong26*. Six upregulated proteins were randomly selected and primers for the mRNAs of these proteins were specifically designed (shown in Table 2). The mRNA levels for all the six proteins were significantly increased in at least one sampling time in wheat *L699*. These changes reflected the increases in proteins and the differentially expressed proteins identified by two-dimensional electrophoresis were validated (Figure 7). However, the mRNAs of four proteins, i.e. fructose-bisphosphate aldolase, phosphoglycerate kinase, 5-methyltetrahydropteroyltriglutamate-homocysteine methyltransferase, and the Germin-like protein (GLP) 8-14, were upregulated earlier than proteins in wheat *L699*. The protein level of fructose-bisphosphate aldolase increased at 12 hpi, but the mRNA level did not exhibit a significant increase at any sampling time point in wheat *Neimai836*.

Table 2. List of primers used for qRT-PCR amplification.

Spot	Protein Name	Accession No.	Primer Sequence 5′-3′
Reference gene	*18S rRNA*	AY049040	Sense: 5′-GTGACGGGTGACGGAGAATT-3′ Antisense: 5′-GACACTAATGCGCCCGGTAT-3′
9	Vacuolar proton-ATPase	ABD85016	Sense: 5′-TATGAACGTGCTGGGAAGGT-3′ Antisense: 5′-GGGTTGCAGAGGTAACAGGA-3′
18	Fructose-bisphosphate aldolase	EMS47455	Sense: 5′-TCTTGTCTGGTGGTCAGTCG-3′ Antisense: 5′-CGTCTTGAGGCAGGTGTTCT-3′
13	Phosphoglycerate kinase	CAA51931	Sense: 5′-AATGGTGCTGTTTTGCTCCT-3′ Antisense: 5′-TGTTCCGAATGCATCGTTTA-3′
36	Ribulose-1,5-bisphosphate carboxylase activase	AAP72270	Sense: 5′-ACGGACCAGTGACCTTTGAG-3′ Antisense: 5′-ACCAGTCTTCATCGCATCCT-3′
34	5-methyltetrahydropteroyltriglutamate-homocysteine methyltransferase	EMS51950	Sense: 5′-TGTGTTCTGGTCCAAGATGG-3′ Antisense: 5′-CTCAAACCTCGGTTGGTCAT-3′
61	Germin-like protein 8-14	EMS51159	Sense: 5′-TGCAGATCACCGACTACGC-3′ Antisense: 5′-CACGGACTTGAGCTTCTTGAC-3′

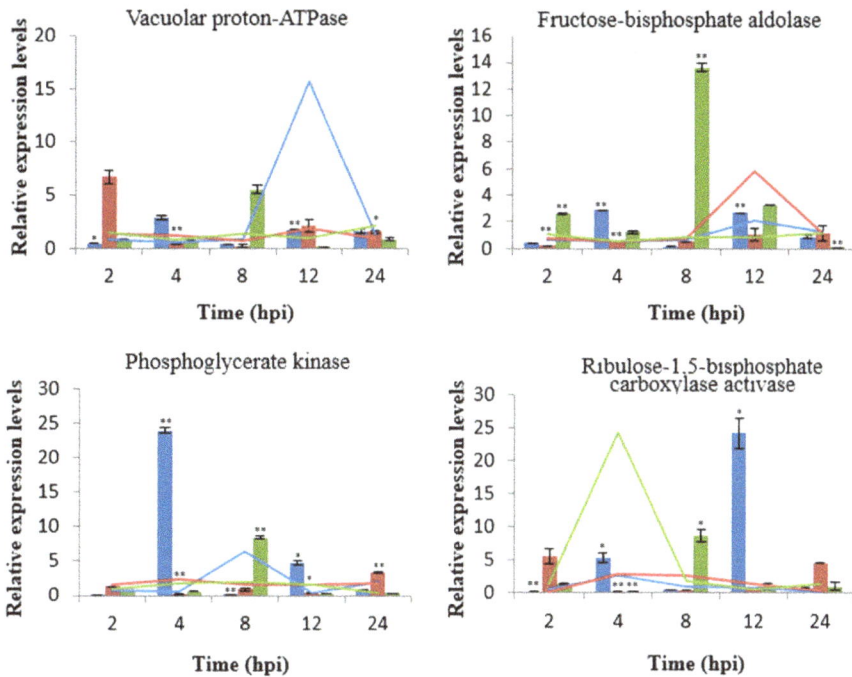

Figure 7. *Cont.*

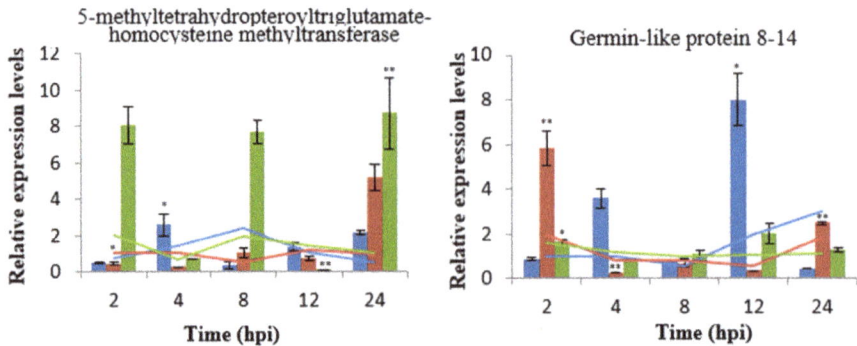

Figure 7. Quantification of six gene transcripts and protein levels at different time points post inoculation with *Bgt* in wheat. The bar graph shows the fold changes of the mRNA expression levels in inoculation vs control samples at five time points. The blue, red, and green columns are representatives of wheat *L699*, *Neimai836*, and *Chuannong26*, respectively. The lines show the fold changes of the protein expression levels in inoculation vs control samples at five time points. The blue, red, and green lines are representatives of wheat *L699*, *Neimai836*, and *Chuannong26*, respectively. The mRNA expression levels were quantified by qRT-PCR normalized against *18S rRNA*. Asterisks indicate statistically significant differences (*, $p < 0.05$; **, $p < 0.01$) of mRNA expression levels between the inoculation and control samples.

3. Discussion

Plants employ two levels of immunity to encounter pathogen invasion: Pathogen-associated molecular pattern (PAMP)-triggered immunity (PTI) and effector-triggered immunity (ETI). In the early phase of defense, PAPMs are recognized as 'non-self' molecules by the host plants. This induces downstream defense signaling, such as the generation of reactive oxygen species (ROS) and the transcription of genes encoding pathogenesis-related proteins (PRs). The pathogens release effector proteins to oppose PTI, and then the plant resistance proteins recognize the effector, which stimulates the plant's ETI, leading to the hypersensitive response (HR) and activating other plant defense pathways [32–34].

However, not only the specific signaling mediated by resistance genes, but also the other basic cellular processes, are involved in the effective defense to support the plant innate immune system [35]. In our results, the differentially expressed proteins, including both resistance proteins against *Bgt* and other proteins related to the direct and indirect defensive processes. The potential roles of these proteins in the defense response are discussed below.

3.1. Stress- and Defense-Related Proteins

Plants experience a variety of biotic and abiotic stresses during the growth and development periods. Studies on the plant stress response found many stress response related proteins. For example, the Germin-like proteins are important stress-related proteins.

Protein spot 61, with an increasing expression level 24 h after *Bgt* infection, was identified as GLPs. GLPs as extracellular glycoproteins are important components of the plant PRs [36]. Recently, GLPs were reported to be involved in the stress responses of Arabidopsis, pepper, barely, and rice [37–40]. GLPs can remove excess ROS generated by plants in the form of enzymes, receptors, or structural proteins in various physiological and biochemical processes. The expression of GLPs increased significantly and potentially catalyzing the production of H_2O_2, in plants infected by fungi, bacteria, viruses, or other pathogens [41–43]. H_2O_2 can selectively participate in the signaling cascade pathway, which can stimulate plant self-defense reactions. In addition, H_2O_2 is able to use the cellulose crosslinking action to strengthen the structure of plant cell walls, which is very important in plant defense against oxidative stress. GLPs play an important role in wheat *L699* resistance to powdery mildew. This result is consistent with the previous study [43].

Heat shock proteins, as chaperones during the stress response, are very important for the correct folding of newly synthesized proteins [44]. Heat shock proteins were first discovered in Drosophila and were a class of proteins expressed by organisms under high temperature stimulation [45,46]. Recently, heat shock proteins were found to have very important roles in the innate immune response and are indispensable for the function of other defense-related proteins [47,48]. Mandal found that heat shock protein expression was significantly increased in wheat *N0308* 72 h after *Bgt* infection [29]. Protein spot 32 was identified as heat shock proteins with an expression level that increased after 8 h with *Bgt* inoculation. Heat shock proteins are closely related to wheat resistance of powdery mildew, as reported previously [29].

Plant lipoxygenases are members of a class of nonheme iron-containing dioxygenases that catalyze the addition of molecular oxygen to fatty acids containing a cis, cis-1,4-pentadiene system, which produces an unsaturated fatty acid hydroperoxide [49]. Currently, an increasing number of studies show that there are many similarities between the plant defense mechanisms and the animal defense mechanisms under adverse conditions [50]. Lipoxygenases in animals and plants play an important role in withstanding adverse environments. Protein spots 3, 4, 5, and 29, which were identified as lipoxygenase, were upregulated 8 h and 12 h after *Bgt* inoculation compared to non-inoculated wheat. In the lipoxygenase pathway, the polyunsaturated fatty acids are catalyzed by lipoxygenases to generate hydrogen peroxide and subsequently form compounds with a specific mass of physiological functions by the catalytic reaction of other enzymes, such as jasmonic acid and guaiac acid, which induces the synthesis of resistance proteins against stresses [51,52].

3.2. Proteins Related to Photosynthesis

Plant defense reactions are closely related to photosynthesis. It is generally believed that plant photosynthesis-related protein biosynthesis is reduced and resources are allocated to the defense response when plants are infected by a pathogen. The plant defense responses to pathogens is known as the "hidden costs" defense [53]. Plants affected by pathogens active the HR response, which is considered as another reason for weakening the plant photosynthesis after the original infection. However, protein spots 11, 28, 33, 36, 44 comprise ribulose carboxylase, which is an indicator of photosynthesis. These spots were upregulated 4 h, 8 h, and 12 h after powdery mildew infection, which could indicate that photosynthesis is increased. It was reported that photosynthesis is enhanced in early plant pathogen infections and weakened on later stage during the infection [54].

3.3. Proteins Involved in Carbohydrate Metabolism and Energy Pathways

The expression of several proteins involved in glucose metabolism, including β-D-glucose hydrolase (spot 6), phosphoglycerate kinase (spot 13, 55), glycerol phosphodiester enzyme (spot 14), diphosphate aldolase (spot 18), ribulose kinase (spot 38), glucose phosphate mutase (50, 51), and six-glucose phosphate decarboxylase (spot 54), were increased in response to wheat powdery mildew in wheat *L699*. Previous studies have shown that hexose can provide extra energy and serve as a signal for activating resistant response. For instance, in response to barley powdery mildew infection, the expression of hexose metabolizing enzymes significantly increased [55].

3.4. Proteins Involved in Gene Expression and Protein Turnover

Each step in the flow of genetic information is very strict, so the error rate of protein synthesis in this process is very low. However, protein synthesis has a certain error rate that is the net result of several processes. Aminoacyl-tRNA synthetases and ribosomes play important roles in protein synthesis [56,57]. Studies have shown that aminoacy-tRNA synthetase is not only involved in protein synthesis, but also participates in other activities, including the regulation of transcription and translation, RNA splicing, signal transduction, and immune response [58]. Current research is focused on the relationship between the function and structure of new amino acid-tRNA synthetases, especially the aminoacyl-tRNA synthetase, which is related to diseases. After powdery mildew infection in wheat *L699*, the expression levels of alanyl-tRNA synthetase (spot47), lysyl-tRNA synthetase (spot49), and ribosomal protein (spot27) increased, suggesting that these proteins may be important in the wheat anti-powdery mildew responses.

3.5. Proteins Associated with Cell Organization

In powdery mildew-infected wheat *L699*, actin (spot 12) was upregulated. The actin cytoskeleton is an essential dynamic component for cells and is highly conserved in eukaryotic cells. The cytoskeleton is closely linked with the membrane and is involved in various cellular processes, including defense signaling based on actin cytoskeletal structures after pathogen infection [59].

3.6. The Correlation of mRNA and Protein Expression

The analysis of six proteins and the expression of the protein-coding regions of their genes showed that the proteins and mRNA levels had a certain uniformity in our study. However, there exists post-transcriptional regulation after translational regulation in wheat, which may lead to differences between protein expression levels and mRNA levels.

3.7. The Novel Proteins Potentially Involed in the Response of Wheat Against Bgt

Some of the identified proteins, such as the lipoxygenase, glucan exohydrolase, glucose adenylyltransferasesmall, phosphoribulokinase, and phosphoglucomutase, are first reported during the interaction of wheat-*Bgt* in this study. These proteins are potentially very critical for the wheat-*Bgt* interaction at early stage. For future research, the defense functions of these novel proteins deserve further investigation by using the integrative approaches, such as the comparative metabolomics, gene overexpression, and silencing methods. The related study will lead to deeper understanding of the detailed functions of these important proteins and more efficient disease control strategies.

4. Materials and Methods

4.1. Plant Materials and Inoculation

L699, *Neimai836*, and *Chuannong26* were used in this study. *L699* carries the resistance gene *Pm40* and shows resistance to most powdery mildew isolates in China. *Neimai836* carries the resistance gene *Pm21*, but not *Pm40*, and also shows resistance to most powdery mildew isolates in China. Nevertheless, *Chuannong26* is highly susceptible to powdery mildew without any effective resistance gene. Plants were cultivated in 30 cm pots in a growth chamber at 18 °C under a 12 h/12 h dark photoperiod. These pots were divided into the *Bgt*-inoculated group and the mock-inoculated group with nonopaque and breathable hoods. Seedlings of the *Bgt*-inoculated group were artificially inoculated by dusting with *Bgt* conidia from sporulating seedlings of *Chuannong26* at two to three leaf stages. Leaf samples were harvested at 2, 4, 8, 12, and 24 hpi with liquid nitrogen and immediately stored at −80 °C. Samples were collected from three biological replicates at each time point and every sample protein was run on three gels. For the analysis, one best gel was selected from three gels.

4.2. Cytological Observation of the Interaction between Wheat and Bgt

Leaves of the *Bgt*-inoculated and mock-inoculated wheat were sampled at 2, 4, 8, 12, and 24 hpi and cut into 2–3 cm leaf fragments. The leaf fragments collected at 2, 4, 8, and 12 hpi were destained using isopropanol fumigation. The leaf fragments collected at 24 hpi were destained with AA solution (ethanol:glacial acetic acid = 1:1, v/v). Then the sample were stained with Coomassie blue staining solution (0.15% trichloroacetic acid aqueous solution:0.6% Coomassie brilliant blue R-250 methanol solution = 1:1, v/v) for 4 h. After rinsing with distilled water, the leaves were saved in a mix solution (glacial acetic acid:glycerol:distilled water = 1:4:15, $v/v/v$). The infection structures including germ tubes, appressoria and haustoria of *Bgt* were observed under electron microscope (40×, Nikon Eclipse 80i, Nikon Corporation, Tokyo, Japan).

4.3. Protein Extraction

Leaf tissue (1 g) was ground in a prechilled mortar with liquid nitrogen. Then, the powder was transferred to a 1.5 mL centrifuge tube with the addition of 1 mL acetone containing 10% trichloroacetic acid (TCA) and 0.07% β-mercaptoethanol. The samples were vortexed and chilled for 1 h at −20 °C. Then the homogenate was centrifuged at 13,000× *g* for 30 min at 4 °C. After the supernatant was gently decanted, the pellet was washed four times with chilled acetone containing 0.07% β-mercaptoethanol, then dried until all the acetone was removed by a vacuum drying instrument. The resulting powder was dissolved in 1 mL of IEF buffer (7 M urea, 2 M thiourea, 4% CHAPS, 20 mM DTT, 0.001% bromophenol blue, and 0.5% ampholyte (pH 3–10)). After centrifugation at 13,000× *g* for 20 min twice, the leaf proteins were obtained from the supernatant, and their concentration was determined using a Bradford dye binding assay [60].

4.4. Two-Dimensional Electrophoresis, Protein Visualization, and Image Analysis

The protein mixture was loaded onto an IPG strip (17 cm, pH 4-7, linear gradient (Bio-Rad, California, CA, USA)) by active rehydration at 50 V for 14 h (20 °C) on a Protein Isoelectric Focusing (IEF) Cell (Bio-Rad). The following conditions were used for the IEF: 20 °C, 50 µA/strip, 250 V for 1 h, 1000 V for 1 h, 10,000 V for 5 h, and 10,000 V, with a total of 60,000 vhs. The focused IPG strips were equilibrated in buffer containing 5 mL 6 M urea, 2% SDS, 20% glycerol, 375 mM Tris-HCl (pH 8.8), and 200 mM DTT for 15 min, then re-equilibrated in a similar buffer whose 200 mM DTT was replaced by 250 mM iodoacetamide for 15 min. Proteins were separated on the second dimension on vertical 12% sodium dodecyl sulfate-polyacrylamide (SDS-PAGE) gel in a Protean II XI Cell (Bio-Rad) at 25 mA/gel. Proteins in the gels were stained by the "blue silver" protocol as described by Candiano et al. [61]. Gels were scanned by a GS-800 scanner (Bio-Rad) and the proteins in the images were analyzed using PDQuest software with version 8.0 (Bio-Rad). There were variations due to sample loading, the 2-DE techniques and staining. To minimize these variations, each spot intensity was normalized according to its percent volume of all protein spots on the gel. The proteins showing at least a twofold change in abundance were considered as differentially expressed proteins (DEPs).

4.5. MS and Database Searches

Protein slices in fresh blue silver-stained gel were excised and plated into a 96-well microtiter plate. Excised slices were first distained twice with 60 µL 50 mM NH_4HCO_3 and 50% acetonitrile, then dried twice with 60 µL acetonitrile. Afterwards, the dried pieces of gels were incubated in ice-cold digestion solution (12.5 ng/µL trypsin and 20 mM NH_4HCO_3) for 20 min, then transferred into a 37 °C incubator for digestion overnight. Finally, peptides in the supernatant were collected after extraction twice with 60 µL extraction solution (5% formic acid in 50% acetonitrile).

The peptide solution described above was dried under the protection of N_2. A 0.8 µL matrix solution (5 mg/mL α-cyano-4-hydroxy-cinnamic acid diluted in 0.1% TFA, 50% ACN) was pipetted to dissolve the peptides. Then, the mixture was spotted on a MALDI target plate (AB SCIEX, Framingham, Massachusetts, MA, USA). MS analysis of the peptides was performed on an AB SCIEX 5800 TOF/TOF. The UV laser was operated at a 400 Hz repetition rate with a wavelength of 355 nm. The accelerated voltage was operated at 20 kV, and the mass resolution was maximized at 1600 Da. The mass instrument with internal calibration mode was calibrated by myoglobin digested with trypsin. All acquired spectra of samples were processed using TOF/TOF Explorer™ Software (AB SCIEX) in default mode. The data were searched by GPS Explorer (V3.6) with the search engine MASCOT (V2.3, Matrix Science, London, UK). The search parameters were as follows: dates were compared against the NCBI nr database, trypsin was digested with one missing cleavage, MS tolerance was set at 100 ppm, and MS/MS tolerance was set to 0.6 Da. Functional annotation of identified proteins based on gene ontology was performed using the Protein Information Resource (https://proteininformationresource.org).

4.6. RNA Isolation and qRT-PCR Assays

Total RNA from *Bgt*-inoculated or mock-inoculated wheat leaves was sampled at 2, 4, 8, 12, and 24 hpi and extracted using Trizol reagent (Tiangen Biotech, Beijing, China). First strand cDNA was synthesized with Transcript One-Step gDNA removal and cDNA Synthesis Supermix (Transgen Biotech, Beijing, China). Primers were specifically designed to anneal to each of the selected genes and the endogenous reference gene *18S rRNA* (GenBank accession No. AY049040) [62]. The expression patterns of selected genes were analyzed with a Bio-Rad iQ5 system. Relative gene quantification was calculated by the comparative $2^{-\Delta\Delta Ct}$ method [63] and normalized to the corresponding expression level of the *18S rRNA*. All reactions were performed in triplicate, including three no-template controls.

5. Conclusions

In summary, we identified 46 differentially expressed proteins in wheat in response to *Bgt* inoculation using 2-DE and mass spectrometry. Among these identified proteins, the lipoxygenase, glucan exohydrolase, glucose adenylyltransferasesmall, phosphoribulokinase, and phosphoglucomutase are first reported during the interaction of wheat-*Bgt*. We inferred that these proteins are not only involved in defense response but also physiology and cellular process for wheat to confer resistance against *Bgt*. The wheat resistance gene products potentially mediate the immune response and coordinate other physiological and cellular processes during the resistance response to *Bgt*.

Author Contributions: Y.L., M.Z., G.G. and X.C. designed the research. Y.L., J.Y., Y.H., L.L., T.L. and L.H. performed the research. Y.L., X.C. and Y.H. analyzed the data. Y.L., Y.X., X.C., M.C. and M.Z. wrote the paper. All authors read and approved the final version of the manuscript.

Funding: The work was supported by the special funds from National Key R&D Program of China, No. 2018YFD0200508; Science and Technology Planning Project of Sichuan Province of China, No. 2016NYZ0053-4; and Science and Technology Department of Sichuan Province of China, No. 2017JY0012.

Acknowledgments: We thank Peigao Luo for providing the wheat cultivar *L699*.

Conflicts of Interest: The authors declare no conflict of interest.

References

1. Wang, Z.; Li, L. Seedling and adult plant resistance to powdery mildew in Chinese bread wheat cultivars and lines. *Plant Dis.* **2005**, *89*, 457–463. [CrossRef]
2. Duveiller, E.; Singh, R. The challenges of maintaining wheat productivity: Pests, diseases, and potential epidemics. *Euphytica* **2007**, *157*, 417–430. [CrossRef]
3. Hao, Y.; Parks, R. Molecular characterization of a new powdery mildew resistance gene *Pm54* in soft red winter wheat. *Theor. Appl. Genet.* **2015**, *128*, 465–476. [CrossRef] [PubMed]
4. Zhang, R.; Sun, B. *Pm55*, a developmental-stage and tissue-specific powdery mildew resistance gene introgressed from Dasypyrum villosum into common wheat. *Theor. Appl. Genet.* **2016**, *129*, 1975–1984. [CrossRef] [PubMed]
5. Hsam, S.; Huang, X. Chromosomal location of genes for resistance to powdery mildew in common wheat (*Triticum aestivum* L. em Thell.). 5. Alleles at the *Pm1* locus. *Theor. Appl. Genet.* **1998**, *96*, 1129–1134. [CrossRef]
6. Singrün, C.; Hsam, S. Powdery mildew resistance gene *Pm22* in cultivar *Virest* is a member of the complex *Pm1* locus in common wheat (*Triticum aestivum* L. em Thell.). *Theor. Appl. Genet.* **2003**, *106*, 1420–1424. [CrossRef] [PubMed]
7. Hao, Y.; Liu, A. *Pm23*: A new allele of *Pm4* located on chromosome 2AL in wheat. *Theor. Appl. Genet.* **2008**, *117*, 1205–1212. [CrossRef] [PubMed]
8. Xie, W.; Ben-David, R. Suppressed recombination rate in 6VS/6AL translocation region carrying the *Pm21* locus introgressed from *Haynaldia villosa* into hexaploid wheat. *Mol. Breed.* **2012**, *29*, 399–412. [CrossRef]

9. Zou, S.; Wang, H. The NB-LRR gene *Pm60* confers powdery mildew resistance in wheat. *New Phytol.* **2018**, *218*, 298–309. [CrossRef] [PubMed]

10. Wiersma, A.T.; Pulman, J.A. Identification of *Pm58* from *Aegilops tauschii*. *Theor. Appl. Genet.* **2017**, *130*, 1123–1133. [CrossRef]

11. Tan, C.; Li, G. Characterization of *Pm59*, a novel powdery mildew resistance gene in Afghanistan wheat landrace PI 181356. *Theor. Appl. Genet.* **2018**, *131*, 1145–1152. [CrossRef] [PubMed]

12. Sun, H.; Hu, J. *Pm61*: A recessive gene for resistance to powdery mildew in wheat landrace Xuxusanyuehuang identified by comparative genomics analysis. *Theor. Appl. Genet.* **2018**, *131*, 2085–2097. [CrossRef] [PubMed]

13. Zhang, R.; Fan, Y. *Pm62*, an adult-plant powdery mildew resistance gene introgressed from *Dasypyrum villosum* chromosome arm 2VL into wheat. *Theor. Appl. Genet.* **2018**, *131*, 2613–2620. [CrossRef] [PubMed]

14. Huang, X.Q.; Hsam, S.L.K. Molecular mapping of the wheat powdery mildew resistance gene *Pm24* and marker validation for molecular breeding. *Theor. Appl. Genet.* **2000**, *101*, 407–414. [CrossRef]

15. Consortium, I.W.G.S. A chromosome-based draft sequence of the hexaploid bread wheat (*Triticum aestivum*) genome. *Science* **2014**, *345*, 1251788.

16. Yahiaoui, N.; Srichumpa, P. Genome analysis at different ploidy levels allows cloning of the powdery mildew resistance gene *Pm3b* from hexaploid wheat. *Plant J.* **2004**, *37*, 528–538. [CrossRef] [PubMed]

17. Cao, A.; Xing, L. Serine/threonine kinase gene *Stpk-V*, a key member of powdery mildew resistance gene *Pm21*, confers powdery mildew resistance in wheat. *Proc. Natl. Acad. Sci. USA* **2011**, *108*, 7727–7732. [CrossRef]

18. Hurni, S.; Brunner, S. Rye *Pm8* and wheat *Pm3* are orthologous genes and show evolutionary conservation of resistance function against powdery mildew. *Plant J.* **2013**, *76*, 957–969. [CrossRef]

19. Sánchez-Martín, J.; Steuernagel, B. Rapid gene isolation in barley and wheat by mutant chromosome sequencing. *Genome Biol.* **2016**, *17*, 221. [CrossRef]

20. Xing, L.; Hu, P. NLR1-V, a CC-NBS-LRR encoding gene, is a potential candidate gene of the wheat powdery mildew resistance gene *Pm21*. *bioRxiv* **2017**, 114058. [CrossRef]

21. Luo, P.G.; Luo, H. Characterization and chromosomal location of *Pm40* in common wheat: A new gene for resistance to powdery mildew derived from *Elytrigia intermedium*. *Theor. Appl. Genet.* **2009**, *118*, 1059–1064. [CrossRef]

22. Zhong, S.; Ma, L. Collinearity analysis and high-density genetic mapping of the wheat powdery mildew resistance gene *Pm40* in PI 672538. *PLoS ONE* **2016**, *11*, e0164815. [CrossRef] [PubMed]

23. Liu, Z.H.; Xu, M. Registration of the novel wheat lines *L658*, *L693*, *L696*, and *L699*, with resistance to Fusarium Head blight, stripe rust, and powdery mildew. *J. Plant Registrat.* **2015**, *9*, 121–124. [CrossRef]

24. Li, J.; Yang, X. Proteomic analysis of the compatible interaction of wheat and powdery mildew (*Blumeria graminis* f. sp. *tritici*). *Plant Physiol. Biochem.* **2017**, *111*, 234–243. [CrossRef] [PubMed]

25. Lim, M.S.; Elenitoba-Johnson, K.S.J. Proteomics in pathology research. *Lab. Investig.* **2004**, *84*, 1227–1244. [CrossRef] [PubMed]

26. Marra, R.; Ambrosino, P. Study of the three-way interaction between *Trichoderma atroviride*, plant and fungal pathogens by using a proteomic approach. *Curr. Genet.* **2006**, *50*, 307–321. [CrossRef] [PubMed]

27. Li, Q.; Chen, X. Differences in protein expression and ultrastructure between two wheat near-isogenic lines affected by powdery mildew. *Russ. J. Plant Physiol.* **2011**, *58*, 686. [CrossRef]

28. Wang, B.; Xie, C. Comparative proteomic analysis of wheat response to powdery mildew infection in wheat *Pm30* near-lsogenic lines. *J. Phytopathol.* **2012**, *160*, 229–236. [CrossRef]

29. Mandal, M.S.N.; Fu, Y. Proteomic analysis of the defense response of wheat to the powdery mildew fungus, *Blumeria graminis* f. sp. *tritici*. *Protein J.* **2014**, *33*, 513–524. [CrossRef]

30. Fu, Y.; Zhang, H. Quantitative proteomics reveals the central changes of wheat in response to powdery mildew. *J. Proteom.* **2016**, *130*, 108–119. [CrossRef]

31. Luo, L.Y. *Defense Responses Induced by Bgt in The Novel Wheat Line L699 Carrying a Pm40 Gene*; Sichuan Agricultural University: Ya'an, China, 2016.

32. Spoel, S.H.; Dong, X. How do plants achieve immunity? Defence without specialized immune cells. *Nat. Rev. Immunol.* **2012**, *12*, 89–100. [CrossRef] [PubMed]

33. Chisholm, S.T.; Coaker, G. Host-microbe interactions: Shaping the evolution of the plant immune response. *Cell* **2006**, *124*, 803–814. [CrossRef] [PubMed]

34. Hammond-Kosack, K.E.; Jones, J.D. Resistance gene-dependent plant defense responses. *Plant Cell* **1996**, *8*, 1773–1791. [CrossRef] [PubMed]
35. Rojas, C.M.; Senthil-Kumar, M. Regulation of primary plant metabolism during plant-pathogen interactions and its contribution to plant defense. *Front. Plant Sci.* **2014**, *5*, 1–12. [CrossRef] [PubMed]
36. Bernier, F.; Berna, A. Germins and germin-like proteins: Plant do-all proteins. But what do they do exactly? *Plant Physiol. Biochem.* **2001**, *39*, 545–554. [CrossRef]
37. Membré, N.; Bernier, F. Arabidopsis thaliana germin-like proteins: Common and specific features point to a variety of functions. *Planta* **2000**, *211*, 345–354. [CrossRef] [PubMed]
38. Park, C.-J.; An, J.-M. Molecular characterization of pepper germin-like protein as the novel PR-16 family of pathogenesis-related proteins isolated during the resistance response to viral and bacterial infection. *Planta* **2004**, *219*, 797–806. [CrossRef]
39. Zimmermann, G.; Bäumlein, H. The multigene family encoding germin-like proteins of barley. Regulation and function in basal host resistance. *Plant Physiol.* **2006**, *142*, 181–192. [CrossRef]
40. Deeba, F.; Sultana, T. Involvement of WRKY, MYB and DOF DNA-binding proteins in interaction with a rice germin-like protein gene promoter. *Acta Physiol. Plant.* **2017**, *39*, 189. [CrossRef]
41. Camejo, D.; Guzmán-Cedeño, Á. Reactive oxygen species, essential molecules, during plant-pathogen interactions. *Plant Physiol. Biochem.* **2016**, *103*, 10–23.
42. Dunwell, J.M.; Gibbings, J.G. Germin and germin-like proteins: Evolution, structure, and function. *Crit. Rev. Plant Sci.* **2008**, *27*, 342–375. [CrossRef]
43. Schweizer, P.; Christoffel, A. Transient expression of members of the germin-like gene family in epidermal cells of wheat confers disease resistance. *Plant J.* **1999**, *20*, 541–552. [CrossRef] [PubMed]
44. Feder, M.E.; Hofmann, G.E. Heat-shock proteins, molecular chaperones, and the stress response: Evolutionary and ecological physiology. *Annu. Rev. Physiol.* **1999**, *61*, 243–282. [CrossRef] [PubMed]
45. Ritossa, F. A new puffing pattern induced by temperature shock and DNP in Drosophila. *Experientia* **1962**, *18*, 571–573. [CrossRef]
46. Tissiéres, A.; Mitchell, H.K. Protein synthesis in salivary glands of Drosophila melanogaster: Relation to chromosome puffs. *J. Mol. Biol.* **1974**, *84*, 389–398. [CrossRef]
47. Neckers, L.; Tatu, U. Molecular chaperones in pathogen virulence: Emerging new targets for therapy. *Cell Host Microbe* **2008**, *4*, 519–527. [CrossRef] [PubMed]
48. Bakthisaran, R.; Tangirala, R. Small heat shock proteins: Role in cellular functions and pathology. *Biochim. Biophys. Acta (BBA) Proteins Proteom.* **2015**, *1854*, 291–319. [CrossRef]
49. Siedow, J.N. Plant lipoxygenase: Structure and function. *Annu. Rev. Plant Biol.* **1991**, *42*, 145–188. [CrossRef]
50. Apel, K.; Hirt, H. Reactive oxygen species: Metabolism, oxidative stress, and signal transduction. *Annu. Rev. Plant Biol.* **2004**, *55*, 373–399. [CrossRef]
51. Vick, B.A.; Zimmerman, D. Oxidative systems for modification of fatty acids: The lipoxygenase pathway. *Biochem. Plants* **1987**, *9*, 53–90.
52. Feussner, L.; Wasternack, C. The lipoxygenase pathway. *Annu. Rev. Plant Biol.* **2002**, *53*, 275–297. [CrossRef] [PubMed]
53. Bilgin, D.D.; Zavala, J.A. Biotic stress globally downregulates photosynthesis genes. *Plant Cell Environ.* **2010**, *33*, 1597–1613. [CrossRef] [PubMed]
54. Scholes, J.; Lee, P. Photosynthetic metabolism in leaves infected with powdery mildew. *Curr. Res. Photosynth.* **1990**, *4*, 219–222.
55. Swarbrick, P.J.; Schulze-lefert, P. Metabolic consequences of susceptibility and resistance (race-specific and broad-spectrum) in barley leaves challenged with powdery mildew. *Plant Cell Environ.* **2006**, *29*, 1061–1076. [CrossRef] [PubMed]
56. Ling, J.; So, B.R. Resampling and editing of mischarged tRNA prior to translation elongation. *Mol. Cell* **2009**, *33*, 654–660. [CrossRef] [PubMed]
57. Zaher, H.S.; Green, R. Quality control by the ribosome following peptide bond formation. *Nature* **2009**, *457*, 161–166. [CrossRef] [PubMed]
58. Park, S.G.; Ewalt, K.L. Functional expansion of aminoacyl-tRNA synthetases and their interacting factors: New perspectives on housekeepers. *Trends Biochem. Sci.* **2005**, *30*, 569–574. [CrossRef] [PubMed]
59. Henty-Ridilla, J.L.; Shimono, M. The plant actin cytoskeleton responds to signals from microbe-associated molecular patterns. *PLoS Pathog.* **2013**, *9*, e1003290. [CrossRef] [PubMed]

60. Bradford, M.M. A rapid and sensitive method for the quantitation of microgram quantities of protein utilizing the principle of protein-dye binding. *Analy. Biochem.* **1976**, *72*, 248–254. [CrossRef]
61. Candiano, G.; Bruschi, M. Blue silver: A very sensitive colloidal Coomassie G-250 staining for proteome analysis. *Electrophoresis* **2004**, *25*, 1327–1333. [CrossRef]
62. Balaji, B.; Bucholtz, D.B. Barley yellow dwarf virus and Cereal yellow dwarf virus quantification by real-time polymerase chain reaction in resistant and susceptible plants. *Phytopathology* **2003**, *93*, 1386–1392. [CrossRef] [PubMed]
63. Livaka, K.J.; Schmittgen, T.D. Analysis of relative gene expression data using real-time quantitative PCR and the $2^{-\Delta\Delta CT}$ method. *Methods* **2001**, *25*, 402–408. [CrossRef] [PubMed]

© 2019 by the authors. Licensee MDPI, Basel, Switzerland. This article is an open access article distributed under the terms and conditions of the Creative Commons Attribution (CC BY) license (http://creativecommons.org/licenses/by/4.0/).

International Journal of
Molecular Sciences

MDPI

Article

Genome-Wide Genetic Diversity and Population Structure of Tunisian Durum Wheat Landraces Based on DArTseq Technology

Cyrine Robbana [1,2,*], Zakaria Kehel [3], M'barek Ben Naceur [2], Carolina Sansaloni [4], Filippo Bassi [3] and Ahmed Amri [3]

1   Faculté des Sciences de Bizerte, Jarzouna, 7021 Bizerte, Tunisia
2   Banque Nationale de Gènes, Boulevard du Leader Yasser Arafat Z. I Charguia 1, 1080 Tunis, Tunisia;
     nour3alanour@yahoo.com
3   International Center for Agricultural Research in the Dry Areas (ICARDA), 10112 Rabat, Morocco;
     Z.Kehel@cgiar.org (Z.K.); F.Bassi@cgiar.org (F.B.); A.Amri@cgiar.org (A.A.)
4   Centro Internacional de Mejoramiento de Maíz y Trigo (CIMMYT), El Batán, 56237 Texcoco, Mexico;
     C.Sansaloni@cgiar.org
*   Correspondence: cyrine_rob@yahoo.fr; Tel.: +216-52-578-958

Received: 13 December 2018; Accepted: 7 February 2019; Published: 18 March 2019

Abstract: Tunisia, being part of the secondary center of diversity for durum wheat, has rich unexploited landraces that are being continuously lost and replaced by high yielding modern cultivars. This study aimed to investigate the genetic diversity and population structure of 196 durum wheat lines issued from landraces collected from Tunisia using Diversity Array Technology sequencing (DArTseq) and to understand possible ways of introduction in comparing them to landraces from surrounding countries. A total of 16,148 polymorphic DArTseq markers covering equally the A and B genomes were effective to assess the genetic diversity and to classify the accessions. Cluster analysis and discriminant analysis of principal components (DAPC) allowed us to distinguish five distinct groups that matched well with the farmer's variety nomenclature. Interestingly, Mahmoudi and Biskri landraces constitute the same gene pool while Jenah Zarzoura constitutes a completely different group. Analysis of molecular variance (AMOVA) showed that the genetic variation was among rather than within the landraces. DAPC analysis of the Tunisian, Mediterranean and West Asian landraces confirmed our previous population structure and showed a genetic similarity between the Tunisian and the North African landraces with the exception of Jenah Zarzoura being the most distant. The genomic characterization of the Tunisian collection will enhance their conservation and sustainable use.

Keywords: durum wheat; Tunisian landraces; center of diversity; genetic diversity; population structure; DArTseq technology

1. Introduction

Durum wheat is the tenth most important crop worldwide, grown on 13 M ha with a production of 39.1 Mt in 2017 (International Grain Council, Grain Market reports, 2017) and is mainly used for pasta production. It is mainly grown in the countries around the Mediterranean basin and its cultivation has extended to India, Mexico, North America, and Russia [1,2]. North Africa grows around 2.5 M ha of durum wheat with around 1 M ha in Tunisia. Landraces are still grown under traditional farming systems and are highly appreciated for local dishes such as frikeh, burghul, couscous, but their acreage is reduced significantly with the large adoption of new cultivars released since the 1970s [3,4].

Durum wheat (*Triticum turgidum* ssp. *durum*) is a self-pollinated allotetraploid cereal (2n = 4x = 28, AABB) originating from a cross between *Aegilops speltoides* (SS) and *Triticum urartu* (AA) and domesticated from primitive wheat (*Triticum turgidum* ssp. *dicoccum*) in the Fertile Crescent around 8000 years ago [5–9]. North Africa and Abyssinian regions are considered as secondary centers of diversity for durum wheat [10]. Tunisia, being part of the secondary center of diversity for durum wheat, has a rich diversity in terms of landraces and wild relatives [11,12].

Durum wheat landraces are valuable parental germplasm for wheat improvement programs around the world. Genesys database (www.genesys-pgr.com) contains 60,488 durum wheat accessions of which 22,600 are landraces. The International Center for Agricultural Research in the Dry Areas (ICARDA) genebank holds one of the largest collections of durum wheat accounting more than 20,531 accessions of which more than 65% are landraces. The Tunisian genebank (NGBT) holds a total of 4000 accessions of durum wheat most of which are pure lines collected from the predominant remaining landraces.

The Tunisian landraces are well adapted to a broad range of environments, genetically diverse, and are considered as an important reservoir of useful genes that could be exploited in wheat breeding programs [3,13,14]. A high genetic diversity of local Tunisian durum wheat has been reported using morphological, agronomic, physiological, and biochemical traits [4–15]. Sourour et al. [16] has shown important phenotypic diversity of the Tunisian germplasm estimated to 0.77 using Shannon–Weaver indices. Several molecular markers techniques were used to assess the genetic diversity in wheat including RFLP, RAPD, ISSR, AFLP, EST–SSR [17], DArT, and SNP [18,19]. A significant genetic difference using SSR markers was detected between landraces of durum wheat originated from Morocco and Syria using the Bayesian method and the Eigen analysis [20]. Medini et al. [21] reported high genetic diversity among 33 old Tunisian cultivars using 15 SSR markers. More recently, progress has been made for wheat in genomic research and genetic diversity analysis methods through the construction of a high-density SNP-based consensus map for durum wheat and the development of several next-generation sequencing (NGS) platforms [22,23]. An Axiom 35K array used to genotype 370 entries of durum wheat panel allowed us to differentiate among improved varieties and landraces and to show that Middle East and Ethiopia had the lowest level of allelic diversity compared to other regions [24]. Many studies have shown that genotyping-by-sequencing (GBS) has been increasingly adopted as a rapid and low-cost molecular technique for whole-genome SNP coverage [25], SNP discovery, genotyping, and genetic variability analysis for various crop species including durum wheat landraces [26–28]. The development of DArTseq™ technology has been applied successfully for large genomes such as barley [29] and polyploid or/and complex genomes such as tetraploid and hexaploid wheat [18,29,30]. The power of the DArTseq™ approach based on Illumina short read sequencing have been proven to be effective when used to study the genetic diversities of Syrian and Turkish durum wheat landraces [31,32], bread wheat [33], watermelon, and common bean landraces [34,35].

The present study aimed at (1) characterizing for the first time the population structure and the genetic diversity within and among the Tunisian durum wheat collection, (2) looking at potential mis-classification by linking off types of landraces to local modern durum wheat varieties and/or other landraces, and (3) comparing the Tunisian durum wheat landraces to landraces from the countries around the Mediterranean basin and West Asia region using high throughput DArTseq™ technology.

2. Results

2.1. DArTseq Marker Characteristics

In this study, a total of 110,856 DArTseq markers were identified in a set of 196 lines issued from six Tunisian durum wheat landraces, of which 16,148 markers were found to be of high quality and polymorphic, after removing markers with more than 20% missing data and with less than 5% minor allele frequency (MAF). Around 10% of the markers had 0 heterozygote alleles. The average of the polymorphism information content (PIC) value of the DArTseq markers was 0.165, and the median was 0.105. The distribution of the majority of MAFs was between 0.05 and 0.15 (Figure 1).

A

Histogram of minor allele frequency (MAF)

B

Histogram of polymorphic information content (PIC)

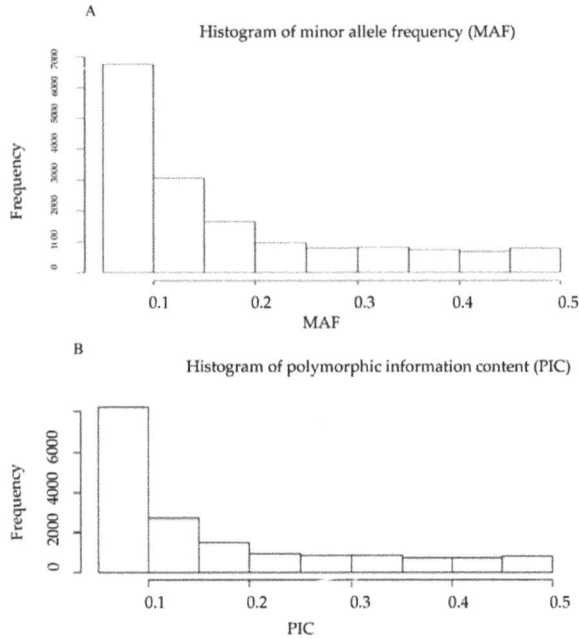

Figure 1. Frequency distribution of (**A**) minor allele frequency (MAF) and (**B**) polymorphic information content (PIC) of 16,148 DArtseq markers.

The DArTseq markers are well distributed across all the 14 chromosomes of durum wheat genomes based on the consensus bread wheat genetic map obtained from the International Wheat Genome Sequencing Consortium database (https://urgi.versailles.inra.fr/download/iwgsc/). The distribution is almost equal between A (4201 markers) and B (4659 markers) genomes with the highest number of markers observed on the chromosomes 7A and 2B. However, 7324 markers are still unassigned to any of the chromosomes (Table 1). A similar range of genetic diversity values (H_e) is observed for both A and B genomes with an average of around 0.25. The maximum value of the expected heterozygosity (H_e) was observed at chromosome 6B (0.28) and the minimum value was seen for chromosomes 3A, 4A, 5A, and 5B (0.24). For all chromosomes, the expected heterozygosity values (H_e) were higher than the observed heterozygosity values (H_0).

Table 1. Number of the selected DArTseq markers (n), the observed heterozygosity (H_0), and the expected heterozygosity (H_e) across the 14 chromosomes in 196 Tunisian durum wheat landraces lines based on the consensus bread wheat genetic map.

	A Genome			B Genome		
Chromosome	n	H_e	H_0	n	H_e	H_0
1	562	0.27	0.15	658	0.26	0.14
2	569	0.25	0.14	864	0.27	0.13
3	633	0.24	0.14	777	0.26	0.14
4	511	0.24	0.16	429	0.25	0.16
5	587	0.24	0.18	639	0.24	0.16
6	452	0.25	0.15	573	0.28	0.15
7	797	0.25	0.14	719	0.26	0.15
Total	4201			4659		
Unassigned	7324	0.25	0.18			

2.2. Genetic Distance and Clustering of the Tunisian Landraces

The allele sharing distance (ASD) among the 196 Tunisian landraces lines ranged from 0 to 0.79 with an average of 0.46 (Figure 2). However, different distance patterns were observed for different landraces. Jenah Zarzoura shows the lowest distance with an average of 0.1 and the lowest variability among its lines ranging from 0 to 0.16. For the Bidi landrace, the average distance is 0.17 with a range of 0.02–0.48. For Rommani, the average distance is 0.25 with values ranging between 0.02 and 0.63 and for Mahmoudi the average distance is 0.30 with a range of 0.35–0.58. Both Biskri and Jenah Khotifa show the highest average distances respectively of 0.43 and 0.44 with the largest variabilities reaching 0.78 and 0.61 respectively.

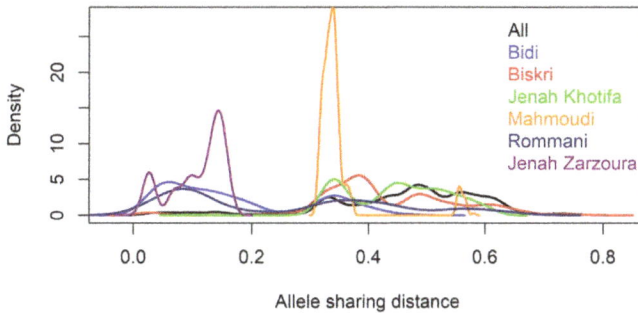

Figure 2. Distribution of allele sharing distance of 196 pure lines derived from six Tunisian durum wheat landraces using DArTseq markers.

Cluster analysis of the 196 lines derived from the six Tunisian landraces was performed using the allele sharing distance (ASD) method and the results allowed us to group them into five clusters and to identify lines wrongly assigned to some landraces. The first cluster comprised all the lines of Jenah Zarzoura (Zar). The second cluster contained mainly Bidi lines (Bid). The third cluster is subdivided into two sub-groups, one containing Rommani lines (Rom) and the other having a mixture of lines from other landraces (seven from Biskri (Bis), three from Jenah Khotifa (jkf), and two from Bidi (Bid)). Cluster 4 had the majority of Jenah Khotifa lines (jkf) but contained also four lines of Rommani, two of Biskri, and one of Mahmoudi. The last cluster constitutes the largest one and gathered almost all lines of Biskri (Bis) and Mahmoudi (Mah) (Figure 3).

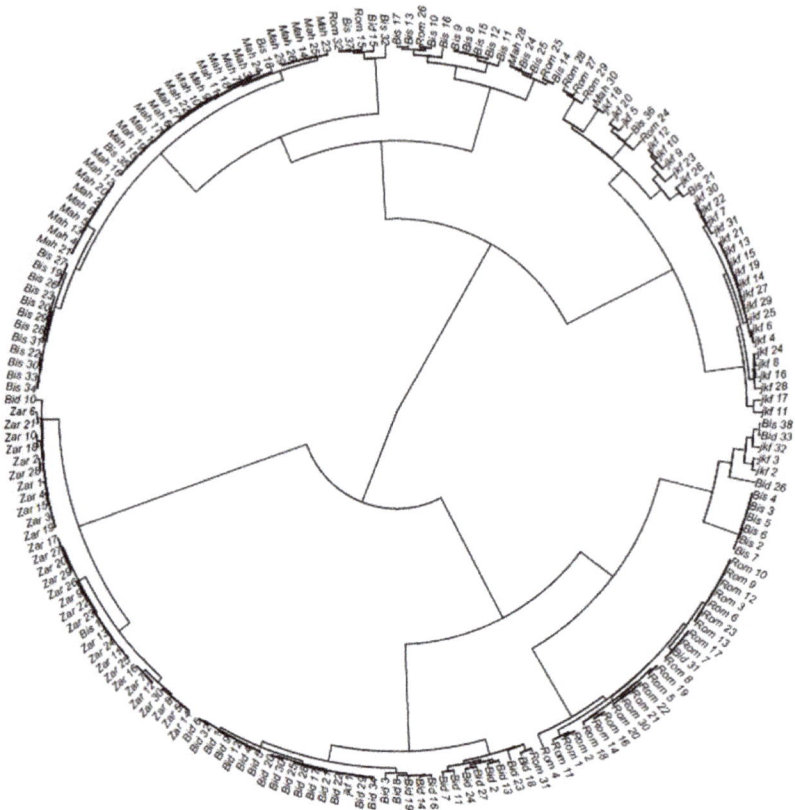

Figure 3. Cluster tree of 196 lines derived from six Tunisian durum wheat landraces based on allele sharing genetic distance. Bid: Bidi; Mah: Mahmoudi; Rom: Rommani; Bis: Biskri; jkf: Jenah Khotifa; Zar: Jenah Zarzoura.

2.3. Population Structure of the Tunisian Landraces

The DAPC (discriminant analysis of principal components) results showed that 88 Principal Components (PCs) explain 82% of the total molecular variance. The optimum number of clusters was obtained with K = 6 using the Bayesian Information Criterion (BIC), which divided the lines into six sub-populations (Figure 4). The first subdivision level using the hierarchical population structure K = 2 separated Jenah Zarzoura lines (Zar) clearly from all the lines of the other landraces (Figure 5A). When K = 3, in addition to the separate sub-population of Jenah Zarzoura, two other sub-populations were identified, one combining most Mahmoudi (Mah) and Biskri (Bis) lines and the other included lines of landraces Bidi (Bid), Jenah Khotifa (jkf), and Rommani (Rom) (Figure 5B). With K = 4, a separate sub-population containing Jenah Khotifa lines appeared (Figure 5C); when K = 5, all the landraces were assigned to different sub-populations, except for Biskri and Mahmoudi, which remained grouped in the same sub-population (Figure 5D). The number of sub-population K = 5 was then chosen to differentiate between different landraces and the sub-populations were given the following names: BID for Bidi, BIS+MAH for Biskri and Mahmoudi, JKF for Jenah Khotifa, ROM for Rommani, and ZAR for Jenah Zarzoura.

Increasing K to 6 resulted into the formation of an additional small group composed by 13 lines mainly of Biskri (7) and Jenah Khotifa (4) showing that these lines could be off-types within their respective landraces, which need further characterization (Figure 5E). The global population structure image was maintained for the values of K equal to 7 and 8 with all different landraces assigned to different sub-populations except the landraces Biskri and Mahmoudi that are still grouped in the same sub-population (Figure 5F).

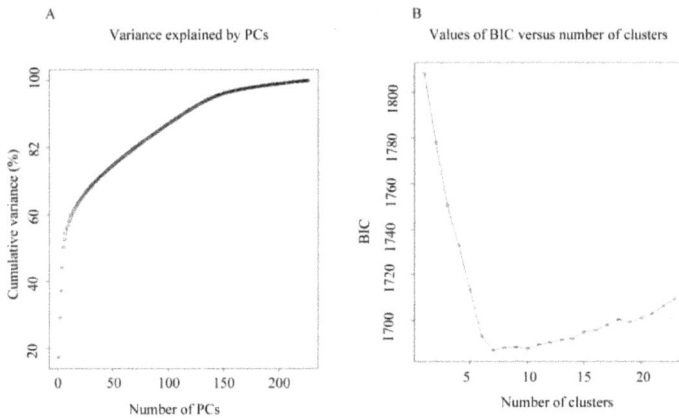

Figure 4. Optimal number of sub-populations using discriminant analysis of principal components (DAPC). (**A**) Cumulative variance explained by PCs; (**B**) BIC versus number of clusters.

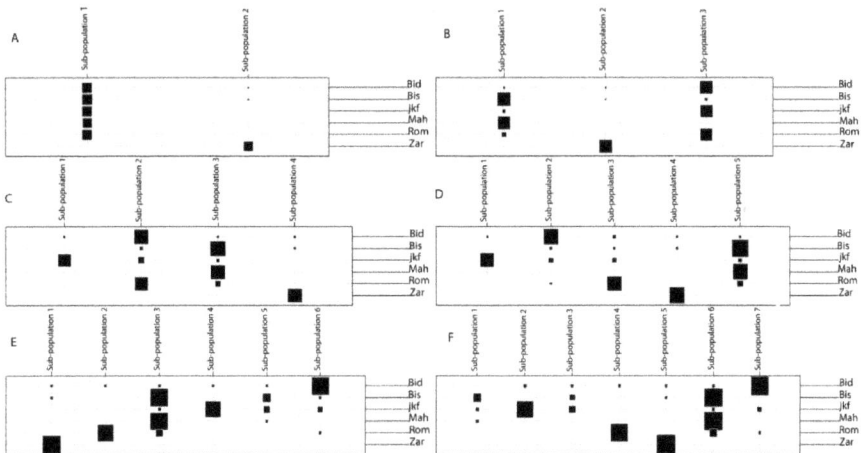

Figure 5. Table graphs comparing distribution of the original landrace classification to the sub-populations using DAPC with (**A**) K = 2; (**B**) K = 3; (**C**) K = 4; (**D**) K = 5; (**E**) K = 6; (**F**) K = 7. K: Number of sub-populations; Bid: Bidi; Mah: Mahmoudi; Rom: Rommani; Bis: Biskri; jkf: Jenah Khotifa; Zar: Jenah Zarzoura.

The results of analysis of molecular variance (AMOVA) for K = 2 to K = 7 showed that the variance between populations was high for K = 2 (31.92%), and it decreased for K = 3 and K = 4 (28.75% and 31.29%, respectively) to increase and reach a plateau for K = 5 with a slightly higher value (31.77%). The variance between lines within populations decreased from 11.32% for K = 2 to reach values close to 0% for K = 5. Comparing AMOVA results between populations with K = 5 and the real groups using the landrace's name clearly support the hypothesis that K = 5 better differentiates genetically the lines than the characterization of landraces by name, resulting in higher variance between populations (31.77% for K = 5 and 28.79% for real groups), lower variance between lines 68.23% for K = 5 and 70.39%

for real groups, and lower variance between lines within populations (Table 2). Thus, these findings indicated a higher genetic variation among rather than within Tunisian landraces.

Table 2. AMOVA summary for comparison between different numbers of sub-populations K and the real groups of lines issued from six Tunisian durum wheat landraces.

Number of Sub-Populations	Variance		
	Among Landraces (%)	Among Lines within Landrace (%)	Among Lines (%)
K = 2	31.92	11.32	56.76
K = 3	28.75	5.60	65.65
K = 4	31.29	1.59	67.12
K = 5	31.77	0.00	68.23
K = 6	31.71	0.00	68.29
K = 7	31.51	0.00	68.49
Real groups	28.79	0.82	70.39

Results from hierarchical AMOVA using hierarchical subdivision strata from K = 2 up to K = 5 indicated that most of the molecular variance was explained with the first level K = 2 separating Jenah Zarzoura from the other landraces (17.18%) and the fifth level K = 5 within the four groups (26.62%) separating Bidi, Rommani, and Jenah Khotifa and the large group constituted by Biskri and Mahmoudi. The remaining molecular variance (variance between lines) was also low compared to non-hierarchical AMOVA results (60.55%) using K = 5 (Table 3).

Table 3. Hierarchical AMOVA results from K = 2 to K = 5 using hierarchical subdivision strata of 196 lines derived from the six durum wheat Tunisian landraces.

Subdivision Strata	Variance Components	Percentage
Variations between K = 2	271.63	17.18
Variations between K = 3 and K = 2	0	0
Variations between K = 4 and K = 3	13.46	0.85
Variations between K = 5 and K = 4	420.82	26.62
Variations between lines and K = 5	957.20	60.56
Variations between lines	0	0
Total variations	1580.63	100

Finally, the resulted variance components from AMOVA analysis with K = 5 were significant when they were tested using permutations (Figure 6). These results showed that the structure found by number of groups with K = 5 was valid and not a result of a random effect.

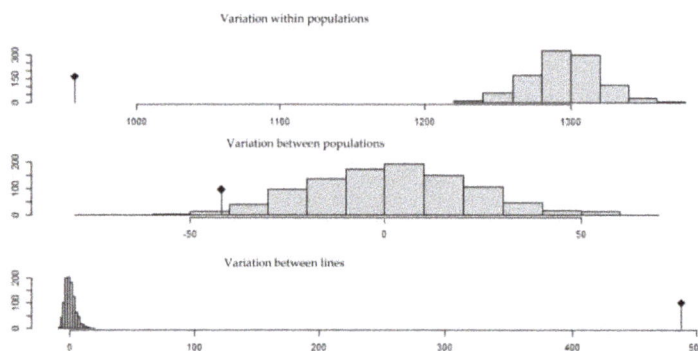

Figure 6. AMOVA permutations test using 1000 permutations for number of sub-population K = 5.

2.4. Genetic Diversity and Genetic Distance between Tunisian Landraces

The genetic distance between landrace populations showed that Jenah Zarzoura sub-population is the farthest from all the other landraces, from Rommani (0.820), from Biskri and Mahmoudi (0.730), from Jenah Khotifa (0.682), and from Bidi (0.816) (Table 4). Furthermore, the Jenah Khotifa sub-population was equally distant to the landraces Bidi and Rommani (0.557). These results confirmed the population subdivision found by DAPC, which separated the Jenah Zarzoura sub-population from the other landraces when K = 2 and grouped the sub-populations Jenah Khotifa, Bidi, and Rommani when K = 3.

Table 4. Reynolds genetic distance between Tunisian durum wheat landrace populations based on sub-populations number K = 5.

Landraces	BID	ZAR	BIS + MAH	JFK
ZAR	0.816			
BIS+MAH	0.624	0.730		
JFK	0.557	0.682	0.466	
ROM	0.729	0.820	0.645	0.577

BID: Bidi; JKF: Jenah Khotifa; ZAR: Jenah Zarzoura; BIS: Biskri; ROM: Rommani; MAH: Mahmoudi.

Results of genetic diversity estimates in each sub-population obtained based on DAPC with K = 5 show that the highest genetic diversity was observed within the Jenah Zarzoura and Rommani populations (H_e = 0.27). The lowest genetic diversity was observed within the Bidi population (H_e = 0.12). Biskri and Mahmoudi forming the same group with 73 lines showed a moderate genetic diversity (H_e = 0.23). Based on F_{ST} values, Jenah Zarzoura showed a low F_{ST} value of 0.05. Thus, this population presents a genetic drift compared to the others, having a fixation of alternate alleles with F_{ST} values superior to 0.25 (Table 5). The results from genetic diversity confirmed the results obtained using the allele sharing distance (ASD) between lines within the same population as shown in Figure 1.

Table 5. Genetic diversity among Tunisian durum wheat landrace populations based on sub-populations number K = 5.

Landraces	n	H_e	F_{ST}
ZAR	32	0.27	0.05
JFK	25	0.20	0.47
BID	34	0.12	0.57
ROM	32	0.27	0.30
BIS+MAH	73	0.24	0.26

n: Number of lines; H_e: Expected heterozygosity (genetic diversity); F_{ST}: Measure of genetic differentiation. BID: Bidi; JKF: Jenah Khotifa; ZAR: Jenah Zarzoura; BIS: Biskri; ROM: Rommani; MAH: Mahmoudi.

2.5. Linking the Mis-Classified Lines to Local Landraces and Improved Varieties or/and to the ICARDA/CIMMYT Elite Lines

Cluster analysis based on the ASD method was performed using a set of 27 Tunisian durum wheat landrace populations, six local improved varieties, and seven ICARDA/CIMMYT inbred elite lines, along with lines which formed the additional sub-population when K = 6 from the previous set. This study aimed at showing the relationships among a larger number of Tunisian landraces, the differences with improved varieties, and germplasm and at shedding more light on the mis-classified lines included in the last sub-population where K = 6.

Results of cluster analysis showed three main clusters (Figure 7). Cluster 1 contained most of the improved varieties released in Tunisia and the advanced lines from ICARDA and CIMMYT, along with three landraces; Azizi (Aziz P), Chetla 1(Chet1 P), and Jenah Khotifa 2 (JK2 P). The six mis-classified lines of Biskri (Bis) from the additional sub-population when K = 6 in the analysis of Set 1 are included in this cluster and are grouped with the improved variety Khiar (Khia T). Cluster 2 contained the majority of the Tunisian landraces along with two improved varieties: INRAT69 (INRA T) and Karim (Kari T). This cluster could be subdivided into two groups: a small group composed of six local landraces (Arbi (Arbi P), Rommani 3 (Rom3 P), Biskri 1(Bis1 P), Chetla 2 (Chet2 P), Richi (Richi P), and Agili (Agil P)) and the two improved varieties, Karim (Kari T) and INRAT 69 (INRA T) and a large group that can be further divided into two sub-groups, one of which constitutes a separate group with only Jenah Zarzoura lines (Zar1 L, Zar2 L, and Zar3 L) and the other of which is composed of the majority of local landraces: Jenah Khotifa 1(jkf1 P), Sbei (Sbei P), Mahmoudi (Mah P), Souri (Sour P), Bayadha (Baya P), Swabei Algia (SA P), Aouadi (Aoua P), Hamira (Hmir P), Rommani (Rom1 P), Bidi 1(Bid1 P), Bidi 3 (Bid3 P), Rommani 2 (Rom2 P), Wardbled (WB P), Derbessi (Derb P), Chili (Chil P), Biskri 2 (Bis2 P), and Bidi 2 (Bid2 P). This last sub-group showed that the two lines of Jenah Khotifa (Jk2 L and Jk4 L) from Set 1 are clustered with Jenah Khotifa population 1(jkf1 P). The Rommani line (Rom1 L) is grouped with the Rommani population (Rom1 P) and the line Bidi 25 (Bid25 L) is grouped with Bidi populations (Bid1 P and Bid3 P). Cluster 3 was divided into two groups: the first group contained a mixture of lines including Jenah Khotifa 11 line (Jk11 L), Biskri 12 line (Bis12 L), and Mahmoudi 8 line (Mah8 L) and the second group showed that the mis-classified lines of Jenah Khotifa 19 (Jk19 L) and Mahmoudi 30 (Mah30 L) included in the additional sub-population when K = 6 of Set 1 analysis are clustered with the ICARDA/CIMMYT inbred line (MCHCB-102). These results showed that most of the lines included in the mixed sub-population of the DAPC analysis with K = 6 are grouped with improved varieties and germplasm and showed that some landraces having the same local name are classified in different clusters.

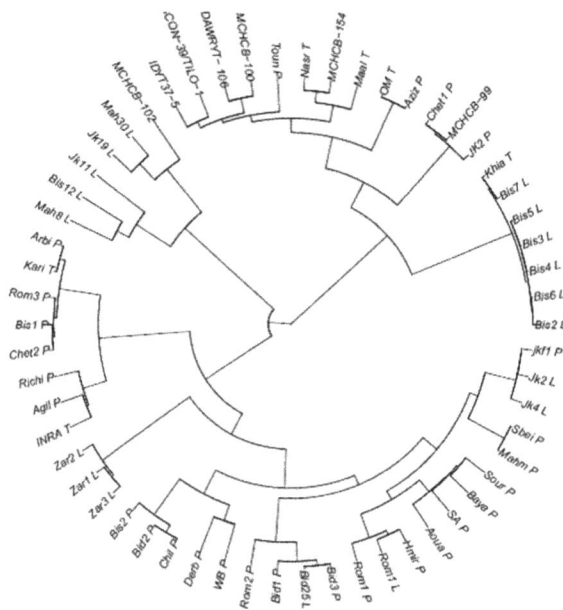

Figure 7. Dendrogram of Tunisian durum wheat landraces, local improved varieties, and ICARDA/ CIMMYT elite lines based on allele sharing genetic distance. L: Line; P: Population. Local landraces: Bid: Bidi; Mah: Mahmoudi; Rom: Rommani; Bis: Biskri; jkf: Jenah Khotifa; Zar: Jenah Zarzoura; Aziz: Azizi; Chet: Chetla; Sbei; Sour: Souri; Baya: Bayadha; SA: Swabei Algia; Aoua: Aouadhi; Hmir: Hamira; WB: WardBled; Derb: Derbessi; Chil: Chili; Agil: Agili; Richi; Arbi; Toun: Tounsia. Local improved varieties: OM: OmRabii; Nasr; Maal: Maali; Khia: Khiar; INRAT: INRAT69; Kari: Karim. ICARDA/CIMMYT elites lines: MCHCB-102: ICARDA inbred line; MCHCB-100: IcaJoudy1; DAWRYT-106: Nachit; MCHCB-154: Zeina4; Con39/Tilo-1: Louiza; IDYT37-5: Ammar 6; MCHCB-99: Ammar 10.

2.6. Comparison of Tunisian Landraces to Landraces from Mediterranean and West Asia Regions

The population structure of the Mediterranean and West Asian landraces along with four lines representing each of six Tunisian landraces included in Set 1 was assessed using the DAPC method. The optimum number of groupings is determined with K = 12 based on the BIC criterion using 66 PCs, which explained 82% of the total molecular variation. Sub-population 6 showed that the Tunisian landraces Rommani (Rom), Jenah Khotifa (jkf), and Bidi (Bid) were grouped with the majority of the Algerian landraces (8); Sub-population 12 included all Ethiopian landraces (11), along with one accession from Afghanistan and two accessions from Yemen; Sub-population 9 included the Tunisian landraces Biskri (Bis) and Mahmoudi (Mah) that were grouped with the remaining five landraces from Tunisia, with the majority of Algerian and Lebanese landraces (6), and with some Moroccan landraces (3). Sub-population 11 included exclusively the Tunisian landrace Jenah Zarzoura (Zar) with two accessions from Jordan (Figure 8). The population structure results confirmed that the newly collected landrace Jenah Zarzoura constitutes a new gene pool and that the Tunisian landraces are genetically closer to North African landraces.

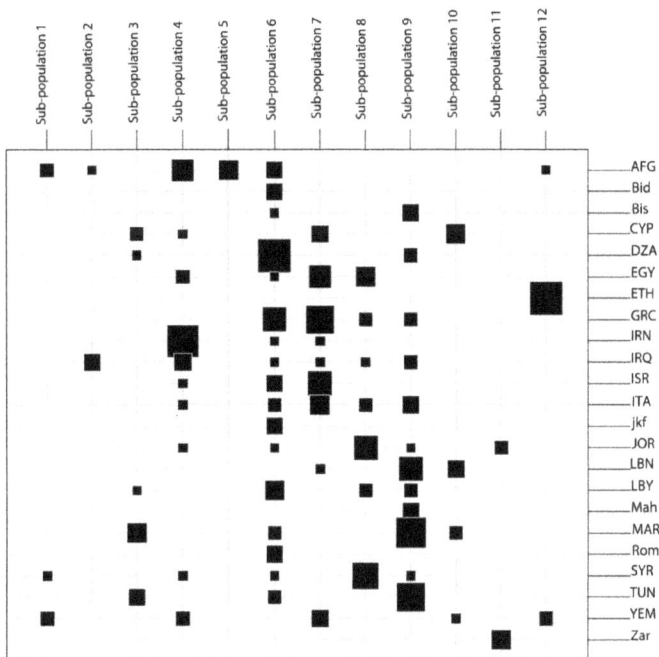

Figure 8. Table graphs comparing Mediterranean, West Asian, and Tunisian landraces classification to the sub-populations using DAPC with K = 12. AFG: Afghanistan; CYP: Cyprus; DZA: Algeria; EGY: Egypt; ETH: Ethiopia; GRC: Greece; IRN: Iran, IRQ: Iraq; ISR: Palestine/Israel; ITA: Italy; JOR: Jordan; LBN: Lebanon; LBY: Libya; MAR: Morocco; SYR: Syria; TUN: Tunisia; YEM: Yemen.

3. Discussion

The molecular markers techniques are important tools for better understanding genetic diversity, undertaking association mapping and ensuring efficient conservation and management of genetic resources. This study demonstrated the relevance of DArTseq technology as a reliable and cost-effective tool for assessing the diversity within and between landraces and for comparing them with other germplasm of durum wheat. This technique yielded a large number of polymorphic and informative markers equally spread in the A and B genomes allowing high coverage of the genomes of the Tunisian durum wheat germplasm compared to other molecular techniques used previously. Similar genomes coverage was found on a panel of 170 durum wheat entries [27]. The large coverage of the genomes can serve to undertake association mapping in the studied germplasm, including finding new allelic variations for major breeders sought traits such as QTLs found by other studies in chromosomes 7A and 2B linked to protein content [36], gluten strength and yellow pigment [37,38], and salinity tolerance and yield components [39,40]. DArTseq technology along with other high throughput and genotyping by sequencing molecular techniques are increasingly used to study the genetic diversity of different crops as they allow to study the genetic diversity of large number entries and complex genomes [41,42].

3.1. DArTseq Polymorphism of Tunisian Durum Wheat Landraces

Polymorphic information content (PIC) values revealed by DArTseq markers averaged 0.165 with an asymmetric distribution skewed towards low values. The same distribution tendency was found using the same approach for 91 durum wheat landraces from the Fertile Crescent and for 138 wheat germplasm from Southwest China [32,41]. Ren et al. [43] have shown the same PIC value for North African durum accessions (0.168) using 946 SNP markers as part of a genetic diversity study in a worldwide durum wheat germplasm collection. Recent studies using DArTseq markers reported higher PIC values for worldwide durum wheat accessions (0.35) and for accessions originating from central Fertile Crescent (0.26) [27,32]. A previous study on 34 Tunisian durum wheat old varieties reported a PIC value of 0.68 [25]. This difference is explained by the bi-allelic nature of DArT markers for which the maximum value for PIC is 0.5 compared to multi-allelic SSR markers with the maximum PIC value of 1 [44].

The lines of the six Tunisian durum wheat landraces showed a moderate level of genetic diversity ($H_e = 0.25$), which is higher than the one exhibited by a set of world-wide durum wheat collection ($H_e = 0.224$) using SNP markers [43]. Our results confirmed previous findings showing that the Fertile Crescent and the eastern Mediterranean durum landraces are more diverse than those from the Western Mediterranean regions [45,46]. Although several studies revealed high genetic diversity in the Tunisian landraces based on phenotypic characterization [15,16], the moderate level of genetic diversity could be explained by the fact that the Tunisian durum wheat lines included in this study are derived from six landraces collected from a limited geographic area in the center and the south parts of Tunisia.

3.2. Genetic Diversity and Population Structure of the Tunisian Durum Wheat Landraces

Allele sharing distance using DArTseq markers allowed us to differentiate among the six landraces Jenah Zarzoura, Biskri, Jenah Khotifa, Mahmoudi, Bidi, and Rommani by showing different distance patterns. Two methods, clustering analysis based on ASD (allele sharing distance) and DAPC (discriminant analysis of principal components) were used for depicting the genetic relationship and structure of the Tunisian collection. The first method classified the panel into five groups matching mostly with the farmer's landraces names, with Jenah Zarzoura being the most distant and Mahmoudi and Biskri included in the same cluster. The closeness of Mahmoudi and Biskri is for a long time reported by Boeuf [11] based on glume and spike color, which could be due to the exchange of these landraces among farmers from Algeria and Tunisia [47]. The clear distinction of Jenah Zarzoura could be due to its confinement to a geographically limited area of the Mareth oasis or to a different pattern of introduction.

The use of the multivariate method DAPC to evaluate the population structure showed better performance and allowed for a population subdivision similar to other studies [24,48]. Several molecular approaches were used to assess the population structure in durum wheat landraces from SSRs to DArTs [49,50]. More recently, GBS-SNPs and DArTseq approaches were mainly used for hexaploid wheat population structure studies [28,51], and some reports have described a durum wheat population structure based on the DArTseq technique that allowed to classify the Turkish and Syrian durum wheat landraces in the same gene pool [32]. The DAPC analysis results were in concordance with those of clustering analysis, despite minor differences. Both showed a good fit between the grouping and the names of the varieties, which reflects the ability of farmers to differentiate among the landraces. However, a small group composed of a mixture of landraces appeared, and some lines of the differentiated landraces are included in different clusters, which could be due mainly to mis-naming of the landraces during the collecting missions and to possible mixtures in the landraces. These findings confirm the multiline nature of landraces that is also found through morphological and agronomic characterization [52]. This heterogeneity offers an important buffering capacity of landraces in drought-prone and fluctuating environments.

Hierarchical AMOVA analysis based on hierarchical subdivision strata from K = 2 up to K = 5 agreed with DAPC analysis results and supported the high level of molecular variance to K = 2 (17.18%) and to K = 5 (26.62%). AMOVA analysis results on the basis of the landraces' names indicated a higher genetic variation among (28.79%) rather than within (0.82%) Tunisian landraces. Taking in consideration the structure of the population based on the optimal number of grouping K = 5, AMOVA results showed an increase in percent of the explained genetic variance among landraces (~31.77%) and a decrease of the genetic variance within them. Soriano et al. [49] revealed much variability within sub-populations (83%) than between them (17%) and Mangini et al. [53] found higher genetic diversity within the two Italian durum wheat landraces "Bianchetta" and "Grano Ricco" (9.5 and 9.4%, respectively) and low genetic diversity within "Dauno III.".

This study allowed us to assess for the first time the genetic diversity within and among the Tunisian durum wheat landraces using the DArTseq technique, which allowed us to show different levels of genetic diversity within landraces. Jenah Zarzoura and Rommani populations showed the highest genetic diversity (H_e = 0.27), and the Bidi landrace showed the lowest genetic diversity (H_e = 0.12). The low genetic variation within the landraces could be explained by the selection from farmers for desirable traits and/or from environmental conditions pressure. Compared to the other landraces, the Jenah Zarzoura landrace showed a high expected heterozygosity (H_e = 0.268) and a low fixation index (F_{ST} = 0.05). Thus, this small differentiation could be explained by the confinement of this landrace to a specific environment in the oasis of Mareth, which reflects the geographic isolation of the oasis of Mareth, limited seed exchange, and selection pressure by farmers. Most often, farmers are selecting the best representative spikes from a landrace to form the seed lots.

3.3. Origin of Tunisian Durum Wheat Landraces

The cluster analysis extended to 27 other Tunisian landraces, improved varieties, and germplasm along with the lines included in the mixed cluster when K = 6 from Set 1 showed large genetic diversity among the germplasm studied. The six lines of landrace Biskri mis-classified in the additional sub-population when K = 6 in the Set 1 analysis were grouped with the modern cultivar Khiar, and the two mis-classified lines of Mahmoudi and Jenah Khotifa were grouped with the elite germplasm (MCHCB-102). These results confirm the possibility of mixture in some landraces, which could be due to seed exchange and threshing practices as suggested by other studies [52,54]. This extended genetic study confirms the uniqueness of the Jenah Zarzoura landrace and the classification of Jenah Khotifa, Rommani, and Bidi lines with their respective landraces, but not with the other populations with the same name. This could be due to the mis-naming of landraces during seed exchange among farmers or during the collecting missions undertaken by the genebank teams.

Tunisia is considered as a secondary center of diversity for durum wheat, and the introduction patterns of durum wheat landraces into Tunisia and North Africa are still under discussion. The DAPC analysis including landraces from Tunisia and landraces from the countries around the Mediterranean and West Asia regions allowed us to define 12 distinct groups, which can be used to highlight the relationships between Tunisian landraces and other landraces. Most Tunisian landraces held at the ICARDA genebank as well as the lines derived from landraces collected recently in Tunisia were grouped with landraces from North Africa neighboring countries and with landraces from Greece, Italy, and Lebanon. Jenah Zarzoura remained distinct from all landraces studied except for the two landraces from Jordan. These results suggest that most Tunisian landraces could be obtained through Lebanon via Greece and Italy, while Jenah Zarzoura was obtained through another introduction pathway. Previous reports have demonstrated two dispersal patterns of durum wheat in the Mediterranean Basin from the Fertile Crescent: over the north side via Turkey, Greece, and Italy and the south side via North Africa [55]. Moragues et al. [56] supported this hypothesis and classified a collection of durum wheat landraces originating from different Mediterranean countries in two groups: (i) landraces from the North and East Mediterranean basin and (ii) landraces from North Africa and the Iberian Peninsula. More recently, Soriano et al. [49] showed an eastern–western dispersal of the Mediterranean durum landraces, which have been classified into four sub-populations: (i) Eastern Mediterranean, (ii) the Eastern Balkans and Turkey, (iii) the Western Balkans, and (iv) Egypt and the Western Mediterranean. The grouping of landraces from North Africa with Italy and Greece was also confirmed by Olivera et al. [57] and Nazco et al. [58], and could be explained by the Roman influence on durum wheat cultivation in North Africa. The early development of Carthage trade maritime activity in the Mediterranean sea enhanced seed exchanges between Tunisia and the Mediterranean countries [59], which might explain the grouping of the Bidi, Jenah Khoutifa, and Rommani lines with the majority of Algerian landraces (Sub-population 6) and that of the Biskri and Mahmoudi lines with Lebanese and Moroccan landraces (Sub-population 9). Moreover, our work confirmed that Biskri and Mahmoudi lines constitute the same gene pool and that Jenah Zarzoura lines constitute a new gene pool distant from the other Tunisian and foreign landraces, which were grouped with only two accessions from Jordan (Sub-population 11). A possible explanation is that the Jenah Zarzoura population, which was collected from the oasis of Mareth, located in the south of Tunisia, near the Mediterranean Sea, might have been introduced from the east through different paths, probably from Egypt to neighboring countries and possibly received from Palestine and Jordan [60,61], or through the introduction by the Phoenicians coming from Lebanon to Carthage between the 9th and 2nd centuries B.C [59]. During the Roman period (7th to 3rd centuries B.C.), Tunisia became the breadbasket of the Italian peninsula and the source of the excellent semolina quality from durum wheat grown in North-African countries [62]. The landraces of Tunisia and North Africa have also be traced to the introductions by Romans, who contributed to the modernization of the irrigation systems and extended wheat cultivation to Southern Tunisia [59].

3.4. Implications on Conservation and Use of Genetic Resources

The results of this study can be used to direct future activities of collection, conservation, and the use of durum wheat genetic resources in Tunisia. In terms of adding new diversity to the existing collections, more collection is needed in Tunisia mainly in the oasis areas to collect different landraces like Jenah Zarzoura. Future studies of this kind of germplasm will shed more light on the specific nature of this germplasm as in the case of durum wheat germplasm from Ethiopia [24]. Additionally, more landraces from other regions, even if they have the same local names, should be collected and given different identifiers within the genebank database and considered as different accessions in the ex situ collection. When collecting, the team should avoid plants with characteristics of improved varieties to avoid mixtures. For ex situ conservation, the genebank in Tunisia should conserve the bulk seeds of each landrace instead of seeds of many individual lines constituting each landrace. This will reduce the cost of conservation and avoid conserving several copies of the same line. DArTseq markers have also allowed us to identify outlier lines within the landraces and can therefore be eliminated during multiplication and characterization. For landraces still prevailing under traditional farming systems and under harsh conditions, on-farm conservation could be promoted to conserve a larger genetic base and the associated local knowledge. The need for ensuring long-term conservation of Tunisian durum wheat landraces is dictated by the on-going genetic erosion due to the spread of newly released cultivars and by their richness in genes for tolerance to drought, heat, and salinity and their quality attributes for different end uses [63].

4. Materials and Methods

4.1. Plant Material

This study used three sets of germplasm:

- Set 1: A total of 196 pure lines issued from six Tunisian durum wheat landraces known as Mahmoudi, Rommani, Jenah Zarzoura, Bidi, Jenah Khotifa, and Biskri, collected from 5 regions between 2009 and 2010 and conserved by the National Genebank of Tunisia (NGBT), were used for assessing the intra and inter genetic diversity (Table 6). The landraces Mahmoudi and Jenah Zarzoura are constituted by 30 lines each, and Rommani and Bidi are constituted by 33 lines each. The landraces Biskri and Jenah Khotifa are represented by two populations each: Biskri1 (31), Biskri2 (7), Jenah Khotifa1 (29), and Jenah Khotifa2 (3).
- Set 2: A total of 40 accessions composed of six improved varieties released in Tunisia, seven ICARDA/CIMMYT elite inbred lines and 27 Tunisian durum wheat landraces (15 landraces are represented by one accession; Jenah Khotifa, Biskri, and Chetla are represented by two accessions each; Bidi and Rommani are represented by three accessions each) were used for identification of potential mis-classified lines from Set 1 and for comparison between Tunisian durum wheat landraces and improved germplasm (Table 7).
- Set 3: A total of 207 durum wheat landraces collected from Mediterranean and West Asia countries—Morocco (17), Algeria (14), Tunisia (13), Libya (9), Egypt (12), Lebanon (10), Syria (11), Jordan (11), Israel/Palestine (10), Iraq (11), Iran (12), Afghanistan (16), Yemen (10), Greece (18), Cyprus (10), Italy (12), and Ethiopia (11)—randomly chosen from the ICARDA genebank collection along with four lines representing each Tunisian landraces from Set 1 were used for a genetic relationship study (Table 8).

The seeds for Sets 1 and 2 were taken from single plant spikes of each line, landrace, and germplasm grown at the Mornag INRA-Tunisia experiment station during the 2014–2015 growing season, while the seeds for Set 3 were send to CIMMYT from the ICARDA genebank within the joint effort to genotype wheat genetic resources.

Table 6. List and sample size of the Tunisian durum wheat landraces collected from the center, the south and the Oasis of Tunisia (Set 1).

Landraces	Number of Lines	Province
Bidi	33	Kairouan
Biskri1	31	Gafsa (Djebel ouled ouhiba)
Biskri2	7	Medenine (Zarzis)
Jenah Khotifa1	29	Kairouan
Jenah Khotifa2	3	Tozeur (El Frid)
Jenah Zarzoura	30	Matmata (Oasis of Mareth)
Mahmoudi	30	Gafsa (Snad)
Rommani	33	Gafsa (Djebel Ouled Ouhiba)
Total	196	

Table 7. List of the Tunisian durum wheat landraces, Tunisia released varieties, and ICARDA/CIMMYT elite lines (Set 2).

Tunisian Landraces	N	Origin	Accession Identifier	Pedigree
Agili	1	NGBT	IG 23903	-
Aouadi	1	NGBT	IG 23908	-
Arbi	1	NGBT	IG 23903	-
Azizi	1	NGBT	IG 23904	-
Bayadha	1	NGBT	IG 23905	-
Bidi	3	NGBT	IG 19553; IG 23906; IG 23929	-
Biskri	2	NGBT	IG 19551; IG 23907	-
Chetla	2	NGBT	IG 19555; IG 19557	-
Chili	1	NGBT	IG 23908	-
Derbessi	1	NGBT	IG 23909	-
Hmira	1	NGBT	IG 23910	-
Jneh khotifa	2	NGBT	IG 23915; IG 999	-
Mahmoudi	1	NGBT	IG 23911	-
Richi	1	NGBT	IG 23912	-
Rommani	3	NGBT	IG 19552; IG 19554; IG 19558	-
Sbei	1	NGBT	IG 23913	-
Souri	1	NGBT	IG 23914	-
Swabei algia	1	NGBT	IG 23916	-
Tounsia	1	NGBT	IG 19559	-
Ward bled	1	NGBT	IG 23917	-
Tunisian Improved Varieties	**N**	**Origin**	**Accession Identifier**	**Pedigree**
Inrat69 *	1	NGBT	IG 23919	Mahamoudi/Kyperounda
Karim *	1	NGBT	IG 23924	Jori"S"/Anhinga"S"//Flamingo"S"
Khiar *	1	NGBT	IG 23922	Chen/Altar 84
Om Rabii *	1	NGBT	IG 23921	Jori C69/ Hau
Nasr *	1	NGBT	IG 23923	GoVZ512/Cit//Ruff/Fg/3/Pin/Gre//Trob
Maali *	1	NGBT	IG 23920	CMH80A.1060/4/T.TURA/CMH74A.370// CMH77.774/3/YAV79/5/RAZZAK/6/DACK /YEL//KHIAR
ICARDA/CIMMYT Elite Lines	**N**	**Origin**	**Accession Identifier**	**Pedigree**
	1	ICARDA	MCHCB-102	OmRabi3/T.*urartu*500651/ch5//980947/3/ Otb4//Ossl1/Rfm6
IcaJoudy1 *	1	ICARDA	MCHCB-100	Atlast1/961081//Icasyr1
Nachit *	1	ICARDA	DAWRYT-106	Ameddkul1/T. *dicoccoides* Syrian collection//Loukos
Zeina4 *	1	ICARDA	MCHCB-154	GdoVZ512/Cit//Ruff/Fg/3/Src3
Louiza *	1	ICARDA	–	Rscn39/Til1
Ammar 6 *	1	ICARDA	IDYT37-5	ICAMORTA0472/Ammar7
Ammar 10 *	1	ICARDA	MCHCB-99	Lgt3/Bicrecham1

N: Number of accessions; *: Accessions marked with * are advanced lines; –: Inbred line without identifier; -: Accession without pedigree (Landrace), NGBT: National Gene Bank of Tunisia; ICARDA: International Center for Agricultural Research in Dry Areas.

Table 8. Number of durum wheat landraces from the Mediterranean and West Asian countries provided by the ICARDA genebank (Set 3).

Geographical Origin	Number of Landraces
Afghanistan	16
Cyprus	10
Algeria	14
Egypt	12
Ethiopia	11
Greece	18
Iran	12
Iraq	11
Israel	10
Italy	12
Jordan	11
Lebanon	10
Libya	9
Morocco	17
Syria	11
Tunisia	13
Yemen	10

4.2. Genotypic Characterization Using the DArtseq™ Method

Fresh young leaves from a single individual plant per accession have been used for genomic DNA extraction performed through a modified CTAB (cetyltrimethylammonium bromide) method [64]. The DNA quality was determined by electrophoresis using 1% agarose gel and quantified with NanoDrop 8000 spectrophotometer V 2.1.0.

A high-throughput genotyping method using DArTseq™ technology was employed to generate a genomic profile of the germplasm at the Genetic Analysis Service for Agriculture (SAGA) facility at CIMMYT, Mexico. This method used a combination of two restriction enzymes (PstI and HpaII) to reduce the genome complexity and to generate a genomic representation of the samples [31]. The genomic DNA was submitted to a process of digestion and ligation with a specific PstI-RE site-adapter tagged with 96 different barcodes, which allow to multiplex 96 DNA samples in a single lane of Illumina HiSeq2500 instrument (Illumina Inc., San Diego, CA, USA). The amplified fragments were sequenced up to 77 bases, generating around 500,000 unique reads per sample. A FASTQ files (full reads of 77 bp) were quality filtered using a Phred quality score of 30, which represents a 90% base call accuracy for at least 50% of the bases. An additional filter has been applied on barcode sequences. DArTsoft 14 was used to generate Silico-DArT score tables as data (1/0), indicating the presence/absence variation (PAV) markers and SNP markers.

4.3. DArTseq Markers and Cluster Analysis

DArTseq markers were mapped using the consensus map version 4.0 (www.diversityarrays.com) developed by DArT Pty. Ltd., Australia, and the reference wheat genome issued from the International Wheat Genome Sequencing Consortium database (IWGSC WGAv0.4), available online at https://urgi.versailles.inra.fr/download/iwgsc/.

DArTseq raw data was filtered according to markers criterion; minor allele frequency >5% and missing data ≤20%.

The summary statistics of the filtered DArTseq markers such as the expected heterozygosity (He) or genetic diversity (GD), minor allele frequency (MAF), and the polymorphic information content (PIC) [44], were calculated using R-project (http://www.r-project.org/) [65].

For cluster analysis of the collection, allele sharing distance matrix was computed as described by Goa et al. [66]. The distance between individuals i and j was defined as

$$D_{ij} = \frac{1}{L} \sum_{l=1}^{L} d_{ij}(l)$$

where L is the total number of markers, and dij(l) is 0, 1, or 2 if individuals i and j have zero, one, or two allele(s) in common at Locus l. Classification of the individuals into groups was performed using the allele sharing matrix and Ward's minimum variance algorithm [67]. The clustering algorithm was run using the hclust function within the R package [68].

4.4. Population Structure and Genetic Differentiation

Discriminant analysis of principal components (DAPC) was used to infer the number of clusters of genetically related individuals [48], using the *adegenet* package and *popr* in R-project [68]. DAPC is a multivariate analysis requiring three steps; first the data is transformed using principal component analysis, sub-groups are then identified using k-mean clustering, and discrimination between the sub-groups is then optimized using discriminant analysis. For the k-mean clustering, the optimal number of groups was identified using the Bayesian information criterion (BIC) as a measure of goodness of fit. The number of sub-groups (K) was set from 2 to 8 and the *K*-value with the lowest BIC was retained as the optimal number of clusters. A discriminant analysis was then implemented using the groups found by k-mean clustering [69].

For detecting the genetic variation among and within population(s) and supporting the hypothesis of the population structure, analysis of molecular variance (AMOVA) was performed for different hierarchical subdivision levels as well as for the full population structure strata from K = 2 to K = 8. Significance levels for variance component were estimated based on 10,000 permutations using the *randtest* function in the R-project as described by Excoffier et al. [70].

For genetic differentiation and relationships among the sub-populations issued from the population structure analysis, the genetic distance between the sub-populations using Reynolds genetic distance was computed [71], and the genetic indices, such as the observed heterozygosity (H_o, the proportion of loci that are heterozygote for a population), the expected heterozygosity or genetic diversity (*He*, the fraction of all landraces which would be heterozygote for any randomly chosen locus), and the F-statistics (F_{ST}) as developed by Wright [72], were calculated.

Author Contributions: Conceptualization, C.R., Z.K. and A.A.; data curation, C.R., Z.K. and A.A.; formal analysis, C.R., Z.K. and A.A.; funding acquisition, A.A.; investigation, C.R. and A.A.; methodology, C.R., Z.K. and A.A.; project administration, A.A.; resources, M'B.B.N.; Software, Z.K.; supervision, A.A.; Validation, Z.K. and A.A.; visualization, Z.K., C.S., F.B. and A.A.; writing—original draft, C.R., Z.K. and A.A.; writing—review & editing, C.R., Z.K. and A.A.

Funding: This work was supported by the Genebank Platform and the Global Wheat Program, ICARDA.

Acknowledgments: We would like to thank the General Director of the National Gene bank of Tunisia, M. Ben Naceur, for providing the germplasm used in this study. We would also like to thank Athanasios Tsivelikas for field assistance and the technical staff of Mornag INRAT station, Tunisia, for handling field and laboratory activities. The genotyping characterization work was implemented by CIMMYT as part of the MasAgro Biodiversidad project in collaboration with ICARDA, made possible by the generous support of the Mexican Secretariat of Agriculture, Livestock, Rural Development, Fisheries and Food (SAGARPA) and CRP Wheat.

Conflicts of Interest: The authors declare no conflict of interest.

Abbreviations

DarTseq	diversity array technology sequencing
DAPC	discriminant analysis of principal components
AMOVA	analysis of molecular variance
RE	restriction enzyme

References

1. Gioia, T.; Nagel, K.A.; Beleggia, R.; Fragasso, M.; Ficco, D.B.M.; Pieruschka, R.; De Vita, P.; Fiorani, F.; Papa, R. Impact of Domestication on the Phenotypic Architecture of Durum Wheat under Contrasting Nitrogen Fertilization. *J. Exp. Bot.* **2015**, *66*, 5519–5530. [CrossRef] [PubMed]

2. *The World Wheat Book: A History of Wheat Breeding*; Bonjean, A.P.; Angus, W.J.; van Ginkel, M. (Eds.) Lavoisier: Paris, France, 2016; Volume 3, ISBN 9782743020910.

3. Daaloul, A.; Harrabi, M.; Amara, H. Evaluation de la collection nationale de blé dur. In *Revure de l'INAT*; numéro spécial; INAT: Tunis, Tunisia, 1998; pp. 337–358. (In French)

4. Deghaïs, M.; Kouki M, M.; Gharbi M, M.; El Felah, M. *Les Variétés de Céréales Cultivées en Tunisie (blé dur, blé Tendre, orge et Triticale)*; INRAT: Tunis, Tunisia, 2007. (In French)

5. Dvorak, J.; Akhunov, E.D. Tempos of Gene Locus Deletions and Duplications and Their Relationship to Recombination Rate During Diploid and Polyploid Evolution in the *Aegilops-Triticum* Alliance. *Genetics* **2005**, *171*, 323–332. [CrossRef] [PubMed]

6. Feldman, M.; Levy, A.A. Allopolyploidy—A Shaping Force in the Evolution of Wheat Genomes. *Cytogenet. Genome Res.* **2005**, *109*, 250–258. [CrossRef] [PubMed]

7. Salamini, F.; Özkan, H.; Brandolini, A.; Schäfer-Pregl, R.; Martin, W. Genetics and Geography of Wild Cereal Domestication in the near East. *Nat. Rev. Genet.* **2002**, *3*, 429–441. [CrossRef] [PubMed]

8. Matsuoka, Y. Evolution of Polyploid Triticum Wheats under Cultivation: The Role of Domestication, Natural Hybridization and Allopolyploid Speciation in Their Diversification. *Plant Cell Physiol.* **2011**, *52*, 750–764. [CrossRef] [PubMed]

9. Zohary, D.; Hopf, M. *Domestication of Plants in the Old World: The Origin and Spread of Cultivated Plants in West Asia, Europe, and the Nile Valley*, 3rd ed.; Oxford University Press: Oxford, UK, 2000; ISBN 9780198503576 9780198503569.

10. Vavilov, N.I. The Origin, Variation, Immunity and Breeding of Cultivated Plants. *Soil Sci.* **1951**, *72*, 482. [CrossRef]

11. Boeuf, F. *Le blé en Tunisie: La Plante. Le Milieu Physico-Chimique*; Société Anonyme de L'Imprimerie Rapide de Tunis: Tunis, Tunisia, 1932; Volume 1.

12. Lala, S.; Amri, A.; Maxted, N. Towards the Conservation of Crop Wild Relative Diversity in North Africa: Checklist, Prioritisation and Inventory. *Genet. Resour. Crop Evol.* **2018**, *65*, 113–124. [CrossRef]

13. Khoufi, S. Morphological and Molecular Characterization of Six of the Most Frequently Cultivated Hard Wheat Varieties in Tunisia. *J. Plant Breed. Crop Sci.* **2012**, *4*. [CrossRef]

14. Lopes, M.S.; El-Basyoni, I.; Baenziger, P.S.; Singh, S.; Royo, C.; Ozbek, K.; Aktas, H.; Ozer, E.; Ozdemir, F.; Manickavelu, A.; et al. Exploiting Genetic Diversity from Landraces in Wheat Breeding for Adaptation to Climate Change. *J. Exp. Bot.* **2015**, *66*, 3477–3486. [CrossRef]

15. Sourour, A.; Chahine, K.; Youssef, T.; Olfa, S.; Hajer, S. Phenotypic Diversity of Tunisian Durum Wheat Landraces. *Afr. Crop Sci. J.* **2010**, *18*. [CrossRef]

16. Sourour, A.; Hajer, S. Distribution and Phenotypic Variability Aspects of Some Quantitiative Traits among Durum Wheat Accessiions. *Afr. Crop Sci. J.* **2010**, *16*. [CrossRef]

17. Ya, N.; Raveendar, S.; Bayarsukh, N.; Ya, M.; Lee, J.-R.; Lee, K.-J.; Shin, M.-J.; Cho, G.-T.; Ma, K.-H.; Lee, G.-A. Genetic Diversity and Population Structure of Mongolian Wheat Based on SSR Markers: Implications for Conservation and Management. *Plant Breed. Biotechnol.* **2017**, *5*, 213–220. [CrossRef]

18. Sohail, Q.; Manickavelu, A.; Ban, T. Genetic Diversity Analysis of Afghan Wheat Landraces (*Triticum aestivum*) Using DArT Markers. *Genet. Resour. Crop Evol.* **2015**, *62*, 1147–1157. [CrossRef]

19. Khan, M.K.; Pandey, A.; Choudhary, S.; Hakki, E.E.; Akkaya, M.S.; Thomas, G. From RFLP to DArT: Molecular Tools for Wheat (*Triticum* spp.) Diversity Analysis. *Genet. Resour. Crop Evol.* **2014**, *61*, 1001–1032. [CrossRef]

20. Kehel, Z.; Garcia-Ferrer, A.; Nachit, M.M. Using Bayesian and Eigen Approaches to Study Spatial Genetic Structure of Moroccan and Syrian Durum Wheat Landraces. *AJMB* **2013**, *03*, 17–31. [CrossRef]

21. Medini, M.; Hamza, S.; Rebai, A.; Baum, M. Analysis of Genetic Diversity in Tunisian Durum Wheat Cultivars and Related Wild Species by SSR and AFLP Markers. *Genet. Resour. Crop Evol.* **2005**, *52*, 21–31. [CrossRef]

22. Maccaferri, M.; Ricci, A.; Salvi, S.; Milner, S.G.; Noli, E.; Martelli, P.L.; Casadio, R.; Akhunov, E.; Scalabrin, S.; Vendramin, V.; et al. A High-Density, SNP-Based Consensus Map of Tetraploid Wheat as a Bridge to Integrate Durum and Bread Wheat Genomics and Breeding. *Plant Biotechnol. J.* **2015**, *13*, 648–663. [CrossRef]

23. Borrill, P.; Adamski, N.; Uauy, C. Genomics as the Key to Unlocking the Polyploid Potential of Wheat. *New Phytol.* **2015**, *208*, 1008–1022. [CrossRef]

24. Kabbaj, H.; Sall, A.T.; Al-Abdallat, A.; Geleta, M.; Amri, A.; Filali-Maltouf, A.; Belkadi, B.; Ortiz, R.; Bassi, F.M. Genetic Diversity within a Global Panel of Durum Wheat (*Triticum durum*) Landraces and Modern Germplasm Reveals the History of Alleles Exchange. *Front. Plant Sci.* **2017**, *8*, 1277. [CrossRef]

25. Holtz, Y.; Ardisson, M.; Ranwez, V.; Besnard, A.; Leroy, P.; Poux, G.; Roumet, P.; Viader, V.; Santoni, S.; David, J. Genotyping by Sequencing Using Specific Allelic Capture to Build a High-Density Genetic Map of Durum Wheat. *PLoS ONE* **2016**, *11*, e0154609. [CrossRef]

26. He, J.; Zhao, X.; Laroche, A.; Lu, Z.-X.; Liu, H.; Li, Z. Genotyping-by-Sequencing (GBS), an Ultimate Marker-Assisted Selection (MAS) Tool to Accelerate Plant Breeding. *Front. Plant Sci.* **2014**, *5*, 484. [CrossRef]

27. Sieber, A.-N.; Longin, C.F.H.; Würschum, T. Molecular Characterization of Winter Durum Wheat (*Triticum durum*) Based on a Genotyping-by-Sequencing Approach. *Plant Genet. Resour.* **2017**, *15*, 36–44. [CrossRef]

28. Alipour, H.; Bihamta, M.R.; Mohammadi, V.; Peyghambari, S.A.; Bai, G.; Zhang, G. Genotyping-by-Sequencing (GBS) Revealed Molecular Genetic Diversity of Iranian Wheat Landraces and Cultivars. *Front. Plant Sci.* **2017**, *8*, 1293. [CrossRef]

29. Wenzl, P.; Carling, J.; Kudrna, D.; Jaccoud, D.; Huttner, E.; Kleinhofs, A.; Kilian, A. Diversity Arrays Technology (DArT) for Whole-Genome Profiling of Barley. *Proc. Natl. Acad. Sci. USA* **2004**, *101*, 9915–9920. [CrossRef] [PubMed]

30. Akbari, M.; Wenzl, P.; Caig, V.; Carling, J.; Xia, L.; Yang, S.; Uszynski, G.; Mohler, V.; Lehmensiek, A.; Kuchel, H.; et al. Diversity Arrays Technology (DArT) for High-Throughput Profiling of the Hexaploid Wheat Genome. *Theor. Appl. Genet.* **2006**, *113*, 1409–1420. [CrossRef]

31. Sansaloni, C.; Petroli, C.; Jaccoud, D.; Carling, J.; Detering, F.; Grattapaglia, D.; Kilian, A. Diversity Arrays Technology (DArT) and next-Generation Sequencing Combined: Genome-Wide, High Throughput, Highly Informative Genotyping for Molecular Breeding of Eucalyptus. *BMC Proc.* **2011**, *5* (Suppl. 7), P54. [CrossRef]

32. Baloch, F.S.; Alsaleh, A.; Shahid, M.Q.; Çiftçi, V.E.; Sáenz de Miera, L.; Aasim, M.; Nadeem, M.A.; Aktaş, H.; Özkan, H.; Hatipoğlu, R. A Whole Genome DArTseq and SNP Analysis for Genetic Diversity Assessment in Durum Wheat from Central Fertile Crescent. *PLoS ONE* **2017**, *12*, e0167821. [CrossRef] [PubMed]

33. Li, H.; Vikram, P.; Singh, R.P.; Kilian, A.; Carling, J.; Song, J.; Burgueno-Ferreira, J.A.; Bhavani, S.; Huerta-Espino, J.; Payne, T.; et al. A High Density GBS Map of Bread Wheat and Its Application for Dissecting Complex Disease Resistance Traits. *BMC Genom.* **2015**, *16*. [CrossRef]

34. Yang, X.; Ren, R.; Ray, R.; Xu, J.; Li, P.; Zhang, M.; Liu, G.; Yao, X.; Kilian, A. Genetic Diversity and Population Structure of Core Watermelon (*Citrullus lanatus*) Genotypes Using DArTseq-Based SNPs. *Plant Genet. Resour.* **2016**, *14*, 226–233. [CrossRef]

35. Valdisser, P.A.M.R.; Pereira, W.J.; Almeida Filho, J.E.; Müller, B.S.F.; Coelho, G.R.C.; de Menezes, I.P.P.; Vianna, J.P.G.; Zucchi, M.I.; Lanna, A.C.; Coelho, A.S.G.; et al. In-Depth Genome Characterization of a Brazilian Common Bean Core Collection Using DArTseq High-Density SNP Genotyping. *BMC Genom.* **2017**, *18*. [CrossRef]

36. Marcotuli, I.; Gadaleta, A.; Mangini, G.; Signorile, A.; Zacheo, S.; Blanco, A.; Simeone, R.; Colasuonno, P. Development of a High-Density SNP-Based Linkage Map and Detection of QTL for β-Glucans, Protein Content, Grain Yield per Spike and Heading Time in Durum Wheat. *Int. J. Mol. Sci.* **2017**, *18*, 1329. [CrossRef] [PubMed]

37. Patil, R.M.; Oak, M.D.; Tamhankar, S.A.; Sourdille, P.; Rao, V.S. Mapping and Validation of a Major QTL for Yellow Pigment Content on 7AL in Durum Wheat (*Triticum turgidum* L. ssp. *durum*). *Mol. Breed* **2008**, *21*, 485–496. [CrossRef]

38. Colasuonno, P.; Lozito, M.L.; Marcotuli, I.; Nigro, D.; Giancaspro, A.; Mangini, G.; De Vita, P.; Mastrangelo, A.M.; Pecchioni, N.; Houston, K.; et al. The Carotenoid Biosynthetic and Catabolic Genes in Wheat and Their Association with Yellow Pigments. *BMC Genom.* **2017**, *18*. [CrossRef]

39. Turki, N.; Shehzad, T.; Harrabi, M.; Okuno, K. Detection of QTLs Associated with Salinity Tolerance in Durum Wheat Based on Association Analysis. *Euphytica* **2015**, *201*, 29–41. [CrossRef]

40. Mangini, G.; Gadaleta, A.; Colasuonno, P.; Marcotuli, I.; Signorile, A.M.; Simeone, R.; De Vita, P.; Mastrangelo, A.M.; Laidò, G.; Pecchioni, N.; et al. Genetic Dissection of the Relationships between Grain Yield Components by Genome-Wide Association Mapping in a Collection of Tetraploid Wheats. *PLoS ONE* **2018**, *13*, e0190162. [CrossRef]

41. Chen, T.; Tantasawat, P.A.; Wang, W.; Gao, X.; Zhang, L. Population Structure of Chinese Southwest Wheat Germplasms Resistant to Stripe Rust and Powdery Mildew Using the DArT-Seq Technique. *Cienc. Rural* **2018**, *48*. [CrossRef]

42. Nadeem, M.A.; Nawaz, M.A.; Shahid, M.Q.; Doğan, Y.; Comertpay, G.; Yıldız, M.; Hatipoğlu, R.; Ahmad, F.; Alsaleh, A.; Labhane, N.; et al. DNA Molecular Markers in Plant Breeding: Current Status and Recent Advancements in Genomic Selection and Genome Editing. *Biotechnol. Biotechnol. Equip.* **2018**, *32*, 261–285. [CrossRef]

43. Ren, J.; Sun, D.; Chen, L.; You, F.; Wang, J.; Peng, Y.; Nevo, E.; Sun, D.; Luo, M.-C.; Peng, J. Genetic Diversity Revealed by Single Nucleotide Polymorphism Markers in a Worldwide Germplasm Collection of Durum Wheat. *Int. J. Mol. Sci.* **2013**, *14*, 7061–7088. [CrossRef]

44. Botstein, D.; White, R.L.; Skolnick, M.; Davis, R.W. Construction of a Genetic Linkage Map in Man Using Restriction Fragment Length Polymorphisms. *Am. J. Hum. Genet.* **1980**, *32*, 314–331.

45. Soriano, J.M.; Villegas, D.; Sorrells, M.E.; Royo, C. Durum Wheat Landraces from East and West Regions of the Mediterranean Basin Are Genetically Distinct for Yield Components and Phenology. *Front. Plant Sci.* **2018**, *9*, 80. [CrossRef]

46. Oliveira, H.R.; Hagenblad, J.; Leino, M.W.; Leigh, F.J.; Lister, D.L.; Peña-Chocarro, L.; Jones, M.K. Wheat in the Mediterranean Revisited – Tetraploid Wheat Landraces Assessed with Elite Bread Wheat Single Nucleotide Polymorphism Markers. *BMC Genet.* **2014**, *15*, 54. [CrossRef] [PubMed]

47. Miège, E. Les principales espèces et variétés de Blé cultivées en Afrique du Nord (Suite et Fin). *Revue Internationale de Botanique Appliquée et d'Agriculture Tropicale* **1950**, *30*, 203–215. (In French) [CrossRef]

48. Jombart, T.; Devillard, S.; Balloux, F. Discriminant Analysis of Principal Components: A New Method for the Analysis of Genetically Structured Populations. *BMC Genet.* **2010**, *11*, 94. [CrossRef] [PubMed]

49. Soriano, J.M.; Villegas, D.; Aranzana, M.J.; García del Moral, L.F.; Royo, C. Genetic Structure of Modern Durum Wheat Cultivars and Mediterranean Landraces Matches with Their Agronomic Performance. *PLoS ONE* **2016**, *11*, e0160983. [CrossRef]

50. Ruiz, M.; Giraldo, P.; Royo, C.; Villegas, D.; Aranzana, M.J.; Carrillo, J.M. Diversity and Genetic Structure of a Collection of Spanish Durum Wheat Landraces. *Crop Sci.* **2012**, *52*, 2262. [CrossRef]

51. Eltaher, S.; Sallam, A.; Belamkar, V.; Emara, H.A.; Nower, A.A.; Salem, K.F.M.; Poland, J.; Baenziger, P.S. Genetic Diversity and Population Structure of F3:6 Nebraska Winter Wheat Genotypes Using Genotyping-By-Sequencing. *Front. Genet.* **2018**, *9*, 76. [CrossRef]

52. Jaradat, A. Wheat Landraces: A Mini Review. *Emir. J. Food Agric.* **2013**, *25*, 20. [CrossRef]

53. Mangini, G.; Margiotta, B.; Marcotuli, I.; Signorile, M.A.; Gadaleta, A.; Blanco, A. Genetic Diversity and Phenetic Analysis in Wheat (*Triticum turgidum* subsp. *durum* and *Triticum aestivum* subsp. *aestivum*) Landraces Based on SNP Markers. *Genet. Resour. Crop Evol.* **2017**, *64*, 1269–1280. [CrossRef]

54. Sahri, A.; Chentoufi, L.; Arbaoui, M.; Ardisson, M.; Belqadi, L.; Birouk, A.; Roumet, P.; Muller, M.-H. Towards a Comprehensive Characterization of Durum Wheat Landraces in Moroccan Traditional Agrosystems: Analysing Genetic Diversity in the Light of Geography, Farmers' Taxonomy and Tetraploid Wheat Domestication History. *BMC Evol. Biol.* **2014**, *14*. [CrossRef]

55. Mac Key, J. Wheat: Its concept, evolution and taxonomy. In *Durum Wheat Breeding: Current Approaches and Future Strategies*; Royo, C., Nachit, M., Di Fonzo, N., Araus, J.L., Pfeiffer, W.H., Slafer, G.A., Eds.; Food Products Press: Binghamton, NY, USA, 2005; pp. 3–61. ISBN 9781482277883.

56. Moragues, M.; Moralejo, M.; Sorrells, M.E.; Royo, C. Dispersal of Durum Wheat [*Triticum turgidum* L. ssp. *turgidum* Convar. *durum* (Desf.) MacKey] Landraces across the Mediterranean Basin Assessed by AFLPs and Microsatellites. *Genet. Resour. Crop Evol.* **2007**, *54*, 1133–1144. [CrossRef]

57. Oliveira, H.R.; Campana, M.G.; Jones, H.; Hunt, H.V.; Leigh, F.; Redhouse, D.I.; Lister, D.L.; Jones, M.K. Tetraploid Wheat Landraces in the Mediterranean Basin: Taxonomy, Evolution and Genetic Diversity. *PLoS ONE* **2012**, *7*, e37063. [CrossRef] [PubMed]

58. Nazco, R.; Villegas, D.; Ammar, K.; Peña, R.J.; Moragues, M.; Royo, C. Can Mediterranean Durum Wheat Landraces Contribute to Improved Grain Quality Attributes in Modern Cultivars? *Euphytica* **2012**, *185*, 1–17. [CrossRef]

59. Essid, M.Y. Chapter 2. History of Mediterranean food, CIHEAM ed. In *MediTERRA 2012 (english): The Mediterranean Diet for Sustainable Regional Development*; Presses de Sciences Po (P.F.N.S.P.): France, Paris, 2012; pp. 51–69.

60. Perrino, P.; Porceddu, E. Wheat Genetic Resources in Ethiopia and the Mediterranean Region. In *Wheat Genetic Resources: Meeting Diverse Needs*; John Wiley & Sons: Chichester, UK, 1990; Volume 27, pp. 161–178, 364–365. ISBN 0471928801.

61. Feldman, M. Origin of Cultivated Wheat. In *The World Wheat Book: A History of Wheat Breeding*; Bonjean, A.P., Angus, W.J., Eds.; Lavoisier: Paris, France, 2001; pp. 3–56.

62. Thompson, D.A.W. Wheat in Antiquity - Naum Jasny: The Wheats of Classical Antiquity. (Johns Hopkins University Studies in Historical and Political Science, Series LXII, No. 3.) Pp. 176; 2 Plates. Baltimore: Johns Hopkins Press, 1944. Paper, $1.75. *Class. Rev.* **1946**, *60*, 120–122. [CrossRef]

63. Zaharieva, M.; Bonjean, A.; Monneveux, P. Saharan Wheats: Before They Disappear. *Genet. Resour. Crop Evol.* **2014**, *61*, 1065–1084. [CrossRef]

64. Hoisington, D.; Khairallah, M.; Gonzalez-de-Leon, D. *Laboratory Protocols, CIMMYT Applied Molecular Genetics Laboratory*, 2nd ed.; CIMMYT: Texcoco, Mexico, 1994; Available online: http://repository.cimmyt.org/xmlui/bitstream/handle/10883/1333/91195.pdf (accessed on 20 April 2015).

65. R core team. *R: A Language and Environment for Statistical Computing*; R Foundation for Statistical Computing: Vienna, Austria, 2015; Available online: http://www.R-project.org/ (accessed on 3 November 2018).

66. Gao, X.; Martin, E.R. Using Allele Sharing Distance for Detecting Human Population Stratification. *Hum. Hered* **2009**, *68*, 182–191. [CrossRef] [PubMed]

67. Ward, J.H. Hierarchical Grouping to Optimize an Objective Function. *J. Am. Stat. Assoc.* **1963**, *58*, 236–244. [CrossRef]
68. Jombart, T. Multivariate Analysis of Genetic Data: Exploring Group Diversity. Available online: http://adegenet.r-forge.r-project.org/files/PRstats/practical-MVAgroups.1.0.pdf (accessed on 18 August 2016).
69. Schwarz, G. Estimating the Dimension of a Model. *Ann. Stat.* **1978**, *6*, 461–464. [CrossRef]
70. Excoffier, L.; Smouse, P.E.; Quattro, J.M. Analysis of Molecular Variance Inferred from Metric Distances among DNA Haplotypes: Application to Human Mitochondrial DNA Restriction Data. *Genetics* **1992**, *131*, 479–491.
71. Reynolds, J.; Weir, B.S.; Cockerham, C.C. Estimation of The Coancestry Coefficient: Basis For A Short-Term Genetic Distance. *Genetics* **1983**, *105*, 767–779.
72. Wright, S. Coefficients of Inbreeding and Relationship. *Am. Nat.* **1922**, *56*, 330–338. [CrossRef]

© 2019 by the authors. Licensee MDPI, Basel, Switzerland. This article is an open access article distributed under the terms and conditions of the Creative Commons Attribution (CC BY) license (http://creativecommons.org/licenses/by/4.0/).

International Journal of
Molecular Sciences

MDPI

Article

Transcriptional Dynamics of Grain Development in Barley (*Hordeum vulgare* L.)

Jianxin Bian [1,†], Pingchuan Deng [1,†], Haoshuang Zhan [1], Xiaotong Wu [1],
Mutthanthirige D. L. C. Nishantha [1], Zhaogui Yan [2], Xianghong Du [1], Xiaojun Nie [1,*] and
Weining Song [1,3,*]

1   State Key Laboratory of Crop Stress Biology in Arid Areas, College of Agronomy and Yangling Branch of
     China Wheat Improvement Center, Northwest A&F University, Yangling, Shaanxi 712100, China;
     brian1791@nwafu.edu.cn (J.B.); dengpingchuan@nwsuaf.edu.cn (P.D.);
     zhanhaoshuang@nwsuaf.edu.cn (H.Z.); wuxiaotong@nwsuaf.edu.cn (X.W.);
     mdlcnishantha@gmail.com (M.D.L.C.N.); xianghongdu@nwsuaf.edu.cn (X.D.)
2   Huazhong Agricultural University, Wuhan 430070, China; gyan@mail.hzau.edu.cn
3   Joint Research Center for Agriculture Research in Arid Areas, Yangling, Shaanxi 712100, China
*   Correspondence: small@nwsuaf.edu.cn (X.N.); sweining2002@yahoo.com (W.S.);
     Tel.: (+86)-29-87082984 (W.S.); Fax: (+86)-29-87080191 (W.S.)
†   These authors contributed equally to this work.

Received: 24 December 2018; Accepted: 19 February 2019; Published: 22 February 2019

Abstract: Grain development, as a vital process in the crop's life cycle, is crucial for determining crop quality and yield. However, the molecular basis and regulatory network of barley grain development is not well understood at present. Here, we investigated the transcriptional dynamics of barley grain development through RNA sequencing at four developmental phases, including early prestorage phase (3 days post anthesis (DPA)), late prestorage or transition phase (8 DPA), early storage phase (13 DPA), and levels off stages (18 DPA). Transcriptome profiling found that pronounced shifts occurred in the abundance of transcripts involved in both primary and secondary metabolism during grain development. The transcripts' activity was decreased during maturation while the largest divergence was observed between the transitions from prestorage phase to storage phase, which coincided with the physiological changes. Furthermore, the transcription factors, hormone signal transduction-related as well as sugar-metabolism-related genes, were found to play a crucial role in barley grain development. Finally, 4771 RNA editing events were identified in these four development stages, and most of the RNA editing genes were preferentially expressed at the prestore stage rather than in the store stage, which was significantly enriched in "essential" genes and plant hormone signal transduction pathway. These results suggested that RNA editing might act as a 'regulator' to control grain development. This study systematically dissected the gene expression atlas of barley grain development through transcriptome analysis, which not only provided the potential targets for further functional studies, but also provided insights into the dynamics of gene regulation underlying grain development in barley and beyond.

Keywords: Barley; Grain development; Transcriptional dynamics; RNA editing; RNA-seq

1. Introduction

Grain development is one of the most essential processes for the life cycle of plants, and especially for crops, which is not only critical for colonization of the environment for plant survival, but also provides the main food source for human beings [1–4]. It is a very complex biological process involving multiple metabolic regulation pathways, which can be further divided into two major phases, embryogenesis and maturation. During the embryogenesis process, embryonic cells divide and go through characteristic stages of development, while maturation mainly involves accumulation of organic materials such as protein, lipids, and carbohydrates [4–7]. A better understanding of the molecular mechanisms of grain development will contribute to crop improvement, including seed formation, morphogenesis, and storage reserve accumulation. During the last few years, a larger number of studies have been conducted to investigate the specific genes or pathways controlling grain development in different species, particularly in model organisms such as rice [8] and Arabidopsis [3,5,9]. These studies have provided important information on both of the regulators of transcription and the genes involved in grain development.

Barley (*Hordeum vulgare*), a member of the Poaceae family, was domesticated thousands of years ago and has been utilized mainly for animal feed, malting, and the brewing industry, and as a human food source [1,10]. Because of its diploid nature and rich genetic diversity, barley is also a well-studied crop regardless of genetics or genomics and once acted as a model plant for Triticeae research. Since the grain is the agriculturally most important organ of crops, many studies have been carried out on barley grain development [1,6,7,11–13]. Preliminary transcriptomic analyses of barley seeds have provided important information regarding tissue-specific metabolic pathways and the regulatory network controlling grain development, including storage reserve accumulation [6], photosynthesis [7], hormone biosynthesis [6,14], programmed cell death [12], sucrose transport [13], and zinc trafficking [15]. Particularly, a comparative study of transcript and metabolite profiling at both temporal and spatial levels dissected that glumes acted as important transitory resource buffers during barley endosperm filling [16]. Furthermore, the differentially expressed genes (DEGs) involved in barley grain development through RNA-seq are also reported and available in barleyGenes—Barley RNA-seq Database (https://ics.hutton.ac.uk/barleyGenes/about_material.html). However, the transcriptomic dynamics of barley grain development are not well understood, and in particular, there is little known about the regulatory network at present. Recently, the high-quality reference sequence of the barley genome was completed, which provides the unprecedented opportunity to investigate the dynamic of the barley grain development at the whole transcriptome level [17]. Here, we performed a study of mRNA-seq analysis at four stages spanning important developmental phases of seed filling with the purpose to the transcriptional dynamics and regulatory process of grain development in barley.

2. Results

2.1. RNA-Seq Analysis of Grain Development in Barley

In this study, the dynamics of mRNA abundance were systematically investigated at four important stages (3, 8, 13, and 18 DPA) of grain development in barley (Figure S1). Totally, 52.69 Gb of raw data was obtained for all of the samples, with the average of about 65 million pair-end reads with 100 bp in size for each sample (Table S1). After removing adaptor sequences and low-quality reads (reads with N and empty reads), there were around 521 million clean reads remaining, which represented 98.97% of the raw data. After quality filter, all of the clean reads were mapped to the barley reference genome [17]. Results showed that on average, 55.10 million paired-end reads (84.53%) could be uniquely mapped to the reference genome [17], with the range from 53.86 (82.75%) to 56.28 million reads (86.29%) (Figure S2A). Out of 39,734 annotated protein-coding genes in the barley genome, 19,165 (48.23%) genes were found to be expressed during grain development (Table S1), which was consistent with a previous study in Arabidopsis (~44%) and hexaploid wheat (~55%) [2,18].

To determine whether this gene expression correlated with developmental stages, reads numbers were firstly normalized to RPKM value, and then were subjected to the usual correlation coefficient (R^2) and hierarchical clustering analysis [19]. We found that two biological replicates of all samples showed consistent determinations of transcript abundance with a coefficient (R^2) greater than 0.93 (Figure 1A). In addition, gene expression correlations between different stages were compared. Results found that the coefficients of 3 DPA (stage01) and 8 DPA (stage02) were 0.73 ($R^2 > 0.73$) and those of 13 DPA (stage03) and 18 DPA (stage04) were higher with the value of 0.89. However, the coefficient between 13 DPA (stage03) and 8 DPA (stage02) was only 0.10, indicating a large change in gene expression pattern occurred during the process. It may be the transition from prestorage phase to storage phase, which was consistent with a previous study finding that the most dramatic transcriptional and physiological changes occurred during transition from the prestorage to the storage phase [6]. Similar stage expression patterns were supported by correlation dendrogram analysis (Figure 1B). The correlation dendrogram demonstrated that the four stages were clearly clustered into two groups, namely 3 DPA (stage01) and 8 DPA (stage02) grouped together, and 13 DPA (Stage03) and 18 DPA (Stage04) clustered into another group. Then, we also used the publicly available barley grain RNA-seq data (5 DPA and 15 DPA) to perform correlation analysis. Results found that 3 DPA and 8 DPA showed high correlation coefficient with 5 DPA and 13 DPA and 18 DPA showed high correlation coefficient with 15 DPA, which also could be clustered into two groups (Figure S3). Overall, the correlation analyses results were in agreement with the developmental order as expected, which were consistent with the previous study [7].

Furthermore, we compared the changes of gene expression abundance at different development stages. The total number of expressed genes in each sample varied from 13,290 (stage04) to 16,532 (stage01) (Table S2) and approximately 63.79% (11,826/18,540) of them were expressed across all of these four stages (Figure 1C). Interestingly, the majority of the genes (80.35%) exhibited peak expression at the prestorage phase (stage01 and stage02) rather than the storage phase (stage03 and stage04) (Figure 1D). GO (Gene ontology)analysis identified the peak expressed genes were significantly enriched in molecular functions at each stage. For example, at the prestorage phase, the peak genes were enriched in terms of cell cycle, DNA, RNA, peptidase biosynthesis, and modification such as cell cycle process (GO: 0022402), cell division (GO:0051301), nuclease activity (GO:0004518), DNA metabolic process (GO: 0006259), tRNA processing (GO: 0008033), double-stranded DNA binding (GO:0003690), microtubule motor activity(GO:0003777), lipoprotein biosynthetic process (GO:0042158), fatty acid biosynthetic process (GO:0006633) (Figure 1E) (Table S3). However, at the storage phase, gene sets were enriched in terms of matter accumulation, such as nutrient reservoir activity (GO: 0045735), catalytic activity (GO: 0003824), 6-phosphofructokinase activity (GO: 0003872), fructose 6-phosphate metabolic process (GO: 0006002), response to temperature stimulus (GO:0009266), response to heat(GO:0009408), disaccharide metabolic process (GO:0005984), protein transporter activity (GO:0008565), carbohydrate derivative catabolic process (GO:1901136) (Figure 1F) (Table S4).

Genes displaying enriched expression patterns in a given developmental stage or tissue are important for understanding the specialized processes within these stages or tissues. As described previously, an empirical cutoff value for positively expressed genes (FPKM $\geq$ 1) was used to detect stage-specific expression candidates [20]. In our study, we identified 2158 stage-enriched genes, of which stage01 possessed the highest number (1183), followed by stage02 (538), stage04 (254), and stage03 (183) (Figure 1C). GO enrichment results showed the stage-specific expression genes involved in DNA metabolic process (GO:0006259), DNA repair (GO:0006281), reproductive process (GO:0022414), RNA processing (GO:0006396) for stage01, transport (GO:0006810), single-organism transport (GO:0044765), lipid metabolic process (GO:0006629), carbohydrate binding (GO:0030246) for stage02, and aromatic compound biosynthetic process (GO:0019438), organic cyclic compound biosynthetic process (GO:1901362), biosynthetic process (GO:0009058), organic substance biosynthetic process (GO:1901576) for stage03, as well as serine-type endopeptidase inhibitor activity (GO:0004867), transferase activity, transferring hexosyl groups (GO:0016758) for stage04 (Table S4), which represented the grain developmental program and biological process.

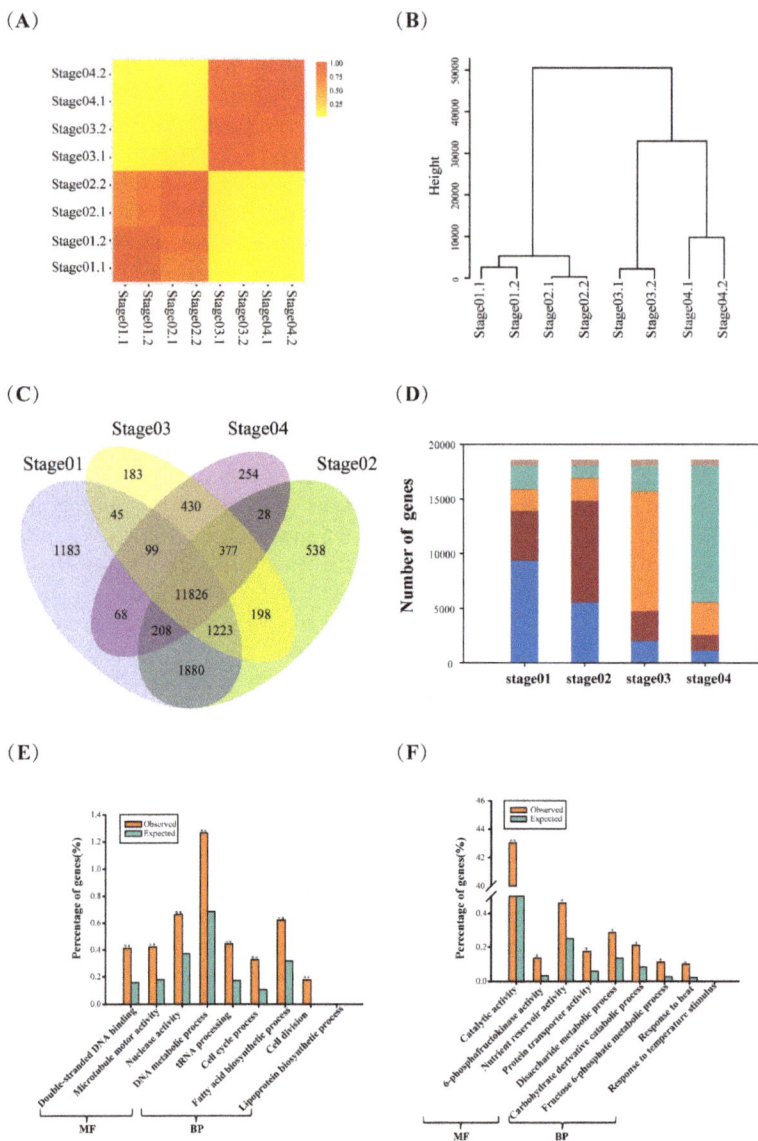

Figure 1. Global characterization of gene expression patterns during the four stages of barley grain development. (**A**) Correlation coefficients between gene expression data sets from two biological duplicates and stages. (**B**) Cluster dendrogram showing global relationships of gene expression in different stages. The branch length indicates the degree of variance. (**C**) Venn diagram analyses of stage-specific genes in barley. (**D**) Overview of genes with different activity degree among the four stages. The different number suggests different activity ranking at each stage and the activity degree decreases as the number increases. For example, '1' indicates gene showed the most activity at this stage and '4' suggests the most inactive stage. (**E**) to (**F**) Functional categories of genes showing peak expression at the prestorage phase (stage01 and stage02) (**E**) and storage phase (stage03 and stage04). (**F**) Padj-adjusted *p* values, * $p < 0.05$ and ** $p < 0.01$. Observed, numbers of genes observed in this study; Expected, numbers of genes in this same category of annotated barley gene models.

2.2. Transcriptome Dynamics of Grain Development in Barley

To capture temporal changes during barley grain development, we compared the gene expression levels between these four stages. Using the Padj < 0.05 and | log2Ratio | ≥1 as threshold, approximately 51.38% of the expressed genes (9527/18,541) were identified as significantly differential expressions, with values ranging from 1532 (stage04 vs. stage03) to 6380 (stage 04 vs. stage01) (Figure 2A, Figure S4, Table S5). When comparing adjacent time points (for example, stage02 vs. stage01 and stage03 vs. stage02), we found that the largest variation in differentially expressed genes occurred between stage03 and stage02, which was consistent with the original correlation analyses. In addition, obvious variations were found on the composition of the differentially expressed genes between any two adjacent time points (Figure 2B). A total of 312 genes were significant differentially expressed at all of the four developmental stages, suggesting the specialized functions and differentiation during barley grain development that might be involved in different sets of genes with distinct patterns of expression. All the differentially expressed genes were further used to define clusters displaying distinct temporal patterns during development. A K-mean clustering method was conducted with the squared Euclidean distance measure, and all the genes potentially involved in the regulation of grain filling were classified into nine categories (Figure S5 and Table S6) determined using the Calinski–Harabasz (CH) index.

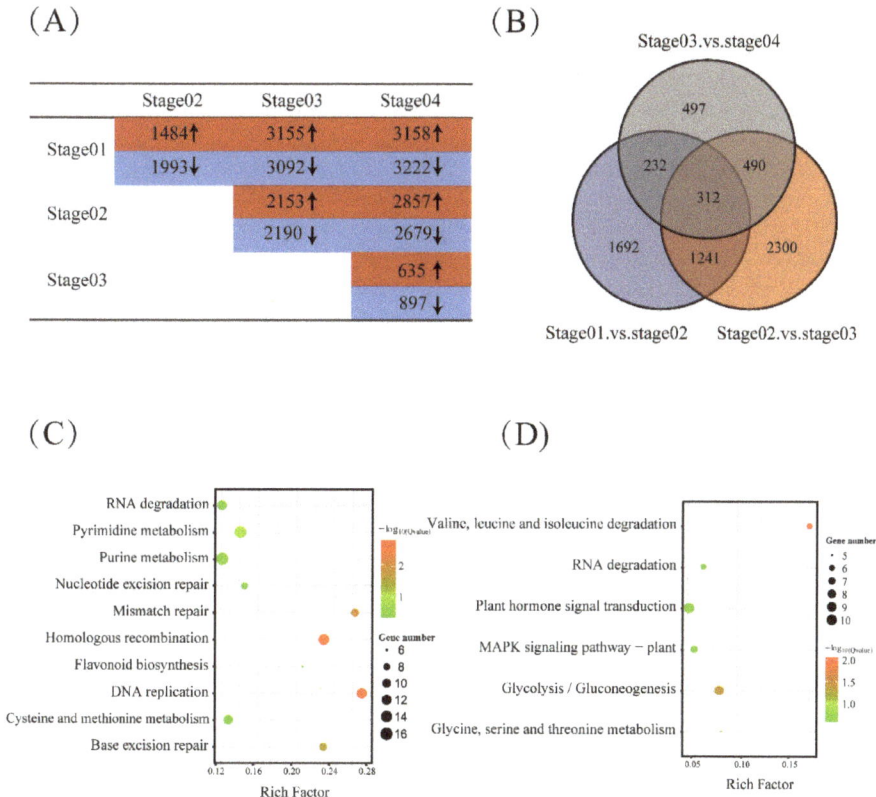

Figure 2. Differential expression genes during barley grain development. (**A**) Number of genes showing up- or downregulated expression during barley grain development (Padj < 0.05 and | log2Ratio | ≥1). (**B**) Venn diagram analyses of genes that were differentially expressed between any two consecutive stages in barley. (**C**) KEGG enrichment analysis of DEGs in cluster V during barley grain development. (**D**) KEGG enrichment analysis of DEGs in cluster VII.

Based on the growth characteristics, starch accumulation patterns, and metabolite profiles, the major events of barley endosperm development were classified into four main stages: prestorage phase (0–5 DPA), late prestorage or transition phase (6–10 DPA), early storage phase (11–15 DPA), and levels off stages (16–20 DPA) [6]. The first four clusters (I–IV) of differentially expressed genes were preferentially expressed at these four development stages, respectively (Table S6). Cluster I comprised 1823 genes with the peak expression at the first stage and then declined sharply at later stages. This cluster showed a clear enrichment of genes involved in protein, chromosome, and cell wall pathways such as protein dimerization activity (GO:0046983), chromosome organization (GO:0051276), cell wall organization (GO:0071555). The transition phase is characterized mainly by an initial accumulation of starch. Expecting for some pathways involved in preparing for starch accumulation such as protein and RNA, genes preferentially expressed at this phase (cluster II, 1311 genes) were also enrichment for some carbohydrate metabolism pathways such as photosynthesis (GO:0015979), transporter activity (GO:0005215), and carbohydrate binding (GO:0030246). Expression of genes in cluster III group (472 genes) activated significant enrichment of genes involved in intracellular transport (GO:0046907), macromolecule localization (GO:0033036), phosphorelay signal transduction system (GO:0000160), pyrophosphatase activity (GO:0016462), and nutrient reservoir activity (GO:0045735). At the levels off stage, there was a significant increase in expression (cluster IV, 746 genes) of genes related to multicellular organism development (GO:0007275) and developmental process (GO:0032502). Cluster V group contained the largest number of differentially expressed genes (2,217 genes), which showed an increased expression at the prestorage phase (stage01 and stage02). In preparation for seed filling, genes related to protein, RNA, cell organization, and cell division were activated and significantly enriched for cell cycle (GO:0007049), cytokinesis (GO:0000910), lipid metabolic process (GO:0006629), cellulose synthase activity (GO:0016759), RNA processing (GO:0006396), nucleosome organization (GO:0034728), chromosome organization (GO:0051276), cellular component biogenesis (GO:004085), protein metabolic process (GO:0019538), organic acid metabolic process (GO:0006082). The KEGG pathway analysis found they were significantly enriched in the DNA replication, mismatch repair pathway (Figure 2C). Transcripts belonging to cluster VII with a peak of expression at the storage phases (stage03 and stage04) were mainly involved in nutrient reservoir activity (GO:0045735), enzyme inhibitor activity (GO:0004857), peptidase inhibitor activity (GO:0030414). Compared to cluster V, genes in this cluster were significantly enriched in glycolysis/gluconeogenesis and plant hormone signal transduction pathway (Figure 2D). Other functional classes, such as stress, transport, cell wall, protein, and RNA, were also over-represented in cluster VI, cluster VIII, and cluster IX (Table S6).

2.3. Functional Analyses of Differentially Expressed Transcription Factors during Barley Grain Development

Transcription factors (TFs) are one of the main factors for signal transduction and regulating gene expression, which play important roles in controlling the growth, development, and stress response in plants. We further identified the transcription factor gene among the identified differentially expressed genes. Results found that there were 606 transcription factors in these DEGs, which could be assigned into 61 families based on the presence of conserved domains. Among them, MYB family members were the most prominent (50), followed by AP2/ERF-ERF (44), NAC (42), bHLH (38), bZIP (35), and MYB-related (30) (Figure 3A and Table S7). MYB proteins are key factors in regulatory networks controlling development and metabolism under biotic and abiotic stress [21]. In this dataset, 17 (34%) MYB TFs members abundantly expressed at the prestorage phase, while another 10 genes (20%) were preferentially expressed at the storage phase (Figure 3B). The differential function of MYBs contributed to controlling development and stress response, which might be the reason the MYB family holds the largest TFs family during grain development. In addition, MADS transcription factor family, known to play crucial roles in flower organ development [22], were found to be abundantly expressed (20 members), implying an unexplored distinct role of MADS box TFs during barley grain development. On the other hand, only 22 TFs (36.73%) were expressed at the level of FPKM $\geq$50, indicating that most TFs were not highly expressed in grain (Table S7). Interestingly, two groups of differently expressed TFs with the highest number exhibited opposite dynamic expression trends from early to late stages: expression levels of 150 gene (cluster V, 24.75%) were highly expressed at prestage (stage01 and stage02), while in contrast, 72 genes in cluster VIII (11.88%) showed low expression at the prestage but highly expressed at the storage phases (stage03 and stage04). Some transcription factors related to meristem function and cell division such as HB (Homeobox), CPP, FAR1 were found in prestage (Figure 3B). Furthermore, 44.44% (4) of the ARFs (auxin response factors), responding to phytohormone auxin and important in determining plant architecture, were identified and their expression levels decreased gradually from early to late stages. This suggests that ARFs play important roles in early stages of barley grain development.

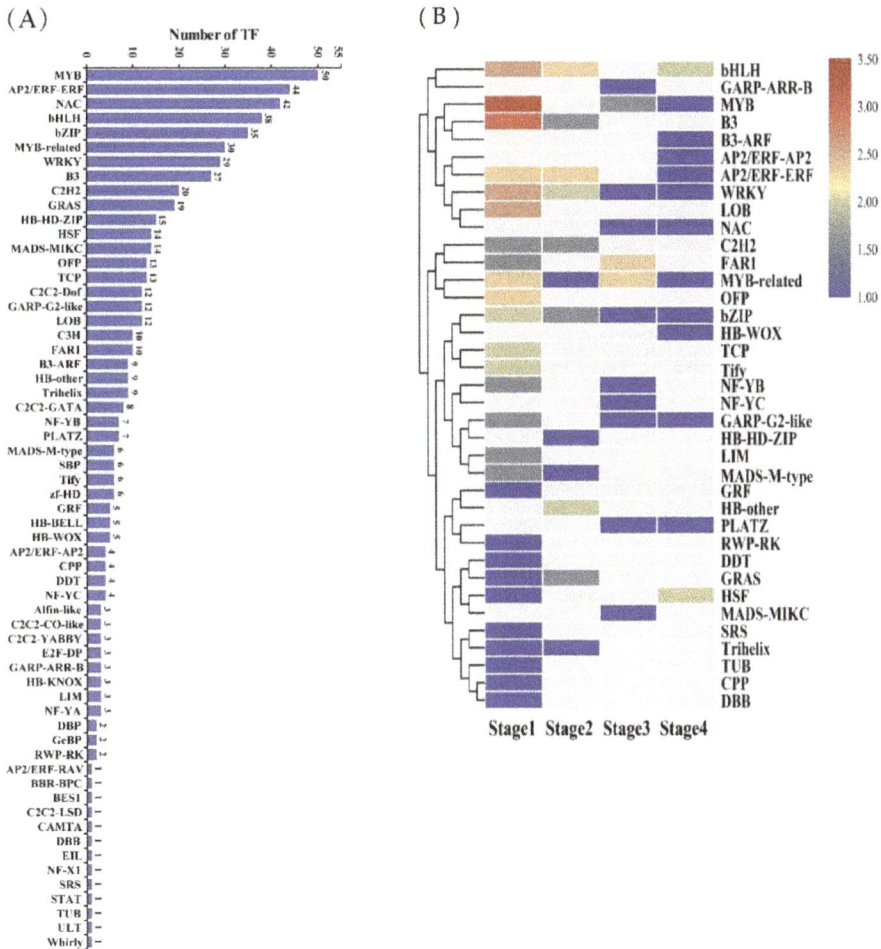

Figure 3. Characterization of transcription factors (TFs) differentially expressed during barley grain development. (**A**) Number of transcription factors (TFs) differentially expressed between any two stages of barley grain development. (**B**) Characterization of TF in each expression pattern.

2.4. Prediction of RNA Editing Sites

RNA editing is an important post-transcription mechanism to enrich genetic information and regulate diverse biological processes. After filtering the lower coverage sites as described in Method section, a total of 4711 editing sites were identified during the barley grain development, with 12 base change types, of which four editing types (A->G; C->T; G->A; T->C) were significantly higher than the other eight types (Figure 4A, Table S8). This result showed the transition occurred more abundantly than transversion in RNA editing, which was consistent with previous studies [23]. Then, the stage-specific and overlapped editing sites of these four stages were investigated. Results showed that stage01 owned the highest unique sites, with the number of 1048, followed by stage02, stage03, and stage04, with the numbers of 847, 707, and 356, respectively (Figure 4B). Among these editing events, a total of 3410 sites were found in 1795 protein-coding genes (Table S8), of which 2412 were found in CDS region, and 356 and 642 sites were located in 5′ and 3′ UTR regions (Figure 4C), respectively, suggesting that most editing could change the genetic information to result in the change of the encoded protein sequence. The gene HORVU7Hr1G000580 showed the most abundant editing sites with the total value of 15, and most of the other genes were observed to have one editing site (Figure 5A). Totally, 783 editing sites in 542 genes were observed to result in the amino acid change, of which 139 editing events (25.65%) led to a conversion from hydrophilic amino acid to hydrophobic amino acid, and 102 (18.82%) changing from hydrophobic amino acid to hydrophilic amino acid (Figure S6). A series of genes related to starch and sugar metabolic processes were found to be edited, such as beta-amylase 5 (HORVU4Hr1G089510), fructose-1,6-bisphosphatase class 1 (HORVU4Hr1G060630, HORVU1Hr1G074330) and low-molecular-weight glutenin subunit (HORVU1Hr1G001350, HORVU1Hr1G001420, HORVU1Hr1G001120, HORVU1Hr1G001120, HORVU1Hr1G000930, HORVU1Hr1G000930, HORVU1Hr1G001080) (Table S9). What is more, 4 terminators were found to be converted to 1 hydrophobic and 3 hydrophilic amino acids, and 1 hydrophobic and 11 hydrophilic amino acids were converted to terminators due to RNA editing. The embryo-defective related gene (HORVU2Hr1G037470) had 4 C to T editing events occur during barley grain development, and two of them caused the amino acid change (H>Y, A>V), one site with no amino acid change (C>C), and one resulted in the termination of the coding transformation (Q>*) (Figure S7). The stop code of this gene caused by editing may lead to the defect development of the embryo (Figure 5B). Furthermore, combined with the expression patterns, 308 DEGs were found to have RNA editing occur and the edited genes were more abundant in cluster I, cluster II, and cluster V, compared to cluster III, cluster IV, and cluster VII (Figure 5C). Then, KEGG analysis of them showed that they were enriched in metabolic pathways, biosynthesis of secondary metabolites, carbon fixation in photosynthetic organisms, carbon metabolism, base excision repair, pentose phosphate pathway, and glycolysis/gluconeogenesis (Figure 5D). These results demonstrated that RNA editing may function as the crucial regulator to control the complex process of barley grain development.

(A)

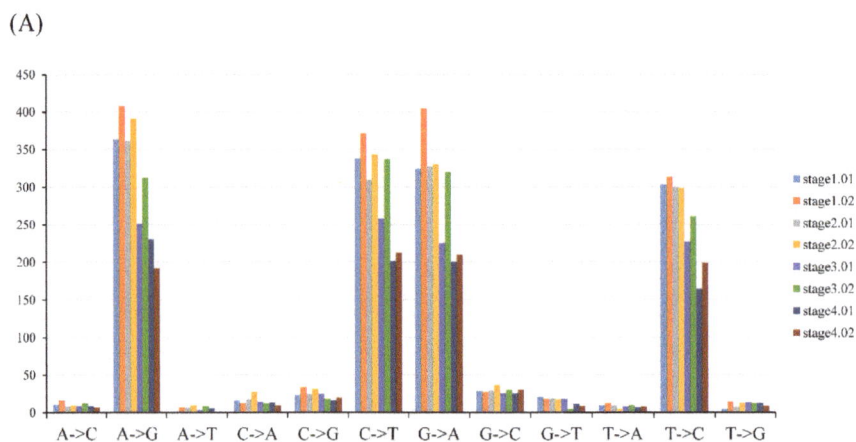

Figure 4. *Cont.*

(B)

(C)

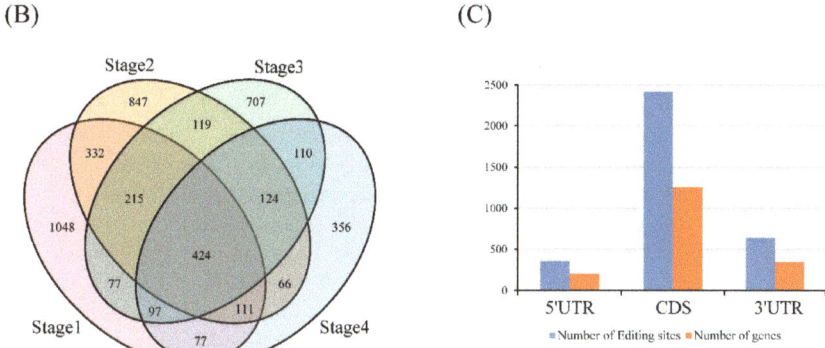

Figure 4. Identification of the RNA editing events during barley grain development. (**A**) The frequency of different types of nucleotide editing at four stages of barley grain development. (**B**) Venn diagram analysis of the abundance of RNA editing events among four stages of barley grain development. (**C**) The localization of RNA editing events in genic regions.

(A)

(C)

(B)

(D)

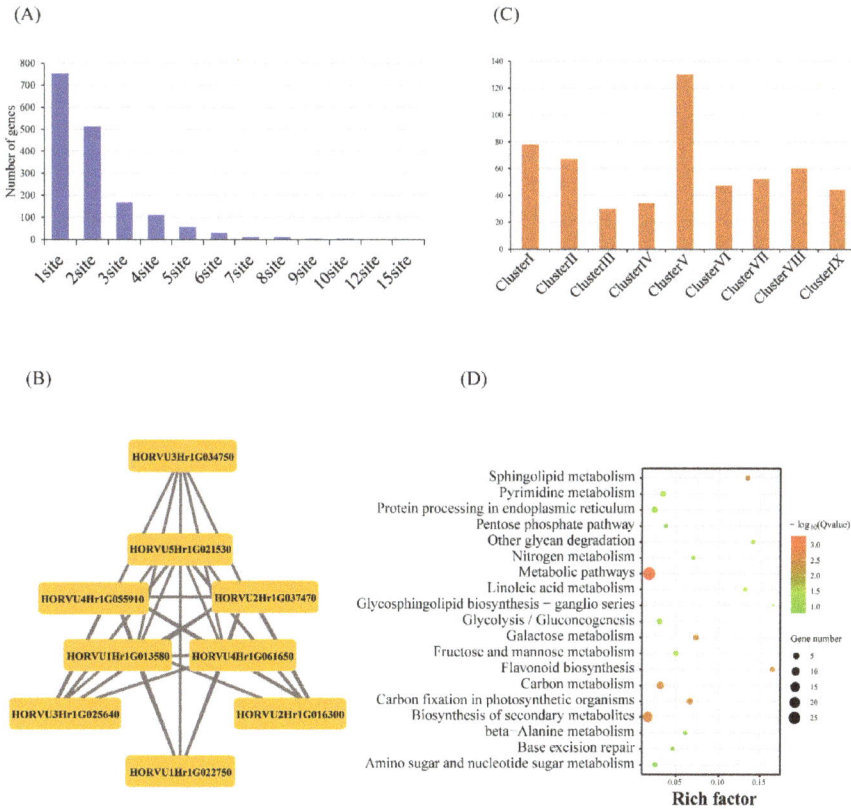

Figure 5. Frequency and functional enrichment of RNA-edited differentially expressed genes (DEGs). (**A**) The abundance of genes harbored the edit sites ranging from 1 to 15. (**B**) The regulatory network of the identified DEGs with RNA editing sites involved in embryonic defects. (**C**) Frequency of RNA-edited DEGs in nine expression clusters. (**D**) KEGG enrichment analysis of the RNA-edited DEGs identified in this study.

3. Discussion

3.1. mRNA Reveals Insights into Dynamic Regulation of Grain Development in Barley

Grain development represents an elaborate and complex phase in the plant life cycle and has been extensively studied in recent years [6,7,13]. It has been suggested that differences in the molecular regulation of grain development may arise from spatial and temporal variation in gene expression [1,2,18]. Hence, interpretation of transcriptomic data providing some clues for the development of grain in plants, and barley in particular, becomes imperative. In this study, we performed a comprehensive analysis of transcriptional dynamics during barley grain development using a transcriptomic sequencing strategy. By pairwise comparisons of the data between four stages, differential transcript abundance and sequence variation were found. The levels of transcripts showed close correlation with the physiological changes during grain development. The majority of the genes exhibited peak expression at the prestorage phase (stage01 and stage02), and relatively low expression at the storage and mature phases (stage03 and stage04). We demonstrated that a dramatic change took place during the transition of the seeds from a prestorage phase to maturation, which enhanced our understanding of the developmental shifts during the filling process of barley grain. Previous studies have identified several pathways or genes involved in regulating barley grain development based on microarray hybridization [1,7,12,14]. In contrast to cDNA chip hybridization, the RNA-seq technology has obvious advantages for transcriptome dynamics analysis, which could generate millions of sequence reads with high reproducibility, especially for the lowly expressed transcripts. In the current study, 48.23% (19,165/39,734) of the annotated barley genes were detected to have the RPKM value more than 1 in at least one stage. These results were similar with that of hexaploid wheat (~55%) [2] and Arabidopsis (~44%) [18], but significantly higher than previous studies in barley [1,6,7,12,14]. Comparison analysis revealed that a total of 9527 genes were considered to be potentially involved in the regulation of barley grain development. These genes were significantly enriched in 16 KEGG pathways (Figure S8A), which provided the clues to explore the global activation of metabolic pathways and gene regulatory networks activated during barley seed development.

3.2. Identification of Genes Required for Grain Development in Barley

Seed development is a complex process in which most of the essential developmental processes are initiated, including embryo and endosperm development, storage essence accumulation, desiccation tolerance forms, and seed dormancy. To obtain the information concerning the genes governing this complex developmental process, several studies have been focused on the identification of "essential" genes (lethal-phenotype genes) for seed development [3,9]. At present, a total of 481 genes were characterized as "essential" genes required for normal embryo development in Arabidopsis and presented in the SeedGenes database (http://seedgenes.org/index.html). The majority (74.43%) of these genes have been confirmed by function analysis [9]. In this study, we identified 3172 seed development-related genes with significant similarity to Arabidopsis based on bidirectional best BLAST hits and an e-value threshold below 1e-30. Among them, 887 genes were significantly differentially expressed during barley grain development and could be considered as potential barley EMB genes (Table S10). Mutations of the majority of these genes (85.34%, 757/887) will cause defects in embryo development. While the other 100 EMB genes are essential to seed pigmentation, mutation of these genes will lead to morphologically normal seeds with color changes such as albino and pale green [3,9]. To define their detailed functional categories, KEGG enrichment analysis of the potential EMB genes for barley used KOBAS (version 3.0) software [24]. Enrichment analysis found that the nucleotide, protein, lipid, and carbohydrate metabolism pathways were significantly enriched (Figure S8B). While these results demonstrate that the majority of genes related to seed development of Arabidopsis [3,9] are also present in barley grains, the definitive functions of these genes in barley seed development remain to be determined.

Developing barley grains are green and contain functional chloroplasts capable of photosynthesis during seed filling in barley. The prestorage phase (0–5 DAP) and transition phase (6–10 DAP) are characterized mainly by cell division and absence (or initial accumulation) of starch in the endosperm [6]. Previous studies in Arabidopsis indicated that the level of photosynthesis was high in the early and middle stages of grain development, when the majority of photosynthesis-related genes showed peak expression [5]. Our findings showed that a total of 29 genes were clustered as photosynthesis-related genes and 68.97% (20/29) of them peaked at stage02, represented by photosynthesis, photosystem I, and photosystem II. For example, several genes encoding photosynthetic light reaction (such as light-harvesting complex II, including Lhca1, Lhca2, Lhca3, Lhca4, Lhcb1, Lhcb2, Lhcb3, Lhcb5, Lhcb6, and Lhcb7) were low in the youngest seeds, increased sharply by 8 DAP (stage02), and then decreased afterwards.

3.3. Plant Hormone Signal Transduction and Sugar Signaling Interaction Networks During the Transition from Pre-Storage to Storage Phase

Plant hormone plays a critical role in the entire growth period in plants. For instance, auxin contributes mainly to cell enlargement and division, and ABA (Abscisic acid) is believed to be involved in the initiation of storage compounds synthesis and linked to seed desiccation tolerance level and seed dormancy [1,4,5,7,14,25,26]. The plant hormone signal transduction pathway bridges the hormone and its response genes. Our transcriptome analysis indicated that auxin signal transduction-related genes were related to cell enlargement and division and their expression showed peak at prestore stage (Figure S9A), but the ABA signal transduction pathway showed a peak at store stage (Figure S9B).

Extensive studies have elucidated that ABA has an important role in the genetic control of seed development. There are four TFs known as main regulators of late embryogenesis in A. thaliana. These regulators are ABA-dependent genes, including LEC1 (HAP3 subunit), LEC2 (B3-domain), ABI3 (ABA-insensitive 3), and FUSCA/FUS3 (fused cotyledon3) [4,14,26]. FUS3 regulates endogenous ABA synthesis, which will activate the ABA-insensitive protein phosphatases 2C (PP2C) and sucrose nonfermenting 1 (Snf1)-related protein kinases 2 (SnRK2). SnRK2 subsequently initiates downstream targets, including ABF, AREB, and ABI5 [4,26]. In this study, we found several ABA signal transduction-related genes were differentially expressed in barley during grain development, and HvPP2C, HvSnRK, and HvABI were preferentially expressed at the storage phase (Figure S9B).

It has been demonstrated that both sugars and ABA played critical roles in controlling the transition from the cell division phase to the cell enlargement phase as storage reserves were accumulated [4,26]. SnRK, known for the regulation of key starch biosynthesis genes, such as sucrose synthase (SUS) and ADP-glucose pyrophosphorylase (AGPase), has been shown to act as a bridge between ABA and sugar signaling [4,26]. During legume seed development, glucose and sucrose acted in almost opposite fashions, with glucose promoting cell division and sucrose being associated with cell expansion [4]. Importantly, previous studies regarding metabolite contents in developing barley endosperm showed that hexoses were highest during early development whereas sucrose peaked at 8 DPA [13,16]. In the current study, our data showed some sugar-related genes were significantly expressed to maintain different contents of glucose and sucrose in distinct stages, thus medicating the phase transition from prestorage to storage phase. In potato tubers, for instance, the expression of invertase or sucrose phosphorylase bypasses sucrose synthase and decreases sucrose levels [4]. Two genes encoding sucrose phosphorylase were significantly expressed in barley and all of them showed a peak expression level at the prestorage phase (stage01 and stage02). It appears that the critical role these sugar-related genes played was to maintain a relatively low content of sucrose at the early stages (Figure S9C). In addition, one sucrose synthase gene, together with an enzyme that hydrolyzes sucrose to glucose and fructose, was preferentially expressed at the early stage, contributing to the high content of glucose at this stage [14] (Figure S9C). Starch biosynthesis was detected at high levels at the storage phase (stage03, stage04) (Figure S9D). A similar pattern was found for one sucrose phosphate and two sucrose synthases genes, indicating their important roles in starch biosynthesis at the storage phase. To conclude, we might hypothesize that the extensive interactions between sugar and ABA signaling have critical roles in mediating the phase transition of barley grain development according to the results obtained in this study.

3.4. RNA Editing Regulated Barley Grain Development

RNA editing is an important post-transcription mechanism to enrich genetic information and regulate diverse biological processes by altering the genetic identity between RNA and genomic DNA. Although the RNA editing also occurs in the noncoding regions, such as 3′-UTR and 5′-UTR, rRNA and tRNAs, it mostly occurs in protein-coding regions, resulting in the change of encoded protein as well as protein function [27–29]. A series of studies have evidenced that the RNA-editing events have important roles in regulating organ formation, architecture growth, and development [30–33]. In this study, during grain development, the editing DEGS were more observed in the prestore stages (cluster V) than store stages (cluster VII) (Figure 5C). In addition, RNA editing was found to occur in some of the genes involved in the embryonic defects development and breakdown of the macromolecule organic matter, and editing has caused the stop coding or amino acid change (Figure S7; Tables S8 and S9), consistent with the empirical evidence that mostly nucleic acid activity and cell cycle occur in embryonic development stages. Combined with the DEGs, KEGG pathway enrichment analysis showed that the editing genes were significantly enriched in metabolic pathways, carbon fixation in photosynthetic organisms, base excision repair, and glycolysis/gluconeogenesis, which is consistent with the physiological process of spike development (Figure 5D). Thus, the RNA editing may function as a 'regulator' to ensure the plant's normal growth and development.

4. Materials and Methods

4.1. Sample Preparation and Sequencing Library Construction

Barley plants (cv. Clipper) were grown in field trial of Northwest A&F University (Yangling, Shaanxi, China) under normal agricultural management. The time of anthesis and developmental stage of caryopses were determined for each head based on the dissection of the middle spikelet as previously described [13]. Individual spikes were tagged at the time of anthesis and harvested in the morning (08:30–09:30 h) at 3, 8, 13, and 18 DPA [16]. Four biological replicates were sampled. For each sample, tissues were collected directly into liquid nitrogen by pooling the middle parts of the spike from 10 to 15 individual plants for the purposes of homogeneity.

Then, two of the four replicates were randomly selected to isolate total RNAs using TRIzol reagent (Invitrogen, Waltham, MA, USA) and were treated with RNase-free DNase I to remove any contaminating genomic DNA. The purified RNAs were used to construct the RNA sequencing library following the Illumina's standard pipeline (Illumina, San Diego, CA, USA). In brief, mRNAs were extracted from purified RNA by Dynabeads oligo (dT) (Dynal, Invitrogen). Double-stranded cDNAs were synthesized using reverse transcriptase (Superscript II, Invitrogen) and random hexamer primers. After size selection and purification through agarose gel electrophoresis, the selected fragments with the length of 500 bp were ligated to sequencing adaptors. Then, the ligated RNAs were reverse transcribed and amplified for Illumina sequencing. The high-throughput sequencing was performed by pair-end (PE) approach with the read length of 100 bp on Illumina HiSeq2000 platform following the standard protocol at BGI Comp (Shenzhen, China). Each sample was sequenced with no less than 60 million reads. All the high-throughput sequencing data were deposited into the Sequence Read Archive (SRA) database in NCBI with the accession number of SUB2126402.

4.2. Expression Profiles Analysis

After removing the adaptor sequences, empty reads, and low-quality reads (Q < 20) with FASTX-toolkit (FASTX-toolkit 0.0.14, http://hannonlab.cshl.edu/fastx_toolkit/), all the clean reads were aligned to barley genome [17] using the default parameters for HISAT2 (version 2.0.5) [34]. Reads counts statistic was calculated using HTSeq (Version 0.7.1) [35] with htseq-count command. Then, gene expression levels were estimated using RPKM values (Reads Per Kilobase transcriptome per million reads). Each sample was treated individually. Pearson correlations between biological replicates were conducted based on the RPKM values of all expressed genes using the function cor in R, which were further used to generate a dendrogram of samples.

4.3. Identification of Differentially Expressed Genes (DEGs)

Differential expression analysis was performed using R packages of DESeq2 (version 1.20.0) between any two stages with two biological replicates [8]. Genes with an adjusted *p*-value < 0.05 and at least twofold changes (| Log2 (treatment/control) | ≥ 1) were considered as differentially expressed. Then, a K-means clustering was used to extract the fundamental patterns of gene expression inherent in the development of barley caryopses [36]. The following three additional steps were performed to pretreat the dataset using the standard tools in cluster 1.52 [37]: (i) Log-transform the data and all data values (FPKM) were replaced by log2(FPKM); (ii) for center arrays [mean], subtract the row-wise mean from the values in each row of data; (iii) normalize arrays: specifies to normalize each row in the data so that the sum of the squares of the values in each row is 1.0. The Venn diagrams in this study were prepared using the Vennerable package (version 3.0) in R based on the gene list for each tissue type.

4.4. Identification of the Transcript Factors in DEGs

All the nucleotide sequences for DEGs were extracted from the barley genome [17] and then used for prediction of the potential transcript factors by the iTAK (version 17.09) online software [38].

4.5. RNA Editing Analysis

All potential RNA editing sites were predicted using SPRINT (SnP-free Rna editing IdeNtification Toolkit) [39] software with default parameters. To support the prediction, we also identified the SNP between RNA reads and reference genomic DNA using SNP calling tools Samtools (version 1.3.1) (http://www.htslib.org) to analyze the mapping BAM files. The overlapped results obtained by both methods remained for further analysis. Then, the editing sites were further filtered using the following parameters: (1) the edited sites had more than five mapped reads; (2) the ratio of editing reads and total mapped reads more than 50%; (3) the editing sites were found in both biological replications.

4.6. Gene Ontology (GO) and KEGG Enrichment Analysis

GO terms that are significantly overrepresented in each cluster were determined by the AgriGO (version 2.0) [40]. Singular Enrichment Analysis (SEA) in AgriGO was used to detect over-represented GO categories in each cluster compared to the whole genes. GO terms with corrected Padj less than 0.05 were taken as significant ones. KEGG was used to identify processes and pathways of specific gene sets [41]. The annotations on genes were carried out using a simple unidirectional BLAST (blastx) search against already classified proteins from Arabidopsis thaliana (thale cress), Oryza sativa japonica (Japanese rice) (RefSeq), and Zea mays (maize) by KAAS (KEGG Automatic Annotation Server) software [41]. Then, KOBAS (version 3.0) software was used to enrich the pathways of all the genes [24].

5. Conclusions

In this study, we systematically investigated the gene expression atlas and RNA editome during barley grain development. Results showed that the expression profiles of transcripts involved in both primary and secondary metabolism were significantly changed between the transitions from prestorage phase to storage phase, which could reflect the physiological changes of grain development. Furthermore, the essential gene and regulatory networks involved in this process were investigated. Results showed that plant hormone signal transduction and sugar signaling interaction networks could play a crucial role in regulating grain development. Finally, 4771 RNA editing sites were detected in these four development stages, and most RNA editing genes were preferentially expressed at the prestore stage compared to store stage, suggesting that RNA editing might act as a 'regulator' to control grain development. This is the first study to report RNA edit events involved in grain development in barley at the whole transcriptome level. Our study systematically dissected the transcriptional dynamics of barley grain development, which will contribute to improved understanding of the molecular mechanism and dynamics of gene regulation underlying grain development in barley and beyond.

Supplementary Materials: Supplementary materials can be found at http://www.mdpi.com/1422-0067/20/4/962/s1; Figure S1: The photos of four stages during barley grain development; Figure S2: RNA-seq read mapping counts against the barley reference genome; Figure S3: The correlation relationship of the gene expression level between four stages reported in this study and two stages in barley RNA-seq database; Figure S4: The volcano map of the number of differently expressed genes between two adjacent stages in barley grain development. A: stage 1 vs. stage 2; B: stage 2 vs. stage 3; C: stage 3 vs. stage; Figure S5: Clusters analysis of genes showing representative expression patterns during barley grain development; Figure S6: The abundance of change type among hydrophobic and hydrophilic amino acid as well as terminator caused by RNA editing; Figure S7: The aligment of the normal and edited protein sequences of the embryonic defect-related gene HORVU2Hr1G037470; Figure S8: Functional characterization of the differentially expressed genes during barley grain development; Figure S9: Identification of the DEGs involved in hormone signal transduction and sugar metabolism processes; Table S1: RNA-seq read mapping counts against the barley reference genome; Table S2: The expression of genes of all the stages(FPKM>=1); Table S3: The GO enrichment of pre-store stage genes; Table S4: The GO enrichment of store stage genes; Table S5: The set of genes showing stage-specific expression during barley grain development (FPKM < 1 in all other stages); Table S6: Go enrichment of the DEGs clustered in different clusters; Table S7: The detail list of transcription factors (TFs) differentially expressed during barley grain development; Table S8: RNA editing sites in coding regions of morex genome predicted by SPRINT tool; Table S9: The function annotation and expression level of editing DEGs; Table S10: List of putative EMB genes for barley grain development.

Author Contributions: Data curation, J.B., X.D., and X.N.; Formal analysis, J.B., P.D., H.Z., X.W., N.M.D.L.C., and X.N.; Funding acquisition, W.S.; Writing—original draft, J.B. and P.D.; Writing—review & editing, Z.Y., X.N., and W.S.

Funding: This work was mainly supported by the National Natural Science Foundation of China (Grant No. 31571647 and 31401373), and partially supported by the State's Key Project of Research and Development Plan of China (Grant No. 2016YFD0101004 and 2016YFD0100302).

Acknowledgments: We thank Kewei Feng and Licao Cui for technical assistance on RNA editing analysis. We are also grateful to the High-Performance Computing (HPC) platform of Northwest A&F University for providing computing resources.

Conflicts of Interest: The authors declare no conflict of interest.

References

1. Druka, A.; Muehlbauer, G.; Druka, I.; Caldo, R.; Baumann, U.; Rostoks, N.; Schreiber, A.; Wise, R.; Close, T.; Kleinhofs, A.; et al. An atlas of gene expression from seed to seed through barley development. *Funct. Integr. Genomics* **2006**, *6*, 202–211. [CrossRef] [PubMed]

2. Pfeifer, M.; Kugler, K.G.; Sandve, S.R.; Zhan, B.; Rudi, H.; Hvidsten, T.R.; Mayer, K.F.X.; Olsen, O.-A. Genome interplay in the grain transcriptome of hexaploid bread wheat. *Science* **2014**, *345*, 1250091. [CrossRef] [PubMed]

3. Tzafrir, I.; Pena-Muralla, R.; Dickerman, A.; Berg, M.; Rogers, R.; Hutchens, S.; Sweeney, T.C.; McElver, J.; Aux, G.; Patton, D.; et al. Identification of Genes Required for Embryo Development in Arabidopsis. *Plant Physiol.* **2004**, *135*, 1206–1220. [CrossRef]

4. Weber, H.; Borisjuk, L.; Wobus, U. Molecular physiology of legume seed development. *Annu. Rev. Plant Biol.* **2005**, *56*, 253–279. [CrossRef] [PubMed]

5. Ruuska, S.A.; Girke, T.; Benning, C.; Ohlrogge, J.B. Contrapuntal networks of gene expression during Arabidopsis seed filling. *Plant Cell* **2002**, *14*, 1191–1206. [CrossRef] [PubMed]

6. Sreenivasulu, N.; Altschmied, L.; Radchuk, V.; Gubatz, S.; Wobus, U.; Weschke, W. Transcript profiles and deduced changes of metabolic pathways in maternal and filial tissues of developing barley grains. *Plant J.* **2004**, *37*, 539–553. [CrossRef] [PubMed]

7. Sreenivasulu, N.; Usadel, B.; Winter, A.; Radchuk, V.; Scholz, U.; Stein, N.; Weschke, W.; Strickert, M.; Close, T.J.; Stitt, M.; et al. Barley grain maturation and germination: Metabolic pathway and regulatory network commonalities and differences highlighted by new MapMan/PageMan profiling tools. *Plant Physiol.* **2008**, *146*, 1738–1758. [CrossRef] [PubMed]

8. Wang, L.; Xie, W.B.; Chen, Y.; Tang, W.J.; Yang, J.Y.; Ye, R.J.; Liu, L.; Lin, Y.J.; Xu, C.G.; Xiao, J.H.; et al. A dynamic gene expression atlas covering the entire life cycle of rice. *Plant J.* **2010**, *61*, 752–766. [CrossRef] [PubMed]

9. Devic, M. The importance of being essential: EMBRYO-DEFECTIVE genes in Arabidopsis. *C R Biol.* **2008**, *331*, 726–736. [CrossRef] [PubMed]

10. Feuillet, C.; Langridge, P.; Waugh, R. Cereal breeding takes a walk on the wild side. *Trends Genet.* **2008**, *24*, 24–32. [CrossRef] [PubMed]

11. Hansen, M.; Friis, C.; Bowra, S.; Holm, P.B.; Vincze, E. A pathway-specific microarray analysis highlights the complex and co-ordinated transcriptional networks of the developing grain of field-grown barley. *J. Exp. Bot.* **2009**, *60*, 153–167. [CrossRef] [PubMed]

12. Sreenivasulu, N.; Radchuk, V.; Strickert, M.; Miersch, O.; Weschke, W.; Wobus, U. Gene expression patterns reveal tissue-specific signalling networks controlling programmed cell death and ABA-regulated maturation in developing barley seeds (vol 47, pg 310, 2006). *Plant J.* **2006**, *47*, 987. [CrossRef] [PubMed]

13. Weschke, W.; Panitz, R.; Sauer, N.; Wang, Q.; Neubohn, B.; Weber, H.; Wobus, U. Sucrose transport into barley seeds: Molecular characterization of two transporters and implications for seed development and starch accumulation. *Plant J.* **2000**, *21*, 455–467. [CrossRef] [PubMed]

14. Seiler, C.; Harshavardhan, V.T.; Rajesh, K.; Reddy, P.S.; Strickert, M.; Rolletschek, H.; Scholz, U.; Wobus, U.; Sreenivasulu, N. ABA biosynthesis and degradation contributing to ABA homeostasis during barley seed development under control and terminal drought-stress conditions. *J. Exp. Bot.* **2011**, *62*, 2615–2632. [CrossRef] [PubMed]

15. Tauris, B.; Borg, S.; Gregersen, P.L.; Holm, P.B. A roadmap for zinc trafficking in the developing barley grain based on laser capture microdissection and gene expression profiling. *J. Exp. Bot.* **2009**, *60*, 1333–1347. [CrossRef] [PubMed]

16. Kohl, S.; Hollmann, J.; Erban, A.; Kopka, J.; Riewe, D.; Weschke, W.; Weber, H. Metabolic and transcriptional transitions in barley glumes reveal a role as transitory resource buffers during endosperm filling. *J. Exp. Bot.* **2015**, *66*, 1397–1411. [CrossRef] [PubMed]

17. Mascher, M.; Gundlach, H.; Himmelbach, A.; Beier, S.; Twardziok, S.O.; Wicker, T.; Radchuk, V.; Dockter, C.; Hedley, P.E.; Russell, J.; et al. A chromosome conformation capture ordered sequence of the barley genome. *Nature* **2017**, *544*, 427–433. [CrossRef] [PubMed]

18. Schmid, M.; Davison, T.S.; Henz, S.R.; Pape, U.J.; Demar, M.; Vingron, M.; Scholkopf, B.; Weigel, D.; Lohmann, J.U. A gene expression map of Arabidopsis thaliana development. *Nat. Genet.* **2005**, *37*, 501–506. [CrossRef] [PubMed]

19. Mortazavi, A.; Williams, B.A.; Mccue, K.; Schaeffer, L.; Wold, B. Mapping and quantifying mammalian transcriptomes by RNA-Seq. *Nat. Methods* **2008**, *5*, 621–628. [CrossRef] [PubMed]

20. Gan, Q.A.; Schones, D.E.; Eun, S.H.; Wei, G.; Cui, K.R.; Zhao, K.J.; Chen, X. Monovalent and unpoised status of most genes in undifferentiated cell-enriched Drosophila testis. *Genome Biol.* **2010**, *11*, R42. [CrossRef]

21. Dubos, C.; Stracke, R.; Grotewold, E.; Weisshaar, B.; Martin, C.; Lepiniec, L. MYB transcription factors in Arabidopsis. *Trends Plant Sci.* **2010**, *15*, 573–581. [CrossRef] [PubMed]

22. De Folter, S.; Immink, R.G.H.; Kieffer, M.; Parenicova, L.; Henz, S.R.; Weigel, D.; Busscher, M.; Kooiker, M.; Colombo, L.; Kater, M.M.; et al. Comprehensive interaction map of the Arabidopsis MADS box transcription factors. *Plant Cell* **2005**, *17*, 1424–1433. [CrossRef] [PubMed]

23. Levanon, E.Y.; Eisenberg, E.; Yelin, R.; Nemzer, S.; Hallegger, M.; Shemesh, R.; Fligelman, Z.Y.; Shoshan, A.; Pollock, S.R.; Sztybel, D.; et al. Systematic identification of abundant A-to-I editing sites in the human transcriptome. *Nat. Biotechnol.* **2004**, *22*, 1001–1005. [CrossRef] [PubMed]

24. Wu, J.M.; Mao, X.Z.; Cai, T.; Luo, J.C.; Wei, L.P. KOBAS server: A web-based platform for automated annotation and pathway identification. *Nucleic Acids Res.* **2006**, *34*, W720–W724. [CrossRef] [PubMed]

25. Locascio, A.; Roig-Villanova, I.; Bernardi, J.; Varotto, S. Current perspectives on the hormonal control of seed development in Arabidopsis and maize: A focus on auxin. *Front. Plant Sci.* **2014**, *5*, 412. [CrossRef] [PubMed]

26. Trontin, J.-F.; Klimaszewska, K.; Morel, A.; Hargreaves, C.; Lelu-Walter, M.-A. Molecular Aspects of Conifer Zygotic and Somatic Embryo Development: A Review of Genome-Wide Approaches and Recent Insights. In *In Vitro Embryogenesis in Higher Plants*; Germana, M.A., Lambardi, M., Eds.; Springer: New York, NY, USA, 2016; pp. 167–207. ISBN 978-1-4939-3061-6.

27. Maier, R.M.; Zeitz, P.; Kössel, H.; Bonnard, G.; Gualberto, J.M.; Grienenberger, J.M. RNA editing in plant mitochondria and chloroplasts. In *Post-Transcriptional Control of Gene Expression in Plants*; Filipowicz, W., Hohn, T., Eds.; Springer: Dordrecht, The Netherlands, 1996; pp. 343–365. ISBN 978-94-009-0353-1.

28. Steinhauser, S.; Beckert, S.; Capesius, I.; Malek, O.; Knoop, V. Plant mitochondrial RNA editing. *J. Mol. Evol.* **1999**, *48*, 303–312. [CrossRef] [PubMed]

29. Ichinose, M.; Sugita, M. RNA Editing and Its Molecular Mechanism in Plant Organelles. *Genes* **2017**, *8*, 5. [CrossRef]

30. Marechaldrouard, L.; Ramamonjisoa, D.; Cosset, A.; Weil, J.H.; Dietrich, A. Editing Corrects Mispairing in the Acceptor Stem of Bean and Potato Mitochondrial Phenylalanine Transfer-Rnas. *Nucleic Acids Res.* **1993**, *21*, 4909–4914. [CrossRef]

31. MarechalDrouard, L.; Kumar, R.; Remacle, C.; Small, I. RNA editing of larch mitochondrial tRNA(His) precursors is a prerequisite for processing. *Nucleic Acids Res.* **1996**, *24*, 3229–3234. [CrossRef]

32. Fey, J.; Weil, J.H.; Tomita, K.; Cosset, A.; Dietrich, A.; Small, I.; Marechal-Drouard, L. Role of editing in plant mitochondrial transfer RNAs. *Gene* **2002**, *286*, 21–24. [CrossRef]

33. Grewe, F.; Herres, S.; Viehover, P.; Polsakiewicz, M.; Weisshaar, B.; Knoop, V. A unique transcriptome: 1782 positions of RNA editing alter 1406 codon identities in mitochondrial mRNAs of the lycophyte Isoetes engelmannii. *Nucleic Acids Res.* **2011**, *39*, 2890–2902. [CrossRef] [PubMed]

34. Kim, D.; Landmead, B.; Salzberg, S.L. HISAT: A fast spliced aligner with low memory requirements. *Nat. Methods* **2015**, *12*, 357–360. [CrossRef] [PubMed]

35. Anders, S.; Pyl, P.T.; Huber, W. HTSeq-a Python framework to work with high-throughput sequencing data. *Bioinformatics* **2015**, *31*, 166–169. [CrossRef] [PubMed]

36. Brock, G.; Datta, S.; Pihur, V.; Datta, S. clValid: An R package for cluster validation. *J. Stat. Softw.* **2008**, *25*, 1–22. [CrossRef]

37. de Hoon, M.J.L.; Imoto, S.; Miyano, S. Statistical analysis of a small set of time-ordered gene expression data using linear splines. *Bioinformatics* **2002**, *18*, 1477–1485. [CrossRef] [PubMed]

38. Zheng, Y.; Jiao, C.; Sun, H.H.; Rosli, H.G.; Pombo, M.A.; Zhang, P.F.; Banf, M.; Dai, X.B.; Martin, G.B.; Giovannoni, J.J.; et al. iTAK: A Program for Genome-wide Prediction and Classification of Plant Transcription Factors, Transcriptional Regulators, and Protein Kinases. *Mol. Plant* **2016**, *9*, 1667–1670. [CrossRef] [PubMed]

39. Zhang, F.; Lu, Y.L.; Yan, S.J.; Xing, Q.H.; Tian, W.D. SPRINT: An SNP-free toolkit for identifying RNA editing sites. *Bioinformatics* **2017**, *33*, 3538–3548. [CrossRef]

40. Tian, T.; Liu, Y.; Yan, H.Y.; You, Q.; Yi, X.; Du, Z.; Xu, W.Y.; Su, Z. agriGO v2.0: A GO analysis toolkit for the agricultural community, 2017 update. *Nucleic Acids Res.* **2017**, *45*, W122–W129. [CrossRef] [PubMed]

41. Ogata, H.; Goto, S.; Sato, K.; Fujibuchi, W.; Bono, H.; Kanehisa, M. KEGG: Kyoto Encyclopedia of Genes and Genomes. *Nucleic Acids Res.* **1999**, *27*, 29–34. [CrossRef]

© 2019 by the authors. Licensee MDPI, Basel, Switzerland. This article is an open access article distributed under the terms and conditions of the Creative Commons Attribution (CC BY) license (http://creativecommons.org/licenses/by/4.0/).

International Journal of
Molecular Sciences

MDPI

Article

The Dark Matter of Large Cereal Genomes: Long Tandem Repeats

Veronika Kapustová [1], Zuzana Tulpová [1], Helena Toegelová [1], Petr Novák [2], Jiří Macas [2], Miroslava Karafiátová [1], Eva Hřibová [1], Jaroslav Doležel [1] and Hana Šimková [1,*]

[1]   Institute of Experimental Botany, Centre of the Region Haná for Biotechnological and Agricultural Research, Šlechtitelů 31, CZ-78371 Olomouc, Czech Republic; kapustova@ueb.cas.cz (V.K.); tulpova@ueb.cas.cz (Z.T.); toegelova@ueb.cas.cz (H.T.); karafiatova@ueb.cas.cz (M.K.); hribova@ueb.cas.cz (E.H.); dolezel@ueb.cas.cz (J.D.)
[2]   Biology Centre, Czech Academy of Sciences, Institute of Plant Molecular Biology, Branišovská 31, CZ-37005 České Budějovice, Czech Republic; petr@umbr.cas.cz (P.N.); macas@umbr.cas.cz (J.M.)
*   Correspondence: simkovah@ueb.cas.cz; Tel.: +420-585-238-715

Received: 5 May 2019; Accepted: 16 May 2019; Published: 20 May 2019

Abstract: Reference genomes of important cereals, including barley, emmer wheat and bread wheat, were released recently. Their comparison with genome size estimates obtained by flow cytometry indicated that the assemblies represent not more than 88–98% of the complete genome. This work is aimed at identifying the missing parts in two cereal genomes and proposing techniques to make the assemblies more complete. We focused on tandemly organised repetitive sequences, known to be underrepresented in genome assemblies generated from short-read sequence data. Our study found arrays of three tandem repeats with unit sizes of 1242 to 2726 bp present in the bread wheat reference genome generated from short reads. However, this and another wheat genome assembly employing long PacBio reads failed in integrating correctly the 2726-bp repeat in the pseudomolecule context. This suggests that tandem repeats of this size, frequently incorporated in unassigned scaffolds, may contribute to shrinking of pseudomolecules without reducing size of the entire assembly. We demonstrate how this missing information may be added to the pseudomolecules with the aid of nanopore sequencing of individual BAC clones and optical mapping. Using the latter technique, we identified and localised a 470-kb long array of 45S ribosomal DNA absent from the reference genome of barley.

Keywords: genome assembly; bread wheat; barley; optical mapping; BAC; ribosomal DNA

---

1. Introduction

Small grain cereals, such as bread wheat, durum wheat, barley and rye are crucial crops for the European population and most of them are grown worldwide. Despite their socio-economic importance, reference genomes of these cereals were only recently generated [1–5], which can be attributed to their high genome complexities, ranging from ~5 Gb for barley to ~16 Gb for bread wheat [6], and enormous proportion of repetitive DNA (85–90%). These assemblies are characterized by high contiguity and low proportion of internal gaps. However, a comparison of assembly lengths with genome-size estimates obtained by flow cytometry indicated that the reference genomes of barley, wild emmer wheat, bread wheat and rye represented no more than 98%, 88%, 90% and 90% of the estimated genome sizes, respectively [6]. This gives rise to an obvious question as to what the missing part of the reference genomes is, and stimulates efforts towards complementing it.

Low-copy genome regions are known to be a relatively easy target for genome assemblers and were found well represented even in early cereal genome sequences that were based on low-coverage Roche/454 data [7–9] or Illumina pair-end reads only [10]. On the contrary, large regions of repeats are known to pose a challenge and result in gaps, mis-assemblies and collapsed tandem repeats in a majority of genome sequences [11]. Dispersed repeats, represented by various types of transposable elements, have been largely resolved in the recent assemblies thanks to the combination of pair-end and mate-pair Illumina reads and sophisticated assembling algorithms [1,3] or implementation of long-read PacBio data [12].

On the other hand, tandem repeats organized as arrays of multiple units (microsatellites, macrosatellites and centromeric satellite repeats) tend to collapse in assemblies into fewer copies. Consequently, they are under-represented in reference genomes and pose a significant source of gaps and assembling errors in de novo assemblies, including that of humans [11]. The repeat-associated gaps are abundant in heterochromatic regions, making it impossible to completely assemble these genome parts. This usually results in genome assemblies missing a majority of (peri)centromeric regions and secondary constrictions [11,13]. To resolve the copy numbers, it is essential to use reads longer than the total array length. To some extent, this can be sorted out by using long-read DNA sequencing technologies, such as PacBio or nanopore sequencing, which produce reads of tens and up to hundreds kilobases, respectively. Nevertheless, only short arrays of simpler repeats, such as 5S rRNA multigene loci spanning over several kilobases, can be tackled by these approaches [14]. Arrays spanning over hundreds to thousands kilobases and consisting of units that are several kilobases long, such as loci coding for 45S rRNA, cannot be resolved by any of the current sequencing technologies.

This shortage can be compensated for by other approaches that facilitate identification, positioning and characterization of long arrays of tandem repeats, such as in situ hybridisation, application of dedicated bioinformatics tools, and optical mapping. The initial methods to investigating distribution of tandem repeats in cereal genomes included in situ hybridisation (ISH) [15–17] and fluorescence in situ hybridisation (FISH) [18,19]. Although the cytogenetic techniques provided first insights into the genome evolution [20] and facilitated chromosome identification and construction of molecular karyotypes [16,19,21], they did not provide information at the DNA sequence level. This could be obtained by application of dedicated bioinformatics tools, such as RepeatExplorer [22,23]. This computational pipeline utilizes a graph-based sequence clustering algorithm to de-novo assemble tandem repeats from raw next-generation sequencing data, without the need of a reference database of known elements. It has been used to identify and characterize repetitive elements in several complex plant genomes, including that of rye [24].

Optical mapping in nanochannel arrays, also known as Bionano genome (BNG) mapping, is a high-throughput long-read technology that generates genome maps of a short sequence motif—the recognition site of an enzyme used for labelling [25]. It has been used to support and validate physical-map and genome assemblies of several complex cereal genomes [2,3,26–28]. The ability of optical mapping to size gaps, cover some types of tandem DNA repeats [26] and identify misassemblies due to collapsed duplicated sequences [29] makes this technology a promising tool for identifying and characterizing missing parts of genome assemblies.

In this study, we identified three new tandem repeats specific for the short arm of wheat chromosome 7D (7DS) and interrogated their representation in recently published bread wheat assemblies, including (i) Triticum 3.1 [12], which combines short Illumina and long PacBio reads, (ii) IWGSC RefSeq v1.0 [3], which is based on short reads only, and (iii) Illumina assemblies of physical map-ordered 7DS-specific BAC clones [30]. While these assemblies comprised of arrays of all three repeats, they failed in unravelling organization of a repeat with unit size of 2726 bp, whose genome arrangement could only be resolved after adding information from nanopore sequencing of two BAC clones bearing arrays of this repeat and by using an optical map (OM) of the wheat 7DS arm [26] as a reference. Besides, we employed an OM of barley cv. Morex [31] to investigate a minor 45S rDNA locus in barley chromosome 1H, which was identified by in situ hybridisation in various barley cultivars [32,33] and underrepresented in the 'Morex' BAC-by-BAC genome assembly [2]. Thus, we demonstrated that optical mapping is a suitable tool to identify the missing parts of the assemblies, and, in some cases, can reveal overall organization of the repeat array. Targeted long-read nanopore sequencing was confirmed as a promising approach to complementing the missing sequences to the genome assemblies.

2. Results and Discussion

2.1. Chromosome-Specific Tandem Repeats in Wheat

Using RepeatExplorer pipeline to cluster Illumina raw data from flow-sorted wheat chromosome arm 7DS, we identified four tandem repeats with monomer lengths ranging from 1167 bp to 2726 bp (Table 1, File S1).

Table 1. Monomer sizes and distribution of identified tandem repeats.

Tandem Repeat	Monomer Size	Distribution
TaeCsTr163	1390 bp	7D subtelomere
TaeCsTr230	1242 bp	7D subtelomere
TaeCsTr99	2726 bp	7D subtelomere
TaeCsTr111	1167 bp	All chromosomes Dispersed

Cytogenetic mapping revealed that three of the repeats, TaeCsTr163, TaeCsTr230 and TaeCsTr99, provided unique FISH signals specific for subtelomeric region of wheat 7DS chromosome arm (Figure 1; Figure S1). A probe derived from the TaeCsTr111 sequence provided dispersed hybridization signals on multiple chromosomes, predominantly in pericentromeric and subtelomeric regions (Figure S1). The clustered hybridization signals obtained with three probes are supportive of the tandem organisation of these repeats. Application of RepeatExplorer on raw data obtained from flow-sorted chromosomes thus proved a suitable approach to identify chromosome-specific tandem repeats. The tool worked efficiently, despite using DNA amplified by multiple displacement amplification, which is known to introduce a quantitative amplification bias [34].

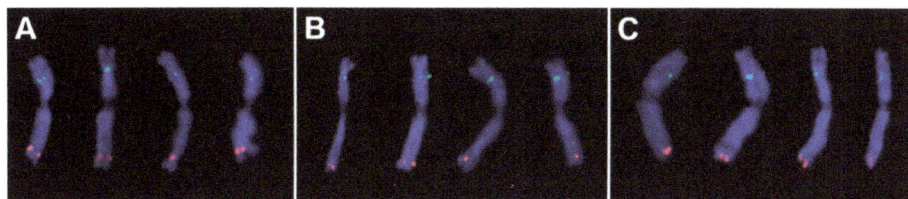

Figure 1. FISH on metaphase 7D chromosomes of bread wheat cv. Chinese Spring with probes for GAA microsatellite (green) and three tandem repeats (red). (**A**) TaeCsTr163, (**B**) TaeCsTr230 and (**C**) TaeCsTr99 repeats localized in subtelomeric region of the 7DS chromosome arm. 7D chromosomes were identified based on the GAA hybridization signal on the 7DL arm. The chromosomes were counterstained by DAPI (blue).

To assess the representation of the tandem repeats in the recently published wheat reference genome, we performed blastn search on IWGSC RefSeq v1.0 assembly [3]. Out of the three repeats assigned specifically to the 7DS arm, only two, TaeCsTr163 and TaeCsTr230, could be reliably identified in the IWGSC RefSeqv1.0 assembly of the 7D chromosome. Using blastn search, we found a cluster of 39 complete and several incomplete units of the TaeCsTr163 repeat, partially tandemly organised, that spanned over 260 kb in the interval of 49.08–49.34 Mb of the 7D pseudomolecule. The TaeCsTr230 repeat was identified as an array of ten complete and three incomplete units located in the interval of 33.254–33.27 Mb of the 7D pseudomolecule (Figure S2). Considering the entire 7DS arm length of 338 Mb [3], the positions of both repeats in the assembly are in agreement with their cytogenetic locations. On the contrary, we failed to find a significant blastn hit in the 7D pseudomolecule of RefSeq v1.0 for the repeat TaeCsTr99 that provided the strongest FISH signal on mitotic chromosomes.

Additional search in the unassigned scaffolds (ChrUn) revealed 12 of them containing TaeCsTr99 units (Figure 2A, Figure S3). Scaffold lengths varied from 3 kb to 163 kb, and they comprised from one to eleven complete TaeCsTr99 units. Altogether, we identified in unassigned scaffolds 36 complete tandemly organised units totalling 98 kb of length, accompanied by several incomplete units and unit fragments. Based on a high sequence homology (Figure S3), some of the scaffolds could be overlapping. Out of the twelve scaffolds, only ChrUn8536 carried an 82.3-kb non-repetitive segment (Figure S4A), which enabled its positioning in the context of the 7D pseudomolecule utilizing the OM of the 7DS chromosome arm. OM contig 77 placed the ChrUn8536 to position 14.4 Mb in the 7D pseudomolecule and revealed an additional 272 kb gap distal of the ChrUn8536 (Figure S5A, Figure 3). The identified position is consistent with the (sub)telomeric location of the repeat indicated by FISH (Figure 1C).

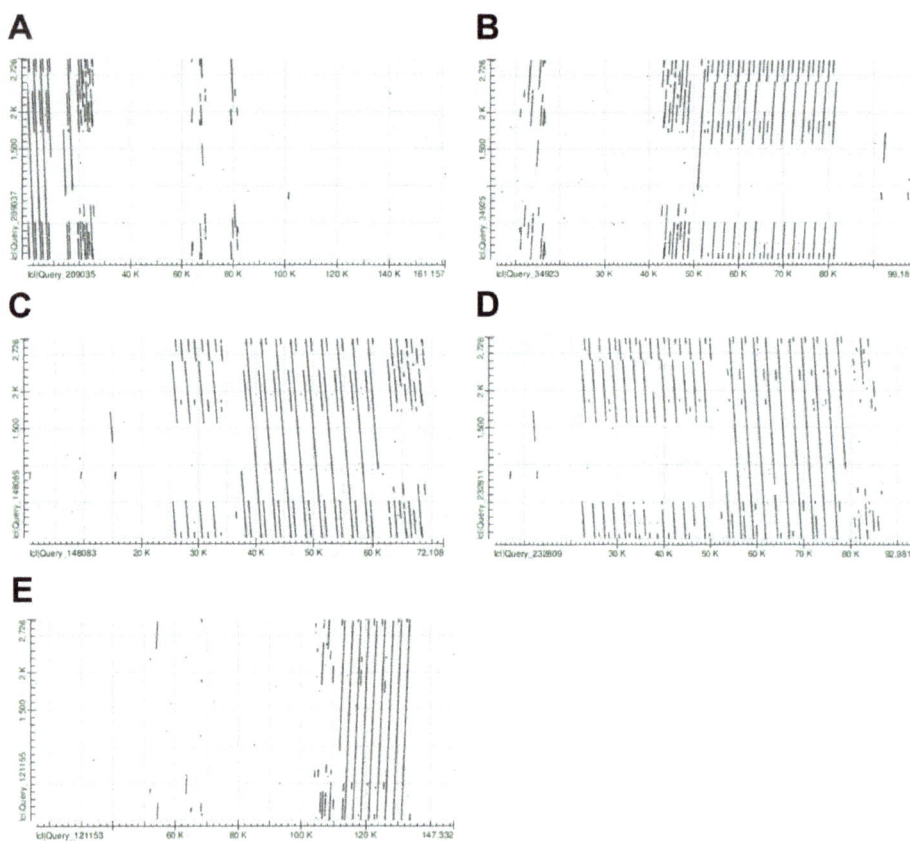

Figure 2. Dot plots showing position and arrangement of TaeCsTr99 repeat in (**A**) ChrUn8536, (**B**) interval 9.06 Mb to 9.16 Mb of 7D pseudomolecule of Triticum 3.1 assembly [12] (**C**) 72-kb Illumina contig of BAC clone 28N04, (**D**) 99.8-kb nanopore read 51ef9015 of BAC clone 28N04, (**E**) 148-kb nanopore read f24cdcf5 of BAC clone 104G18.

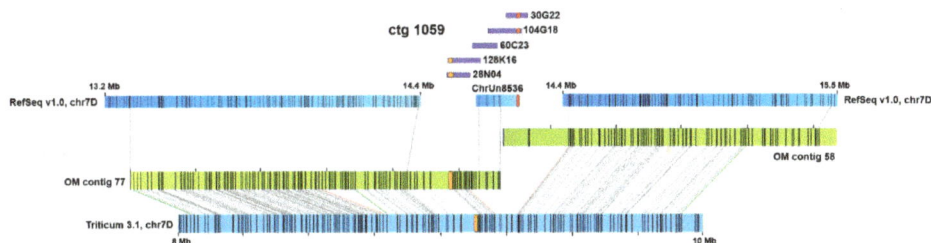

Figure 3. Organization of the TaeCsTr99 region. Positions of two TaeCsTr99 arrays (highlighted in orange and red) and an overall arrangement of the region were obtained by aligning ChrUn8536, IWGSC RefSeq v1.0 [3] and Triticum 3.1 [12] 7D pseudomolecules (blue bars) and BAC clones of 7DS physical-map contig 1059 [30] (violet bars) to 7DS optical map [26] (green bars). Numbers at the 7D pseudomolecules indicate assembly coordinates.

In order to span the entire TaeCsTr99 array, we employed short-read sequence assemblies of physical-map ordered 7DS BAC clones [30]. TaeCsTr99 was identified in four BAC clones belonging to 7DS physical-map contig 1059. Clones 104G18 and 30G22 overlap and show sequence homology with ChrUn8536 (Figure 3). Illumina assemblies of these BAC clones were rather fragmented and did not allow reconstructing the entire TaeCsTr99 array. Surprisingly, an array of the TaeCsTr99 repeat was also found in overlapping BAC clones 28N04 and 128K16. The latter could be aligned to the OM 77 (Figure S5B), but it was separated from the TaeCsTr99 array identified in clones 104G18 and 30G22 by a 225-kb non-repetitive segment (Figure 3, Figures S4 and S5). This suggested the presence of two separate TaeCsTr99 arrays in a close proximity, which we termed distal (covered by clones 128K16 and 28N04) and proximal (covered by ChrUn8536, 104G18 and 30G22), respectively. The size of the distal array was deduced from the assembly of BAC clone 28N04, which appeared to comprise the entire array in one scaffold, and was estimated to be ~44 kb. This array was composed of nine complete units, four units comprising a 1056-bp deletion and an additional cluster of repeat fragments spanning over ~6 kb (Figure 2C). The size of the proximal array in ChrUn8536 was 26 kb and this, likely incomplete array sequence comprised of three complete and three partial units and a ~6-kb cluster of repeat fragments (Figure 2A). Apparently, the organization of the repeat array in BAC clones 28N04 and 128K16 differed from that in ChrUn8536 (Figure 2, Figure S4A), which supported our hypothesis of two spatially separated arrays of TaeCsTr99 repeat in 7DS.

We hypothesized that the difficulties in assembling and incorporating of the TaeCsTr99 arrays into the pseudomolecule could be overcome by employing longer reads, such as those generated by SMRT sequencing (PacBio technology). To verify this, we explored bread wheat Triticum 3.1 assembly [12] that combines short-read Illumina and long-read PacBio data to search the 7D pseudomolecule for the TaeCsTr99 sequence. The blastn search revealed in the position 9.11 - 9.14 Mb a ~30-kb array composed of 14 incomplete TaeCsTr99 units, 13 of which carried the 1056-bp deletion observed in BAC clone 28N04 (Figure 2). We also observed the ~6-kb cluster of repeat fragments located distal of the array. Alignment of this region to 7DS OM placed the array to position ~1.2 Mb in OM contig 77 (Figure S5C), which corresponded to the position of the distal array identified in the short-read assemblies (Figure 3). On the contrary, we did not find any evidence of the proximal array in the Triticum 3.1 assembly.

To resolve the discrepancies in the location and organization of the TaeCsTr99 arrays identified in various assemblies, we made use of the long-read platform of Oxford Nanopore Technologies (ONT) and generated nanopore reads from BAC clones 28N04 and 104G18, which cover the distal and proximal array, respectively. For each of the clones, we obtained two reads that spanned over the entire insert and showed a consistent array structure. ONT read 51ef9015 (File S2) of 99,802 bp covering clone 28N04 confirmed the complex structure of the distal array composed of two sub-arrays with differently organised units (Figure 2D, Figure S4B). The distally located sub-array, approximately 27 kb in size, comprised of 10 complete units of the TaeCsTr99. The proximal sub-array was of similar length and consisted of 12 incomplete units, bearing a distinct deletion between 507 and 1563 bp of the TaeCsTr99 sequence. The ONT read also confirmed the presence of the adjacent ~6-kb cluster consisting of TaeCsTr99 fragments, which was less obvious here than in the Illumina BAC assembly due to the inherent inaccuracy of the nanopore technology (Figure 2C,D). Except for the variation in the number of units, the overall structure of the distal array looked highly similar in the BAC Illumina assembly and the ONT read. On the contrary, the corresponding array in whole-genome Triticum 3.1 assembly differed by the absence of the full-length units (Figure 2B). The proximal array was covered by a 148,009 bp full-length read f24cdcf5 of clone 104G18 (File S2), comprising the entire array that spanned over ~30kb and had a simple structure, including eight complete and one incomplete unit and the cluster of repeat fragments (Figure 2E). The total number of TaeCsTr99 units in the ONT reads covering the distal and the proximal array (18) was smaller than that identified in unassigned scaffolds (36). This could be due to non-recognized overlaps between the scaffolds, which may have resulted in overestimating the number of the repeats. Alternatively, we cannot exclude the presence of additional TaeCsTr99 array(s), missing both in the pseudomolecules and in the 7DS BAC assemblies that might be located in proximity of the confirmed ones.

The data obtained in our study suggest that tandemly organised repeats with unit size of 1–3 kb are not the major contributor to the missing part of the wheat IWGSC RefSeq v1.0 assembly as three of such repeats were well represented in the wheat reference genome obtained from short read data. Nevertheless, a more detailed analysis of a repeat with unit size of 2726 bp revealed that it was completely missing from the 7D pseudomolecule and was found in unassigned scaffolds (ChrUn) of the RefSeq v1.0 only. Thus, we concluded that this type of repeats may cause shrinking of pseudomolecules without impacting size of the entire assembly. Our results are in line with a finding that 27% centromeric sequences, identified by association with a centromere-specific histone H3 variant and highly enriched in centromere-specific repeats, were found in ChrUn of the RefSeq v1.0 [3]. This indicates that advanced assemblers can to some extent assemble shorter arrays of tandemly organised repeats but integration of these arrays in the pseudomolecule context may still pose a substantial challenge.

The TaeCsTr99 repeat was also found underrepresented and likely misassembled in the 7D pseudomolecule of Triticum v3.1, generated from both Illumina and PacBio reads. Moreover, both tested wheat whole-genome assemblies failed in discriminating two similar arrays located 225 kb apart, which could only be resolved after nanopore sequencing of BAC clones. This approach was successful not only because it employed a technology that provides reads exceeding the length of the whole array, but also because it leveraged the separation of the two arrays into the individually sequenced BAC clones. Interestingly, the identification of relatively long arrays of tandem repeats in BAC clones contradicts the finding of [11] that the tandem repeats are underrepresented in BAC libraries because of their toxicity for bacteria. The organisation of the whole tandem arranged region was resolved thanks to the application of the OM of the 7DS arm, which provided a reference for alignment of various sequences and revealed existing gaps and misassemblies. Nevertheless, the full potential of this genomic resource could not be exploited because none of the repeats analysed comprised a *Bsp*QI site (GCTCTTC) labelled in the 7DS OM. Consequently, the repeat arrays could not be recognised in the map through a specific labelling pattern, but appeared as longer regions devoid of labels. This shortage of the method might be overcome by the application of a new approach based on CRISPR-mediated labelling of specific sequences in the context of the optical map [35], which may facilitate straightforward mapping and quantifying of any repeat of interest.

2.2. Minor 45S rDNA Locus in Barley Chromosome 1H

Our second target was a minor 45S ribosomal DNA locus in barley chromosome 1H, identified by in situ hybridisation in various barley cultivars [32,33]. To access this locus, we first reconstructed the 1H-specific rDNA unit from 1H-specific paired-end Illumina reads (File S3). The unit sequence with the length of 8407 bp was then used to search 1H pseudomolecule of the barley 'Morex' reference genome [2]. Fragments of the unit were found between 139.05 Mb and 139.33 Mb of the 1H pseudomolecule, which fits well with the cytogenetic location of the rDNA locus at ~60% of the short arm of 1H [32], but we did not identify a regular rDNA array at this locus. To investigate completeness of the sequence in this region, we aligned it to available OM of barley cv. Morex [2], which identified OM contig 310 spanning over the region (Figure 4). Central part of the contig 310 did not align to the pseudomolecule and showed a regular labelling pattern with label spacing of approximately 5 kb. We compared it with the label pattern predicted for tandemly organised rDNA units. The reconstructed rDNA unit sequence comprised of three *Bsp*QI sites (File S3), but two of them were located just 1133 bp apart, which is too close for them to be discriminated in optical maps generated on the Bionano Genomics Irys platform. Thus, the labels associated with the *Bsp*QI sites were predicted to generate a composed pattern alternating ~3.5- and ~4.9-kb units (Figure S6A). This roughly corresponded to the pattern seen in the OM, with a discrepancy relating to the predicted ~3.5 kb "restriction fragment", which was not apparent in the optical map. This fragment covers the intergenic spacer (IGS) that comprises of two types of shorter tandem repeats with 78- and 135-bp unit length, respectively. It is likely that these repeats are collapsed in our consensus sequence and their real number is larger, extending the proposed IGS size by as much as 1.5 kb. This hypothesis was supported by the analysis of several partial rDNA units found in the 1H pseudomolecule, which were showing for both spacer repeats a higher number than included in our rDNA consensus sequence. Thus, we suggest that the complete size of the 1H rDNA unit is ~9.9 kb, which is supported by findings of [36] who identified in barley ribosomal DNA units of two sizes, 9.9 kb and 9 kb. Co-localisation of the blastn hits for 45S rDNA with the array in the optical map and the reported cytogenetic position lead us to the conclusion that the ~470-kb long array comprising ~47 putative rDNA units represents the minor 1H rDNA locus detected by in situ hybridisation. Our copy number estimate is close to that of [32] who quantified rRNA genes in the 1H chromosomes by in situ hybridization and proposed 50-100 copies in this locus. The slight discrepancy could be due to using a different barley cultivar (Morex vs. Sultan).

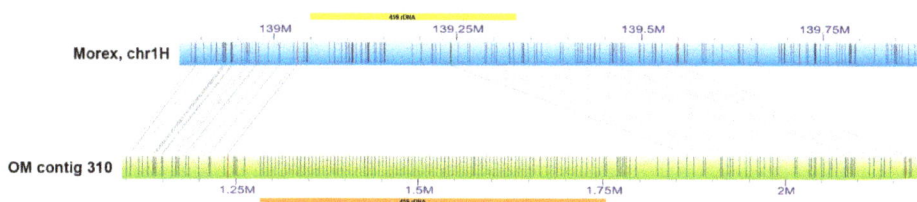

Figure 4. Positioning of the 45S rDNA locus in barley chromosome1H. Alignment of barley 'Morex' 1H pseudomolecule [2] (blue bar) to 'Morex' OM contig 310 (green bar) revealed a tandemly organized repeat with ~5 kb label spacing (highlighted by orange bar) missing in the pseudomolecule. Co-localisation with a cluster of 45S rDNA fragments in the sequence assembly (highlighted by yellow bar) indicates that the array represents the 1H rDNA locus.

Using the optical map, we identified a 470-kb segment that is absent from the Morex 1H pseudomolecule. We also performed blastn search for the consensus 45S rDNA sequence in chromosomes 5H and 6H, which are known to harbour major barley rDNA loci comprising thousands of genes [32,36], and in unassigned scaffolds of the 'Morex' assembly [2]. The search failed in revealing rDNA arrays in any of the datasets and identified fragments of the rDNA units only. This indicates that the missing rDNA loci contribute significantly to the dark matter of the cereal genomes.

3. Materials and Methods

3.1. De Novo Identification of Wheat Tandem Repeats

In order to identify new tandem repeats specific for the short arm of wheat chromosome 7D (7DS), we randomly selected 3.6 million reads obtained by Illumina sequencing multiple-displacement-amplified (MDA) DNA of flow-sorted 7DS [37]. Raw reads were examined and filtered by quality using FastQC and Trimmomatic tool. Repeat identification was performed employing similarity-based clustering of paired-end (2 × 100 nt) Illumina reads using local installation of the RepeatExplorer pipeline [23]. The pipeline employs graph representation of read similarities to find clusters of frequently overlapping reads corresponding to various repetitive elements or their parts. Putative tandem repeats were identified based on circular topology of their graphs [22] and tandem structure of contigs assembled from the reads within individual clusters. Sequences of the assembled contigs were then used to design PCR primers to verify the presence of corresponding sequences in the wheat genome (Table S1). In addition, the amplified fragments were cloned using the TOPO-TA Cloning Kit for Sequencing (Invitrogen, Carlsbad, CA, USA), selected clones were verified by sequencing and used as probes for in situ hybridisation experiments.

3.2. In Situ Hybridisation

For the in situ hybridisation experiment, we employed seeds of *Triticum aestivum* L., cv. Chinese Spring, kindly provided by Dr. Pierre Sourdille (INRA, Clermont-Ferrand, France). Seed germination, cell cycle synchronisation, metaphase accumulation and squash preparations were performed from wheat root tip meristems according to [38] with minor modifications. Metaphase accumulation was done by incubating root tips in 2.5 μM amiprophos-methyl for 2h in the dark at 25 °C. Inserts of clones bearing particular repeats were amplified using M13 primers and the PCR products were labeled by biotin using BioNick™ Labeling System (Invitrogen, Carlsbad, CA, USA). GAA microsatellite, used for identification of wheat chromosomes, was labelled by digoxigenin. Biotin- and digoxigenin-labeled probes were detected using streptavidin-Cy3 (Invitrogen, Carlsbad, CA, USA) and anti-digoxigenin-fluorescein (Roche, Basel, Switzerland), respectively. Chromosomes were counterstained with 4′,6′-diamidino-2-phenylindole (DAPI) and the preparations were imaged using Axio Imager Z.2 Zeiss microscope (Zeiss, Oberkochen, Germany) equipped with a CCD camera.

3.3. Reconstruction of Barley 1H rDNA Unit

RepeatExplorer pipeline was used to perform reconstruction of 45S rDNA sequence of barley. To do this, whole-genome paired-end (2 × 100 nt) Illumina reads of barley (*Hordeum vulgare*) cv. Morex (SRR490932) were downloaded from the SRA database, trimmed to quality and used for graph-based clustering. The resulting barley consensus 45S rDNA sequence was then used as a guide for reconstruction of a barley 1H chromosome-specific 45S rDNA sequence. This was done using online version of RepeatExplorer pipeline on the Galaxy platform and applying paired-end (2 × 100 nt) Illumina reads from flow-sorted 1H chromosome [39] of *H. vulgare* cv. Morex (SRR490144). The graph-based clustering resulted in five clusters homologous to the barley consensus 45S rDNA. The 1H-specific rDNA unit was then assembled manually utilizing the barley consensus 45S rDNA as a reference.

3.4. Application of Optical Maps

To validate sequences and analyse repeats in wheat 7DS and barley 1H chromosome, we employed available optical (BNG) maps constructed from 7DS chromosome arm of wheat cv. Chinese Spring [26] and the whole genome of barley cv. Morex [31], respectively. Both maps were assembled from single molecule data obtained after labelling molecules at Nt.*Bsp*QI nicking sites (motif GCTCTTC). Comparison of the optical maps with sequences was carried out using the IrysView 2.5.1 software package (Bionano Genomics, San Diego, CA, USA). For the alignment, cmap files were generated from fasta files of particular sequences. Query-to-anchor comparison was done with default parameters and P-value threshold of $1e^{-10}$.

3.5. Nanopore Sequencing

To resolve organization of the TaeCsTr99 repeats in the wheat genome, nanopore sequencing was conducted on 7DS BAC clones TaaCsp7DS028N04 (28N04) and TaaCsp7DS104G18 (104G18) from the 'Chinese Spring' 7DS arm-specific BAC library [40]. BAC DNA was extracted using alkaline lysis method followed by phenol-chloroform extraction and ethanol precipitation. Finally, the DNA was purified by incubating with 1:1 AMPure XP beads (Beckman Coulter, Miami, FL, USA) for 5 min and eluted into 30 μl 10 mM Tris, pH 8.5. Barcoded sequencing libraries were prepared from 700 ng DNA per BAC clone using Rapid Barcoding Sequencing Kit (SQK-RBK004; Oxford Nanopore Technologies, Oxford, UK) and sequenced together with additional ten clones on the MinION platform (Oxford Nanopore Technologies, Oxford, UK). Raw data were basecalled using Poretools 0.6.0 (https://github.com/arq5x/poretools, accessed on: 30 May 2019), demultiplexed using Porechop 0.2.3 (https://github.com/rrwick/Porechop, accessed on: 30 May 2019) and size-filtered >10 kb, which yielded 315 reads ranging from 10,003 to 101,160 bp, and 62 reads ranging from 10,082 to 149, 812 bp for the clone 28N04 and 104G18, respectively. Selected reads of 99,802 bp and 148,009 bp for 28N04 and 104G18, respectively, spanned the entire lengths of the respective clones.

4. Conclusions

Our study on tandem organised DNA repeats with unit sizes of 1.2–2.7 kb suggested that such repeats might be present in genome assemblies of large cereal genomes even if generated from short-read data. Nevertheless, they are typically comprised in short sequence contigs or scaffolds and thus may be difficult to incorporate into the pseudomolecules. We demonstrated that tandem repeats could be identified by a dedicated bioinformatics tool—RepeatExplorer—on a chromosome-specific basis and that nanopore sequencing of BAC clones provided a reliable approach to analysing organization of particular repeat arrays. We showed that an optical map might be useful for anchoring unassigned repeat-bearing scaffolds and for validating sequence assemblies in the problematic regions. The potential of the method was confirmed in our attempt to localise and characterise a minor 45S ribosomal DNA locus, which is missing in the reference genome of barley. Since BAC resources and optical maps are available for many plant species including major crops, the approaches presented in our study are widely applicable.

Supplementary Materials: Supplementary materials can be found at http://www.mdpi.com/1422-0067/20/10/2483/s1.

Author Contributions: Conceptualization, H.Š. and J.M.; FISH experiment, M.K., Repeat assembling P.N., J.M., V.K., E.H.; Repeat analyses V.K., Z.T., H.T.; Nanopore sequencing, Z.T.; Writing—Original Draft Preparation, V.K.; Writing—Review & Editing, H.Š., J.D.; Supervision, H.Š.; Funding Acquisition, H.Š., J.D. All authors have read and approved the manuscript.

Funding: This research was funded by the Czech Science Foundation (grant award 17-17564S) and the ERDF project "Plants as a tool for sustainable global development" (No. CZ.02.1.01/0.0/0.0/16_019/0000827).

Acknowledgments: We acknowledge the excellent assistance of Andrea Koblížková in validating the wheat tandem repeats. Computational resources were provided by the ELIXIR-CZ project (LM2015047).

Conflicts of Interest: The authors declare no conflict of interest.

Abbreviations

7DS	Short arm of wheat chromosome 7D
BAC	Bacterial Artificial Chromosome
BNG mapping	Bionano genome mapping
CRISPR	Clustered Regularly Interspaced Short Palindromic Repeats
DAPI	4,6-diamidino-2-phenylindole
FISH	Fluorescence in situ hybridisation
IGS	Intergenic spacer
ISH	In situ hybridisation
IWGSC	International Wheat Genome Sequencing Consortium
MDA	Multiple Displacement Amplification
OM	Optical map
ONT	Oxford Nanopore Technologies

PacBio	Pacific Biosciences
rDNA	Ribosomal DNA
SMRT	Single-molecule real-time
SRA	Sequence read archive

References

1. Avni, R.; Nave, M.; Barad, O.; Baruch, K.; Twardziok, S.O.; Gundlach, H.; Hale, I.; Mascher, M.; Spannagl, M.; Wiebe, K.; et al. Wild emmer genome architecture and diversity elucidate wheat evolution and domestication. *Science* **2017**, *357*, 93–97. [CrossRef] [PubMed]

2. Mascher, M.; Gundlach, H.; Himmelbach, A.; Beier, S.; Twardziok, S.O.; Wicker, T.; Radchuk, V.; Dockter, C.; Hedley, P.E.; Russell, J.; et al. A chromosome conformation capture ordered sequence of the barley genome. *Nature* **2017**, *544*, 427–433. [CrossRef] [PubMed]

3. The International Wheat Genome Sequencing Consortium (IWGSC). Shifting the limits in wheat research and breeding using a fully annotated reference genome. *Science* **2018**, *361*. [CrossRef]

4. Maccaferri, M.; Harris, N.S.; Twardziok, S.O.; Pasam, R.K.; Gundlach, H.; Spannagl, M.; Ormanbekova, D.; Lux, T.; Prade, V.M.; Milner, S.G.; et al. Durum wheat genome highlights past domestication signatures and future improvement targets. *Nat Genet.* **2019**, *51*, 885–895. [CrossRef] [PubMed]

5. Stein, N. (IPK, Gatersleben, Germany). Personal communication, 2018.

6. Doležel, J.; Čížková, J.; Šimková, H.; Bartoš, J. One major challenge of sequencing large plant genomes is to know how big they really are. *Int. J. Mol. Sci.* **2018**, *19*, 3554. [CrossRef]

7. Brenchley, R.; Spannagl, M.; Pfeifer, M.; Barker, G.L.; D'Amore, R.; Allen, A.M.; McKenzie, N.; Kramer, M.; Kerhornou, A.; Bolser, D.; et al. Analysis of the bread wheat genome using whole-genome shotgun sequencing. *Nature* **2012**, *491*, 705–710. [CrossRef]

8. Martis, M.M.; Zhou, R.; Haseneyer, G.; Schmutzer, T.; Vrána, J.; Kubaláková, M.; König, S.; Kugler, K.G.; Scholz, U.; Hackauf, B.; et al. Reticulate evolution of the rye genome. *Plant Cell* **2013**, *25*, 3685–3698. [CrossRef]

9. Mayer, K.F.X.; Martis, M.; Hedley, P.E.; Šimková, H.; Liu, H.; Morris, J.A.; Steuernagel, B.; Taudien, S.; Roessner, S.; Gundlach, H.; et al. Unlocking the barley genome by chromosomal and comparative genomics. *Plant Cell* **2011**, *23*, 1249–1263. [CrossRef]

10. Mayer, K.F.X.; Rogers, J.; Doležel, J.; Pozniak, C.; Eversole, K.; Feuillet, C.; Gill, B.; Friebe, B.; Lukaszewski, A.J.; Sourdille, P.; et al. A chromosome-based draft sequence of the hexaploid bread wheat (*Triticum aestivum*) genome. *Science* **2014**, *345*, 1251788. [CrossRef]

11. Chaisson, M.J.; Wilson, R.K.; Eichler, E.E. Genetic variation and the *de novo* assembly of human genomes. *Nat. Rev. Genet.* **2015**, *16*, 627–640. [CrossRef]

12. Zimin, A.V.; Puiu, D.; Hall, R.; Kingan, S.; Clavijo, B.J.; Salzberg, S.L. The first near-complete assembly of the hexaploid bread wheat genome, *Triticum aestivum*. *Gigascience* **2017**, *6*, 1–7. [CrossRef]

13. Handa, H.; Kanamori, H.; Tanaka, T.; Murata, K.; Kobayashi, F.; Robinson, S.J.; Koh, C.S.; Pozniak, C.J.; Sharpe, A.G.; Paux, E.; et al. Structural features of two major nucleolar organizer regions (NORs), Nor-B1 and Nor-B2, and chromosome-specific rRNA gene expression in wheat. *Plant J.* **2018**, *96*, 1148–1159. [CrossRef]

14. Symonová, R.; Ocalewicz, K.; Kirtiklis, L.; Delmastro, G.B.; Pelikánová, Š.; Garcia, S.; Kovařík, A. Higher-order organisation of extremely amplified, potentially functional and massively methylated 5S rDNA in European pikes (*Esox* sp.). *BMC Genom.* **2017**, *18*, 391. [CrossRef] [PubMed]

15. Appels, R.; Gerlach, W.L.; Dennis, E.S.; Swift, H.; Peacock, W.J. Molecular and Chromosomal Organization of DNA Sequences Coding for the Ribosomal RNAs in Cereals. *Chromosoma* **1980**, *78*, 293–311. [CrossRef]

16. Rayburn, A.L.; Gill, B.S. Use of biotin-labeled probes to map specific DNA sequences on wheat chromosomes. *Heredity* **1985**, *76*, 78–81. [CrossRef]

17. Mukai, Y.; Endo, T.R.; Gill, B.S. Physical mapping of the 18S.26S rRNA multigene family in common wheat: Identification of a new locus. *Chromosoma* **1991**, *100*, 71–78. [CrossRef]

18. Leitch, I.J.; Leitch, A.R.; Heslop-Harrison, J.S. Physical mapping of plant DNA sequences by simultaneous in situ hybridization of two differently labelled fluorescent probes. *Genome* **1991**, *34*, 329–333. [CrossRef]

19. Mukai, Y.; Nakahara, Y.; Yamamoto, M. Simultaneous discrimination of the three genomes in hexaploid wheat by multicolor fluorescence in situ hybridization using total genomic and highly repeated DNA probes. *Genome* **1993**, *36*, 489–494. [CrossRef] [PubMed]

20. Jiang, J.; Gill, B.S. New 18S. 26S ribosomal RNA gene loci: Chromosomal landmarks for the evolution of polyploid wheats. *Chromosoma* **1994**, *103*, 179–185. [CrossRef]

21. Brandes, A.; Röder, M.S.; Ganal, M.W. Barley telomeres are associated with two different types of satellite DNA sequences. *Chromosome Res.* **1995**, *3*, 315–320. [CrossRef]

22. Novák, P.; Neumann, P.; Macas, J. Graph-based clustering and characterization of repetitive sequences in next-generation sequencing data. *BMC Bioinform.* **2010**, *11*, 378. [CrossRef]

23. Novák, P.; Neumann, P.; Pech, J.; Steinhaisl, J.; Macas, J. RepeatExplorer: A Galaxy-based web server for genome-wide characterization of eukaryotic repetitive elements from next-generation sequence reads. *Bioinformatics* **2013**, *29*, 792–793. [CrossRef]

24. Martis, M.M.; Klemme, S.; Banaei-Moghaddam, A.M.; Blattner, F.R.; Macas, J.; Schmutzer, T.; Scholz, U.; Gundlach, H.; Wicker, T.; Šimková, H.; et al. Selfish supernumerary chromosome reveals its origin as a mosaic of host genome and organellar sequences. *Proc. Natl. Acad. Sci. USA* **2012**, *109*, 13343–13346. [CrossRef]

25. Lam, E.T.; Hastie, A.; Lin, C.; Ehrlich, D.; Das, A.K.; Austin, M.D.; Deshpande, P.; Cao, H.; Nagarajan, N.; Xiao, M.; et al. Genome mapping on nanochannel arrays for structural variation analysis and sequence assembly. *Nat. Biotechnol.* **2012**, *30*, 771–777. [CrossRef]

26. Staňková, H.; Hastie, A.R.; Chan, S.; Vrána, J.; Tulpová, Z.; Kubaláková, M.; Visendi, P.; Hayashi, S.; Luo, M.C.; Batley, J.; et al. BioNano genome mapping of individual chromosomes supports physical mapping and sequence assembly in complex plant genomes. *Plant Biotechnol. J.* **2016**, *14*, 1523–1531. [CrossRef]

27. Luo, M.C; Gu, Y.Q.; Puiu, D.; Wang, H.; Twardziok, S.O.; Deal, K.R.; Huo, N.; Zhu, T.; Wang, L.; Wang, Y.; et al. Genome sequence of the progenitor of the wheat D genome *Aegilops tauschii*. *Nature* **2017**, *551*, 498–502. [CrossRef]

28. Zhu, T.; Wang, L.; Rodriguez, J.C.; Deal, K.R.; Avni, R.; Distelfeld, A.; McGuire, P.E.; Dvorak, J.; Luo, MC. Improved genome sequence of wild emmer wheat Zavitan with the aid of optical maps. *G3 (Bethesda)* **2019**, *9*, 619–624. [CrossRef]

29. Tulpová, Z.; Toegelová, H.; Lapitan, N.L.V.; Peairs, F.B.; Macas, J.; Novák, P.; Lukaszewski, A.J.; Kopecký, D.; Mazáčová, M.; Vrána, J.; et al. Accessing a Russian wheat aphid resistance gene in bread wheat by long-read technologies. *Plant Genome* **2019**, *12*, 1–11. [CrossRef]

30. Tulpová, Z.; Luo, M.C.; Toegelová, H.; Visendi, P.; Hayashi, S.; Vojta, P.; Paux, E.; Kilian, A.; Abrouk, M.; Bartoš, J.; et al. Integrated physical map of bread wheat chromosome arm 7DS to facilitate gene cloning and comparative studies. *N. Biotechnol.* **2019**, *48*, 12–19. [CrossRef]

31. Beier, S.; Himmelbach, A.; Colmsee, C.; Zhang, X.Q.; Barrero, R.A.; Zhang, Q.; Li, L.; Bayer, M.; Bolser, D.; Taudien, S.; et al. Construction of a map-based reference genome sequence for barley, *Hordeum vulgare* L. *Sci. Data* **2017**, *4*, 170044. [CrossRef]

32. Leitch, I.J.; Heslop-Harrison, J.S. Physical mapping of the 18S-5.8S-26S rRNA genes in barley by in situ hybridization. *Genome* **1992**, *35*, 1013–1018. [CrossRef]

33. Szakács, É.; Kruppa, K.; Molnár-Láng, M. Analysis of chromosomal polymorphism in barley (*Hordeum vulgare* L. ssp. *vulgare*) and between *H. vulgare* and *H. chilense* using three-color fluorescence in situ hybridization (FISH). *J. Appl. Genet.* **2013**, *54*, 427–433. [CrossRef]

34. Shoaib, M.; Baconnais, S.; Mechold, U.; Le Cam, E.; Lipinski, M.; Ogryzko, V. Multiple displacement amplification for complex mixtures of DNA fragments. *BMC Genom.* **2008**, *9*, 415. [CrossRef]

35. Zhang, D.; Chan, S.; Sugerman, K.; Lee, J.; Lam, E.T.; Bocklandt, S.; Cao, H.; Hastie, A.R. CRISPR-bind: A simple, custom CRISPR/dCas9-mediated labeling of genomic DNA for mapping in nanochannel arrays. *bioRxiv* **2018**, preprint. [CrossRef]

36. Gerlach, W.L.; Bedbrook, J.R. Cloning and characterization of ribosomal RNA genes from wheat and barley. *Nucleic Acid Res.* **1979**, *7*, 1869–1886. [CrossRef]

37. Berkman, P.J.; Skarshewski, A.; Lorenc, M.T.; Lai, K.; Duran, C.; Ling, E.Y.; Stiller, J.; Smits, L.; Imelfort, M.; Manoli, S.; et al. Sequencing and assembly of low copy and genic regions of isolated *Triticum aestivum* chromosome arm 7DS. *Plant Biotechnol. J.* **2011**, *9*, 768–775. [CrossRef]

38. Karafiátová, M.; Bartoš, J.; Doležel, J. Localization of low-copy DNA sequences on mitotic chromosomes by FISH. In *Plant cytogenetics. Methods and Protocols*; Kianian, S.F., Kianian, P.M.A., Eds.; Humana Press: New York, NY, USA, 2016; Volume 1429, pp. 49–64.

39. Muñoz-Amatriaín, M.; Lonardi, S.; Luo, MC.; Madishetty, K.; Svensson, J.T.; Moscou, M.J.; Wanamaker, S.; Jiang, T.; Kleinhofs, A.; Muehlbauer, G.J.; et al. Sequencing of 15 622 gene-bearing BACs clarifies the gene-dense regions of the barley genome. *Plant J.* **2015**, *84*, 216–227. [CrossRef]

40. Šimková, H.; Šafář, J.; Kubaláková, M.; Suchánková, P.; Číhalíková, J.; Robert-Quatre, H.; Azhaguvel, P.; Weng, Y.; Peng, J.; Lapitan, N.L.V.; et al. BAC Libraries from wheat chromosome 7D: Efficient tool for positional cloning of aphid resistance genes. *J. Biomed. Biotechnol.* **2011**, *2011*, 302543. [CrossRef]

© 2019 by the authors. Licensee MDPI, Basel, Switzerland. This article is an open access article distributed under the terms and conditions of the Creative Commons Attribution (CC BY) license (http://creativecommons.org/licenses/by/4.0/).

International Journal of
Molecular Sciences

MDPI

Brief Report

New ND-FISH-Positive Oligo Probes for Identifying *Thinopyrum* Chromosomes in Wheat Backgrounds

Wei Xi [1,2,†], Zongxiang Tang [1,2,†], Shuyao Tang [1,2], Zujun Yang [3], Jie Luo [1,2] and Shulan Fu [1,2,*]

1   Provincial Key Laboratory of Plant Breeding and Genetics, Sichuan Agricultural University, Wenjiang, Chengdu 611130, Sichuan, China; xiwei923700915@163.com (W.X.); zxtang@sicau.edu.cn (Z.T.); tangshuyao705708@sina.com (S.T.); luojie2010@sina.com (J.L.)
2   College of Agronomy, Sichuan Agricultural University, Wenjiang, Chengdu 611130, Sichuan, China
3   Center for Informational Biology, University of Electronic Science and Technology of China, Chengdu 610054, Sichuan, China; yangzujun@uestc.edu.cn
*   Correspondence: fushulan@sicau.edu.cn
†   These authors contributed equally to this work.

Received: 4 April 2019; Accepted: 22 April 2019; Published: 25 April 2019

Abstract: *Thinopyrum* has been widely used to improve wheat (*Triticum aestivum* L.) cultivars. Non-denaturing fluorescence in situ hybridization (ND-FISH) technology using oligonucleotides (oligo) as probes provides a convenient and efficient way to identify alien chromosomes in wheat backgrounds. However, suitable ND-FISH-positive oligo probes for distinguishing *Thinopyrum* chromosomes from wheat are lacking. Two oligo probes, Oligo-B11 and Oligo-pThp3.93, were designed according to the published *Thinopyrum ponticum* (*Th. ponticum*)-specific repetitive sequences. Both Oligo-B11 and Oligo-pThp3.93 can be used for ND-FISH analysis and can replace conventional GISH and FISH to discriminate some chromosomes of *Th. elongatum*, *Th. intermedium*, and *Th. ponticum* in wheat backgrounds. The two oligo probes provide a convenient way for the utilization of *Thinopyrum* germplasms in future wheat breeding programs.

Keywords: wheat; *Thinopyrum*; chromosome; ND-FISH; oligo probe

---

1. Introduction

Thinopyrum intermedium (*Th. intermedium*), *Thinopyrum ponticum* (*Th. ponticum*) and *Thinopyrum elongatum* (*Th. elongatum*) are important reservoirs of elite genes for wheat (*Triticum aestivum* L.) breeding programs. Genomic in situ hybridization (GISH) and fluorescence in situ hybridization (FISH) technologies can be used to differentiate and localize *Th. intermedium*, *Th. ponticum* and *Th. elongatum* chromosomes in wheat backgrounds [1–8]. However, GISH and FISH are time-consuming because of the preparation and labeling of probe sequences, and denaturing of the probes and chromosomes [9,10]. Oligonucleotide (oligo) probes combined with non-denaturing fluorescence in situ hybridization (ND-FISH) can be used to discriminate alien chromosomes in wheat backgrounds conveniently [10,11]. Suitable ND-FISH-positive oligo probes for distinguishing *Thinopyrum* chromosomes from wheat are lacking. In this study, *Thinopyrum* chromosome-specific oligo probes were developed.

2. Results

2.1. Production of Thinopyrum Chromosome-Specific Oligo Probes

Th. ponticum species-specific repetitive sequences B11 [3] and pThp3.93 [6] were aligned using the DNAMAN (Ver. 4.0, Lynnon Corp., Quebec, QC, Canada). The sequence of pThp3.93 has a 59.78% similarity with the 1138–1622 bp segment of the reverse_complement B11 sequence. The 1304–1348 bp segment of the reverse_complement B11 sequence and the 158–216 bp segment of the pThp3.93

sequence were used as oligo probes, and were named as Oligo-B11 and Oligo-pThp3.93, respectively. The two oligo probes are listed in Table 1.

Table 1. Oligonucleotide sequences of new oligo probes developed in this study.

Probe	Amount for Each Slide (ng/slide)	Oligonucleotide Sequence (5′-3′)
Oligo-B11	72.5–96.6	TCCGCTCACCTTGATGACAACATCAGGTGGAATTC CGTTCGAGGG
Oligo-pThp3.93	69.5–92.6	GGACTCCCACTAGATGTATCCGTCAAGGTGAATCCA GAGGAATCACCCTCGATGGCATT

2.2. ND-FISH Analysis Using Oligo-B11 and Oligo-pThp3.93

The Oligo-B11, Oligo-pThp3.93, Oligo-pSc119.2-1, and Oligo-pTa535-1 probes combined with the ND-FISH assay could distinguish *Thinopyrum* chromosomes from wheat in Xiaoyan 68, 8802, Xaioyan 7430, and Zhong 3. Both Oligo-B11 and Oligo-pThp3.93 probes produced dispersed signal patterns on 12 and 14 chromosomes in Xiaoyan 68 and 8802, respectively (Figures 1 and 2). A pair of wheat-*Th. ponticum* translocation chromosomes TTh-4DS·4DL in Xiaoyan 68 could also be detected by Oligo-B11, Oligo-pThp3.93, Oligo-pSc119.2-1, and Oligo-pTa535-1 (Figure 1).

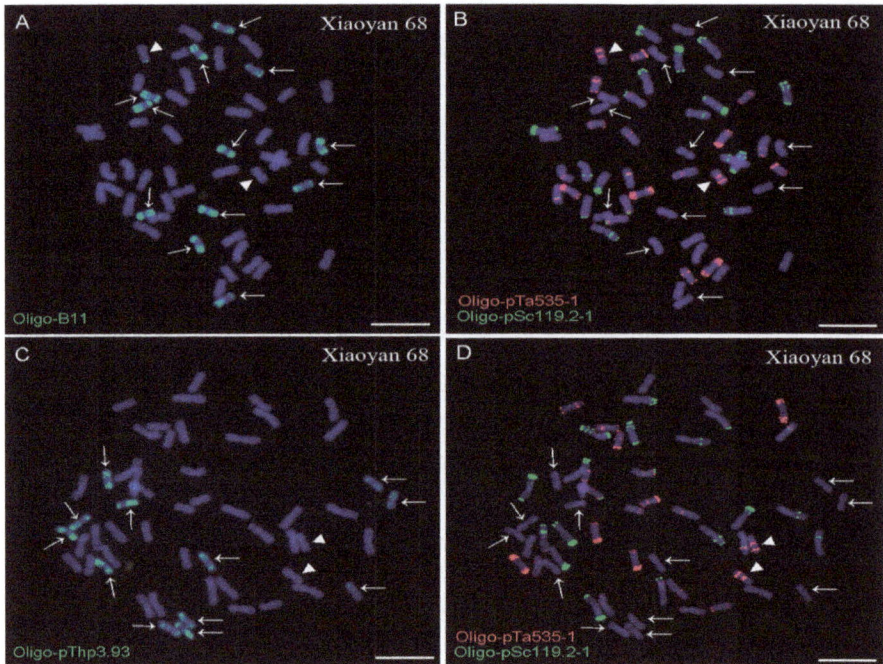

Figure 1. ND-FISH analysis of root tip metaphase chromosomes of Xiaoyan 68 using the Oligo-B11 (green), Oligo-pThp3.93 (green), Oligo-pTa535-1 (red), and Oligo-pSc119.2-1 (green) as probes. (**A**) and (**B**) are the same cell, and (**C**) and (**D**) are the same cell. Arrows indicate the *Th. ponticum* chromosomes that carry signals of Oligo-B11 (**A**) and Oligo-pThp3.93 (**C**). Triangles indicate the wheat-*Th. ponticum* translocation chromosomes TTh-4DS·4DL. Chromosomes were counterstained with DAPI (blue). Scale bar: 10 μm.

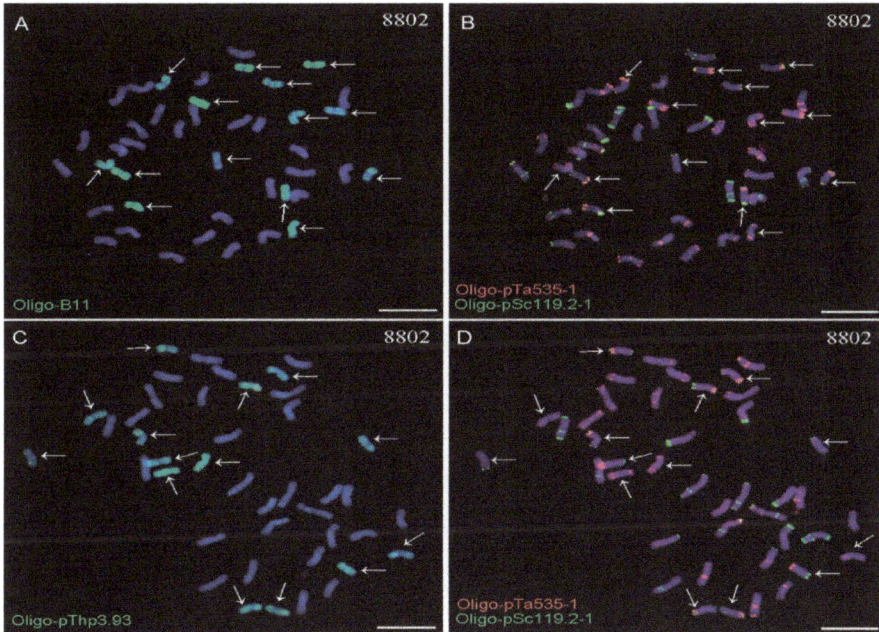

Figure 2. ND-FISH analysis of the root tip metaphase chromosomes of 8802 using the Oligo-B11 (green), Oligo-pThp3.93 (green), Oligo-pTa535-1 (red) and Oligo-pSc119.2-1 (green) as probes. (**A**) and (**B**) are the same cell, and (**C**) and (**D**) are the same cell. Arrows indicate the *Th. elongatum* chromosomes that carry signals of Oligo-B11 (**A**) and Oligo-pThp3.93 (**C**). Chromosomes were counterstained with DAPI (blue). Scale bar: 10 µm.

The Oligo-B11 and Oligo-pThp3.93 probes also produced whole-chromosome signal patterns on 12 and 14 chromosomes in Xiaoyan 7430 and Zhong 3, respectively (Figures 3 and 4). Therefore, both of the ND-FISH-positive Oligo-B11 and Oligo-pThp3.93 probes produced whole-chromosome signal patterns on 12, 14, 12, and 14 chromosomes in Xiaoyan 68, 8802, Xaioyan 7430, and Zhong 3, respectively. Additionally, no obvious signals of Oligo-B11 and Oligo-pThp3.93 were observed on wheat chromosomes in the materials used in this study (Figures 1–4).

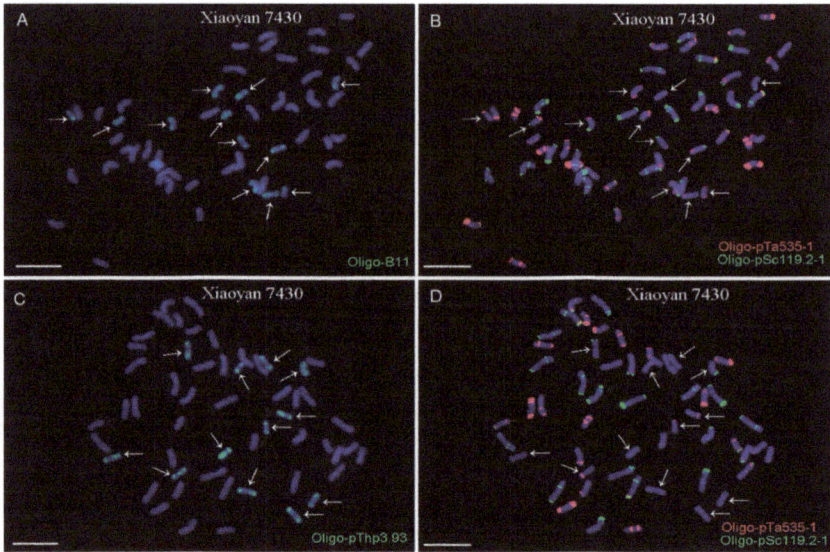

Figure 3. ND-FISH analysis of the root tip metaphase chromosomes of Xiaoyan 7430 using the Oligo-B11 (green), Oligo-pThp3.93 (green), Oligo-pTa535-1 (red) and Oligo-pSc119.2-1 (green) as probes. (**A**) and (**B**) are the same cell, and (**C**) and (**D**) are the same cell. Arrows indicate the *Th. ponticum* chromosomes that carry signals of Oligo-B11 (**A**) and Oligo-pThp3.93 (**C**). Chromosomes were counterstained with DAPI (blue). Scale bar: 10 μm.

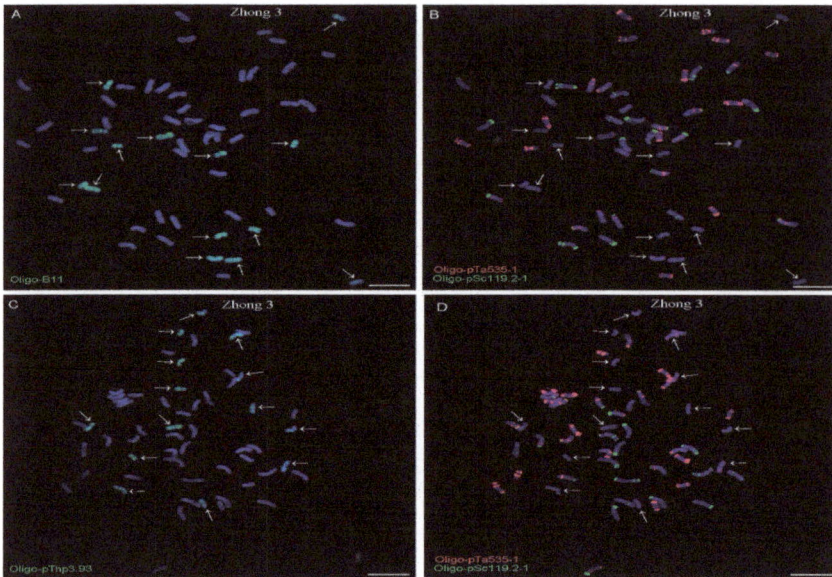

Figure 4. ND-FISH analysis of the root tip metaphase chromosomes of Zhong 3 using the Oligo-B11 (green), Oligo-pThp3.93 (green), Oligo-pTa535-1 (red) and Oligo-pSc119.2-1 (green) as probes. (**A**) and (**B**) are the same cell, and (**C**) and (**D**) are the same cell. Arrows indicate the *Th. intermedium* chromosomes that carry signals of Oligo-B11 (**A**) and Oligo-pThp3.93 (**C**). Chromosomes were counterstained with DAPI (blue). Scale bar: 10 μm.

3. Discussion

3.1. Using Oligo Probes and ND-FISH Assay to Identify Alien Chromosomes

Since Cuadrado et al. used the ND-FISH assay to detect plant telomeres [12], this technology has been widely used to analyze chromosomes of Triticeae species because of its convenience and high efficiency [4,10,11,13–26]. Using oligo probes combined with the ND-FISH assay, rye (*S. cereale*) and *Dasypyrum villosum* chromosomes can be effectively and accurately distinguished from common wheat chromosomes [4,10,11]. In addition, two ND-FISH-positive oligo probes, Oligo-B and Oligo-D, can replace multicolor GISH to identify wheat A-, B-, and D-genome chromosomes [23]. During the ND-FISH procedure, oligo probes can be commercially synthesized and the denaturing of probes and chromosomes is not necessary [4,10,11,23]. Therefore, compared with GISH and FISH, ND-FISH analysis with suitable oligo probes can efficiently distinguish the chromosomes between two distant genera and within the same genus of the Triticeae tribe.

3.2. ND-FISH-Positive Oligo Probes for Identifying Thinopyrum Chromosomes

For a successful ND-FISH assay, finding suitable oligo probes is the key initial step. However, the necessary ND-FISH-positive oligo probes for distinguishing *Thinopyrum* chromosomes in wheat backgrounds are still lacking. Some disease-resistance and stress-resistance genes from *Thinopyrum* have been introduced into wheat backgrounds [1–8,18,27]. Now, conventional GISH and FISH are the main methods to distinguish and localize *Thinopyrum* chromosomal segments in wheat backgrounds [1–8,18,27]. In this study, two suitable ND-FISH-positive oligo probes, Oligo-B11 and Oligo-pThp3.93, were designed based on *Th. ponticum* species-specific repetitive sequences [3,6]. Both of these oligo probes produced whole-chromosome signal patterns on 12, 14, 12, and 14 chromosomes in Xiaoyan 68, 8802, Xaioyan 7430m and Zhong 3, respectively (Figures 1–4). The numbers of the *Thinopyrum* chromosomes identified by Oligo-B11 and Oligo-pThp3.93 in each of these materials were consistent with those previously reported [1,6,28,29].

In addition, the Oligo-B11, Oligo-pThp3.93, and Oligo-pTa535-1 probes could also identify the TTh-4DS·4DL translocation chromosomes in Xiaoyan 68 (Figure 1). In the materials used in this study, the *Thinopyrum* chromosomes came from *Th. ponticum*, *Th. elongatum*, and *Th. intermedium* [1,6,28,29]. Therefore, Oligo-B11 and Oligo-pThp3.93 combined with the ND-FISH assay can replace conventional GISH and FISH to conveniently discriminate the *Th. elongatum*, *Th. intermedium*, and *Th. ponticum* chromosomes in Xiaoyan 68, 8802, Xaioyan 7430, and Zhong 3 from wheat chromosomes.

4. Materials and Methods

4.1. Plant Materials

Wheat-*Th. ponticum* partial amphiploids Xiaoyan 68 and Xaioyan 7430, hexaploid *Trititrigia* 8802, and wheat-*Th. intermedium* partial amphiploid Zhong 3 were kindly provided by Professor Fangpu Han, Institute of Genetics and Developmental Biology, Chinese Academy of Science, Beijing, China. The 8802 contains chromosomes of *Th. elongatum* [28].

4.2. Cytological Analysis

Root-tip metaphase chromosomes were prepared following the methods described by Han et al. [30]. Oligo-pSc119.2-1 and Oligo-pTa535-1 were also used in this study [31]. Oligo-B11, Oligo-pThp3.93, Oligo-pSc119.2-1, and Oligo-pTa535-1 probes were synthesized by Tsingke Biological Technology Co. Ltd. (Beijing, China). Both the Oligo-B11 and Oligo-pThp3.93 probes were 5′-end-labelled with 6-carboxyfluorescein (6-FAM), the Oligo-pTa535-1 probe was 5′-end-labeled with 6-carboxytetramethylrhodamine (TAMRA), and the Oligo-pSc119.2-1 probe was 5′-end-labeled with Cyanine Dye 5 (Cy5). The synthesized oligo probes were diluted by 1 × TE solution (pH 7.0). For Oligo-pSc119.2-1 and Oligo-pTa535-1, probe dilution and the probe amounts per slide were carried out according to the methods described by Tang et al. [31]. For Oligo-B11 and Oligo-pThp3.93, 100 μL 1 × TE was used to dissolve each 1OD probe. Then, the original solution was diluted five times and were used as the working solutions. Two × SSC and 1 × TE buffers (pH 7.0) were mixed as a 1:1 (volume) ratio. The target probes were added into the 2 × SSC + 1 × TE buffer and uniformly mixed. For each slide, 10 μL of probe mixture was used. When the probe mixture was dropped onto the cell spreads, the room temperature around the slide were kept at above 28 °C, then the slides were covered with glass coverslips and immediately put in a moist box that was incubated at 42 °C in advance, and then stored in the moist box at 42 °C for 1–2 h. After incubation, the slides were washed for 15–20 s in 2 × SSC at 42 °C. When the slide washing was completed, the slides were dried with a rubber suction bulb, and the slides were mounted with Vectashield mounting medium (Vector Laboratories, Burlingame, CA, USA) with DAPI (4′,6-diamidino-2-phenylindole). An epifluorescence microscope (BX51, Olympus Corporation, Tokyo, Japan) equipped with a cooled charge-coupled device camera operated with HCIMAGE Live software (Hamamatsu Corporation, Sewickley, PA, USA) was used to take images.

5. Conclusions

In conclusion, the Oligo-B11 and Oligo-pThp3.93 probes combined with the ND-FISH assay can replace GISH and FISH to conveniently discriminate the *Th. elongatum*, *Th. Intermedium*, and *Th. ponticum* chromosomes in Xiaoyan 68, 8802, Xaioyan 7430, and Zhong 3 from wheat chromosomes. Therefore, the two oligo probes provide a convenient way for the utilization of *Thinopyrum* germplasms in Xiaoyan 68, 8802, Xaioyan 7430, and Zhong 3 in future wheat breeding programs.

Author Contributions: S.F. and Z.T. designed the study and the oligo probes, analyzed the data, and wrote the manuscript. W.X., S.T., and J.L. performed the experiments. Z.Y. analyzed the data.

Acknowledgments: This project was supported by the National Key Research and Development Program of China (No. 2016YFD0102001) and the "13th Five-Year" Crops, Livestock and Poultry Breeding Program of Sichuan Province (No. 2016NY0030).

Conflicts of Interest: These authors have no conflicts of interest. The funding sponsors had no role in the experiment design, material creation, data analysis, manuscript writing and the decision to publish the results.

References

1. Han, F.; Liu, B.; Fedak, G.; Liu, Z. Genomic constitution and variation in five partial amphiploids of wheat–*Thinopyrum intermedium* as revealed by GISH, multicolor GISH and seed storage protein analysis. *Theor. Appl. Genet.* **2004**, *109*, 1070–1076. [CrossRef]

2. Fu, S.; Lv, Z.; Qi, B.; Guo, X.; Li, J.; Liu, B.; Han, F. Molecular cytogenetic characterization of wheat-*Thinopyrum elongatum* addition, substitution and translocation lines with a novel source of resistance to wheat Fusarium Head Blight. *J. Genet. Genomics* **2012**, *39*, 103–110. [CrossRef] [PubMed]

3. Yao, H.; Tang, C.G.; Zhao, J.; Zheng, Q.; Li, B.; Hao, C.Y.; Li, Z.S.; Zhang, X.Y. Isolation of *Thinopyrum ponticum* genome specific repetitive sequences and their application for effective detection of alien segments in wheat. *Sci. Agric. Sin.* **2016**, *49*, 3683–3693.

4. Li, J.; Lang, T.; Li, B.; Yu, Z.; Wang, H.; Li, G.; Yang, E.; Yang, Z. Introduction of *Thinopyrum intermedium* ssp. *trichophorum* chromosomes to wheat by trigeneric hybridization involving *Triticum*, *Secale* and *Thinopyrum* genera. *Planta* **2017**, *245*, 1121–1135. [CrossRef]

5. Lang, T.; La, S.; Li, B.; Yu, Z.; Chen, Q.; Li, J.; Yang, E.; Li, G.; Yang, Z. Precise identification of wheat-*Thinopyrum intermedium* translocation chromosomes carrying resistance to wheat stripe rust in line Z4 and its derived progenies. *Genome* **2018**, *61*, 177–185. [CrossRef]

6. Liu, L.; Luo, Q.; Teng, W.; Li, B.; Li, H.; Li, Y.; Li, Z.; Zheng, Q. Development of *Thinopyrum ponticum*-specific molecular markers and FISH probes based on SLAF-seq technology. *Planta* **2018**, *247*, 1099–1108. [CrossRef] [PubMed]

7. Li, D.; Li, T.; Wu, Y.; Zhang, X.; Zhu, W.; Wang, Y.; Zeng, J.; Xu, L.; Fan, X.; Sha, L.; et al. FISH-based markers enable identification of chromosomes derived from tetraploid *Thinopyrum elongatum* in hybrid lines. *Front. Plant Sci.* **2018**, *9*, 526. [CrossRef] [PubMed]

8. Mago, R.; Zhang, P.; Xia, X.; Zhang, J.; Hoxha, S.; Lagudah, E.; Graner, A.; Dundas, I. Transfer of stem rust resistance gene *SrB* from *Thinopyrum ponticum* into wheat and development of a closely linked PCR-based marker. *Theor. Appl. Genet.* **2019**, *132*, 371–382. [CrossRef] [PubMed]

9. Du, W.; Wang, J.; Lu, M.; Sun, S.; Chen, X.; Zhao, J.; Yang, Q.; Wu, J. Characterization of a wheat-*Psathyrostachys huashanica* Keng 4Ns disomic addition line for enhanced tiller numbers and stripe rust resistance. *Planta* **2014**, *239*, 97–105. [CrossRef] [PubMed]

10. Fu, S.; Chen, L.; Wang, Y.; Li, M.; Yang, Z.; Qiu, L.; Yan, B.; Ren, Z.; Tang, Z. Oligonucleotide probes for ND-FISH analysis to identify rye and wheat chromosomes. *Sci. Rep.* **2015**, *5*, 10552. [CrossRef]

11. Xiao, Z.; Tang, S.; Qiu, L.; Tang, Z.; Fu, S. Oligonucleotides and ND-FISH displaying different arrangements of tandem repeats and identification of *Dasypyrum villosum* chromosomes in wheat backgrounds. *Molecules* **2017**, *22*, 973. [CrossRef]

12. Cuadrado, Á.; Golczyk, H.; Jouve, N. A novel, simple and rapid nondenaturing FISH (ND-FISH) technique for the detection of plant telomeres. Potential used and possible target structures detected. *Chromosome Res.* **2009**, *1*, 755–762. [CrossRef]

13. Cuadrado, Á.; Jouve, N. Chromosomal detection of simple sequence repeats (SSRs) using nondenaturing FISH (ND-FISH). *Chromosoma* **2010**, *119*, 495–503. [CrossRef]

14. Cuadrado, Á.; Carmona, A.; Jouve, N. Chromosomal characterization of the three subgenomes in the polyploids of *Hordeum murinum* L.: New insight into the evolution of this complex. *PLoS ONE* **2013**, *8*, e81385. [CrossRef]

15. Carmona, A.; Friero, E.; Bustos, A.D.; Jouve, N.; Cuadrado, A. Cytogenetic diversity of SSR motifs within and between Hordeum species carrying the H genome: *H. vulgare* L. and *H. bulbosum* L. *Theor. Appl. Genet.* **2013**, *126*, 949–961. [CrossRef]

16. Cabo, S.; Carvalho, A.; Martin, A.; Lima-Brito, J. Structural rearrangements detected in newly-formed hexaploid tritordeum after three sequential FISH experiments with repetitive DNA sequences. *J. Genet.* **2014**, *93*, 183–188. [CrossRef]

17. Delgado, A.; Carvalh, A.; Martín, A.C.; Martín, A.; Lima-Brito, J. Use of the synthetic Oligo-pTa535 and Oligo-pAs1 probes for identification of *Hordeum chilense*-origin chromosomes in hexaploid tritordeum. *Genet. Resour. Crop Evol.* **2016**, *63*, 945–951. [CrossRef]

18. Li, G.; Wang, H.; Lang, T.; Li, J.; La, S.; Yang, E.; Yang, Z. New molecular markers and cytogenetic probes enable chromosome identification of wheat-*Thinopyrum intermedium* introgression lines for improving protein and gluten contents. *Planta* **2016**, *244*, 865–876. [CrossRef]

19. Delgado, A.; Carvalho, A.; Martín, A.C.; Martín, A.; Lima-Brito, J. Genomic reshuffling in advanced lines of hexaploid tritordeum. *Gene. Resour. Crop Evol.* **2017**, *64*, 1331–1353. [CrossRef]

20. Jiang, M.; Xiao, Z.Q.; Fu, S.L.; Tang, Z.X. FISH karyotype of 85 common wheat cultivars/lines displayed by ND-FISH using oligonucleotide probes. *Cereal Res. Commun.* **2017**, *45*, 549–563. [CrossRef]

21. Duan, Q.; Wang, Y.Y.; Qiu, L.; Ren, T.H.; Li, Z.; Fu, S.L.; Tang, Z. Physical location of new PCR-based markers and powdery mildew resistance gene(s) on rye (*Secale cereale* L.) chromosome 4 using 4R dissection lines. *Front. Plant Sci.* **2017**, *8*, 1716. [CrossRef]

22. Du, P.; Zhuang, L.; Wang, Y.; Yuan, L.; Wang, Q.; Wan, D.; Dawadondup; Tan, L.; Shen, J.; Xu, H.; Zhao, H.; et al. Development of oligonucleotides and multiplex probes for quick and accurate identitication of wheat and *Thinopyrum bessarabicum* chromosomes. *Genome* **2017**, *60*, 93–103. [CrossRef]

23. Tang, S.; Tang, Z.; Qiu, L.; Yang, Z.; Li, G.; Lang, T.; Zhu, W.; Zhang, J.; Fu, S. Developing new oligo probes to distinguish specific chromosomal segments and the A, B, D genomes of wheat (*Triticum aestivum* L.) using ND-FISH. *Front. Plant Sci.* **2018**, *9*, 1104. [CrossRef]

24. Ren, T.; He, M.; Sun, Z.; Tan, F.; Luo, P.; Tang, Z.; Fu, S.; Yan, B.; Ren, Z.; Li, Z. The polymorphisms of oligonucleotide probes in wheat cultivars determined by ND-FISH. *Molecules* **2019**, *24*, 1126. [CrossRef]

25. Lang, T.; Li, G.; Wang, H.; Yu, Z.; Chen, Q.; Yang, E.; Fu, S.; Tang, Z.; Yang, Z. Physical location of tandem repeats in the wheat genome and application for chromosome identification. *Planta* **2019**, *249*, 663–675. [CrossRef]

26. Du, H.; Tang, Z.; Duan, Q.; Tang, S.; Fu, S. Using the 6RL^{Ku} minichromosome of rye (*Secale cereale* L.) to create wheat-rye 6D/6RL^{Ku} small segment translocation lines with powdery mildew resistance. *Int. J. Mol. Sci.* **2018**, *19*, E3933. [CrossRef]

27. Nie, L.; Yang, Y.; Zhang, J.; Fu, T. Disomic chromosome addition from *Thinopyrum intermedium* to bread wheat appears to confer stripe rust resistance. *Euphytica* **2019**, *215*, 56. [CrossRef]

28. Guo, X.; Shi, Q.; Wang, J.; Hou, Y.; Wang, Y.; Han, F. Characterization and genome changes of new amphiploids from wheat wide hybridization. *J. Genet. Genomics* **2015**, *42*, 459–461. [CrossRef]

29. He, F.; Wang, Y.; Bao, Y.; Ma, Y.; Wang, X.; Li, X.; Wang, H. Chromosomal constitutions of five wheat–*Elytrigia elongata* partial amphiploids as revealed by GISH, multicolor GISH and FISH. *Comp. Cytogenet.* **2017**, *11*, 525–540. [CrossRef]

30. Han, F.; Lamb, J.C.; Birchler, A. High frequency of centromere inactivation resulting in stable dicentric chromosomes of maize. *Proc. Natl. Acad. Sci. USA* **2006**, *103*, 3238–3243. [CrossRef]

31. Tang, Z.; Yang, Z.; Fu, S. Oligonucleotides replacing the roles of repetitive sequences pAs1, pSc119.2, pTa-535, pTa71, CCS1, and pAWRC.1 for FISH analysis. *J. Appl. Genet.* **2014**, *55*, 313–318. [CrossRef] [PubMed]

© 2019 by the authors. Licensee MDPI, Basel, Switzerland. This article is an open access article distributed under the terms and conditions of the Creative Commons Attribution (CC BY) license (http://creativecommons.org/licenses/by/4.0/).

MDPI

St. Alban-Anlage 66

4052 Basel

Switzerland

Tel. +41 61 683 77 34

Fax +41 61 302 89 18

www.mdpi.com

International Journal of Molecular Sciences Editorial Office

E-mail: ijms@mdpi.com

www.mdpi.com/journal/ijms

www.ingramcontent.com/pod-product-compliance
Lightning Source LLC
Chambersburg PA
CBHW051719210326
41597CB00032B/5538